Hans Joachim Blaß, Carmen Sandhaas

Timber Engineering

Principles for Design

Timber Engineering

Principles for Design

by
Hans Joachim Blaß
Carmen Sandhaas

KIT Scientific Publishing

Translation: Richard Mort

Impressum

KIT Scientific Publishing

Karlsruher Institut für Technologie (KIT)
KIT Scientific Publishing
Straße am Forum 2
D-76131 Karlsruhe
KIT Scientific Publishing is a registered trademark
of Karlsruhe Institute of Technology.
Reprint using the book cover is not allowed.
www.ksp.kit.edu

Print on Demand 2017 – Gedruckt auf FSC-zertifiziertem Papier

ISBN 978-3-7315-0673-7
DOI 10.5445/KSP/1000069616

Foreword

Successfully designing timber structures requires an in-depth knowledge and a fundamental understanding of wood and wood-based materials as modern and sustainable construction materials, of structural elements, systems and joints using wood and wood-based materials. Although consideration of design rules set out in standards is still important for structural design, successful and efficient constructions require more than simply applying calculation standards alone. Accordingly, this book should contribute to a more in-depth understanding of the principles used to design timber structures.

The first step towards explaining the Eurocode 5 rules came in 1995 with the STEP anthology – Structural Timber Education Programme. STEP, which was compiled by approximately 50 authors from 14 European countries, was intended both to educate those studying timber engineering and provide further training for structural engineers in the European countries concerned. Since it was released however, knowledge has advanced significantly in areas such as construction materials, structural elements and joints, as is also reflected in the relevant Eurocode 5 rules. This book embraces and builds on the STEP concept, to establish a timber engineering textbook tailored to the national situation. This involved expanding the timber engineering topic scope to accommodate additional contributions, while also omitting some of the original STEP material, given its limited relevance to timber construction in German-speaking countries. Contributions on individual topics were expanded, deepened and updated beyond the scope of the original STEP books, while derivations on individual topics were included in an annex and the authors of the original STEP contributions are cited in the relevant places. The scientific employees of the KIT Timber Structures and Building Construction section also provided invaluable assistance to revise and supplement various contributions.

There will always be room to improve a book as comprehensive as this one. We kindly request that anyone reading this who comes across an error informs us accordingly, so that this can be rectified for future editions.

Hans Joachim Blass and Carmen Sandhaas

General note: Most standards and additional codes of practice are included without specifying the date, given frequent changes. Exception: Excerpts from standards are dated, so that the reader knows which edition was consulted. For this edition, DIN EN 1995-1-1:2010, DIN EN 1995-1-1/A2:2014 and DIN EN 1995-1-1/NA:2013 have all been considered.

Contents

A Introduction 1
A1 Codes of practice in Germany 3
A2 Sustainability 11

B Material wood 19
B1 Wood anatomy 21
B2 Wood physics 37
B3 Wood formation 53
B4 Durability 75
B5 Wood drying and strength grading 85
B6 Wood products 99

C Principles for design 133
C1 Safety concept 135
C2 Actions on structures 145

D Structural members and systems 159
D1 Basic stresses 161
D2 Stability 177
D3 Influence of volume and stress distribution 195
D4 Members with variable cross-sections or curved shapes 205
D5 Non-reinforced notches and holes 219
D6 Glued composite members 233
D7 Mechanically jointed members and CLT elements 249
D8 Reinforcements 263
D9 Diaphragms and bracings 279
D10 Timber-concrete composite structures 309
D11 System strength 317

E Joints 323

E1 Joints in timber structures 325
E2 Johansen model 347
E3 Joints with nails and staples 365
E4 Joints with bolts and dowels 375
E5 Joints with self-tapping screws 383
E6 Joints with connectors 393
E7 Punched metal plate fasteners 407
E8 Cold-formed steel connectors 417
E9 Contact joints (Carpentry joints) 423
E10 Glued-in steel rods 435
E11 Joints loaded perpendicular to the grain 441
E12 Reinforced joints 459
E13 Joints with multiple fasteners 477
E14 Moment-resisting joints 491
E15 Joints with multiple shear planes 509

F Serviceability 515

F1 Deformations 517
F2 Vibrations 525

G Accidental loads and additions 535

G1 Reaction to fire and fire design 537
G2 Joints subject to seismic loads 561
G3 Earthquake-compliant structural details 573
G4 Damages in hall structures 585

Annexes 609

Annex 1: Dynamic modulus of elasticity 611
Annex 2: Stress interactions 613
Annex 3: Elastic buckling load and buckling lengths 619
Annex 4: Derivations lateral torsional buckling 623
Annex 5: Mechanically jointed beams 629
Annex 6: Derivation Johansen equations 639

A

Introduction

A1 Codes of practice in Germany

Author: Rainer Görlacher

Above all, the building authority centres on preventing dangers to the public, which may arise from the construction process. In Germany, the general conditions governing the building, alteration or demolition of structures have been a federal responsibility since the 19th century. Even so, individual tasks were transferred to the Deutsche Institut für Bautechnik (DIBt) in an agreement and this allocation of responsibilities, alongside European rules which have to be implemented in Germany make it far from easy for structural engineers to decipher the current legal situation.

The following article will provide an overview (including online references) of the relevant codes of practice for structural engineers. It is important to remember that regulations on national and European levels are prone to change and since reflecting all such changes in this article is impossible, only the basic context will be explained in this case. We advise structural engineers to check the current status of regulations (particularly the current list of Technical Building Rules, Section A1.2 and the current Construction Products Lists, Section A1.3) on the specified web pages.

A1.1 Model Building Code and Federal Building Codes

The building codes of the individual German federal state dictate the form of construction supervision imposed, although these State Building Codes (LBOs) tend to follow the jointly compiled Model Building Code (MBO). The Model Building Code and the model ordinances of ARGEBAU can be viewed and downloaded on the ARGEBAU homepage (www.is-argebau.de; there: Mustervorschriften (model rules) / Mustererlasse (model decrees) / Bauaufsicht (construction supervision) / Bautechnik (construction technology)). However, the Model Building Code and model ordinances are not laws, but merely a framework providing guidance for the legal building regulations of each state. With simplicity in mind, the model regulations are adopted more or less uniformly across different federal states for construction products and types of construction. Even so, the regulations of individual federal states should still be taken into consideration.

§ 3 of the LBOs sets out the general requirements for a construction:

(1) Built structures must be laid out, built, altered and maintained such as to ensure there is no threat to public order and security; particularly life, health and natural resources.

(2) Construction products and types of construction may only be used if the built structures in question are subject to proper maintenance for an appropriate duration commensurate with the purpose, meet requirements pursuant to or based on this law and are fit for purpose.

(3) The technical rules stipulated by the highest building authority as publicly disclosed Technical Building Rules must be complied with.

These general requirements imposed on construction result in the following provisions, with which structural engineers must comply:

- Compliance with technical rules is mandatory to verify structural safety. These are included in the list of Technical Building Rules of German federal states (LTB).
- The construction products and types of construction to be used in built structures, are detailed in the so-called Construction Products Lists A, B and C.

A1.2 Technical Building Rules

The list of Technical Building Rules (LTB) of federal states is set out separately based on a sample list (M-LTB) for each state, meaning the content may vary slightly between individual states. Moreover, the individual lists are not released at the same time. The Deutsches Institut für Bautechnik (DIBt) is tasked with preparing the introduction, on behalf of the states (https://www.dibt.de/de/Geschaeftsfelder/BRL-TB.html):

- Part I: Technical rules for planning, design and construction of built structures and their parts.
- Part II: Regulations governing the use of construction products and kits in accordance with European Technical Assessments and harmonised standards pursuant to the Construction Products Regulation.
- Part III: Regulations governing the use of construction products and kits in accordance with European Technical Assessments and harmonised standards pursuant to the Construction Products Regulation and Sections § 17 para. 4 and § 21 para. 2 of the Model Building Code. This part concerns the regulation to assess the suitability of construction products and types of construction in accordance with water law through verifications in accordance with the Model Building Code (WasBau-PVO).

Part I, as specified on the DIBt homepage, is a model application, from which the state regulations may differ. Parts II and III, conversely, include valid usage regulations in Germany. The regulations for the state of Baden-Württemberg can be retrieved from the homepage of the Ministry of the Environment, Climate Protection and the Energy Sector. (http://um.baden-wuerttemberg.de/de/umwelt-natur/berg-und-baurechtsbehoerde/bautechnik-und-bauoekologie/technische-baubestimmungen/).

Part I

The technical rules for planning, design and construction are generally set out as standards (national or European) imposed as part of building regulations by inclusion in the LTB. In certain cases (e.g. technical rules for health protection or to protect against fire), directives may also be issued to meet building legislation requirements.

For timber structures, Part I of the LTB for EC 5 (DIN EN 1995) and its parts,

- Part 1-1: General – Common rules and rules for buildings,
- Part 1-2: General – Structural fire design,
- Part 2: Bridges.

are introduced to meet building legislation requirements.

Regulation is also imposed through respective National Annexes (NA) and the Nationally Determined Parameters (NDP, such as partial safety factors or load duration classes). The NAs also contain non-contradictory complementary information (NCI) and member states cannot amend European standards.

However, the application of EC 5 also requires compliance with additional provisions, contained in the annexes of the LTB. The relevant annexes for EC 5 are 2.5/1E and 2.5/2, which specify regulations to be complied with when using construction products in accordance with harmonised standards for timber structures. Here, reference is made to application standards (DIN V 20000-X), which also apply when using specific construction products in accordance with harmonised standards, although such use may also be excluded.

Relevant examples:

- Wood-based panels are governed by the harmonised standard EN 13986:2004, which is also set out in the Building Rules Lists of the DIBT (see next section). However, when using wood materials in accordance with EN 13986, compliance with the related application standard DIN V 20000-1:2005 is also required.
- Dowel-type fasteners (nails, screws, dowels) are governed by the harmonised standard EN 14592:2008, which is also set out in the Building Rules Lists of the DIBT (see next section). However, Annex 2.5/1E specifies that this standard is only applicable for bolts, dowels and smooth nails and other fasteners require additional approvals in line with building legislation requirements (Note: In this case, however, an application standard DIN V 20000-6 is compiled).

Finally, the list of Technical Building Rules Part 1 for timber construction also introduces

- DIN 1052-10: Design of timber structures – Part 10: Additional provisions

to meet building legislation requirements. This standard was necessary, since EC 5 and the related reference standards did not include all applications regulated to date in DIN 1052 (including resined staples, glued-in rods, steel rods with wood thread), or the European product standards did not meet the statutory requirements imposed in Germany to meet building legislation requirements (e.g. proof of suitability for gluing from an accredited test centre).

Part II

Since 01.07.2013, the European Construction Products Regulation (CPR), which regulates the certification procedure for construction products and kits, has come into force and differs in some aspects compared to the previous Construction Products Directive, which applied up to 30.06.2013. For example, there are no more guidelines, so-called ETAGs, and without guidelines, via so-called CUAPs, no further construction products and kits can be approved. The old ETAGs have been transferred to the new European Assessment Document (EAD), while existing European Technical Approvals (ETAs, whether or not a guideline is included) remain valid until the date of expiry.

In the regulations governing the use of construction products and kits in accordance with European Technical Assessments (ETAs) and harmonised standards in accordance with the Construction Products Regulation, the LTB of federal states make direct reference to the list of Technical Building Rules, as published in the "DIBt Official communication" (www.bauministerkonferenz.de or www.dibt.de). The applicable rules in Part II are classified in accordance with the procedure when creating the rules:

- Regulations governing the use of **construction products** within the applicable scope of European Assessment Documents (EAD) for European Technical Assessments or for construction products with European Technical Approval, which were compiled in accordance with a guideline prior to 01.07.2013.
- Regulations governing the use of **construction kits** within the applicable scope of EADs for European Technical Assessments or for construction kits with European Technical Approval, which were compiled in accordance with a guideline prior to 01.07.2013.
- Regulations governing the use of **construction products**, for which European Technical Approvals were compiled **without** a guideline prior to 01.07.2013.
- Regulations governing the use of **construction kits**, for which European Technical Approvals were compiled **without** a guideline prior to 01.07.2013.
- Regulations governing the use of **construction products** in accordance with **harmonised standards**.

This means any user wishing to identify applicable regulations in the Technical Building Rules must know whether the relevant object is a construction product or a construction kit and also whether the ETA was created in accordance with an EAD, prior to 01.07.2013 based on a guideline (ETAG) or without a guideline (CUAP). Finally, there are also construction products which correspond to harmonised standards.

For construction products subject to European regulations (via ETAs or harmonised standards), no additional regulations apply and they can be used, subject to inclusion in the Construction Products Lists (see next section) and taking into consideration any existing application standards (DIN 20000-X), as specified in Part I.

A1.3 Construction Products Lists

In § 17, the Model Building Code (MBO) regulates the use of construction products, which must correspond to one of the following specifications and be identified:

National

- Construction products regulated in accordance with Construction Products List A Part 1 (products for which technical rules (standards) exist.
- Non-regulated construction products, for which the following is required:
  - A general (national) technical approval or
  - A general (national) test certificate (Construction Products List A, Part 2) or
  - Consent in an individual case ("Zustimmung im Einzelfall").

Excluded are construction products of only minor importance for meeting the requirements of or based on this law and which the Deutsches Institut für Bautechnik (DIBt) has disclosed in a list C, by agreement with the highest building authority.

Technical rules (standards) apply to regulated construction products, with which the construction product must comply. If a construction product deviates significantly from the technical rules, or if no technical rule exists for the same, a general technical approval can be issued for this product (List of approvals for timber construction: https://www.dibt.de/de/Fachbereiche/Referat_I5.html).

If the construction products used do not meet key structural safety requirements or if the requirements can be assessed by generally accepted test procedures, instead of the general **technical approval**, a general **test certificate** (allgemeines bauaufsichtliches Prüfzeugnis) applies. These construction products are set out in the Construction Products List A, Part 2 and for construction kits in Part 3. Finally, **consent in an individual case** (Zustimmung im Einzelfall) can also be issued for the use of construction products in a specific construction, even if not all the above requirements are met.

European

Construction products in accordance with Construction Products List B:

- Construction products within the applicable scope of harmonised standards in accordance with the Construction Products Regulation (CPR).
- Construction products within the applicable scope of European Assessment Documents (EAD) for European Technical Assessments (ETA).
- Construction products, for which European Technical Approvals (ETA) were compiled prior to 01.07.2013 with or without a guideline.

The following note in Annex 01 of the Construction Products List is important for all construction products subject to European regulations: "The levels, classes and conditions of use specified in the Federal Building Codes (LBO) and in the provisions set out under the same shall apply." This means additional regulations often apply for a construction product in Germany subject to European regulations, which must be taken from LBOs or LTBs (or annexes thereof). The current Construction Products Lists are published on the DIBt homepage: https://www.dibt.de/de/Geschaeftsfelder/BRL-TB.html

If there are no additional regulations, construction products for which an ETA has been compiled can be used in Germany, even when not explicitly stated in the Construction Products List B. An overview of granted and valid ETAs is included on the EOTA page: http://valideta.eota.eu/pages/valideta/

A1.4 Marking

According to Federal Building Code § 22 Attestation of Conformity, construction products require an attestation of conformity with the technical rules, the general technical approvals, the general test certificates or consent in an individual case; any minor deviations are also classed as compliance. In this case, national regulated (A, Part 1) and non-regulated (A, Part 2) construction products are marked with an Ü-symbol (compliance mark), see Figure A1-1 on the left. This mark indicates the technical specification (standard or approval) with which the construction product complies and the nature of the Attestation of Conformity (declaration of the manufacturer or certificate of a certification body). A CE-mark, as shown in Figure A1-1 on the right, indicates the harmonised European standard or ETA with which the construction product complies and the reference number of the **Declaration of Performance** (**DOP**). This declaration of performance includes all values required for the planner, e.g. characteristic stiffness and strength properties of wood products, but also formaldehyde emission values for glued products. Figure A1-2 and Figure A1-3, respectively, show examples of Ü- and CE-marks.

Ü-mark in accordance with the compliance regulations of the federal states (e.g. ÜZVO)	CE-mark in accordance with the European Construction Products Regulation (CPR)
Letter „**Ü**“	Letter „**CE**“
Manufacturer and manufacturing plant or supplier and manufacturing plant	**Manufacturer** (Name and registered address or a mark)
	Last two figures of the **year**, in which the CE-mark was first applied
	Clear **identity code** of the product type
	Reference number of the declaration of performance (**DoP No.**)
	Declared performance by level or **class**
Technical **specification** and key features	Technical **specification**
Label of the **certification body** (only for construction products requiring certification)	Identification number of the **notified body** (if required for the construction product)
	Purpose of use (defined in the technical specification)

Figure A1-1 Right: Ü-mark. Left: CE-mark.

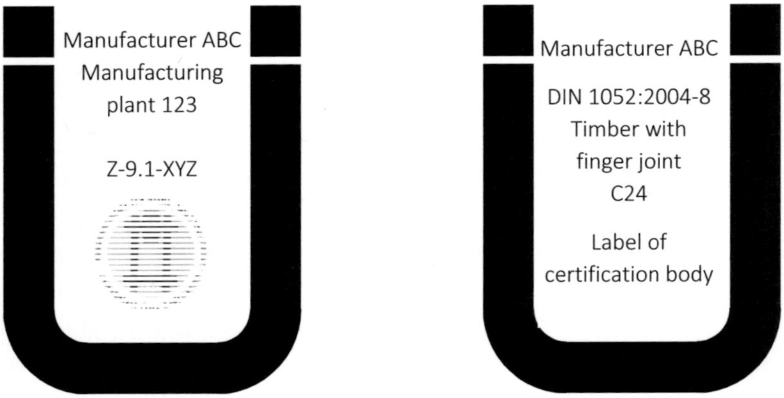

Figure A1-2 Example of Ü-mark, technical specification is the national technical approval (left) or national standard (right).

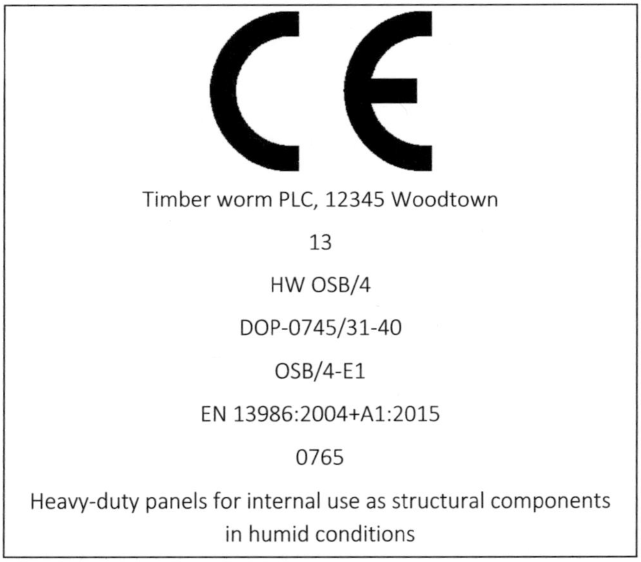

Figure A1-3 Example of CE-mark.

A2 Sustainability

Original article: T. Vihavainen

An important point to consider when using timber in constructions is the sustainability of wood and wood products. As a renewable resource, wood has a key advantage compared to alternative construction materials. For example, when biomass is formed (e.g. leaves, wood), CO_2 is extracted from air via photosynthesis, carbon (C) is incorporated in the biomass and oxygen (O) is released into the air. Thanks to this effect, timber is considered a CO_2-neutral construction material and it is only when biomass breaks down when wood is burnt or when wood rots in the forest that the same amount of CO_2 is released as was originally absorbed. One further advantage is the fact that wood requires far less primary energy than other materials when being processed into construction products. Wood grows naturally, need not be manufactured and is far easier to work with than e.g. steel or aluminium. Figure A2-1 shows the ratio of primary energy consumption required to produce a cubic metre of construction material from various raw materials. However, different mechanical properties and larger cross-sections mean larger volumes may be required to create wooden structures compared to those made of steel or concrete. Conversely, when it comes to other physical features like heat conduction, wood outperforms steel with a lower heat conductivity λ between 0.13 and 0.20 W/(m·K) (steel: λ = 60 W/(m·K)).

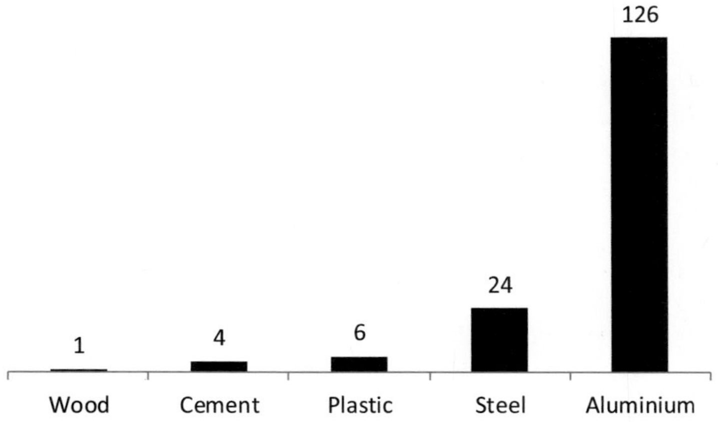

Figure A2-1 Ratio of primary energy consumption required to produce a cubic metre of construction material. (Forstabsatzfonds (Forestry Sales Fund): Ich, der Wald, bin mehr als Sie denken, Bonn, 1994)

A2.1 Life Cycle Analysis (LCA)

Sustainability is defined in numerous ways, but the original definition came from Hans Carl von Carlowitz, who requested as early as 1713 that "We should only take as much wood out of the forest, as will grow back." Nowadays, various aspects have to be considered when assessing the sustainability of a system. The forest example includes focusing on sociocultural aspects, such as its recreational value, economic aspects like the monetary value of the wood and ecological aspects like the need to preserve biodiversity. If we assess the sustainability of a building, we soon realise just how complex a task it really is. As well as the need to take the energy consumption and emissions of all materials into consideration, including their production and the building construction themselves, the assessment also includes expenditures during the useful life, for example the heat energy required. At the end of its useful life, the building is demolished and the left-over material is either recycled or used for energy recovery. Balance sheets, including sustainability assessments for these aspects, also have to be compiled. All the above points refer solely to the ecological aspects, so an ecological balance sheet for the building in question is compiled. To perform a complete sustainability assessment however, criteria for economic and sociocultural sustainability also need to be defined. Here, feasible economic criteria would include, for example, how well the building retains its value or the costs of the products used. Living comfort, the urban integration of the building or infrastructural criteria meanwhile, like public transport connections or distance from the supermarket, would constitute typical sociocultural aspects of a sustainability assessment.

Within the scope of this chapter, only the first section of a sustainability analysis mentioned, the ecological balance, is relevant and this kind of analysis, featuring the development of a system right up to the end of its life, is known as a life cycle analysis, LCA. Figure A2-2 schematically illustrates the life cycle of raw material over production, construction and usage phases right up to demolition and recycling or energy recovery from the waste material.

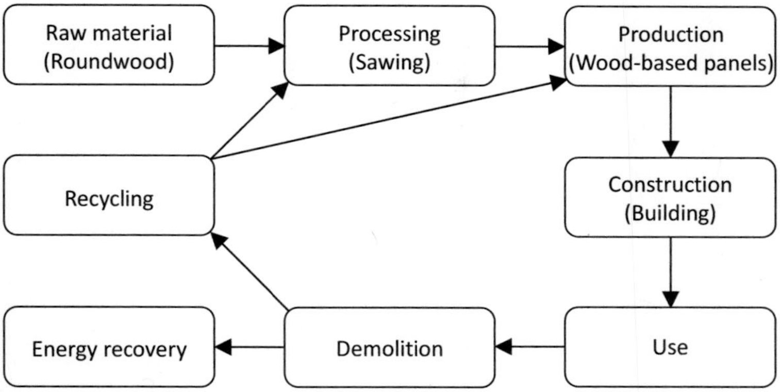

Figure A2-2 Life cycle. (Frühwald, 2007)

A life cycle analysis comprises four steps:

- Definition of system limits (goal and scope definition):
  The goal and frame of reference have to be defined. Is the goal to consider an individual product or the ecological balance of the interaction of multiple individual products? When there is a need to compare products made of various materials, consideration of a functional unit is often more significant. For example, static/dynamic reasons dictate that over a one-kilometre stretch of power line, more wooden power poles are used than those made of steel or concrete. Another example is the ecological assessment to determine the origin of the wood; do you start with the distribution of the seeds, the harvesting of the logs or only after the cutting? In response, the system limits and assessment scope must be configured in a way that facilitates comparison of the data determined.

- Life cycle inventory (LCI):
  The life cycle analysis determines all flows of materials and energy. The resource consumption (raw materials, auxiliary materials, energy, water,...) and emissions (end products, by-products, waste, air emissions,...) of the system observed are arranged, while a life cycle inventory is normally performed for the functional unit, namely per cubic metre or kilogram of product.

- Impact assessment (LCIA):
  The material and energy flows described in the previous step are evaluated in terms of their environmental impact, which involves considering a range of impact categories introduced after the following step.

- Interpretation:
  The data obtained in steps 2 and 3 is evaluated and recommendations are made for the client. This is mainly dictated by the quality of the available data and involves assessing the data quality itself. The parameters used may include comparability, completeness and the accuracy or consistency of the obtained data.

Some selected impact categories are:

- Global warming potential GWP (in kg CO_2-equivalent):
  The GWP specifies the extent to which a specific quantity of a greenhouse gas contributes to the greenhouse effect, where the comparable figure is carbon dioxide, namely the CO_2-equivalent. The scope usually involves observing the average heating effect over a century. All emissions recorded in the life cycle inventory are evaluated for their GWP and converted to a CO_2-equivalent figure.

- Acidification potential AP (in kg SO_2-equivalent):
  All emissions which exacerbate acidification are expressed in terms of sulphur dioxide equivalent.

- Human toxicity potential HTP (in kg of body weight):
  Recording all emissions impacting on health.

- Photochemical ozone formation potential POFP (in kg C_2H_4-equivalent):
  Emissions contributing to the formation of ozone are expressed in terms of ethylene equivalent.

- Eutrophication potential EP (in kg PO_4-equivalent):
  Expressing the contributing emissions as a phosphate equivalent.

The site http://www.oekobaudat.de lets you retrieve complete life cycle analyses for various construction materials. Key here is to ensure comparability of data at all times, given the variance in system limits used.

A2.2 Ecological significance of wood

Around 30% of global land is forested, most of which comprises tropical rainforests and boreal coniferous forests. Figure A2-3 shows an overview of the world's forests, with a global total of 4 billion hectares, more than half of which is distributed over the five most forested countries (the CIS, Brazil, Canada, the USA and China. Source: FAO 2010). The biomass of these forests also holds around 289 gigatons of CO_2 (FAO, 2010) and since CO_2 is the most significant greenhouse gas generated by human activity, this means forests perform a key carbon storage role. Such forests are known as carbon pools. Virgin forests establish an equilibrium and only emit as much carbon as they absorb. The volume of storage depends on the type of forest involved, the soil condition (CO_2 storage in the soil) and management. Forests can also function as carbon sinks if the timber stock rises, namely, where more forest is planted than felled. Forests where harvesting outstrips planting, namely where the net carbon release exceeds the figure incorporated into the growing wood, are deemed carbon sources.

Development of the global forested area (see Figure A2-4) shows a rising volume of forest in Europe (⟶ carbon sink), albeit at a slow rate, while those in South America and Africa are declining rapidly (⟶ carbon source). The high growth rate in Asia is due to the large-scale reforestation programs in China, in contrast to the shrinking forest elsewhere in Asia. This also underlines the need for continued vigilance in checking the origin of tropical woods, to reconfirm that they have been sourced from sustainably managed forests.

Sustainable management means not only minimising the ecological impact of the timber harvest as far as possible (no surface clearance, preserving the ecological function of the forest, including all animals and other plant varieties, preserving the most valuable areas of forest from an ecological perspective, control tools for monitoring), but also ensuring the long-term sustainability of forest management by getting forest farmers involved and clearly defining their claims of ownership and use. As well as these social aspects of sustainable forestry, however, economic aspects also play a part, such as ensuring an equal share of proceeds from the timber trade and the fact that forest management generally has to be economically self-sustaining. Establishing a sustainable forestry can therefore retain the function of the forest as a carbon reservoir and provided the forest is used effectively and sustainably, can help it fulfil its role as a carbon sink, if more biomass is produced than harvested, as applies in Europe, for example.

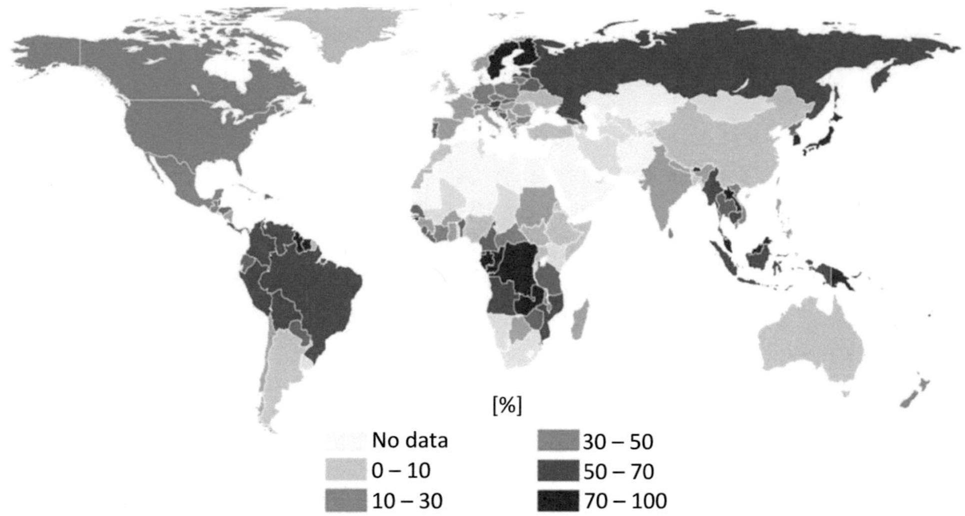

Figure A2-3 Ratio of forest area as a proportion of the overall land area. (FAO, 2010)

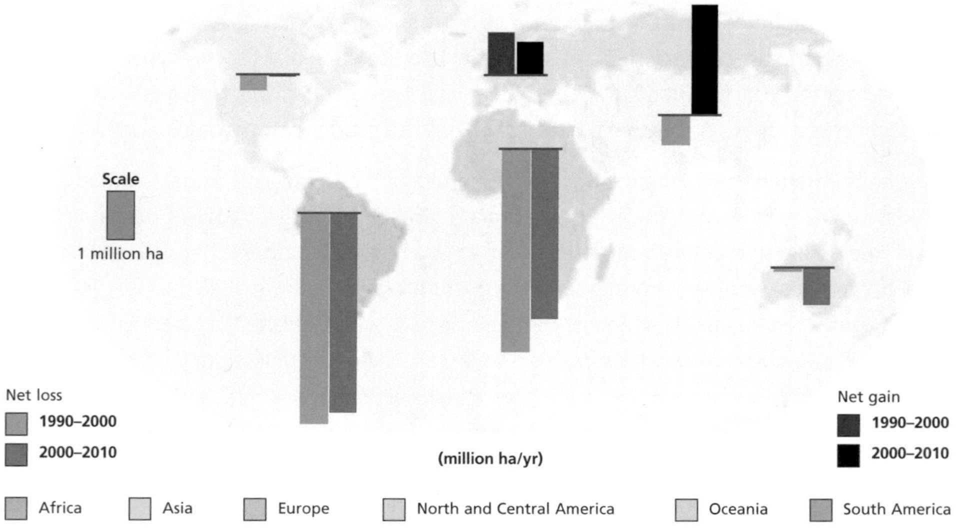

Figure A2-4 Change in forest area per region. (FAO, 2010)

A2.3 Sustainable forest management

Deforestation and silvicultural exploitation are becoming increasingly key aspects of the international environmental debate and the fact that our forests are vanishing threatens the Earth like never before. In response, most countries have committed to using forests sustainably and for a long while, efforts to preserve production capacity as a measure of sustainability came to the fore. Equally important, however, is focusing on wider ecological goals, such as protecting forest ecosystems as a whole, preserving diversity and nurturing cultural, aesthetic and recreational values. This reflects how logging is changing ecological conditions for forests in many areas. Numerous forest plants and animal species require specific conditions in which to flourish. Nowadays in Europe, truly natural forests are very rare and the natural forests or special biotopes that do exist are not generally commercially exploited, or timber production is strictly regulated. Meanwhile, debate continues to rage as to whether sufficient areas or natural parks have been established.

These are all reasons why timber consumers need to verify that the timber they use comes from sustainable forestry, the most well-known certificates for which are the label of the Forest Stewardship Council (www.fsc.org) and PEFC (www.pefc.de). Both prioritise sustainable forestry, which preserves the ecological, social and cultural values of forests and an unbroken "chain of custody" allowing the trajectory of the wood, from felling right up to arrival at the end consumer, to be properly traced.

A2.4 Literature

T. Vihavainen, original Article A16, STEP 1995.

FAO (2010). Global forest resources assessment 2010. FAO Forestry Paper 163, Rome.

Frühwald A. (2007). The ecology of timber utilization, life cycle assessment, carbon management etc.. Kick-off meeting Probos Foundation, Doorn, the Netherlands.

B

Material wood

B1 Wood anatomy

Original article: P. Hoffmeyer

Only with a solid grounding in wood anatomy can mechanical and other physical characteristics of timber be understood and wood used in a way that ensures material compatibility or optimally exploits its potential. Wood can also be observed on a range of levels, as clearly shown in Figure B1-1, while a tree trunk can be observed in terms of structural timber, which involves evaluating knots and other growth-related properties. Observing at a level of detail beyond that for structural timber means examining wood on a macroscopic level, including all features visible to the naked eye. The scope also includes flawless wood without fibre deviation (clear wood). When the mechanical properties depend on the system size, meanwhile, we describe the wood on a mesoscopic level, e.g. in terms of annual rings. The microscopic level includes the wood fibres within the vessels, the large water-conducting cells in hardwood and the smaller tracheids, key to hardwood strength and performing both reinforcement and water-transport functions in softwood. Besides the cell wall structure, the chemical structure can also be examined.

Wood is a natural and organic material, comprising cells. It is also a compound chemical complex of cellulose, hemicellulose, lignin and other constituents. Wood tends to be anisotropic, due to the elongated structure of its cells and the orientation of the cell walls. Anisotropy is also a factor of the variation in cell size during growth and partially due to the preferred direction of specific cell types (e.g. rays). Meanwhile, the three key structural variables impacting on the properties of wood as a construction material are its fine cell wall structure, the collection of cells in clear wood and growth irregularities in timber. For example, the fact that shrinking and swelling perpendicular to the grain is generally 10 to 20 times as large as in the grain direction can be explained by the submicroscopic structure of the cell walls. The microscopic structure of clear wood, meanwhile, is why wood is 20 to 40 times more rigid longitudinally rather than perpendicular to the grain, while its macrostructure (knots, fibre deviation) underlines its tensile strength in the grain direction. This may range from over 100 N/mm^2 in clear wood to under 10 N/mm^2 for low-quality timber, for the same wood species.

This chapter will initially explain the general stem structure and section planes of the wood, before focusing on its anatomy, from the macroscopic level right up to the chemical structure. Wood can be extracted from two main groups of plants; softwoods and hardwoods, where the vast majority used for construction in Central Europe are spruce and fir, two conifers. The differences between conifers (softwoods) and deciduous trees (hardwoods) emerge in their growth pattern, different leaf shapes and also the structure of the wood itself (Figure B1-2).

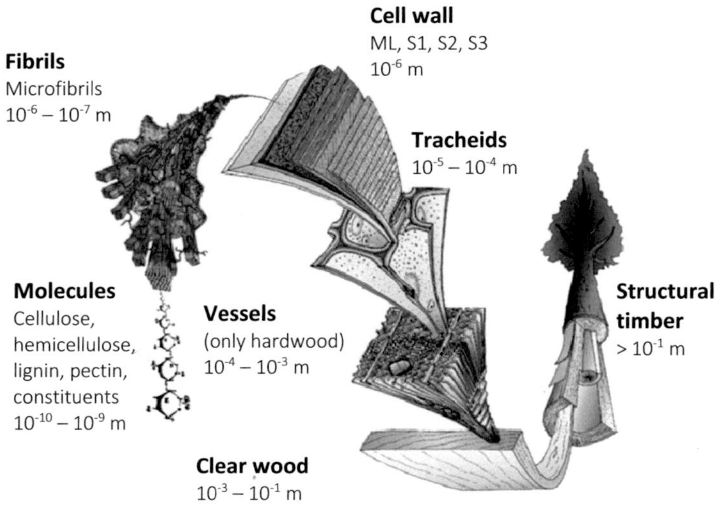

Figure B1-1 Hierarchical levels of wood. (Mark Harrington, University of Canterbury, 1996)

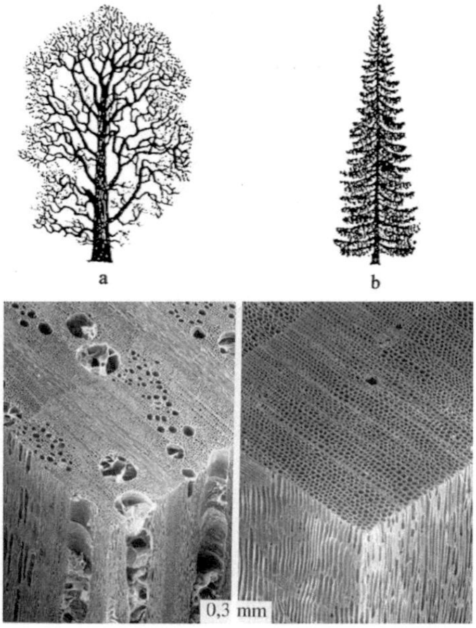

Figure B1-2 Growth pattern and microscopic structure of oak (*Quercus robur*, left) and spruce (*Picea abies*, right). (STEP 1995 Article A4)

B1.1 General structure of a tree stem

The role of a tree's stem or trunk is to support the crown and supply it with water, while at the same time, the nutrients formed in the crown are transported down the stem via its bark. Nutrients not used to form wood are channelled towards the roots and stored there. Accordingly, a tree stem meets the requirements imposed on it, as its structure reflects. Most cell structures (water canals, fibres) follow the axial structure of the stem, with only a small percentage (rays) oriented radially, to guarantee the radial transport of nutrients and water. A tree thus grows both longitudinally as well as around its girth and the concentric rings formed over the course of a year within the so-called cambium are known as annual rings. The annual rings themselves are, in turn, sub-classified into early- and latewood. Wood is thus not homogenous or isotropic, but a naturally grown material, which always adapts to the respective local conditions. Figure B1-3 and Figure B1-4 show the stem structure of a tree.

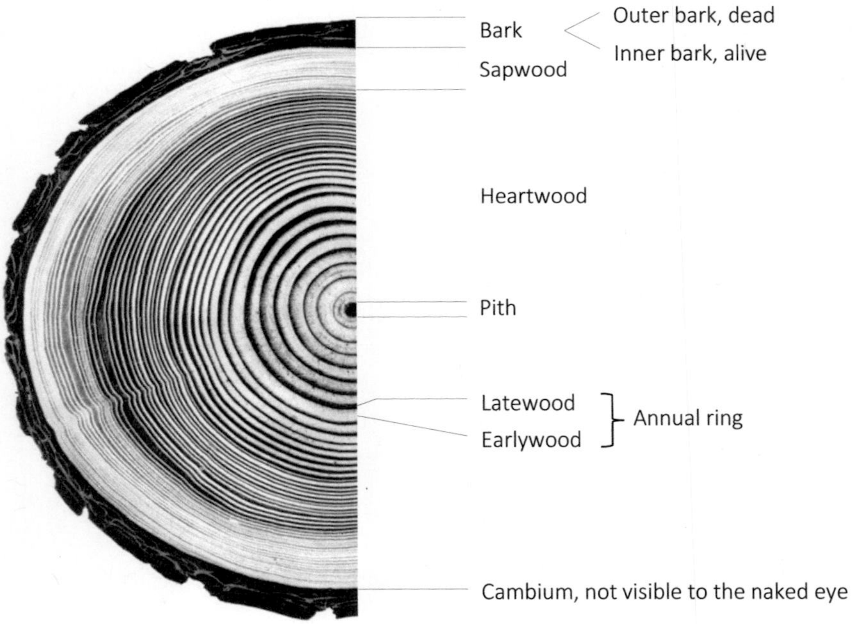

Figure B1-3 Cross-section through a coniferous stem, Douglas fir. (Grosser, 1977)

B1.2 Section planes of wood

The natural section planes of the wood emerge as a result of the clearly axial symmetrical and round cross-section of the stem, as is clearly shown in Figure B1-3. The main section planes are cross-, radial and tangential sections, as shown schematically in Figure B1-4 and a complete picture of the three-dimensional wood structure is only possible when all three sections are combined.

The **cross-section** is performed at right angles to the stem axis and substantially circular as a rule; revealing heartwood, sapwood and annual rings. Since the dominant axial structure has been severed, this cross-section often reveals a reticular texture (see Figure B1-5), while the rays run from inside to outside.

The **radial section** splits the stem longitudinally through the pith, similar to a circle segment and shows the axial and radial cell structures. Here, the rays run horizontally and the annual rings can be clearly seen.

The **tangential section**, meanwhile, runs tangentially to the annual rings, like a circle segment. The annual rings do not show up more easily in this case. The axial cells are easily visible, while the radial rays are visible in the cross-section.

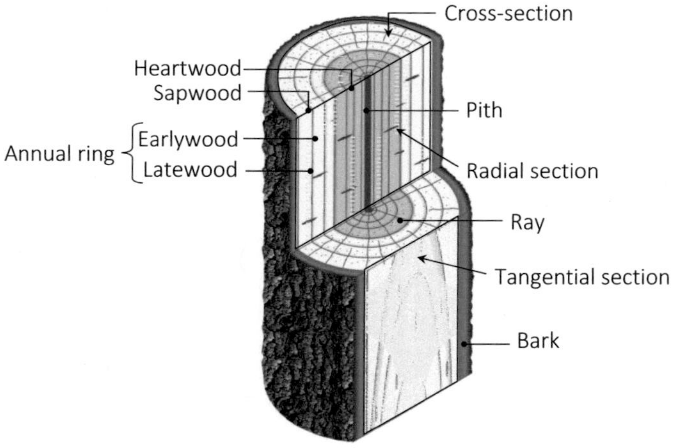

Figure B1-4 Macroscopic structure and section planes.

B1.3 Macroscopic structure

The following section will explain the macroscopic structure from inside to outside. We start with the pith over the rays, to explain the heartwood and sapwood, then spotlight the functions of the cambium and bark.

Pith

The pith comprises parenchymatous – accumulating – tissue comprising so-called parenchyma cells. Its cross-section may be rounded, radial or square and its diameter is only in the order of millimetres. Its main role is to supply the young and developing scion with water and it perishes prematurely in a range of woods.

Rays

Rays, which exist in all deciduous trees and conifers, also include parenchyma cells and facilitate the radial transport of water and nutrients, while their size and frequency vary significantly by wood species. They form bright and fine lines, seldom more than 1 mm wide, which run radially from the centre of the stem to the outside. Only the initially formed and so-called primary rays lead from the pith to the bark, which is why they are called pith rays. All so-called secondary rays are unconnected to the pith. The later they emerge from the cambium, the further out their point of origin in the stem.

Heartwood and sapwood

As the secondary growth of the tree continues, the entire stem cross-section is no longer used to conduct water, but only the external portion and younger annual rings, between 5 and 100 or so, depending on the tree species. This external portion of the stem is known as sapwood and also functions as a storage recipient, characterised by the presence of living and physiologically active wood cells. The inner portion, meanwhile, known as heartwood, generally no longer contains living cells and is solely tasked with stabilising and strengthening the tree. Sapwood transforms into heartwood via biochemical processes, so-called heartwood formation, which is synonymous with the death of parenchyma cells and consumption of their starch deposits. As the heartwood forms, the gas and water balance change, while the pits between the individual cells close off. In addition, constituents are often stored in the cells. Heartwood is often darker in colour, with differences in shade attributable to the chemical structure of the heartwood pulp. Incidentally though, lighter coloured heartwood does not necessarily indicate a lack of heartwood pulp, but merely the fact that no pigmented heartwood pulp has been formed. The fact that heartwood closes itself up and dies often makes **it far more durable than sapwood**.

Cambium

The cambium comprises an area of dividable cells and is an extremely thin layer between the wood and bark. It retains its divisibility up to the point at which the tree dies, forming wood cells inwardly and inner bark cells outwardly and facilitating the secondary growth (growth in girth) of the tree. It is not visible to the naked eye within the cross-section.

Bark

Bark is further sub-classified into outer and inner bark. The outer bark comprises dead cells and functions to protect the stem, while the inner bark is made up of living cells and transports nutrients produced in the crown to the cambium and storage cells.

Resin canals

Many wood species house resin canals, which are macroscopically visible. Resin canals run parallel and perpendicular to the stem axis and transport resin, which is used as sticky protection for the tree against wounds or invaders. The canals are lined with parenchyma cells capable of excreting resin, so-called epithelial cells, although not all wood species have resin canals. Firs and yews are resin-free, although resin is contained in spruce, pine, larch and Douglas fir trees.

B1.4 Mesoscopic structure, annual rings

The secondary growth of the tree is shown in the form of concentric annual rings. In areas exposed to seasonal and temperate climates, the secondary growth is closely linked to the prevailing climatic conditions, namely the water supply and temperature. When the growing season gets underway in spring, the tree must conduct water effectively and swiftly to ensure proper supply to the treetop. Reflecting this, the portion of the wood structure in conifers which grows during spring is characterised by a large pore volume and thin-walled cells, since the tree is focusing on additional transfer of water. In hardwoods, conversely, particularly large and numerous vessels are the stand-out features and the elements responsible for the transfer of water. Wood formed in spring is known as earlywood, but the priorities of the tree change when it comes to late summer and autumn. Key at this time is the strength of the wood and the importance of the water supply declines. Accordingly, conifers tend to form particularly thick-walled cells with a small cell lumen and small cell cavities, whereas deciduous trees form fewer and narrower vessels. This so-called latewood is far thicker than earlywood and in softwood in particular, these differences in thickness result in clear colour changes between early- and latewood (see Figure B1-3).

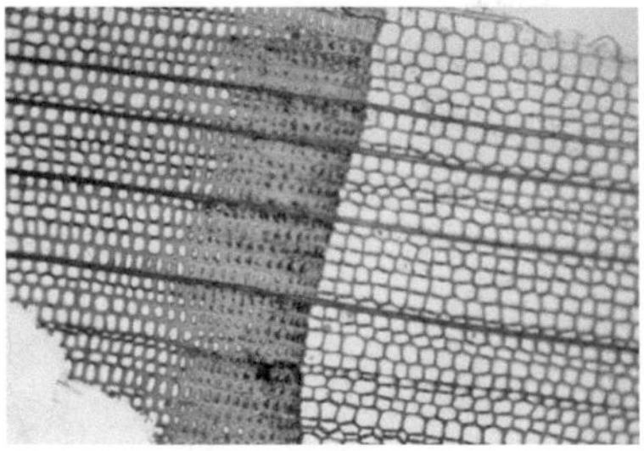

Figure B1-5 Cross-section through softwood, in which early- and latewood are clearly visible. (Schmid, 2002)

In winter, meanwhile, the tree stops growing. In softwood, Figure B1-5, the prominent disparity in cell wall thicknesses leads to characteristic variation in density, colour and hardness within an annual ring. The wide-lumen earlywood is lighter and the thicker latewood is darker in colour.

Figure B1-6b clearly shows that in coniferous trees, earlywood comprises the vast majority of the annual rings. Even so, when exposed to stress, particularly dryness and cold, a tree may only be capable of very limited secondary growth and accordingly only forms minimal earlywood. In particularly bad years, the annual ring may be narrow to the point of being virtually invisible to the naked eye and comprise no more than a few isolated latewood cell lines. In Figure B1-6, the differences in annual ring widths for various tree species and within a single tree species can be clearly seen. Spruce (Figure B1-6a) is a softwood that grows far faster than yew (Figure B1-6c); the same applies to poplar (Figure B1-6d) and robinia (Figure B1-6e, in which the vessels present in hardwood are also easily visible). The yew clearly shows the external impacts on secondary growth, as reflected in the very varied width of annual rings. This leads us to conclude that the width of the annual ring, the proportion of latewood and the regularity of annual growth (particularly based on differences in respective densities) all have a key impact on the strength of the wood.

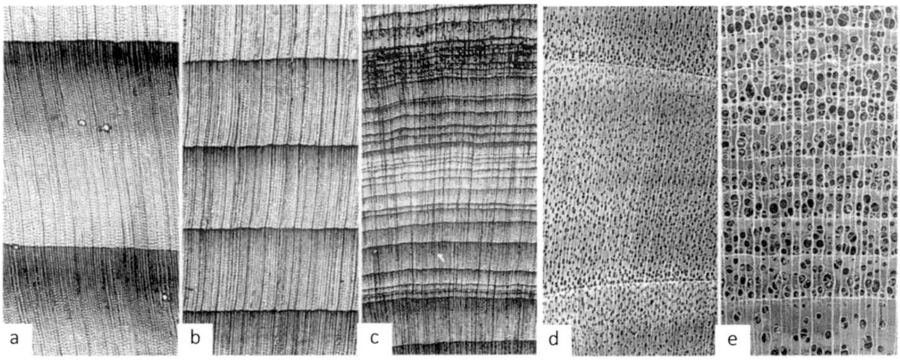

Figure B1-6 Annual ring widths (cross-section): a. wide annual ring in spruce, b. medium-wide annual rings in fir, c. narrow annual rings in yew, d. wide annual rings in poplar, e. narrow annual rings in robinia. (Wagenführ, 1999)

B1.5 Microscopic structure, cell types

Softwood

The wood structure of conifers comprises just two cell types; tracheids and parenchyma cells. Tracheids are long (2 to 5 mm) and thin (10 to 50 µm) cells with tapered or curved sealed ends. As shown in Figure B1-7, the cells are arranged in two criss-crossing systems, so tracheids are distinguished in terms of longitudinal and perpendicular tracheids, while parenchyma comprise longitudinal and ray parenchyma. The longitudinal tracheids take up by far the most space in conifers at around 90 to 95%. Rays (= ray parenchyma), longitudinal parenchyma and resin canals, in contrast, occupy a relatively modest 5 to 10% in the form of remaining cell elements or tissue systems. Some trees, however, such as the yew, have no longitudinal parenchyma and resin canals at all. In firs, meanwhile, there are no resin canals in normal wood and longitudinal parenchyma are only very seldom present. The parenchyma excretion cells, which line the resin canals and from which the resin emerges, are known as epithelial cells.

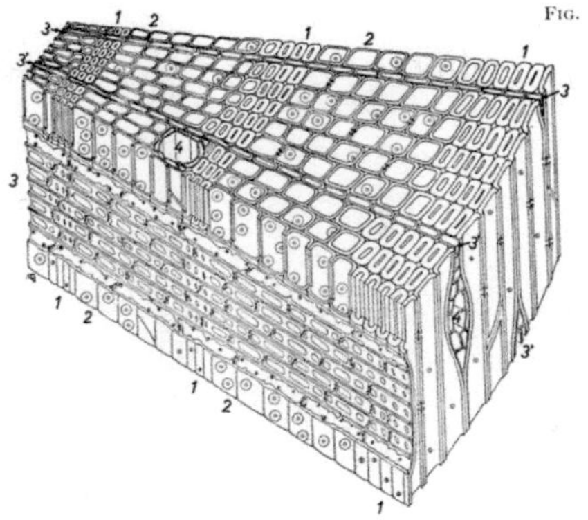

1) Latewood tracheids
2) Earlywood tracheids
3) Rays
4) Resin canals

Figure B1-7 Schematic structure and cell types of softwood. (Nardi-Berti, 1993)

Hardwood

Hardwood has evolved to a higher level than softwood and also includes distinctly more cell types (depending on the species) and a more extensive division of labour, i.e. specialisation of cells, than softwood. See Figure B1-8 for the schematic structure. Hardwood includes a range of tracheids (= dead cells), which is why people tend to refer to hardwood in terms of fibres rather than tracheids. The libriform fibres, for example, are solely for strength, whereas the fibre tracheids also function to transport water as well as providing reinforcement. The vessels or pores, which comprise dead cells and solely conduct water, can be up to several metres long in some tree species. As in softwood, the parenchyma cells store nutrients, but unlike softwoods, the rays in hardwoods may also comprise multiple rows of parenchyma cells and are often clearly visible.

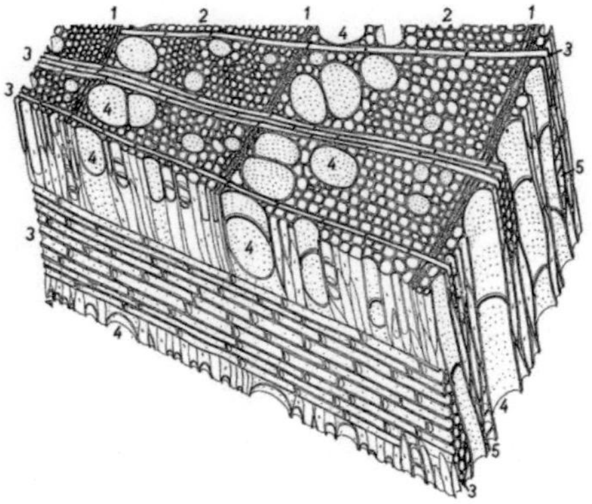

Latewood section
Earlywood section
Rays
Vessels
Longitudinal parenchyma

Figure B1-8 Schematic structure and cell types of hardwood. (Nardi-Berti, 1993)

Pits

The cell-to-cell material exchange takes place via small openings or gaps in the fibre wall, which are known as pits. The main type in softwood is bordered pits, as shown in Figure B1-9. As well as letting water through, they also prevent any ingress of air into sap-filled cells; preventing the collapse of the water columns running from the roots to the crown, which would ultimately mean the tree dying. Since closure of a pit is generally irreversible, this must be borne in mind during wood drying or impregnation. For example, bordered pits in the spruce tree shut very quickly, which often means poor permeability for this species. (Impregnating agent cannot penetrate.)

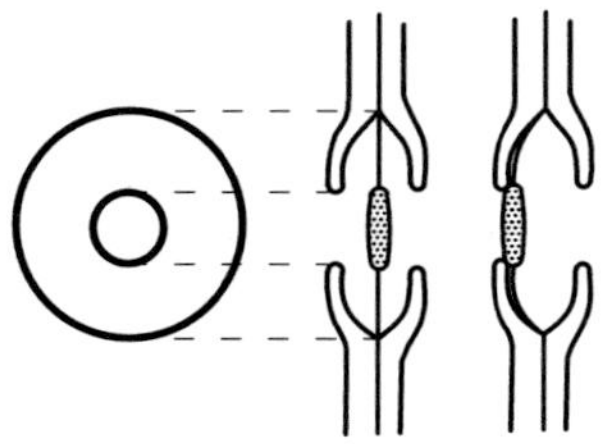

Figure B1-9 Bordered pits of conifers, schematic. Left: top view, centre and right: longitudinal section, right: closure under one-sided pressure. (Strasburger, 1998)

B1.6 Submicroscopic structure, cell wall

The wall of a wood cell generally comprises various cell wall layers, namely middle lamella and the primary, secondary and tertiary walls. The individual layers differ in terms of thickness, chemical composition and orientation of the cellulose micro-fibrils. The structure is schematically shown in Figure B1-10 and the cell wall structure itself dictates many of the mechanical properties. The orientation of the cellulose micro-fibrils, for example, shows the direction where the tensile strength peaks. From an engineering perspective, the cell wall is a particularly ingenious construction and the dominant S2 layer of axially oriented micro-fibril bundles can withstand tensile forces very effectively. When pressure is exerted, long and slender columns emerge from the micro-fibril bundles, but do not buckle thanks to the reinforcing effect of the subtly sloping internal and external S1 and S3 layers.

Middle lamella

The middle lamella interconnects neighbouring cells and adjacent cells possess a common middle lamella; mainly comprising lignin and pectin. The high lignin content gives the middle lamella very high compressive strength, while the pectin acts as a binder. The middle lamella is 0.5 to 1.5 μm thick and appears thinner in earlywood than latewood.

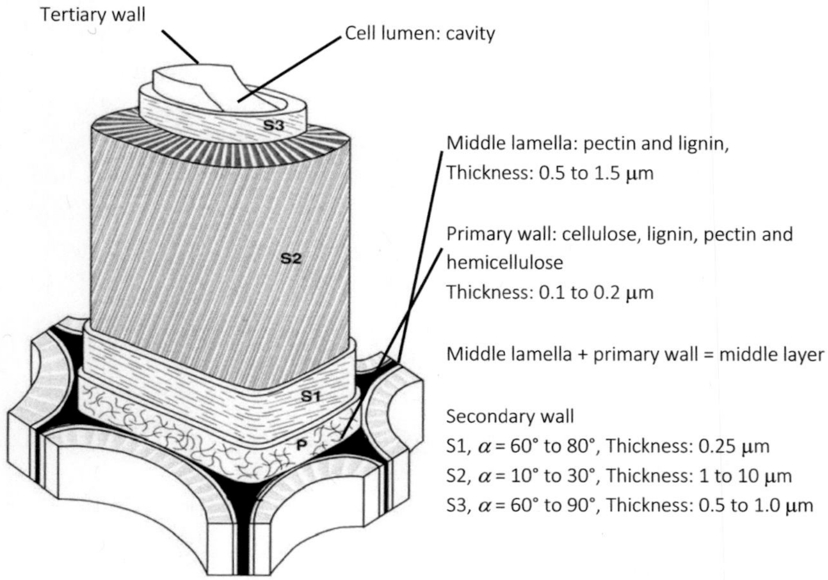

Figure B1-10 Structure of the cell wall. (Booker and Sell, 1998)

Primary wall

The middle lamella forms the middle layer together with the primary wall. The diffuse texture (see Figure B1-11) of the cellulose micro-fibrils, namely their amorphous non-crystalline arrangement, helps ensure the high dimensional stability of the cell. The primary wall is 0.1 to 0.2 μm thick.

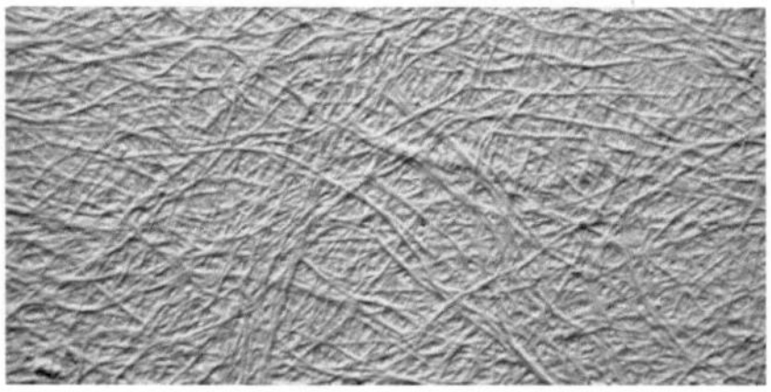

Figure B1-11 Diffuse texture of the primary wall.

Secondary wall

The secondary wall connects to the middle layer and is by far the thickest cell layer, with around 90% of its volume comprising cellulose micro-fibrils. The secondary wall includes three distinct wall layers, which differ in terms of the orientation of the cellulose micro-fibrils and thicknesses. The external secondary wall (S1) is also known as the transition lamella; located on the primary wall and around 0.25 μm thick. The cellulose micro-fibrils feature a parallel texture with a fibril angle of 60° to 80°, meaning they are virtually perpendicular to the cell axis, see Figure B1-10. The central secondary wall (S2) forms the main portion of the cell wall in the early- and latewood with a thickness of 1 to 10 μm. The fibril angle is at around 10° to 30° to the cell axis and therefore basically in axial direction, as also shown in Figure B1-10. The micro-fibrils are tightly packed and run parallel to each other in a spiral shape (screw-texture) in the direction of the cell axis. The internal secondary wall (S3) is around 0.5 to 1.0 μm thick and the micro-fibrils have a parallel texture. The fibril angle to the cell axis is 60° to 90°; and as in the S1 layer, the micro-fibrils run virtually perpendicular to the cell axis.

Tertiary wall

The tertiary wall divides the cell wall from the cell lumen and like the middle layer, contains a high proportion of lignin.

Cell lumen

The internal hollow cell space, which is filled with air or water, is known as the lumen, which is why you will often hear the term wide-lumen cells used to describe cells with a large lumen as opposed to narrow-lumen cells. Earlywood and latewood typically include wide- and narrow-lumen tracheids respectively.

B1.7 Chemical structure

Biomass is formed through the process of photosynthesis, whereby CO_2 is extracted from the air, carbon is incorporated into the biomass ($C_6H_{12}O_6$, carbohydrates) using water (H_2O) and the excess oxygen is discharged back into the air:

$$6H_2O + 6CO_2 \rightarrow C_6H_{12}O_6 + 6O_2 \tag{B1-1}$$

Wood mainly comprises organic compounds:

50%	carbon	*C*
43%	oxygen	*O*
6%	hydrogen	*H*
< 1%	nitrogen	*N*
< 1%	minerals	

These main components cluster to form chain and macro molecules, which then form the wood structure. The molecules formed in this case are cellulose, hemicellulose and lignin. Wood also contains other constituents like tanning agents or colourants and resins. Taking all tree species, the proportion of wood made up by cellulose is around 40%. However, the constituents vary significantly, depending on the tree species in question and also result in dark-coloured heartwood in certain tree species.

Cellulose

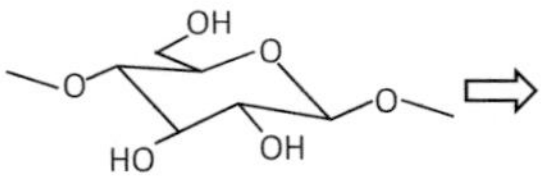

β-1,4-glucose

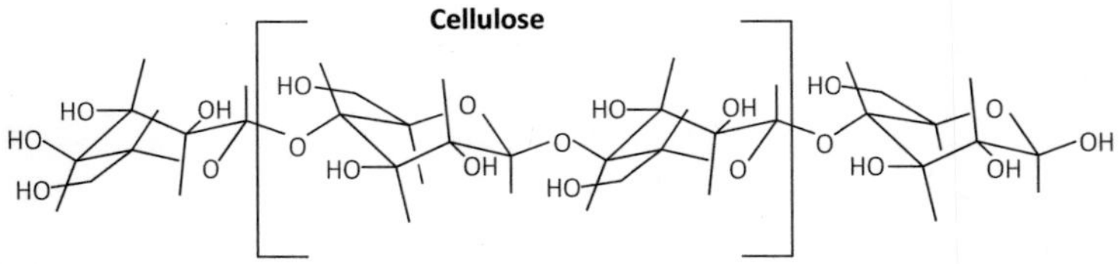

Cellulose is an extremely long-chain unbranched polysaccharide, which is formed from at least 10,000 glucose components, so-called *β*-1,4-glucose. Cellulose is (in the chain direction) a high-polymer chain molecule with very high tensile strength, which can exist in crystalline and amorphous form. The fibrils of the cell wall are formed of cellulose molecules and cellulose ensures the axial strength of the cell. Cellulose is particularly hydrophilic, due to the numerous free hydroxyl groups (*–OH*).

Hemicelluloses

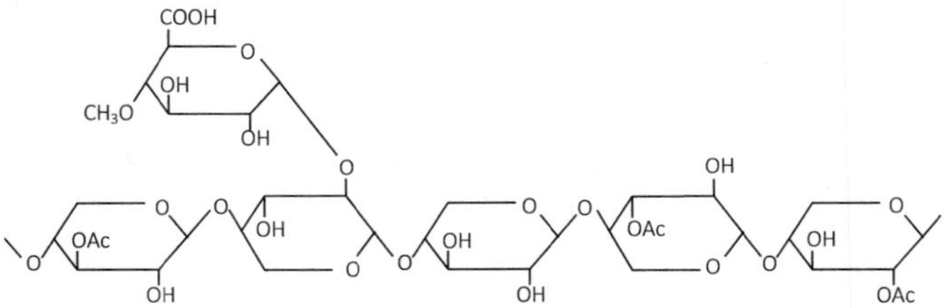

Unlike cellulose, so-called hemicelluloses comprise five different sugar components (glucose, galactose, mannose, arabinose and xylose), for polysaccharides that are short-chain, branched and – unlike cellulose – three-dimensional. Hemicelluloses have many free hydrophilic groups (*–OH, –CHO, –COOH*) and are highly reactive, due to their branched structure. Along with cellulose, hemicelluloses form the framework of a wood cell (although fibrils are exclusively made of cellulose), but the hydrophilic groups mean hemicelluloses have other functions. They absorb or discharge water to control the permeability of the cell membrane or may also act as matrix components and sealant.

Lignin

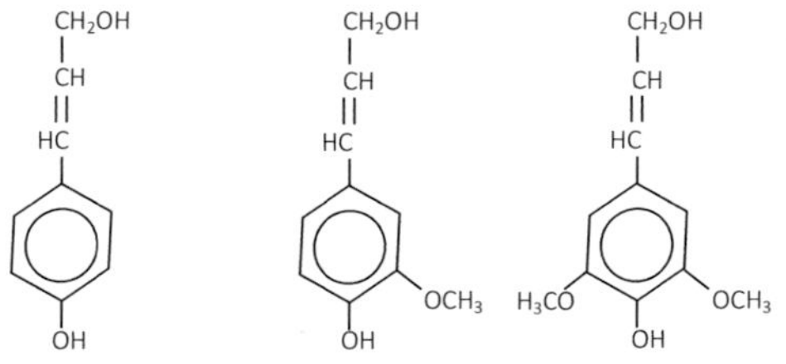

p-coumaryl alcohol   coniferyl alcohol   sinapic acohol

Lignin is a highly complex three-dimensional macromolecule, formed from three molecules of coumaryl alcohol, coniferyl alcohol and sinapic alcohol. Strictly speaking, it should be referred to as lignins in the plural, since the composition of the macromolecules can vary quite considerably and the three components are the product of numerous compositional variations. Lignin molecules may also differ depending on the extraction method, which is why, for example, we talk of Klason lignin using the Klason method. Since lignin is very pressure-resistant and hydrophobic, it gives the cell compressive strength as well as featuring as a fixed and rigid compound for micro-fibrils. In addition, lignin also protects the cell wall against the ingress of water.

Constituents

As previously explained, the constituents depend on the tree species concerned and despite comprising only a modest proportion of the overall wood mass, can exert a major impact on various properties. However, the proportion of constituents not only varies according to the tree species, but also within a tree species itself, or even a single tree. For example, the tree directs far more constituents like resin to damaged areas than those which are unscathed and heartwood generally contains more constituents than sapwood. These constituents, in turn, may exert an impact on various chemical, biological and physical features, such as the pH value, pest resistance, odour or colour. Resins, for instance, are used by the tree to seal up wounds and combat pests; invading insects are "stuck together." Greases repel water; colourants alter the colour and sensitivity to light. Terpenes and phenols have microcidal effects. With this in mind, one interesting finding is that trees from areas with a moderate climate generally have fewer constituents, around 1 to 10% by volume, than those from the tropics, where the equivalent figure ranges between 2 to 30% depending on the tree species.

B1.8 Literature

P. Hoffmeyer, original Article A4, STEP 1995.

Booker R.E. and Sell J. (1998). The nanostructure of the cell wall of softwoods and its functions in a living tree. Holz als Roh- und Werkstoff 56:1-8.

Grosser D. (1977). Die Hölzer Mitteleuropas. Ein mikrofotografischer Lehratlas. Springer Verlag, Berlin, 208 p.

Nardi-Berti R. (1993). La struttura anatomica del legno ed il riconoscimento dei legnami italiani di più corrente impiego. Istituto del Legno, Consiglio Nazionale delle Ricerche, Florenz, 155 p.

Schmid M. (2002). Anwendung der Bruchmechanik auf Verbindungen mit Holz. Dissertation, Universität Karlsruhe (TH).

Strasburger E. (1998). Lehrbuch der Botanik. Gustav Fischer Verlag, Stuttgart, 1007 p.

Wagenführ R. (1999). Anatomie des Holzes. DRW-Verlag, Leinfelden-Echterdingen, 188 p.

B2 Wood physics

Original articles: P. Hoffmeyer, L. D. Andriamitantsoa

Wood physics, a key element of wood sciences in general, draws on findings from wood chemistry, wood anatomy and biology, as well as classical chemistry, physics and mechanics and can be defined as the "Science of physical-mechanical properties of wood and wood-based materials". Fields of wood physics meriting further discussion include how wood behaves when exposed to moisture, wood density and the rheological (visco-elastic) properties of the wood.

B2.1 Wood moisture content

The way wood is arranged structurally means its moisture content determines almost all its other properties, which becomes particularly evident when going below the fibre saturation point. With increasing moisture:

- Stiffness and strength decrease,
- Sustained loading increases creep deformation,
- The thermal conductivity of wood rises,
- Wood becomes increasingly prone to fungal infection, particularly when the moisture content exceeds 20%.

Table B2-1 indicates the impact of humidity on certain mechanical properties of wood, whereby the more moisture is absorbed, the lower the strength observed, as the molecular binding forces decline. Namely, the ingress of water into the intermicellar and interfibrillar cavities forces the cellulose chains far apart. In an oven-dry state, the close proximity of the cellulose chains in the wood helps generate strong intermolecular binding forces, which render the structural and bonding substance stiff and brittle. However, as water penetrates the cell wall and more moisture is absorbed, the forces of attraction decline and the hydrogen bonds holding the cell wall together are weakened. Changes in moisture content exceeding the fibre saturation point have no further effect on mechanical properties, but simply lead to excess water accumulating in the cell cavities, while a combination of high temperature and humidity render the wood far more malleable and prone to permanent deformation. This is precisely the property exploited when making so-called (Thonet) coffeehouse stools from steam-bent beech wood.

The impact of changing levels of moisture content on the various mechanical properties varies. For example, the wood may break down when compressive stress is applied along the grain due to the fibres buckling, a process in which the moisture-sensitive hydrogen bonds play a key role. Conversely, tensile breakdown along the grain signals a break in the covalent bonds as the micro-fibrils of the cell wall are torn apart. Accordingly, compressive strength is more sensitive to humidity than tensile strength.

Table B2-1 Change in the properties of clear wood with a one percent change in wood moisture. The benchmark figure is the properties at a moisture content of 12%.

Property	Change
Compressive strength parallel to the grain	6%
Compressive strength perpendicular to the grain	5%
Bending strength	4%
Tensile strength parallel to the grain	2.5%
Tensile strength perpendicular to the grain	2%
Shear	2.5%
Modulus of elasticity (MOE) parallel to the grain	1.5%

It is crucial to ensure appropriate moisture content with subsequent construction use in mind. This also helps avoid unwanted swelling and shrinking movements of the wood, which result in timber members cracking or warping of panels in the absence of expansion gaps.

Wood is a capillary-porous material, with porosity ranging from around 50 to 70% depending on density and a correspondingly huge internal surface area. Its cavity system is hygroscopic, capable of absorbing airborne moisture, while capillary transport processes in the cell lumen also allow liquid water or other liquids (e.g. wood preservative or adhesives) to be absorbed. Depending on the water content of the wood, three boundary conditions are distinguished:

- Oven-dry
  The wood contains no water, meaning 0% moisture content.
- Fibre saturation point
  The entire microsystem of the wood is filled with water. The fibre saturation point occurs at around 28% and differs slightly depending on the type of wood concerned.
- Water saturation
  The microsystem and macrosystem (cell lumen) of the wood are filled with water.

The term used for the proportion of water in the wood up to the fibre saturation point is bound water, since the stored water is bound in cellulose hydroxyl groups within the micro system, by hydrogen bonds in the cell walls. The remaining water in the macro system up to the water saturation point is termed free water and found in the cell lumen. The wood moisture content u is defined as the ratio of the mass of water contained in the wood ($m_u - m_{dtr}$) and the mass of the dry wood (m_{dtr}), equation (B2-1), where the oven-dry state is the reference variable, meaning moisture content exceeding 100% is possible.

$$u = \frac{m_u - m_{dtr}}{m_{dtr}} \cdot 100 \tag{B2-1}$$

where

u Moisture content of the wood

m_u Mass of the moist wood

m_{dtr} Mass of the oven-dry wood

When freshly cut timber is dried, the water initially emerges from the cell cavities. This water has no molecular link to the wood and is known as free water, as opposed to the water within the cell walls, which is linked to the cell wall via hydrogen bonds (to hydroxyl groups) and Van der Waals forces and thus known as bound water. This is why far more energy is required to drive out water from cell walls compared to eliminating free water.

The moisture content at which the cell walls are deemed saturated with water, but prior to the presence of any free water in the cell cavities, is known as the **fibre saturation point**, which means all "free (*–OH*) groups are occupied" and the only further water that can be absorbed is free water. For most wood species, the fibre saturation point is between 25 and 35%, making 28% a reasonable mean for most practical applications. For engineers, knowing the fibre saturation point is crucial, given the significant changes to most physical and mechanical properties before it is attained. Above the fibre saturation point, however, most of the properties are virtually constant.

As a hygroscopic material, wood constantly discharges and absorbs moisture from its surroundings. At any one combination of temperature T and humidity of the surrounding air ψ, a corresponding level of moisture content applies, where the moisture diffusing into the wood balances out the moisture escaping from the same. This moisture content is known as the equilibrium moisture content ω, although prevailing climatic changes mean wood is seldom actually in this state. The moisture content and even the extent and speed of the moisture transport have a key impact on almost all relevant engineering properties of the wood.

The way the relative humidity at constant temperature dictates the self-adjusting equilibrium moisture content via so-called sorption isotherms is shown using a typical S-shaped course in Figure B2-1. The incongruent nature of the isotherms for moisture absorption (adsorption) and moisture release (desorption) is also clear and this hysteresis effect explains why the adjusted moisture content of the wood for desorption exceeds that of adsorption by around 1 to 2%. Accordingly, under uniform ambient conditions, if you dry a piece of wood from a wet state to the equilibrium moisture content, the equilibrium moisture content will exceed that for adsorption. Engineers can leverage the sorption hysteresis of the wood: since wood exposed to an alternating climate shows fewer changes in moisture content with given humidity changes than would otherwise be expected without hysteresis, sorption hysteresis lowers the effective gradient of current sorption isotherms. This means, in turn, that any changes in e.g. wood dimensions linked to varying humidity levels may be smaller than expected.

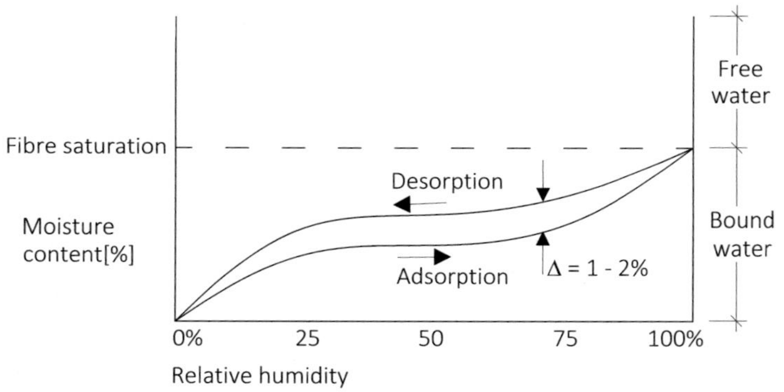

Figure B2-1 Sorption isotherms with the hysteresis effect. Sorption and desorption isotherms reveal a so-called hysteresis loop.

The **equilibrium moisture content** referred to previously adjusts itself within the wood when sufficient time is allowed and under specific ambient conditions (relative humidity, temperature, air pressure). It varies by wood species and is a vital parameter for timber engineers, since the calculations of the relative moisture content of the timber elements to be assembled depend on the ambient conditions expected when assembling. The aim is to minimise any moisture differences in the wood as far as possible (→ reduce swelling and shrinking). Wood moisture content is also a key variable when assessing vulnerability to pests and durability.

In larger cross-sections, reaching the equilibrium moisture content is a slow process, since the capillary and diffusion processes only trigger swifter changes in moisture in the edge regions. Very considerable time is required for the wood moisture to attain equilibrium, under constant ambient conditions. For example, four weeks would be required for a cross-sectional piece of spruce wood measuring 50 mm x 100 mm and with a moisture content of 20% to attain an internal equilibrium moisture content of 10%, at 20°C and relative humidity of 54%. This means that the equilibrium moisture content of a timber member will be obtained at an earlier stage with longer exposure to the relevant average temperature and relative humidity than with shorter cycles involving higher or lower humidity. Figure B2-2 shows how the equilibrium moisture content is determined for a wood species under specific ambient conditions. In this example, the adjusted equilibrium moisture content of around 11% is marked at a relative humidity of 50% and a temperature of 20°C.

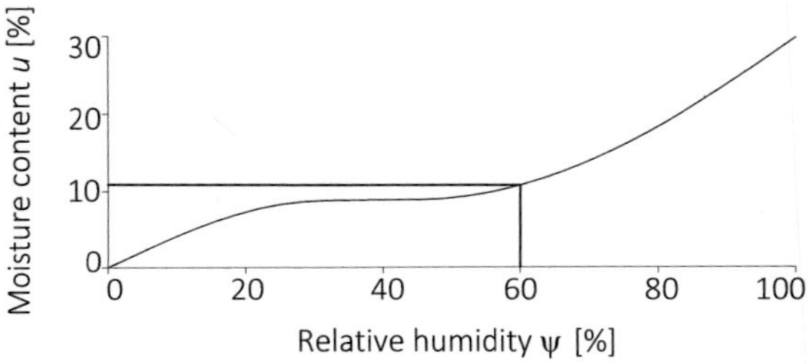

Figure B2-2 Schematic equilibrium moisture content u of 11% with T = 20°C and 50% relative humidity ψ.

Swelling and shrinking behaviour of wood

The water molecules stored in the intermicellar and interfibrillar spaces over the hydroxyl groups lead to the cell walls expanding, as shown schematically in Figure B2-3 in a process known as swelling. Conversely, the volume contraction that occurs when moisture is released is termed shrinking. Since the expanding/contracting of cell walls dictates whether water molecules are accumulated or released, swelling and shrinking is limited to the hygroscopic area alone. There is no further swelling and shrinking above the fibre saturation point, since only free water can be absorbed or discharged from this point.

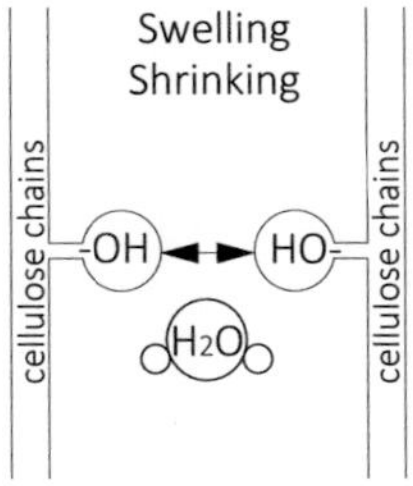

Figure B2-3 Swelling and shrinking. (www.holzfragen.de)

Wood is a roughly polar orthotropic material, although this is usually considered irrelevant for practical construction purposes. The calculation only takes into account two section planes of the wood; parallel and perpendicular to the grain respectively. However, this simplified summary of the radial and tangential section planes to the section plane perpendicular to the grain is infeasible for swelling and shrinking processes. Even so, construction practicalities mean tangential and radial shrinking is, in turn, collectively considered as shrinking in a perpendicular direction, since the type of cut applied to a piece of wood varies immensely and the engineer lacks such details. Figure B2-4 shows the swelling of the common beech as an example. Swelling varies considerably in radial and tangential directions. Longitudinal swelling is very low, since the micro-fibrils of the layer are oriented in the fibre direction with very few fibrils perpendicular to the grain. This also means that, conversely, numerous free hydroxyl groups of the cellulose chains can be found perpendicular to the fibre direction, whereupon far more water can accumulate perpendicular to the grain, intensifying any resulting swelling/shrinking. Figure B2-4 also shows that swelling stabilises from the fibre saturation point (≈ 35% for beech) onwards.

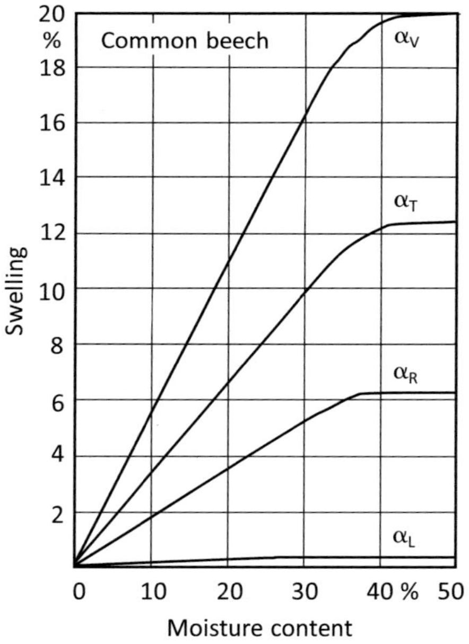

Figure B2-4 Swelling of common beech; α_L longitudinal, α_R radial, α_T tangential, α_V volume. (Kollmann and Coté, 1968)

On average, the degree of longitudinal swelling for European woods is 0.4%, for radial swelling 4.3 and for tangential swelling 8.3%; the level of longitudinal swelling is thus just a tenth of that which occurs in radial or tangential directions. Key parameters influencing the swelling and shrinking behaviour of various wood species include the wood density, latewood proportion, anatomical structure and lignin proportion. The lignin content influences the swelling and shrinking behaviour, since lignin is far more hydrophobic than cellulose and strongly lignified wood species swell far less than more weakly lignified species. However, the key influential parameter is the angle of the cellulose fibril in the S2 layer (see Article B1). In compression wood fibres for example, the fibrils of the S2 layer are at a wider angle to the fibre direction than with normal fibres (compression wood, see Article B3, section on reaction wood), which increases the degree of longitudinal swelling and reduces the level of swelling in radial and tangential directions. In compression wood, the angle of the cellulose fibril in the S2 layer may be up to 45°, in which case equivalent humidity deformations occur in longitudinal and perpendicular directions.

The deformations of cut wood (Figure B2-5) can be effectively clarified with the "axial symmetrical" swelling and shrinking behaviour mentioned previously. The range of swelling and shrinking behaviour in three main directions results in significant deformation of the wood and internal stresses and explains, for example, why the left side of a board, namely that facing away from the pith, is always concave when drying out. If not dried properly (if excessive humidity gradients are present between the outer and inner layers of the wood), this may result in extensive cracking, particularly in wood with a higher density (and particularly in a radial direction), so swift drying is best avoided.

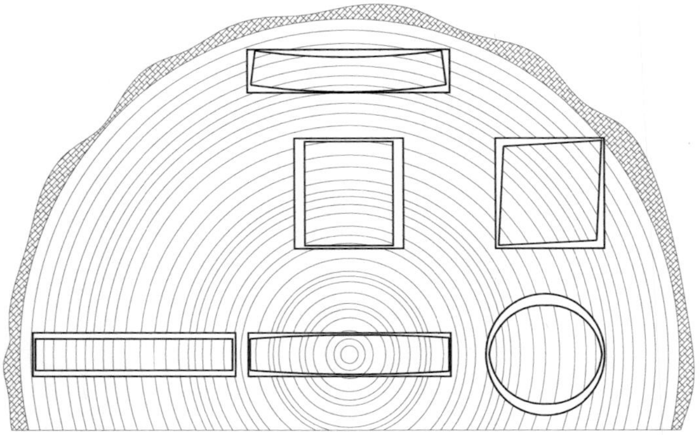

Figure B2-5 Distortion of wood due to varying swelling. (Kollmann and Coté, 1968)

The dimensions of the wood vary linearly with a wood moisture content ranging between 5 and 20%, within which range, humidity deformations can be calculated from the following:

$$h_2 = \frac{h_1 \cdot \left[100 + \beta \cdot \left(\left|u_2 - u_1\right|\right)\right]}{100} \tag{B2-2}$$

where h_1 and h_2 are the dimensions (thickness) for a wood moisture content of u_1 or u_2. β is the degree of swelling (positive) or shrinking (negative) in %/%.

In the absence of specific values for the wood species in question dictating the degree of swelling and shrinking, an approximation equation may be used which means that the degree of volume swelling and shrinking β_V corresponds to the numeric value for the density in g/cm^3. This means that the volume of wood with a density of 0.4 g/cm^3 swells by an additional 0.4% with each 1% increase in moisture content. The implicit and fundamental consideration here is that the degree of volume swelling is equivalent to the volume of the absorbed water. While the degree of swelling and shrinking in a longitudinal direction β_0 may generally be discounted, the level in a perpendicular direction β_{90} equates to half the degree of volume swelling or shrinking.

For most wood species, such as spruce, fir, pine, larch, poplar and oak, engineers can use values β_0 = 0.01%/% and β_{90} = 0.24%/%. For denser wood species such as beech (*Fagus sylvatica*) and ekki/bongossi/azobé (*Lophira alata*), there is more swelling in a perpendicular direction: β_{90} = 0.3%/% should be used for beech and β_{90} = 0.36%/% for ekki.

In plywood, the humidity deformations in the panel plane are of the same order of magnitude as those of wood in a longitudinal direction. For other wooden composites, however, such as particleboards or fibreboards, these deformations depend very strongly on the special panel type and production technique used. Perpendicular to the panel plane meanwhile, the reversible humidity deformations are comparable to those for timber. However, many panel products exposed to high compressive stresses during production show additional, irreversible thickness swelling, which is also known as "spring back".

If the expansion of the wood is hampered (e.g. in joints employing mechanical fasteners), the moisture absorption generates internal forces. The viscoelastic-plastic behaviour of the wood means such stresses decline over time, resulting in irreversible dimensional changes. If the wood reverts to its former humidity, the dimensions will since have shrunk, meaning the joint will no longer fit properly and will be less able to support any load. This underlines the need for full access to any construction details if retightening may be required.

B2.2 Wood density

Density, the ratio of mass to volume, is a key parameter when distinguishing wood. Wood is a capillary-porous, swellable material, which may contain water, water vapour, air or a feeder liquid and the mass and volume of which vary according to the proportions of such substances within the wood. Since the moisture content of wood varies depending on the ambient climate and the volume also changes accordingly when under the fibre saturation point, the density depends on the level of moisture. This is why the density must always be specified for a specific climate (often an ambient temperature of 20°C and 65% relative humidity (normal or indoor room environment) – around 12% wood moisture). The density ρ_u is the quotient from the mass of wood (including the water contained in the pores) and the volume of a wooden body (including cavities) for a defined moisture content u:

$$\rho_u = \frac{m_u}{V_u} \tag{B2-3}$$

where

ρ_u Density at moisture content u
m_u Mass of the wood at moisture content u
V_u External volume of the wood at moisture content u

In wood technology and timber engineering, mainly oven-dry density ρ_{dtr} and density ρ_{12} at 12% moisture content are used. The density values (= ρ_{12}) in accordance with EC 5 or EN 338 refer to mass and volume at the equilibrium moisture content (≈ 12%), which is obtained at a temperature of 20°C and relative humidity of 65%. The density values in accordance with EC 5 or EN 338 either refer to the mean density $\rho_{12,mean}$ or the characteristic density $\rho_{12,k}$, which is defined as the 5% quantile. The assumption made for the strength class of the timber generally involves a normally distributed density with a variation coefficient (COV) of 10%. This results in the following: (factor 1.645 for normal distribution and 0.1 = 10% COV):

$$\rho_{12,k} = \rho_{12,mean} - 1.645 \cdot (0.1 \cdot \rho_{12,mean}) \tag{B2-4}$$

The density ρ_c of the cell wall is around 1500 kg/m³. Accordingly, the density of the wood depends on the porosity, which is defined as a volume fraction of the cell lumen. Timber generally has an oven-dry density in the region of 300 to 550 kg/m³, which corresponds to a void volume proportion of 0.8 to 0.63.

Influence of wood density on wood characteristics

Density is one of the main variables influencing virtually all wood characteristics. Most mechanical properties correlate positively to density. Figure B2-6 schematically shows the effect of density on certain selected properties.

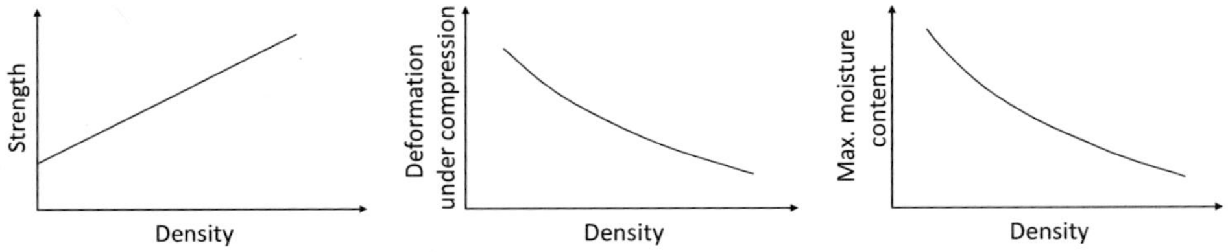

Figure B2-6 Schematic illustration of the influence of density.

Influences on the density and density distribution of wood

Wood species

The ratio between the cell wall and porosity varies considerably among individual wood species and the same applies to density. However, density still varies considerably, even within a single wood species.

Local conditions

The ground, climate and all other ambient conditions (including the interval distance between trees) have a key effect on the growth of trees and with it, the density distribution of wood (see also Article B3). This means, for example, that slower-growing conifers from colder areas of Europe have higher densities than faster-growing conifers of the same species from the Mediterranean region.

Early- and latewood proportion

The early wood of conifers has a lower density than latewood, since the cell walls are considerably thinner and the cell lumen larger.

Annual ring width

Smaller annual rings mean a greater latewood proportion and hence a higher density of softwood, while for deciduous trees, the opposite applies (see also Article B3, Figure B3-2).

B2.3 Influence of temperature

Compared to swelling and shrinking of the wood, the thermal expansion of the wood is far less significant. Any change in temperature, however, also alters the moisture content and thus the swelling and shrinking deformations. The degree of swelling and shrinking perpendicular to the grain is around tenfold greater than the thermal changes. Temperature changes over a wide area affect the properties of the wood, while the strength values of the wood decline with increasing temperature and vice versa (see Table B2-2). Further correlations and the fire behaviour of wood and wood-based products are shown in Article G1.

Table B2-2 Influence of temperature on certain properties with an increase of 20°C to 100°C.

Property	Change
Bending strength	-28%
Tensile strength	-8%
Compression strength	-44%
Modulus of elasticity (MOE) in bending	-17%

B2.4 Rheological properties

The flow and deformation properties of a material are referred to as rheological properties and may apply to a solid material (→ elasticity, plasticity) or a liquid. The term used when referring to liquids is viscosity; the higher the viscosity, the more sluggishly the liquid flows. Wood exhibits both elastic and viscous, time-dependent behaviour, which is why it is known as a viscoelastic material. Its viscoelastic properties are classified into categories including:

- Creep
- Relaxation

These two key aspects of the rheological behaviour of wood are briefly discussed in the following section and creep behaviour is by far the most critical parameter of the two for timber construction engineers. In design meanwhile, creep is taken into account using k_{def} as a coefficient.

Creep

Creep is described as an increase in deformation over time under constant load. Figure B2-7 above shows a load scheme which may result in creep deformations in wood shown subsequently in Figure B2-7. From the time t_0 at which the load is first applied, the elastic deformation of the wood also progresses up to time t_{inst}. When reaching t_{inst}, the elastic deformation u_{inst} under applied load F is completed and immediately followed by creep deformation. The creep curve can be subdivided into two areas. While creep deformation increases rapidly at the start, it is then consolidated with a constant creep speed. When load is removed at time t_{fin}, this results in a recovery, which would revert to the value $u = 0$ if ideal viscoelastic behaviour applied. However, this is usually not the case for wood, depending on the load level and loading duration. A permanent deformation u_{plast} after the entire load is removed indicates that the microscopic area of the wood has been damaged, so it is technically inaccurate to refer to wood as a viscoelastic material.

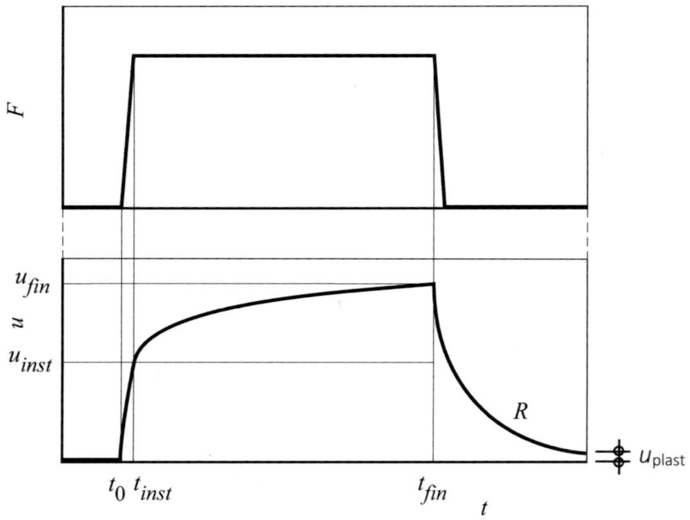

Figure B2-7 Load (above) and creep deformation caused (below), horizontal axis time axis. (STEP 1995 Article A19)

Variables influencing creep

The main parameters affecting creep behaviour are:

- Moisture content and changes in moisture content,
- Duration of load,
- Temperature,
- Level of stress.

The first two factors are what influence creep more than any other. The wood moisture is included in the calculation via the selectable "service class" and the duration of load via the "load-duration class". Creep behaviour also varies considerably among different materials, meaning wood-based panels are generally more prone to creep than solid timber and the smaller the wooden components of the panels, the more pronounced the actual degree of creep. This means that for hardboards in service class 2, a k_{def} value of 3.00 applies, 2.25 for OSB and 0.80 for solid timber, glued laminated timber and laminated veneer lumber (LVL).

Figure B2-8 shows creep curves for wood under various moisture conditions. The curves shown also reveal an additional effect, namely the so-called mechano-absorptive creep, which is a collective term referring to swelling, shrinking and creep. The creep deformation increases during the drying phase (shrinking, curve (b)) and declines in the moisture penetration phase (swelling, curve (d)). Figure B2-8 clearly shows the great extent to which this mechano-absorptive effect affects the creep curves and the swifter and stronger the changes in wood moisture, the greater the impact the wood moisture changes will have. Meanwhile, the influence of the stress level is shown in Figure B2-9. The greater the load applied, the more rapidly and intensely creep progresses. The influence of temperature on creep can also be shown using the chemical structure of wood, where the polymer structure means creep deformations rise with increasing temperature.

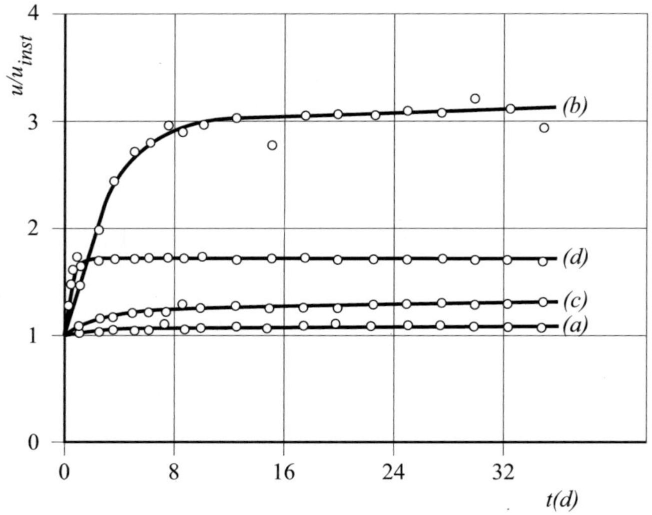

Figure B2-8 Relative deformation-time curves for beams at different moisture conditions. (*a*) green timber kept green; (*b*) green timber drying to 12% moisture content; (*c*) timber kept at 12% moisture content; (*d*) timber initially at 12% moisture content allowed to absorb moisture. *t*(*d*) is time in days. Alpine ash, 24% of average short-term strength, *T* = 25°C. (STEP 1995 Article A19)

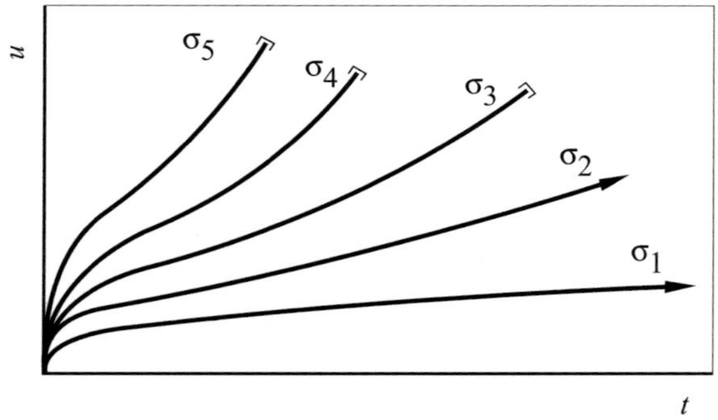

Figure B2-9 Influence of the stress level on creep, $\sigma_1 < \sigma_2 < \sigma_3 < \sigma_4 < \sigma_5$, *u* is deformation, *t* is time. (STEP 1995 Article A19)

Relaxation

The decrease of stresses with constant deformations over time is described as relaxation and particularly significant for prestressed structures. Figure B2-10 shows relaxation under compressive stress perpendicular to the grain over the course of a year. This phenomenon means that prestressed members must be retensioned, which was done twice in Figure B2-10. While the compressive stress perpendicular to the grain declines during the drying phase due to shrinking of the wood, it increases during the moisture adsorption phase. As the number of cycles increases, however, the stress declines dramatically.

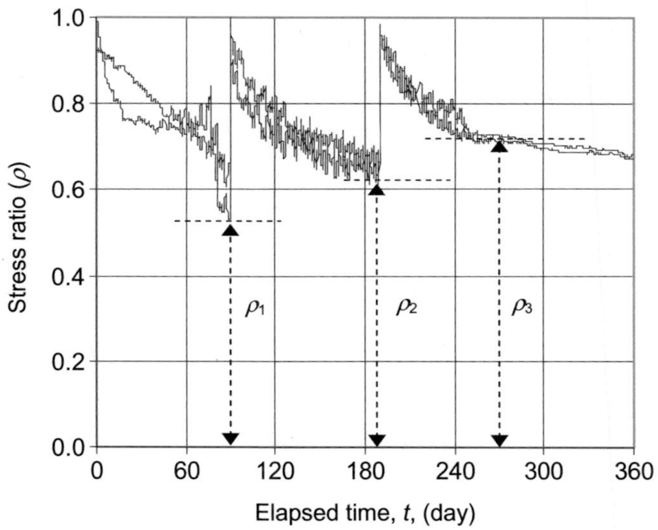

Figure B2-10 Stress curve over a year for a prestressed and twice re-stressed connection. (Awaludin et al. 1999)

B2.5 Literature

P. Hoffmeyer, L.D. Andriamitantsoa, original Articles A4, A19, STEP 1995.

Kollmann F.F.P and Coté W.A. (1968). Principles of wood science and technology. Volume I, Solid wood. Springer Verlag, Berlin, 592 p.

Awaludin A., Hirai T., Hayashikawa T., Sasaki Y. and Oikawa A. (2008). One-year stress relaxation of timber joints assembled with pretensioned bolts. Journal of Wood Science 54(6):456-463.

B3 Wood formation

Original articles: P. Hoffmeyer, L. M. R. Nunes, P. P. de Sousa

The life of the tree is the fundamental basis for understanding wood as a raw and construction material, since it exerts a greater impact on properties which affect the end usage than many other raw materials. This also explains the strong focus on the individual details of a tree when assessing the raw material. The properties of the wood depend on the tree species, the individual tree and the tree section, but the world of nature encompasses innumerable structures, dimensions and shapes, which can even be a plus when special requirements are imposed (e.g. formerly: curved tree stems for hulls). Nevertheless, amid increasing industrialisation and the overwhelming trend to pursue rational large-scale serial production methods, variability in wood characteristics within a particular timber variety is considered undesirable. Knowing details of how wood formation is influenced and possible changes in the wood can help mitigate such problems.

B3.1 Wood formation

The stem

The stem shape, Figure B3-1, is determined by the number and width of annual rings in the wood and the bark thickness. Depending on the yearly shoot lengths, meanwhile, the number of existing annual rings may decline with increasing height. If the shoot lengths and annual ring widths were the same at all stem heights, the debarked stem would be conical in shape. However, since the shoots vary in length and the width of the annual rings is also subject to change, tree stems deviate from the conical shape to a greater or lesser extent. Over the lifetime of the tree, the distribution of annual ring widths may vary at various stem heights, particularly following any change in the prevailing environmental conditions. If, for example, older trees, which were cultivated in dense proximity, are released, the greatest increase in diameter comes in the lower portion of the stem; a phenomenon particularly prominent in coniferous rather than deciduous trees. In free-standing or released trees meanwhile, the annual ring width may decline from the bottom up.

The vertical classification of a tree into the roots, stem and crown or the roots, trunk and branches changes throughout its life, which is why, in young trees for example, the crown initially grazes the ground. The stem and crown first establish separation as the lower branches die, in a phenomenon which intensifies as the branches fall away and the stumps are overgrown with knot-free stem wood. The vertical classification of trees, particularly the stem length and crown size, is influenced by the distance between the trees. Free-standing trees retain abundant branches deep into old age and feature a shorter stem full of branches and a larger treetop. When trees are in close proximity, however, the shadowing of the lower branches tends to result in a proportionally longer stem and smaller crown from a relatively early stage. The most valuable wood in terms of length, diameter and wood quality is obtained from the stem. Varieties of wood which fall from the crown area generally stand out due to the reduced dimensions as well as the inferior internal quality (e.g. larger knots).

Figure B3-1 Stem shapes: left prismatic, right tapered. Tapering tends to occur; particularly in free-standing trees and at leeward forest areas. A significant decline in diameter to the smaller stem end (= tree taper) has a very detrimental effect on usability and yield. (Steuer, 1990)

The annual rings

Variation in ring width is dictated by the wood species and age of the tree, its vitality, its positioning among others, environmental factors such as the supply of water, nutrients, light and heat and silvicultural care measures. Forest fires and insect damage can also impact on the annual ring width. In addition to the change in annual ring width, change in the latewood proportion in conifers and ring-porous hardwoods is also a key pointer indicating variation in annual rings. The water flow is limited to the areas of earlywood, meaning that the stem cross-section features concentric rings of water-conducting tissue with corresponding dry zones. Curves c in Figure B3-2, which indicate the way earlywood width rises with increasing ring width, show that in highly developed ring-porous hardwoods (ash, Figure left), the tissue for the water flow forms independently of the growth rate, while among conifers (pine, Figure right) the opposite applies. In ring-porous hardwood types, meanwhile, the earlywood width is virtually constant, while latewood width increases depending on growth. In contrast, the latewood width shows minimal change with increasing ring width in conifers, which explains differences in annual ring widths as well as the latewood proportion of fine ring-patterned spruces from Northern Scandinavia and of coarse ring-patterned spruces containing a high proportion of earlywood from a good site in Germany. These connections impact on density and thus on mechanical properties.

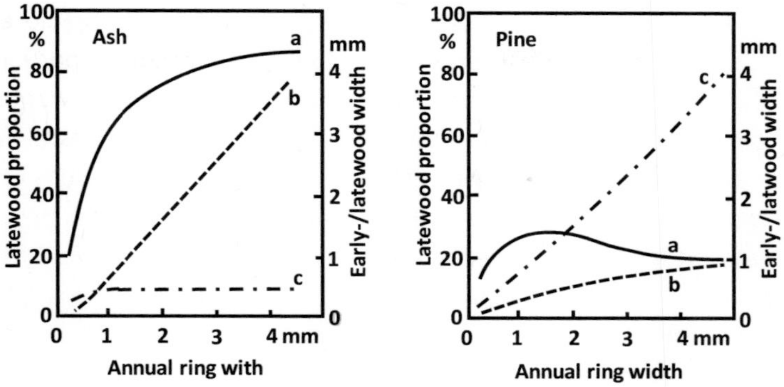

Figure B3-2 Early- and latewood width and latewood proportion depending on the width of annual rings in *Fraxinus excelsior* (ash, left) and *Pinus silvestris* (pine, right); latewood proportion (a), latewood width (b), earlywood width (c). (Knigge and Schulz, 1966)

The branches/knots

When it comes to structural use, the branches of a tree are a key feature of timber, since each knot actually disrupts the wood structure. Most of this disruption involves localised deviations from the straight grain, since the stem fibres have to grow around the knot. As far as vertical branch distribution on the stem is concerned, though, differences emerge between each tree species; namely some in which the branches are locally massed and others in which they are scattered around. The most prominent accumulations of branches occur in tree species in which the branches are arranged in whorl shapes (Figure B3-3) and which feature knot-free or virtually knot-free interim longitudinal sections of annual shoots, such as the pine, spruce and fir. The impact of environmental conditions often means branches are thicker on one stem side (e.g. trees skirting the forest). The diameter of branches having formed at the stem tends to increase on average from the bottom up to the crowning height and then decline within the crown space itself. For coniferous species in particular, however, small branches still proliferate, even in higher areas of the stem.

Figure B3-3 Cross-section with whorling branch.

Forestry measures are a highly significant factor influencing average branch thicknesses, since they determine the growing space available to the individual plants via planting distance. The more spaced out the trees and the greater the incidence of light, the more growth is accelerated and branches at the lower edge of the crown live much longer, which means they are larger in diameter. Pruning is another vital forestry measure, which can allow thicker knot-free layers to emerge as a result. Branch thickness directly impacts on the quality of the wood, based on the extent of disruption to the wood structure and indirectly via the periods of decay and overgrowth, which dictate the condition of the snag. As vertical growth continues and new branches form, the lower branches are increasingly shadowed by younger crown parts and neighbouring trees and eventually die. At the stem surface meanwhile, a protective barrier tends to form within the dying branch, which segregates the healthy and living tree sections from the diseased branch sections dying and dropping off. It is often also macroscopically visible thanks to the dark colouring, as a narrow and sharply delineated zone. After dying, the branch is particularly prone to attack and decay, initiated by fungi but also insects, in a process which usually occurs more rapidly in deciduous trees than most coniferous varieties. The branches of deciduous trees, for example, tend to break off completely, namely in one piece. In many coniferous species, however, they break off in piecemeal fashion. Stumps emerging from the stem thus vary considerably in length, or are preserved for a shorter or longer time respectively which, in turn, influences how the timber is used, since the snag is encircled by the secondary growth of the trunk. In fact, it lacks any form of bond to the surrounding stem wood and resembles a foreign body in otherwise healthy timber (Figure B3-4). Even the healthy portion of the branch near the pith, which is connected to the wood, is seen as a disruption, since the type and orientation of its cells and hence properties clearly differ from those of the surrounding stem wood. The overgrowth of wood, which encircles the snag, generally curves upwards, so that even after the wound has been closed up, a bulging protrusion remains. Within this, the wood fibres deviate, initially to a greater and then lesser extent, from the longitudinal orientation.

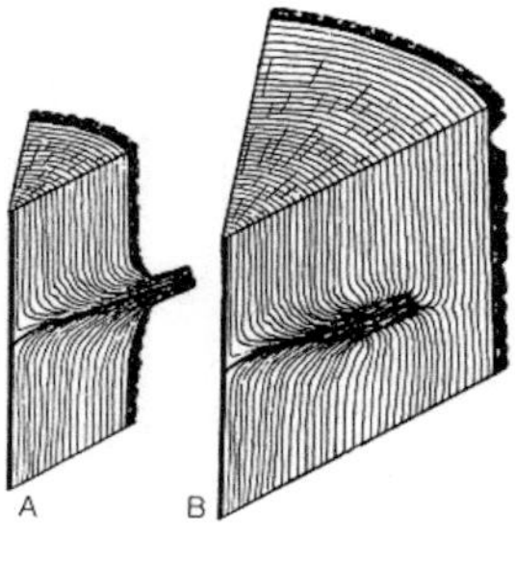

Figure B3-4 Formation of the dead branches or knot holes: The dead branch (A) has not completely fallen off. The subsequent annual rings have completely encircled the branch stub (B). The branch only re-emerges when the stem is cut open. (Steuer, 1990)

B3.2 Wood characteristics

Generically defining the term "wood characteristics" is no easy task. This is because many features of tree growth, such as decreasing diameter, knots etc. are often considered natural from a biological perspective and inevitably subject to specific limits. When it comes to processing or using the wood in question, though, such characteristics frequently prove problematic. Accordingly, there is no one-size-fits-all answer to the question of how exactly to define a wood characteristic; rather, it depends on the intended end use. The growth irregularities and wood characteristics described in this article can be roughly classified into the following categories:

Deviations from the ideal stem shape

For most usage purposes, the ideal stem shape would be as long a cylinder as possible. The following are considered undesirable:

- Unwanted deviations occurring longitudinally. These include: curvatures and excessive decline in diameter (e.g. tree taper).
- Unwanted deviations, which impact on the cross-section. For example, cross-sections which deviate from the circular form (e.g. fluting).

Deviation from average wood formation

- Unwanted deviations from the average chemical and anatomical composition of the wood (e.g. reaction wood).
- Unwanted deviations from the normal orientation of structural elements relative to the longitudinal axis of the scion (e.g. spiral grain or knots).
- Lack of bonding between successively formed wood layers (e.g. in the event of wounded areas being overgrown).
- Ingress of resin, bark etc. (e.g. in the overgrown areas of branches).

Unwanted subsequent changes

- Discolourations, which can have a whole range of causes (e.g. action of light, access to oxygen, fungal infection etc.).
- Cracks, which either run radially or in the form of so-called ring shakes of annual rings, which are separated from each other, in whole or in part.
- Fractures (e.g. treetop/stem breakages).
- Holes, primarily caused by wood-boring insects, as well as by white pocket rot.
- Destruction by fungi.

Many of the so-called fundamental features described in the following section (see Figure B3-5) do not occur in isolation, but alongside other characteristics, namely in the form of combined wood characteristics. Such combined wood characteristics can be found e.g. in the area of former wounds, where a greater or lesser area of the cambium perished in the course of the tree's life cycle. The diversity of the existing fundamental features in the wound area, such as discolourations, ingrowths, structural changes etc. can be seen in Figure B3-6.

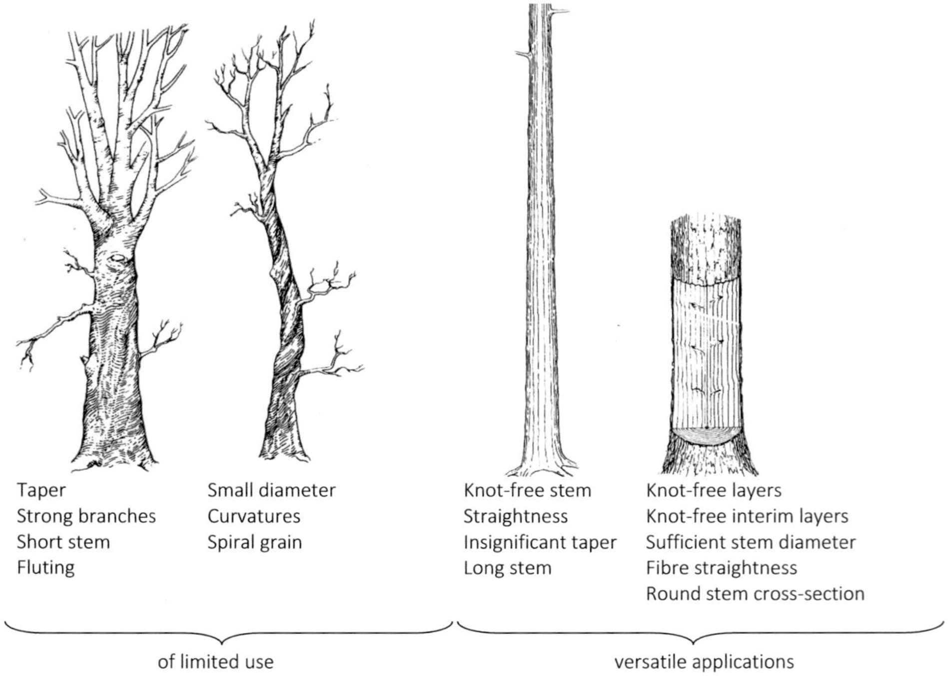

Figure B3-5 Presenting certain features of trees offering limited use and versatile applications respectively. (Knigge and Schulz, 1966)

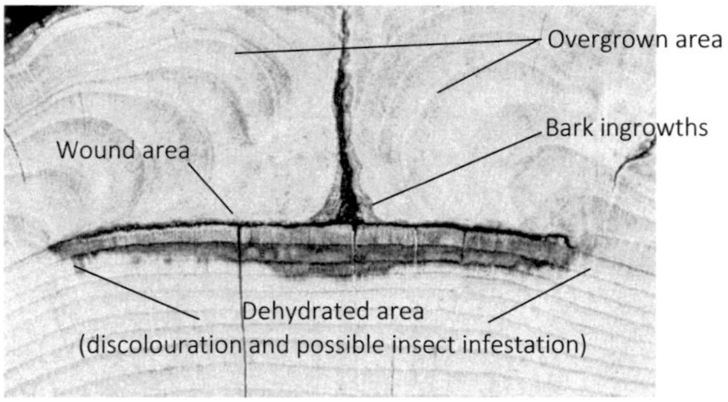

Figure B3-6 Wounds and wound overgrowth in beech wood. (Knigge and Schulz, 1966)

Taper

One key characteristic when assessing stems is the degree of tapering (above-average decrease in diameter with increasing tree height, Figure B3-1. A decline in diameter of up to 1 cm per metre of section is deemed prismatic, stems with any values exceeding this figure are deemed tapered. The taper is particularly apparent when cutting the logs into long sawn timber boards. The smaller stem diameter at the tapered end means a decline in the yield and reduces the overall value. The degree of taper also determines the cutting losses in the sawmill. Goods produced with this wood have also lower strength, since parallel cuts expose more annual rings and fibres than in prismatic sections. Tapering depends significantly on the location and care of the tree, with peripheral trees, for example, particularly prone to tapering.

Curvatures

Curvature is defined as the deviation of the stem from the straight-line axis and is also normally always associated with other characteristics such as eccentric position of the pith, reaction wood, etc.

Deviation from the circular form

Oval stem cross-sections, as shown in Figure B3-7, are usually attributable to the effects of wind, whereby the diameter peaks in the wind direction and the radius for coniferous trees peaks on the side sheltered from the wind. Inclined stems and those featuring curvatures are also prone to deviations from the normal circular cross-section. Additional causes include one-off stresses from the sun, snow push, crown shape and hillside location.

Figure B3-7 Oval stem shape, eccentric growth.

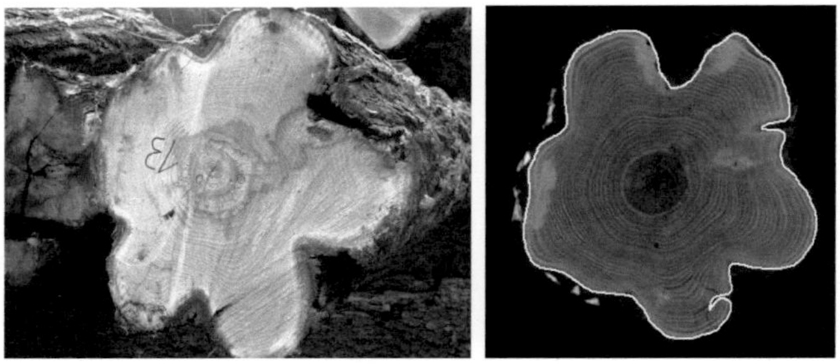

Figure B3-8 Fluting in robinia.

Another form of cross-sectional change is fluting, as shown in Figure B3-8. Here, the annual rings run in undulatory manner in the cross-section, resulting in a stem shape penetrated by channels and meanders which significantly reduce the volume yield. There is also the danger of warp and a loss of strength. However, the fluting may also surface sporadically, e.g. in the form of so-called recesses (proneness to grooves) under strong branches and is attributable to irregular segmentation in the cambium. Triggers in this case may include injuries, genetic disposition or butts.

Forked growth

A tree fork (Figure B3-9) forms due to two trees converging at the stem base or due to bud damage (e.g. in the ash tree, caused by the ash bud moth), browsing by game or genetic causes. The need to cut out forked sections reduces the overall yield. Other prevalent issues include sources of rot and discolourations, e.g. formation of false heartwood in the beech, while forked growth is also often linked to bark ingrowths and intermingling of fibres.

Figure B3-9 Tree fork.

Spiral grain

When the grain does not rise in parallel to the stem axis but assumes a spiral form, this is known as spiral grain and becomes apparent when observing the tangential surface, see Figure B3-10 on the left. Certain tree species (e.g. the horse chestnut) are known to always have spiral grain associated with a specific rotation direction. Many coniferous trees also exhibit a known tendency to rotate to the left while young, whereupon a portion of the stem is subject to straight grain or right spiral grain at an earlier or later stage. Tropical tree species in particular show a rhythmic change in the rotation direction (interlocked grain). What has also become apparent is that numerous tree species have very specific regularities in terms of their spiral graining behaviour and virtually all wood has some degree of spiral graining, Figure B3-10 on the right. Spiral grain thus does not represent any exceptional phenomenon, switching from left to right is generally associated with indigenous conifers. It is also important to underline that the rotation angle of the fibres normally undergoes numerous changes throughout the life cycle of the tree. Spiral grain depends on the wood species, the main wind direction and one-sided foliage.

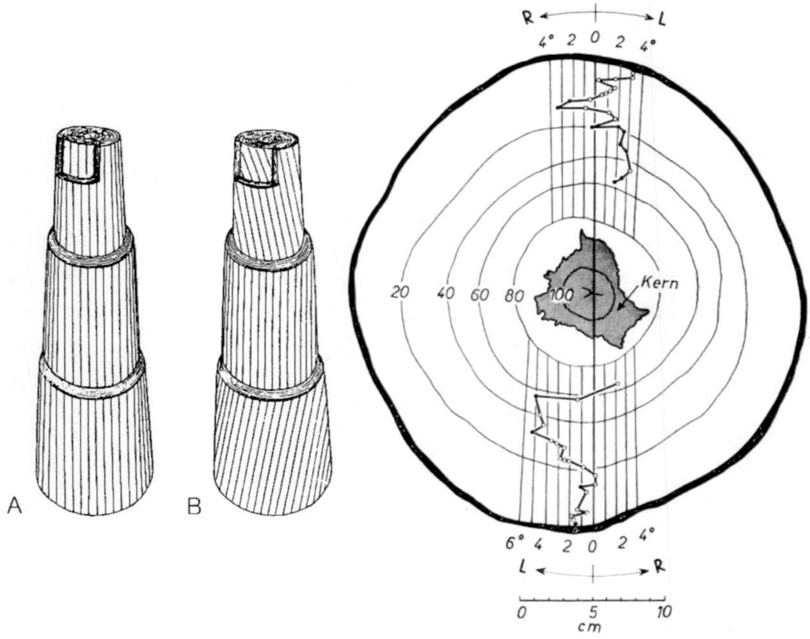

Figure B3-10 Left: Schematic presentation of the straight grain in all wood layers (A) and a change in the grain from left to right (B). (Steuer, 1990) Right: Changes to the rotation angle with tree age in a beech. (Knigge and Schulz, 1966)

In spiral-grained wood layers, more fibres are diagonally bisected following an incision made parallel to the stem than in straight-grown sections, with a corresponding decline in the strength of the processed sections. Changes in moisture trigger greater differences in swelling and shrinking behaviour and spiral-grained wooden bodies are prone to warp, as dictated by the degree and direction of their rotation. This process has to be taken into account during wood drying or in poles, for example, this leads to rotation of the tapered end relative to the embedded base. Cracks in spiral-grained sections look particularly unattractive, since they result in a helical exposure of the section in line with the grain, reducing the value of a larger area of wood than would otherwise apply to cracks in straight-grown stems. Spiral grain also disrupts surface processing, since on the same wood surface, tools come into contact with the wood fibres at different angles.

Reaction wood

Reaction wood is an active form of direction tissue in trees, which involves the tree attempting to bring tree sections (trunk, branches) back into their original position, after they have been moved due to e.g. wind pressure or landslide. The term reaction wood is used for the "compression wood" of conifers and the "tension wood" of deciduous trees, Figure B3-11.

Compression wood forms on the underside of slanted stems and branches of softwood. Increased lignin storage leads to reddish discolouration of the wood and increases the density and hardness. The increased incline of the cellulose micro-fibrils in the S2-cell wall layer causes the level of axial swelling and shrinking to soar (see Article B2). Given its hard and brittle nature, there is limited scope to process compression wood and compression wood is also very prone to longitudinal shrinking. Tension wood develops on the upper side of lop-sided or unbalanced hardwood stems or branches, while the low lignin content gives it a white to silvery colouring. Unlike the compression wood of conifers, tension wood is no harder than the surrounding wood, but does show generally increased axial shrinkage and poorer processability.

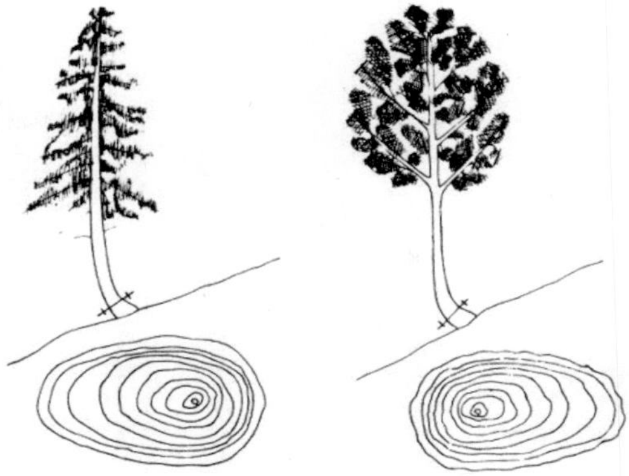

Figure B3-11 Eccentric growth and reaction wood. Compression wood of coniferous trees is formed on the underside of drooping trunks or branches (left), while tension wood of deciduous trees forms on the upper side of drooping trunks and branches (right). (Bosshard, 1984)

Growth stresses

Growth stresses are internal stresses that emerge during the tree growth due to the interaction of growth processes (particularly secondary growth) and the weight of the tree itself. External factors such as the amount of wind and the bending stresses it causes may also impact on how internal stresses are generated. Such growth stresses are clearly visible in a fresh felled and cut stem (undried). Felling or split cuts results in sections of stems which previously held each other in place becoming relaxed and deforming, whereupon cracking may emerge.

When a fresh stem is longitudinally bisected, both halves bend apart (comparable to cut flower stems, Figure B3-12 on the right). This shows that as newly formed wood layers grow in the grain direction, longitudinal tensile stresses form, but pass over towards the pith in longitudinal compressive stresses. This is confirmed if the fresh stem is cut perpendicular to the grain, whereupon the cutting surfaces undulate due to compressive stresses in the stem centre and the tensile stresses show convex movement at the stem edge. These internal stresses have already formed by the time the tree is standing; intensifying with age and stem diameter and capable of exacerbating or triggering incidence of heart shakes, star shakes, ring shakes and schilfer shakes (Figure B3-12 on the left).

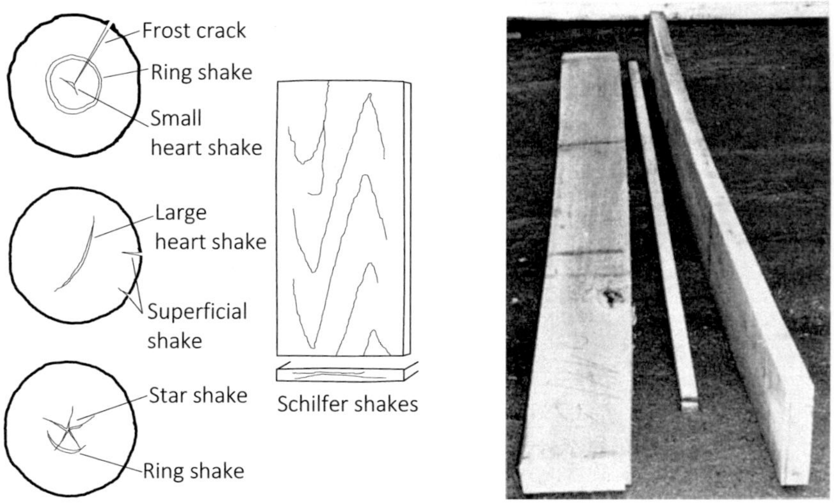

Figure B3-12 Left: Presentation and description of different crack/shake types. Right: Beech planks, subject to significant deformation right after being cut. (Knigge and Schulz, 1966)

Pith

As far as timber applications are concerned, the pith component is often unwanted. It is seen as particularly disadvantageous when it shows a very uneven course. In older stems meanwhile, short cracks often emerge from the pith, which exacerbate the level of defects.

False heartwood

Many older hardwood trees frequently respond to environmental factors by forming so-called false heartwood during the process of transforming sapwood to heartwood. Such discolourations tend to be undesirable, because of the wide-ranging colours produced within the same core area, dark edge zones and irregular shapes. Figure B3-13 shows incidence of false heartwood in a stem slice taken from a pear tree.

Figure B3-13 False heartwood in a pear tree.

Pitch pockets and pitch shakes

Pitch pockets are formed when the cambium sustains damage and remain visible as localised wood discolourations following overgrowth. Pitch pockets occur in spruce, larch and pine, but not in fir trees (since the latter contains no resin) and frequently lead to incomplete heartwood formation, particularly in larch. During processing, they are cut from a range of angles; the end product is considered substandard cut timber, which stands out for its loss of strength, discolouration and the unappealing appearance when installed in a visible area.

Irregularities in annual ring formation

Prominent wood characteristics associated with the anatomical structure of the wood also include all irregularities in annual ring formation, which are attributable to wide-ranging causes, particularly the impact of weathering. The annual ring width reveals pointers to certain wood characteristics, such as its internal knots, density, consistency of quality, surface properties of timber products and geometrical and dimensional stability. Many quality characteristics of the wood are closely linked to the annual ring width. In softwood for instance, narrow annual rings point to more favourable physical features of the wood, since they indicate a higher proportion of latewood. In hardwood, particularly oak, narrow annual rings indicate a relatively soft wood. Wide annual rings in oak, for example, point to hard wood, since wide annual rings indicate a higher latewood proportion, a marker of hardness. It follows therefore that changes in annual ring widths reflect how the wood hardness varies, which impacts on aspects including surface processing. The annual ring width also dictates density in particular, which, in turn, is the key determining factor for many physical wood properties. Accordingly, wood in which the annual ring widths differ is subject to varying degrees of shrinkage or moduli of elasticity. The annual ring width shrinks to a minimum of 1 mm (e.g. in yew) and peaks at the level of a few centimetres (e.g. for poplar or radiata pine).

Knots

We have already explored what may, at times, be a dramatic reduction in the strength properties of timber members caused by knots, where quantity is less important than the diameter and position of individual knots. The effect of an individual knot on defectiveness of the wood is influenced by multiple factors, the relevance of which may increase or decline depending on the purpose of use:

- By knot diameter. The thicker a knot, the greater its disruptive effect on the end use of the wood in question.
- By the length of the intergrown knot. Short snags, which end near the pith, allow thicker knot-free layers to form, which is less disruptive to the wood.
- By branch angle. Assuming unchanged length, steeper branches traverse a longer longitudinal, but shorter cross-sectional area of the respective section of wood. Depending on the purpose of use, either a larger or smaller branch angle may be more advantageous.
- By the proportion of various knot areas in intergrown snags. The section of branch facing the pith, where the wood is tightly intergrown and healthy (living or sound knot area) is more advantageous than the external stump portion, nearer to the point where the branch has broken (dead or unsound knot area).
- By the first cut of the knot. An incision through the centre of the stem often sees knots released following a longitudinal cut in the form of so-called spike knots. A tangential incision, meanwhile, reveals round or even oval knot cross-sections, which tend to only reveal a specific area of the knot. For branches with a large proportion of dead wood, cutting into planks often reveals so-called dead knots, which stand out due to their lack of bonding with the surrounding wood and which easily fall off during the drying process.

Knots represent the key wood characteristic as far as mechanical properties are concerned and are classified based on their surface appearance, Figure B3-14.

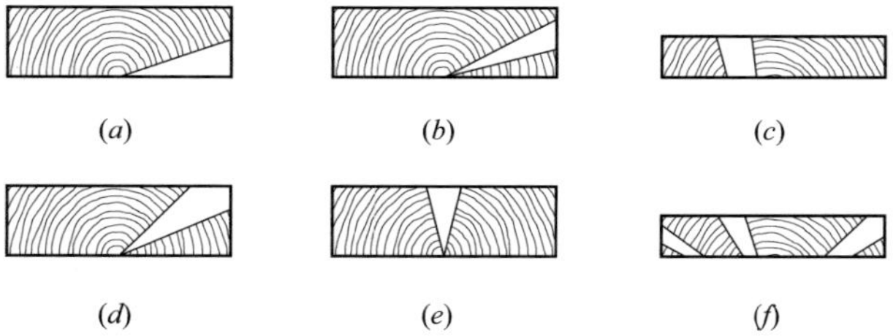

Figure B3-14 Different types of knots. (*a*) spike knot; (*b*) narrow face knot; (*c*) through knot; (*d*) arris knot; (*e*) wide face knot; (*f*) knot cluster. (STEP 1995 Article A4)

Wood characteristics influenced by forestry operations

Trees may also sustain damage due to forestry measures or forest visitors, which mainly involves injuries to the cambium. Noteworthy examples also include felling damage to felled tree stems:

- Stems are particularly prone to rupture, if the upright stem is subject to one-sided pressure and the directional notch lacks sufficient depth. While in storage, crack edges pave the way for secondary damage (e.g. fungal attack).
- Breakages and cracks affect higher stem sections and result in either stem breakage or the branch stubs breaking off. The impact of strong and steep branches (particularly when forked) often leaves the stem splitting downwards.

Cracks which go unnoticed and which are not taken into consideration at the time of cutting can adversely affect timber strength. Moreover, the felling process can also result in damage to neighbouring trees (damage to bark due to rubbing, impact damage) or damage to the undergrowth. Damage akin to felling may also occur when transporting the timber through the forest (skidding damage) and both these forms of damage leave the timber prone to attack by fungi and insects. Felling at the wrong time (summer felling) may result in discolourations, while incorrect storage may result in decay damage (due to moisture or the lack of underlay), intensified crack formation (due to solar radiation) and discolouration.

Wood characteristics caused by extreme wheathering

The impact of weathering can damage a living tree in such a range of ways that potentially all the basic characteristics listed may be at risk. The most frequent examples are breakages, cracks and cambium damage and their consequences, while weather-related structural characteristics and damage to living wood cells also play a role. Weather-related wood characteristics often come into play during extreme temperatures and drought, sudden changes in temperature and under ambient conditions, where the tree lacks sufficient time to adapt. Temperature and moisture influence wood formation to such an extent that they are considered the main cause of any irregularities in the annual ring formation. The irregularities themselves may vary very widely, but particularly include rapidly changing annual ring widths, earlywood widths in coniferous trees and latewood widths in deciduous trees. Additional weather-related characteristics include so-called sunburn, which can lead to much of the cambium dying off due to overheating, or frost cracks. Frost cracks occur radially, in a scope that often goes from the stem edge up to the pith. Lightning can cause radial (lightning) cracks, which usually commence under the crown and extend up to the stem base. These cracks may exacerbate the problem of fungal and insect-related damage. Among young and thin-barked trees in particular, hail can cause damage in the form of pitch pockets and bark ingrowths to the stem. Figure B3-16 shows a schematic diagram of the various basic characteristics.

Storm-related damage can trigger breakages of various kinds. If they affect the stem, the impact on the wood value is dictated by the height, size and area of breakage, the length of the fragmented and cracked areas and the extent of secondary damage. The pace at which the timber is prepared also has a key impact on secondary damage.

Treetop breakages as a wood characteristic are assessed in accordance with the size and smoothness of the area of breakage. Smaller breakage areas in trees that are still relatively young can often be effectively overgrown. Provided no rot sets in, once a branch has been straightened to support the vertical function, the result is a significant bend, which emerges in a curve along the stem. Gradually however, the compensatory growth renders the bend virtually imperceptible, meaning that ultimately, within and above the old breakage area, the only features pointing to the former damage are the eccentric pith position and multiple emergence of reaction wood. At the former breakage site, the internal wood layers lack any fixed bond to the overgrowth, which may render this point weaker. Some examples of treetop breakage are shown in Figure B3-15.

Compression failures occur when the stem is subject to significant warpage, particularly in spruces and firs. On the side facing away from the force application (pressure side), the wood is compressed to such an extent at the point where the load peaks that compression failures occur when elasticity limits are exceeded (moreover, compressive strength in upright trees is lower than that of dry wood). They run perpendicular to the grain and often on top of each other, from the stem edge to the pith, while the stem side under tensile stress remains free of damage. The tree then proceeds to form thick bulges over the compression failures, namely the wound supporting tissue. The damage to the wood is caused by the loss of strength in the area of breakage and the heterogeneity of the wood layers formed before and after the breakage. The damage is hardly discernible in sawn timber, since the tree is not actually broken, but has simply formed pressure folds in a specific area. These pressure folds are hardly visible, even on planed wood.

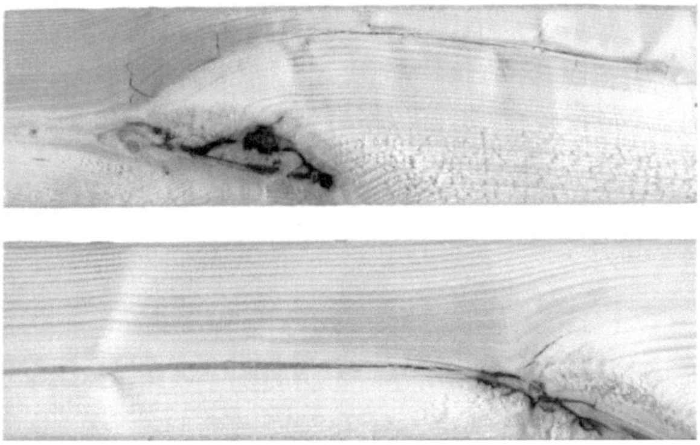

Figure B3-15 Two examples of treetop breakage.

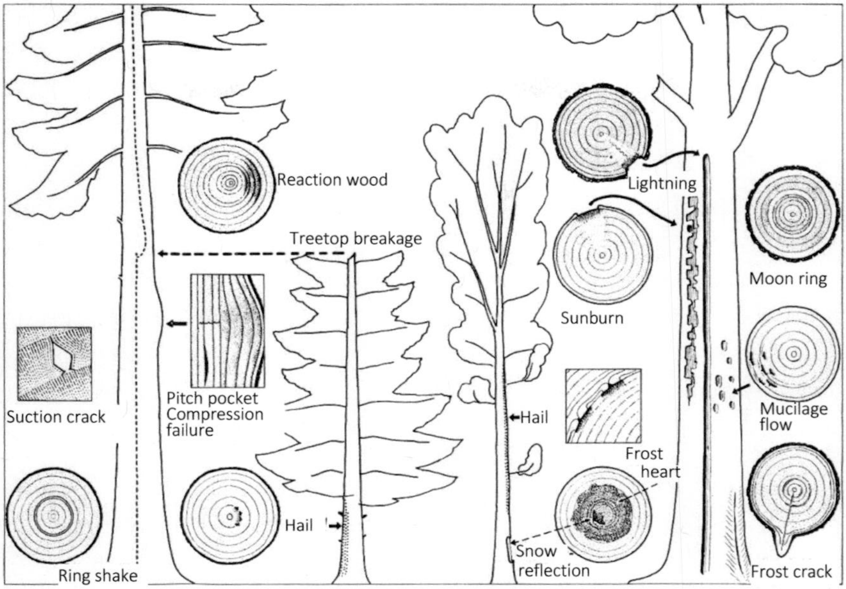

Figure B3-16 Wood characteristics caused by the impact of weathering. (Knigge and Schulz, 1966)

B3.3 Biodegradation

Insects

Insects, particularly beetles, mainly damage wood via the burrowing of their larvae ("woodworms"), which may involve a dramatic reduction in the cross-section and hence load-bearing capacity. A difference is established between greenwood insects, which infest the wood of living trees or freshly felled green wood and dry-wood insects, which infest semi-dry and constructional wood (stakes, fences) or dry timber (attics, furniture). A third group includes decaying wood insects, which gravitate towards old and rotten wood. Greenwood insects (e.g. bark beetles, *Scolytinae*) and decaying wood insects (e.g. long-horned beetle, *Ergaster faber*) can be disregarded in terms of their impact on processed wood.

While wood-destroying fungi can no longer grow when the wood moisture goes below 20% or so, normal larval development in some species of insects is possible with wood moisture of just 8 to 12%, which means purely constructive measures (keeping it dry) to protect the wood against infestation by such insects is infeasible. When the relative humidity is 60% at 20°C, the wood moisture content is between 10 and 12%, which means larval development is still possible. Within the forest, methods to prevent such beetle infestation include prompt removal of the timber, preferably before warmer seasons and before the beetles migrate. The close links between many pests to the average moisture content of the wood and the egg deposition, most of which takes place on the bark, suggest the need, however, to debark the wood promptly and either swiftly cut it up and dry it or, conversely, use wet storage to keep it above the moisture region of danger.

House longhorn beetle (*Hylotropus bajulus Linné*)

The house long-horn is native to the whole of Europe as well as Asia Minor and North Africa and in Central Europe, is by far the most dangerous and economically devastating destroyer of coniferous timber, although hardwood is spared. Since warm and humid climatic conditions promote larval development, the house longhorn beetle thrives in warm and sunlit attics or similarly exposed constructional timber. Typical entrance holes and burrows can be seen in Figure B3-17.

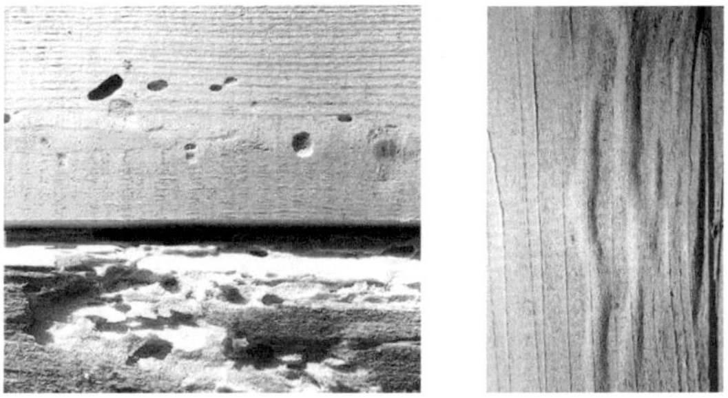

Figure B3-17 House longhorn beetle. Left: entrance holes and destruction under the intact wood surface. Right: Vein-like bulges, which point to the presence of a house long-horn beetle. (Sutter, 1997)

Common furniture beetle (*Anobium punctatum*)

The common furniture beetle is the most significant native beetle and found Europe-wide. However, *A. punctatum* has also migrated to South Africa, the USA, Brazil, South-East Australia and New Zealand. Coniferous and deciduous trees are both attacked, particularly sapwood, while durable heartwoods are only affected after initial fungal growth and tropical wood species like Abachi, Ilomba or Limba are spared. The common furniture beetle proliferates in buildings as a destroyer of processed and constructional wood, whereas outdoor wood and wood exposed to rain remains almost unscathed. Given their overwhelming proliferation as pests, after the house longhorn beetle, the common furniture beetle and the related *Anobiidae*-species are the key wood pests of note. Their philopatric infestation in particular exacerbates the damage they cause and the main issue is often the destruction of furniture items, rather than actual structural timber.

Brown powderpost beetle (*Lyctus brunneus*)

Beetles from the powderpost family (*Lyctidae*) are among the most feared hardwood pests worldwide, with most representatives of the over 60 species in this family populating the warmer regions of the Earth. Having migrated to Germany via the timber trade and since proliferated, the brown powderpost beetle is now one of the most significant destroyers of dry wood. It infests large-pored, high-starch imported woods (Abachi, Limba, Okoumé) and sapwood of indigenous deciduous species (oak, ash, elm, walnut and chestnut). The powderpost beetle attacks only the sapwood portion of woods with dark heartwood. While the light sapwood shows evidence of numerous burrows full of bore dust, the dark heartwood is free of parasites.

Additional wood parasites

Termites are wood pests of paramount economic importance, far exceeding the significance of indigenous species. Our coasts are particularly plagued by *Teredo navalis L.*, a marine woodworm, and *Limnoria lignorum R.*, a gribble, which threaten harbour structures.

Micro-organisms

The microbial breakdown of wood is part of the overall course of life cycles and a natural and necessary process. Using micro-organisms, particularly bacteria and fungi, complex macromolecules like wood cellulose are broken down into smaller and simpler molecules and redirected into the materials cycle (the fungal enzyme attacks the hydroxyl groups of the cellulose). The microbial breakdown takes place in any location where the ambient temperature and humidity allow such micro-organisms to develop. Soil, for example, is an ideal habitat for bacteria and fungi, as reflected in the rotting of branches and tree stumps in the forest or the deliberate composting of organic waste. But even structural timber, which is not in contact with the ground, may be attacked and destroyed by micro-organisms (mainly fungi) since sufficient moisture content is the sole prerequisite. This is why wood exposed to weather and unprotected or which stands moist within a building, unable to dry out, is prone to attack from a range of fungi. Indeed, fungi are a key cause of the breakdown of constructional wood and wooden objects in interior spaces. The optimal range of moisture, in which the most common wood-inhabiting fungi flourish, is between 30 and 60%. During the so-called dry phase, however, certain fungi (mycelium and spores) can withstand extended dry periods and start regrowing when living conditions improve. The temperature can also impact on fungal growth, which – depending on the type of fungus involved – takes place within the range -2.5 and +40°C.

Classification

Two main groups of wood-inhabiting fungi can be distinguished, wood-staining and wood-destroying respectively. The problem with wood-staining fungi is primarily visual, while the presence of wood-destroying fungi signals a widespread loss in the substance and strength of the wood. The fungi are classified in accordance with the damage pattern and nature of the wood components targeted, as shown in Figure B3-18.

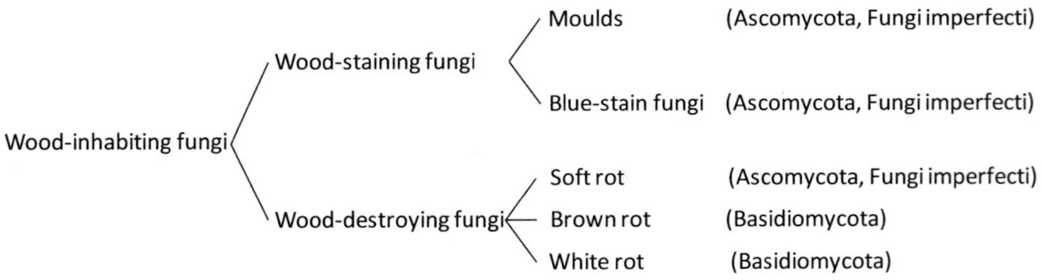

Figure B3-18 Classification of wood-inhabiting fungi in accordance with their damage pattern. (Sutter, 1997)

Moulds

Moulds grow on surfaces, without penetrating the wood to any significant extent and the main giveaway is a discolouration of the surface itself. Moulds do not generally break down the wood components (cellulose, lignin), so the physical features of the wood remain intact. Nevertheless, mould infestation is generally a sign of conditions (optimal temperature from 24 to 28°C, humidity ranging between 30 and 150%), which also promote the growth of other, more dangerous fungi. Moulds may also prove a health risk in interior spaces, particularly if they remain undetected (e.g. under parquet flooring).

Blue-stain fungi

Blue-stain fungi grow from within wood, where the dark inherent colour of the fungi themselves stains the wood a bluish-blackish colour, hence the name blue stain. Blue-stain fungi tend to affect softwood in particular and only target sapwood in trees forming dark heartwood. Blue-stain fungi do not generally break down cell walls, which means they do not trigger rotting or any strength loss of the wood.

Soft rot

Soft rot fungi destroy the secondary walls of cells while growing, mainly breaking down the cellulose, while leaving the lignin intact. Wood affected by soft rot shows tell-tale softness, even before any clear signals of destruction become visible to the naked eye. Wood that has broken down even to a minor extent loses much of its strength. Moist wood gives when thumb pressure is applied and can be depressed slightly with the fingernail. Moist wood is also dark in colour, with a musty and soft surface (name). Cracks in dry wood only emerge at a significantly advanced stage and resemble the cube break typical of brown rot, although not penetrating to the same depth. Deciduous trees are more prone to soft rot fungi than conifers. Ideal conditions for soft rot to develop exist in the soil, namely the ground-air zone. Similarly, wood used in buildings or other objects exposed to water and high levels of moisture (for an extended period), may be damaged by soft rot.

Brown rot

Brown rot is the most significant form of fungal-related wood destruction affecting interiors and is triggered by a number of basidiomycetes, which thrive at varying levels of temperature and moisture. Brown-rot fungi grow in the cell cavities (cell lumen, pith rays, resin canals) and from there, proceed to break down the cellulose of the secondary wall via enzymatic means, although the lignin remains more or less intact. As the cellulose breaks down, the wood loses its strength and weight. The brown-rot wood cracks into deep cubical pieces, with one typical example shown in Figure B3-19. The two key types of brown-rot fungi are true dry rot and cellar fungus.

Figure B3-19 True dry rot with cubical cracks.

Gloeophyllum spp.

Gloeophyllum are brown-rot fungi, which exhibit the unappealing property of growing from inside to out. This means any outbreak is only detected at a very late stage and when the fruiting bodies can be seen from outside, the affected wood will already have been completely destroyed. Gloeophyllum tend to attack softwood and do not affect interior spaces. Sites of affected wooden members include bridge constructions, balconies and windows. Gloeophyllum can tolerate high temperatures (thriving on sun-exposed wood and debarked wood) and sporadic dryness.

B3.4 Literature

P. Hoffmeyer, L.M.R. Nunes, P.P. de Sousa, original Articles A4, A15, STEP 1995.

Bosshard H.H. (1984). Holzkunde. Zur Biologie, Physik und Chemie des Holzes. Birkhäuser Verlag, Basel, 312 p.

Grosser D. (1985). Pflanzliche und tierische Bau- und Werkholzschädlinge. DRW-Verlag, Leinfelden-Echterdingen, 159 p.

Knigge W. and Schulz H. (1966). Grundriss der Forstbenutzung. Entstehung, Eigenschaften, Verwertung und Verwendung des Holzes und anderer Forstprodukte. Verlag Paul Parey, Hamburg.

Steuer W. (1990). Vom Baum zum Holz. Nutzholzarten, Holzschäden, Ausformung, Holzernte, Rundholzsortierung, Verkauf. DRW-Verlag, Leinfelden-Echterdingen, 256 p.

Sutter H.-P. (1997). Holzschädlinge an Kulturgütern erkennen und bekämpfen. Verlag Paul Haupt, Bern, 164 p.

B4 Durability

Original articles: G. Sagot, L. M. R. Nunes, P. P. de Sousa

Durability in the sense of resistance to destructive organisms and thus the ability to guarantee load-bearing capacity and usability throughout the service life of an object is imperative for wood as an organic material. The scope of durability generally also includes other aspects such as corrosion resistance of metallic fasteners. However, our remit in this book is limited to wood-specific matters, namely the biological effects of elements like fungi and insects. Frequently occurring damaging micro-organisms were already covered in Article B3. In the following section, we will focus on natural durability in particular.

B4.1 Assessing natural durability

As was explained in Article B1, wood incorporates other constituents, as well as the main chemical components of cellulose, hemicellulose and lignin, which are either absorbed or formed as sapwood transforms into heartwood and are specific to particular tree species. Some also have microcidal qualities, like terpene or phenols. Trees in temperate climates contain around 1 to 10% constituents, but this percentage can be much higher in the case of tropical woods, ranging from 2 to 30%. The closing mechanisms of the heartwood cells, which vary according to tree species, have a key influence on natural durability, while isolating the heartwood in a virtually "hermetic" manner, via tyloses for example, which is crucial help in warding off the attack of pests and damaging materials. Since heartwood formation is often linked to the storage of constituents, heartwood is clearly more resistant to pests than sapwood. Common tree species are assessed in terms of their biological durability during experiments performed in a laboratory or in the open air. Figure B4-1 shows, for example, specimens from durability tests.

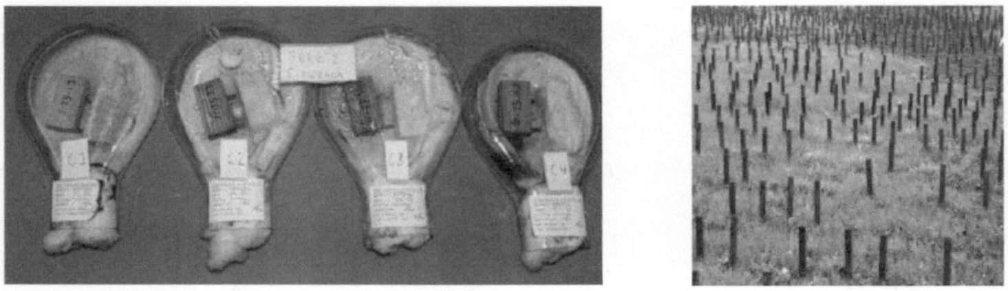

Figure B4-1 Left: Specimens in the laboratory in accordance with EN 113. Right: Outdoor test bodies in accordance with EN 252. (Rapp and Augusta, 2000)

Durability tests featuring various wood pests and untreated wood species show how the resistance of the different wood species varies very considerably, which means they are classified into a range of durability classes. The durability classes are specified in EN 350-2:1994 „Natural durability of solid wood“:

Against fungi

1	Very durable
2	Durable
3	Moderately durable
4	Slightly durable
5	Non-durable

Against insects and wood pests in sea water

D	Durable
M	Moderately durable
S	Vulnerable
SH	Where heartwood is also classed as vulnerable

In EN 350-2, numerous wood species are included in the various durability classes. Spruce is included in durability class 4 and SH, while a tropical wood spieces Greenhart is durability class 1 and D (termites). A European hardwood species with tyloses like robinia achieves a higher durability class 1 to 2 against fungal attack. The classification specified in EN 350-2 of **durability against fungi applies only to heartwood**. Sapwood should generally be classified in durability class 5, unless otherwise specified. The generally high durability of tropical wood species compared to those in Europe is primarily attributable to the constituents. These make up a higher proportion of tropical woods, namely 2 to 30%, a considerable portion of which also exhibit microcidal effects. The higher proportion depends on the tree species involved – since tropical woods tend to be more exposed to wood pests (e.g. termites), which explains why they have developed better defence mechanisms through their evolution. Biological pests rely on specific environmental conditions, key among which by far is wood moisture content. Fungi cannot grow optimally in wood with moisture content underneath the fibre saturation point (due to the lack of free water in the cell lumen). Another key factor is whether the wood is permanently moist or whether it can be dried again. Spruce members can resist fungal attack, provided the moisture content is kept under 20% at all times, with resistance declining if the moisture content fluctuates and the wood becoming non-durable in the event of continual moisture. These fluctuating environmental conditions are taken into consideration by classifying into service classes, as stated in DIN 68800-1 "Wood preservation – Part 1: General", Table B4-1.

Table B4-1 Classification in service classes (SC). (Table from DIN 68800-1:2011)

SC		Moisture content / exposure[ab]	General conditions of use	Threat from				Leaching
				In-sects	Fungi[c]	Soft rot	Wood pests in the sea water	
0		Dry (permanently ≤ 20%), average relative humidity up to 85%	Wood or wood product under roof, not exposed to weathering or moisture, the risk of structural damage caused by insects can be excluded in line with 5.2.1	No	No	No	No	No
1		Dry (permanently ≤ 20%), average relative humidity up to 85%	Wood or wood product under roof, not exposed to weathering or moisture	Yes	No	No	No	No
2		Occasionally humid (> 20%), average relative humidity over 85%[d] or occasional humidification due to condensation	Wood or wood product under roof, not exposed to weathering, high relative humidity can lead to occasional but not permanent humidification	Yes	Yes	No	No	No
3	**3.1**	Occasionally humid (> 20%), accumulation of water in wood, even on a limited basis, not expected	Wood or wood product not under roof, exposed to weathering, but without permanent contact with soil or water, accumulation of water in wood, even on a limited basis, is not expected due to swift drying	Yes	Yes	No	No	Yes
	3.2	Frequently humid (> 20%), accumulation of water in wood, even on a limited basis, expected	Wood or wood product not under roof, exposed to weathering, but without permanent contact with soil or water, accumulation of water in wood, even on a limited basis, expected[e]	Yes	Yes	No	No	Yes
4		Mainly to permanently humid (> 20%)	Wood or wood product in contact with earth or freshwater and hence mainly to permanently exposed to moisture leading to moderate to severe stress[f]	Yes	Yes	Yes	No	Yes
5		Permanently humid (> 20%)	Wood or wood product permanently exposed to sea water	Yes	Yes	Yes	Yes	Yes

a The terms “occasionally”, “frequently”, “mainly” and “permanently” show increasing levels of stress, but do not constitute very precise numerical data, due to wide-ranging input variables.

b The value of 20% includes a safety margin (see 4.2.2, Note 1).

c Wood destroying basidiomycetes (see 4.2.2, Note 2) and wood staining fungi (see 4.2.3).

d The decisive factor for allocating wood components to a service class is the respective moisture content.

e Components expected to accumulate deposits of dirt, earth, foliage etc. over several months and components exposed to particular stress, e.g. due to splashing water, should be classed in SC 4.

f “Moderate” or “severe” stress refers to the hazard potential of a fungal attack (moisture conditions, soil conditions) and the intensity of leaching.

Insect infestation is more difficult to assess. DIN 68800 points out that assuming a normal interior climate, insect attack does not represent any risk except lyctus beetle infestation of starchy hardwood species. The standard also cites constructive wood preservation measures against insect attack, using sealed panelling or an effective means of inspecting the wooden members. The second part of DIN 68800-2 "Wood preservation – Part 2: Preventive constructional measures in buildings" offers hints on preserving constructional wood, since effective construction details can help significantly reduce the vulnerability of wood members to biological pests, as well as the choice of wood species used. As a general rule, wood should be installed with moisture content commensurate with subsequent use and not permanently moist (in other words, covered or allowed to dry out again and water can be effectively drained off). Preserving constructional wood against fungal attack thus equates squarely to humidity protection. The scope of wood preservation also encompasses positive silvicultural practices. Wood is felled in winter, when the nutrient and water transfer processes in the tree are on hold and outside the flight and mating seasons for insects that damage wood. This also means that the felled wood has to be transported away and processed prior to spring (→ when the insects mate and lay eggs). The wet storage or deliberate watering of stems is performed when stems cannot be processed immediately and have to be placed in storage and offers two advantages. Firstly, it imbues the wood with a moisture content which is excessive for wood-destroying fungi or insects. Secondly, it helps prevent shrinkage due to drying out and hence shrinkage cracking, which reduces the access for insects due to fewer microscopic cracks for laying eggs or fungal spores.

B4.2 Wood preservatives

Chemical wood preservatives are applied to wood using various impregnation methods. Paints and coatings are considered physical forms of wood preservation and also represent a means of weathering protection, since they can help mitigate precipitation, high humidity, UV rays and mechanical stresses. Chemical wood preservation should always be a last resort, since it inevitably means using biocides and should be avoided where possible by selecting the right wood species as well as constructive wood preservation measures. Chemical wood preservation may, however, make sense as remedial action following an infestation. The prevailing wood species in our area, spruce, is very difficult to treat, even when using pressure-impregnation processes, since the sapwood can only be penetrated a few millimetres. Heartwood, meanwhile, tends to be untreatable, since its cell walls are saturated with core constituents and the pits close off as part of the drying process. The unpopularity of biocide use notwithstanding, effective chemical wood preservation, namely non-stop treatment of the entire volume of wood, is infeasible for conventional structural elements.

B4.3 Wood modification

European timber species, particularly the number one choices for construction, spruce, are not particularly durable. Since chemical wood preservation is frowned on, various modification measures are preferred to make the timber more durable. One starting point involves reducing the hygroscopicity of wood (= its ability to bind in water). This tendency of wood sees environmental conditions form (lowered equilibrium moisture content), which hinder the growth of fungi and insects and significantly reduce the risk of infestation.

Preventing hygroscopic water absorption, which involves water being bound, particularly to cellulose and hemicellulose hydroxyl groups (*–OH*), ensures that less water can be bound into cell walls, which lowers the equilibrium moisture content of the wood. The final level of moisture content attained is then too low for pests. Moreover, promoting the decomposition of hydroxyl groups and blocking them deactivates the docking sites for the fungal enzyme. Figure B4-2 shows various approaches taken for reducing the hygroscopicity of wood, which equates to minimising the storage of water in cell walls. The hydrophilic hydroxyl groups can be broken down (→ Thermally treated wood), cross-linked (→ Belmadur, Kebony) or grafted (→ Accoya). Absorbing less water achieves the desirable outcome of reduced swelling and shrinking behaviours and a resulting boost in dimensional stability.

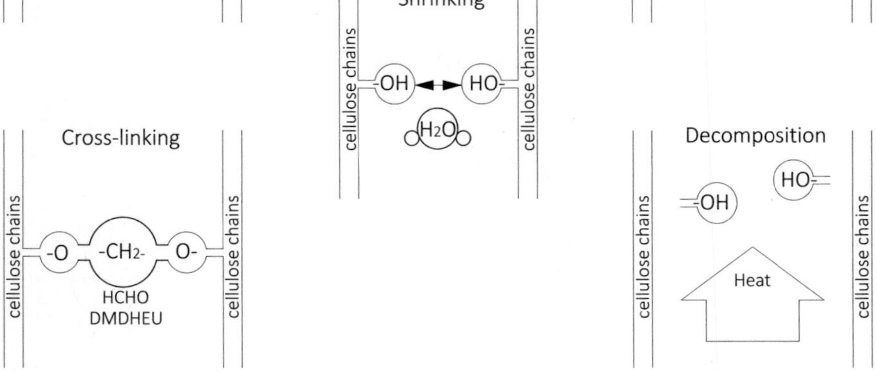

Figure B4-2 Starting points for wood modification. (www.holzfragen.de)

Both the most recently cited methods represent a chemical modification of the timber, since a chemical reaction is triggered when the timber is impregnated with specific materials, which involves the free hydroxyl groups being occupied. For the Belmadur® product, the result is cross-linking with the methylol already in use during the textile cross-linking (non-iron laundry). Kebony®, conversely, uses furfuryl alcohol, an agricultural industry waste product, to cross-link the hydroxyl groups and "artificially soak" the wood in a certain way.

In contrast, acetylation with acetic anhydride, which is the process used in Accoya®, involves occupying (grafting) the hydroxyl group, so that no further water can be stored. The waste product of acetylisation is acetic acid, which must then be recycled.

The final modification option is thermally treated wood, which incidentally has already become widely established. The wood is heated to a temperature of between 150 and 240°C, which causes the *–OH* groups to "decompose". Thermally treated beech may, depending on the intensity of the thermal treatment applied, attain durability class 1, as opposed to thermally treated pine, which can "only" achieve durability class 2 to 3 (see also pre-standard DIN CEN/TS 15679), while thermally untreated pine wood is classified in durability class 3-4 in accordance with EN 350-2.

Up to Accoya, all modified wood products are only used in outdoor and garden areas. Wood having undergone special thermal treatment is not usable for load-bearing purposes, since the mechanical properties decline by up to 30%, while impact energy may drop by up to 60%, rendering the material very brittle. Accoya alone has been used in load-bearing applications to date; the most well-known examples of which include the bridges in Sneek (NL) (see Figure B4-3).

Figure B4-3 Accoya bridge in Sneek, the Netherlands. (Blass and Eberhart, Karlsruhe)

All the chemical modification measures require impregnation, most often pressurised, followed by drying and a reaction period. However, not all wood varieties lend themselves to impregnation (→ pits). Spruce, for example, is very difficult to impregnate, which rules out the possibility of chemical modification. Timber species open to the use of acetylation include pine, beech, maple and particularly Radiata pine, a fast-growing species with large annual ring widths. Not all the workings behind such modification methods are known or understood to date, but to put matters in a clearer chronological context, it is worth noting that the first acetylation took place in 1928 and the first furfurylisation in the 1950s. Even so, several more decades elapsed before the modified wood was first rendered commercially usable.

B4.4 Preventive constructional measures

Unlike metallic parts, which may corrode, wood is exposed to biotic damaging influences. In terms of biotic damage, the two main causes identified are fungal and insect attack, which means rules must be established for using wood members under various climatic conditions without risking the load-bearing capacity and durability of members made of wood or wood-based products. Practical examples should indicate which details are to be avoided and which types of construction are advisable.

Fungal attack

Fungal attack occurs in wood with moisture content above 20%. Fungi require free water and oxygen, to develop, although the optimal moisture content depends on the type of fungus involved. Fungal attack can significantly reduce the load-bearing capacity of wooden constructions, although the strength loss varies and is dictated by the type of fungus and the extent of the infestation. Wood may already be significantly weakened, even if it looks virtually pristine, but merely ensuring a building is designed properly should prevent wooden members from being exposed to excessive moisture and any associated damage caused to the wood by fungi. Designs should take the following requirements into account:

- Ensure humidification of wood is avoided as far as possible,
- Ensure swift water drainage and ventilation, if occasional humidification cannot be prevented,
- Ensure woods with sufficient natural durability are used, if permanent humidification cannot be prevented.

Insect attack

Heat tends to encourage insect attacks, since it promotes the development and reproduction of insects. Those capable of completely destroying wood quicker than any other are termites, which proliferate in warmer areas of the world, although their existence and chances of survival are far slimmer in Central and Northern Europe (as things stand). Central heating systems in buildings tend to establish conditions which favour the survival and proliferation of insects, since they ensure optimal temperatures are reached, even during colder seasons, while at lower temperatures, the insect larvae may perish. Cracks and gaps which form in treated wood and penetrate the external treated wood layer, can signal the basis for egg-laying or the start of infestation, given the fact that they significantly impair the effectiveness of any wood preservative treatment. The natural durability of the individual wood species against insect attack varies considerably. Among most woods, heartwood tends to be durable, although resistance among individual wood species to termite infestations varies. The wood and insect species involved determine whether or not wood from the sapwood range is durable.

Corrosion of metallic parts

Under normal conditions of use, wood is not attacked by acids and bases. However, metallic parts must be protected against corrosion, where the conditions of use could impact on their long-term behaviour. To avoid surface discolouration of visible wood, there may be a need to paint or coat metallic parts.

Preventive constructional measures

Many factors have to be taken into account when considering the durability of wood. As far as possible, the wood should be installed at the equilibrium moisture content present in the building, so that only additional seasonal variations in humidity need be taken into account. If care is not taken here, however, the wood may be prone to cracking after being installed. This risk of cracking peaks in wood members directly exposed to weathering or where there are significant seasonal variations in humidity. Cracks allow the ingress of water and fungal spores and egg-laying by insects within the internal cross-section beyond the protected outer area. The structural engineers must take into consideration the risk of humidity changes due to:

- Direct water influx,
- The hygroscopic behaviour of the wood in response to the moisture and temperature of the surrounding air.

In a liquid state, water is particularly liable to infiltrate the wood in the grain direction. Accordingly, end grain surfaces should be arranged or covered such as to ensure that no water can penetrate via the capillary effect. Certain cases which may cause moisture content to increase are now listed below:

- Warm and moist air e.g. in poorly ventilated attics, in which ventilation shafts are installed, is conducive to moisture absorption of the wood.
- Condensation problems can be prevented by additional heat insulation and a vapour barrier. Condensation at the base of glass walls must also be taken into consideration.
- Wood is exposed to moisture when in contact with the ground, in doors and window frames and in areas, in which snow accumulates due to snow drift. Particular attention should be paid to areas liable to exposure to splashing water, i.e. showers, bathrooms or kitchens.
- Water may accumulate in the soil, in walls behind watertight barriers and in connections, because the natural moisture release is prevented by evaporation in these areas. Any features which may allow water to accumulate in the vicinity of metallic plates must also be avoided.
- There is an increased risk of humidification while wooden members are stored at the building site and during assembly, while no roof covering is in place. Stacks of wood should be covered and when there is no other alternative for assembly reasons, stored for a maximum of one week. Minimising the storage period is particularly important in the event of adverse weather conditions.

One of the most fundamental preventive constructional measures is to cover exposed members, particularly those with end grain. DIN 68800-2 explains precautionary construction measures used in building construction.

Precautionary fungal protection

The risk of any increase in moisture content can be reduced by careful planning of structural details. Conversely however, high temperatures based on the geographical location cannot be influenced and the risk of higher levels of moisture content rises with increasing average temperatures. If it is impossible to prevent the ingress of water into the wood, swift drying must be ensured or the moistened area must be segregated, so that the moisture content does not exceed the limit of 20%. The moisture content results from the equilibrium between water absorption and water release and it can be lowered, if the water absorption is delayed and efforts made to promote water release. One good example is the use of metallic column footing, which allow a minimum distance between the end grain of the column and the ground surface of 300 mm to be established. Even if rain falls, the wood is not exposed to additional moisture from splashing water and can dry out again afterwards.

Precautionary insect protection

The first step should be to determine the natural durability of the selected wood species in terms of resistance to individual and potentially invasive insect species. There is also a need to clarify whether the individual insects are likely to be present in the area in which the wood is installed. As well as wood preservation using wood with natural durability, structural barriers shielding the wood from the ground may also constitute effective protection, in the event of any risk of infestation by termites which exist at ground level. Termites often form tunnel-like routes with canopies between the nest chambers and wood sections attacked. Structural barriers or covers make it easier to detect the movement of such termites. There is a need for regular maintenance inspections to discover and eliminate the relevant locations and care must be taken to ensure that the surfaces between the ground and wood sections are both accessible and visible.

Corrosion protection for fasteners

EC 5 includes examples of minimum requirements or minimum corrosion protection needed for various service classes. Enhanced corrosion protection measures are required, e.g. in storage buildings for chemical products, salt and fertilisers, or also in phosphoric acid factories, in which the use of bolts, dowels and steel plates made of special stainless steels is imperative.

B4.5 Literature

G. Sagot, L.M.R. Nunes, P.P. de Sousa, original Articles A14, A15, STEP 1995.

Rapp A.O. and Augusta U. (2000). Dauerhaftigkeit in den Gefährdungsklassen unter besonderer Berücksichtigung von Lärchen- und Douglasien Kernholz. 22. Holzschutz-Tagung der Deutschen Gesellschaft für Holzforschung, Bad Kissingen.

B5 Wood drying and strength grading

Original article: P. Glos

B5.1 Principles of wood drying

Wood is hygroscopic when its moisture content is underneath the point of fibre saturation, namely its moisture content varies depending on the prevailing climate. In the process, the cross-sectional dimensions also vary by swelling or shrinking. In addition, practically all the technological properties such as strength and stiffness or heat insulation in the hygroscopic area (**below the fibre saturation point**, see Article B2) depend on the moisture content. Moreover, when its moisture content exceeds around 20%, wood becomes particularly vulnerable to pests. Accordingly, avoiding subsequent damage involves drying the wood before end processing of the same to ensure that it is already adapted to the future prevailing climate as far as possible. In central Europe, the equilibrium moisture content for wood placed outside is 12 to 14% in summer and at least 18% in winter. An overview of wood moisture guidance values for installation is shown in Table B5-1.

Table B5-1 Wood moisture guidance values.

Purpose of use	Moisture in %
Structural timber	8-20
External windows and doors	13-16
Furniture, room doors, interior construction with permanent heating	6-9
Plywood, LVL, musical instruments	5-7

As a general rule, wood has to be dried; not only when producing structural and glued laminated timber, but in all areas of woodworking and processing, for producing wood-based panels for example. During the construction period, it is important to strictly limit increase in wood moisture during either transport or storage and deployment of the dried wood, which can be done by storing the wood on covered underlayers and swift covering or, where applicable, provisional covering of members. Absorption of moisture during storage occurs particularly quickly and intensively via end grain surfaces, recesses, boreholes and so on.

Physical and technical principles of wood drying

The relative humidity

Relative humidity denotes the amount of water vapour absorbed by the air and is calculated from the percentage ratio of the actual current quantity of water vapour in the air relative to the maximum absorbable quantity, assuming unchanged air temperature and pressure. If the air contains the maximum possible quantity of water vapour, this is referred to as saturated steam or saturated humid air; whereupon the vapour starts condensing on the surfaces of the space. Underneath the maximum quantity, the vapour is described as unsaturated or overheated. As Figure B5-1 shows, the relative humidity clearly depends on air temperature. Warm air can absorb more humidity than cold air and is relatively drier amid unchanged absolute water content. Increasing the temperature in a specific room triggers a simultaneous decline in relative humidity. This link is significant for wood drying, since falling relative humidity causes the equilibrium moisture content of wood to plummet, while the direct impact of air temperature on moisture content is far lower.

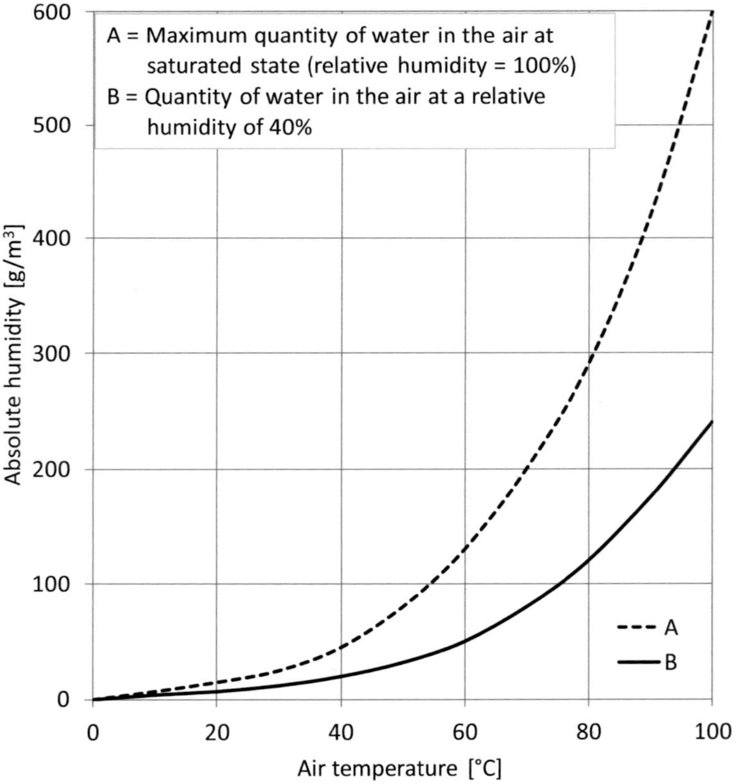

Figure B5-1 Absolute humidity depending on relative humidity and air temperature.

The equilibrium moisture content of wood

The resulting relative moisture content thus depends on the relative humidity, air temperature and pressure and wood species. Given storage for sufficient time under specific climatic conditions, wood will reach the related equilibrium moisture content. The key factor dictating the so-called equilibrium moisture content of wood u_{gl} is the relative humidity of the surrounding air, whereas the direct impact of temperature is less important. Figure B5-2 shows the Keylwerth diagram, which can be used to determine the connection between relative humidity, air temperature and the equilibrium moisture content of wood (here: spruce).

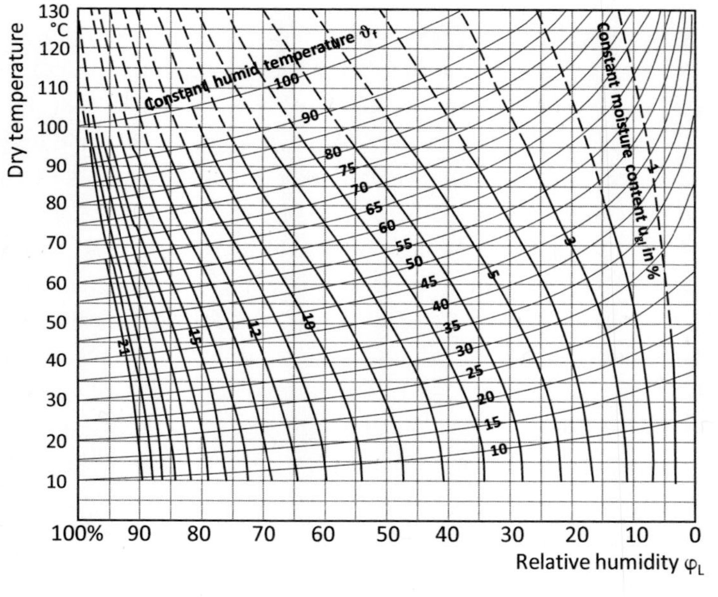

Figure B5-2 Keylwerth diagram for spruce. (Kollmann and Coté, 1968)

Transport of moisture in the wood while drying

The transport of moisture in the wood is based on a complicated system of capillary water movements and diffusion. At the beginning of the drying process, since levels of moisture content exceeding the fibre saturation point and capillary transport processes are accelerated for the drying process, free water is discharged more quickly than bound water. The process of drying from $u = 50\%$ to $u = 40\%$ moisture content proceeds, with the same cross-sectional dimensions, namely far faster than $u = 20\%$ to $u = 10\%$. The speed of the moisture release as regards this capillary water movement is hardly direction-dependent.

The moisture exchange between wood and the surrounding proceeds at different rates for the individual wood species, while the drying speed clearly declines in the following order: pine sapwood, spruce, beech, sipo, oak, iroko and teak. Since natural water transport takes place in sapwood, it is easier to dry than heartwood, while wood with a greater density generally dries more slowly than lightweight wood. Morphological features, such as storage of constituents during heartwood formation or the closure of bordered pits in conifers may considerably hinder water transport and hence the drying process. In such cases, the internal wood layers may still be very humid, while the external layers have already dried out, whereas for freshly sawn timber, the water contained in the wood is still distributed virtually uniformly throughout the cross-sectional portion. With the release of moisture via the surfaces, the external layers are first to dry, hence reducing the disparity in moisture content between the interior and surface, Figure B5-3. This disparity is the decisive factor controlling the movement of moisture within the wood and thus the most important element for an optimal drying process. Excessive wood moisture disparity, e.g. at the start of the drying process, hinders the capillary water movement and means most of the free water from the interior has to travel via the wood in the form of vapour, which is far slower than for liquid water. If the wood moisture disparity is very considerable, drying defects may occur (e.g. casehardening of wood). Conversely, if the disparity is insufficient, the drying times required will be excessive and uneconomical. The term "casehardening" refers to the residual stress state of the wood after the drying process, due to permanent localised tensile or compressive deformations perpendicular to the grain. The external areas dry too quickly and shrink, which exerts compressive stresses perpendicular to the grain accumulating on the inside and tensile stresses perpendicular to the grain on the surface; ultimately resulting in permanent tensile deformations perpendicular to the grain. If the drying process is continued, the internal sections also dry out and shrink and the inability of the external areas to accommodate such deformations leads to internal cracks forming.

Drying methods used include natural outdoor drying, while technical drying methods (kiln-drying) in drying chambers are particularly popular.

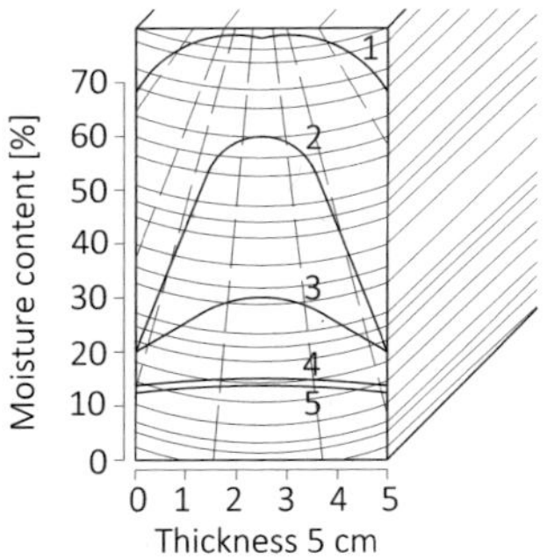

Figure B5-3 Change in wood moisture in a 50 mm thick plank during storage at the lumberyard, Line 1: Green, Line 5: Wood moisture arising.

B5.2 Principles of strength grading

As a natural raw material, wood is produced from a wide range of trees depending on their type, genetic material, growth and environmental conditions. Wood characteristics are also very diverse, from tree to tree, but also within the same stem, stem cross-section and longitudinal stem direction (see Articles B1 and B3). Cutting up logs by sawing destroys the wood structure having developed, for example severing wood fibres in the area of branches or in spiral-grained stems. This also means greater variation in the strength properties of sawn timber, particularly for smaller cross-sectional dimensions, which may vary much more than those of logs subject to no or minimal processing. In fact, the strength properties of unsorted sawn timber within a single wood species may vary so considerably that the strength of the strongest piece may be ten times as much as the weakest, as in Figure B5-4 above. Since the most important factor influencing the possible end use of the wood is always its characteristic strength, namely the 5% quantile of the basic population, using ungraded wood makes it impossible to exploit the high strength of the majority of sawn timber. For economic reasons, however, the wood needs to be classified into different strength classes by ensuring appropriate grading, Figure B5-4 below. However, where the strength of the individual piece is unknown and can only be determined indirectly via visually recognisable approaches or non-destructively assessing the wood properties correlating to strength, the variation in strength within a specific classification cannot be completely narrowed down. For this reason, the classes overlap depending on the quality of grading (see Figure B5-4 below) and the less effective the grading method used, the greater the overlap concerned. This shows how crucial it is to ensure effective strength grading to exploit the economic potential of the wood. In addition, grading is also a prerequisite to ensure the availability of sawn timber in the quality and quantity demanded by consumers and its compliance with all user requirements, particularly meeting the criterion of being a predictable construction material with reliable properties.

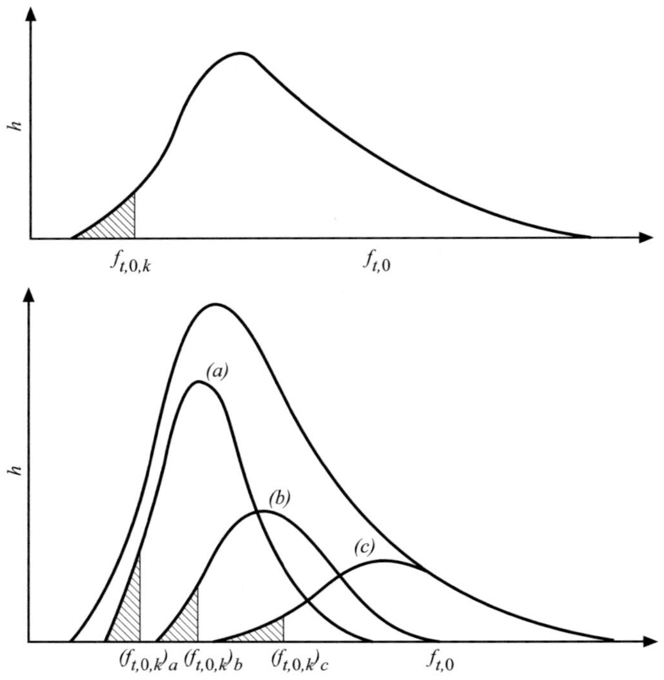

Figure B5-4 Above: Tensile strength ungraded; below: Tensile strength graded into three classes. (STEP 1995 Article A6)

Traditionally, visual assessment has been used as a means of grading wood, which involves determining its quality by assessing visible characteristics, particularly knots and annual ring width. Up to the start of this century, visual grading was primarily conducted based on traditions having been handed down and local experiences. The first detailed standardisation of strength grading took place in 1923 in the USA and subsequently, from the 30s onwards, the process established itself throughout Europe. Given the various species and qualities of wood around Europe and the range of timber construction traditions, for example using sturdy or slender cross-sections, it is no surprise that the last half-century has seen variation in grading regulations for grading criteria; particularly in terms of how knots are classified and classes divided. The common problem, however, for these grading regulations is the visual grading process. Naturally, practical considerations dictate that only visually recognisable wood characteristics can be taken into consideration and simple combination rules defined as a result. Key factors determining strength, such as wood density, cannot be assessed accurately via visual means alone, for example via the annual ring width. The visual strength grading approach thus suffers from inherent and unavoidable ambiguity, which limits the effectiveness of any grading. Moreover, since the classification also depends on the alertness of the grader, it is not completely objective.

As part of efforts towards more reliable strength grading of sawn timber, enhanced usage efficiency for the present timber stock and higher strength classes, machine grading methods were developed in Australia, the USA and the UK from around 1960 onwards, with other countries following suit later on. The increasing industrialisation in woodworking and processing, the growing importance of quality assurance and the increasing demand for premium quality wood have seen interest in such machine grading rocket in recent years, which has catalysed the development of new and more efficient grading machines. During wood grading, there is generally a need to distinguish between so-called appearance grading and strength grading. The former process of appearance grading involves assessing the wood in terms of optical appearance, namely aesthetic criteria. This is particularly important whenever wood is deployed as a visible fixture, whether as wall or ceiling panelling; for structural timber used in permanently visible construction or in furniture-making. Conversely, during strength grading, the wood is assessed solely in accordance with relevant criteria dictating its load-bearing capacity. It follows that under certain circumstances, timber used in high-level and permanently visible members must be graded in accordance with both criteria and not just strength alone. However, the remainder of this article only covers the strength grading of sawn timber.

General strength grading requirements

The strength grading process should ensure confirmed compliance with all key properties of relevance when using timber for load-bearing purposes. To do so, the grading rules for each grade must include limit values for wood characteristics, which correlate sufficiently to the actual strength and stiffness of the wood. For conventional visual grading, these particularly include the annual ring width as a measure of the wood structure relevant for strength and strength-reducing wood characteristics such as knots, slope of grain, cracks, reaction wood, fungi and insect attack and mechanical damage. During machine grading, other properties, which are not visually measurable, such as the dynamic modulus of elasticity (MOE), can be used and allow far more accurate estimation of the strength and stiffness properties. In addition to relevant strength and stiffness grading criteria, limit values for geometric properties, such as the wood wane, curvature and warp must also be determined, given their relevance when the wood is used for construction. The fact that the wood dimensions, as well as its curvature, warp and cracks, depend on the moisture content means the limit values have to be based on a reference moisture content, which is fixed at 20% in Germany (see DIN 4074). In addition, the moisture content also has to be taken into consideration during machine grading, if moisture-related wood parameters are measured.

Visual strength grading

During visual grading, the appearance of the wood is considered and the characteristics that determine its strength are assessed. Accordingly, the result depends on the grader and the key visual grading parameter is knots. The fact that location and climate influence the tree growth means the sample used to determine the correlation of the measured grading parameters with the strength must be representative. The sample results must also, without fail, be applicable to the basic population, since this is the only way to guarantee that the actual non-destructive strength grading leads to a wood piece being reliably classified in a specific strength class using the correlated grading parameter. The first step involves establishing a representative sample, which means the following influences are to be taken into consideration:

- Climate,
- Growth area,
- Moisture content,
- Geometry of specimens,
- Wood species.

The next step is to perform tests to correlate the grading parameter with the targeted strength property, e.g. bending strength. A typical result of bending strength, depending on the grading parameter knots, is shown for softwood in Figure B5-5. Large knots mean a drastic drop in bending strength, which means the wood has to be assigned to a lower strength class. However, knots are only weakly correlated to strength and the end result is a wide and diversified point cloud.

Subsequently, to use knots as a grading parameter, limit values are needed for each strength class. Figure B5-6 shows a schematic view of the process. The characteristic bending strength of a strength class is set to e.g. 24 or 30 N/mm^2, while the horizontal lines of the selected bending strengths intersect with the vertical lines of knot area ratios, so that only 5% of all pieces of wood with equivalent or less knot area ratio have a smaller bending strength (black areas in Figure B5-6). Accordingly, Figure B5-6 shows knot area ratios of 0.4 and 0.2, respectively, as decisive grading criteria for strength classes, with characteristic bending strengths of 24 or 30 N/mm^2. EN 1912 links a visual grade, which, in this case, equates to grade S10 "maximum knot area ratio = 0.4" in accordance with DIN 4074 Part 1, to strength class C24 in accordance with EN 338, where $f_{m,k}$ = 24 N/mm^2. In EN 1912, national visual grades are assigned to a European strength class.

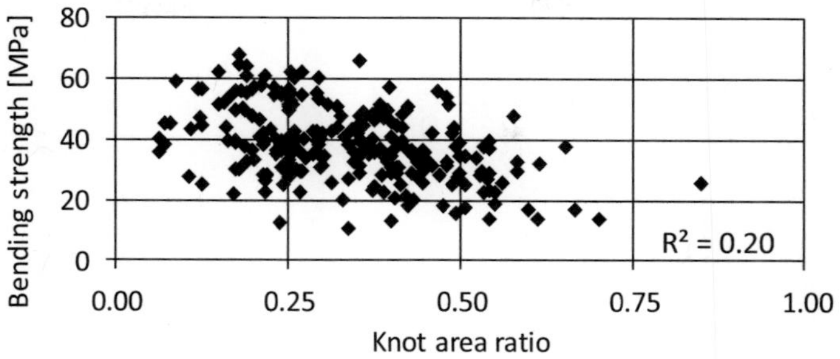

Figure B5-5 Correlation between bending strength and knot area ratio (softwood).

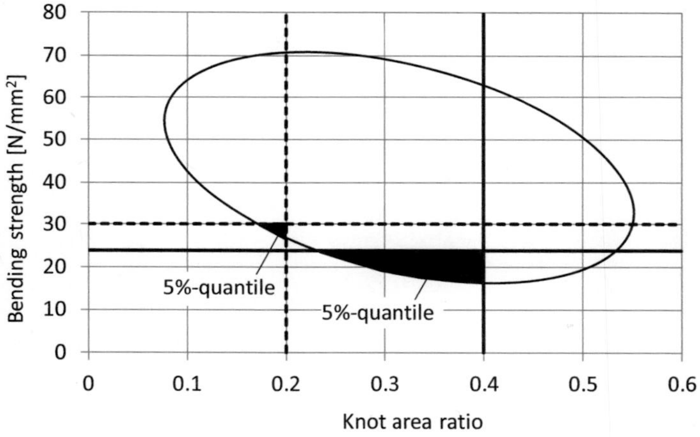

Figure B5-6 Determining grading parameter "knot area ratio" per strength class (softwood).

Visual grading parameter and grading criteria

The key grading parameter "knot area ratio" is shown as an example in Figure B5-7 for squared timber (softwood). Knot parameter A is always the product of geometrical considerations concerning the ratio of knot area to the remaining area of wood. Evaluating boards and planks, however, is somewhat more complex, since the knots are present as individually interspersed on the narrow or wide side, through knots or knot clusters (see also Figure B3-14). Various limits (= grading criteria) are applied to the grading parameter "knot area ratio" depending on the grade. In addition to the knot area ratio, other grading parameters must also be added, such as slope of grain, wane, annual ring width, curvatures, cracks, reaction wood or discolourations. For example, Table B5-2 lists the grading criteria for squared timber in accordance with DIN 4074.

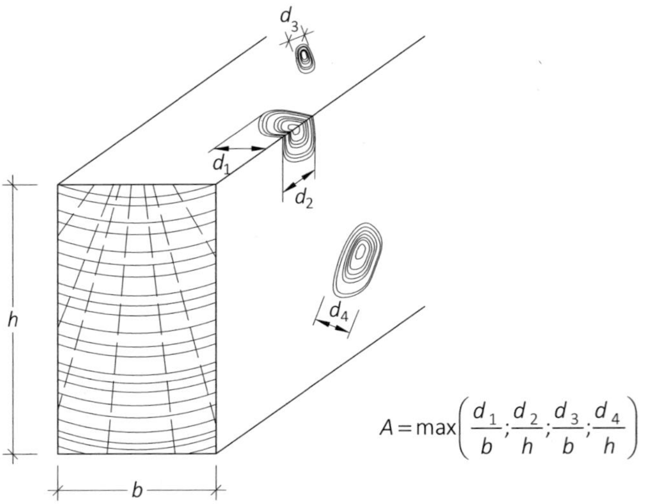

Figure B5-7 Knots in squared timber, Knot parameter A in accordance with DIN 4074-1.

Machine strength grading

When machine strength grading is used, the key grading parameter is not knots, but modulus of elasticity. The measurement accuracy exceeds that of visual grading and the grading result is more easily reproducible. The modulus of elasticity correlates better with strength than knots; Figure B5-8 shows a clearly higher coefficient of determination (R^2) than Figure B5-5.

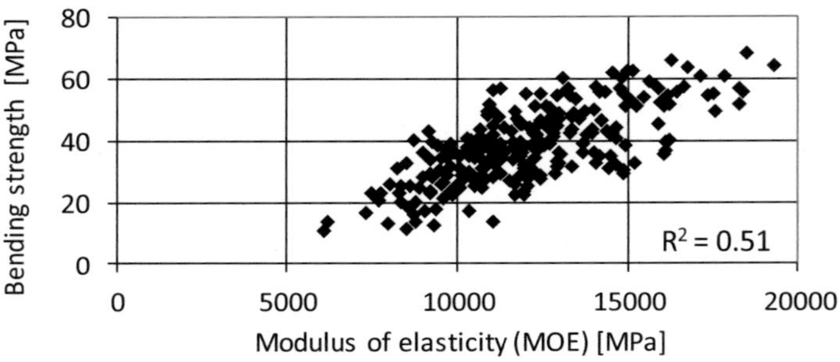

Figure B5-8 Correlation between bending strength and MOE (softwood).

Table B5-2 Grading criteria for square timber (softwood) during visual grading, DIN 4074-1:2008.

Grading parameter	Grade		
	S7	S10	S13
1. Knots	up to 3/5	up to 2/5[a]	up to 1/5
2. Slope of grain	up to 12%	up to 12%	up to 7%
3. Pith	permissible	permissible	not permissible[b]
4. Annual ring width			
- In general	up to 6 mm	up to 6 mm	up to 4 mm
- For Douglas fir	up to 8 mm	up to 8 mm	up to 6 mm
5. Cracks/shakes			
- Shrinkage cracks[c]	up to 1/2	up to 1/2	up to 2/5
- Lightning, ring shake	not permissible	not permissible	not permissible
6. Wood wane	up to 1/4	up to 1/4	up to 1/5
7. Curvature[c]			
- Longitudinal curvature	up to 8 mm	up to 8 mm	up to 8 mm
- Warp	1 mm / 25 mm height	1 mm / 25 mm height	1 mm / 25 mm height
8. Discolouration, rot			
- Blue stain	permissible	permissible	permissible
- Brown and red stripes	up to 2/5	up to 2/5	up to 1/5
- Brown, white rot	not permissible	not permissible	not permissible
9. Compression wood	up to 2/5	up to 2/5	up to 1/5
10. Insect damage	burrows up to 2 mm in diameter by greenwood insects permissible		
11. Other features	are to be taken into consideration correspondingly, in line with the other grading criteria		

[a] For spruce and Douglas fir, up to ½ for annual ring widths of up to 4 mm for spruce and 5 mm for Douglas fir. The proportion within a delivery must not exceed 25%.

[b] Permissible for squared timber with width > 120 mm.

[c] These grading parameters are not taken into consideration for non-dry woods.

Machine grading parameters

As well as the modulus of elasticity, additional parameters may also be mechanically recorded and evaluated, including knots, density, moisture content and the cross-sectional dimensions. What all parameters have in common is the fact that they can be measured via non-destructive methods. The modulus of elasticity is measured via static bending or dynamic measurement, while optical methods, such as surface scanning or even X-rays or gamma rays are used to measure knots. Wood moisture measurement devices reveal the moisture content, while modern strength grading machines measure the dynamic modulus of elasticity via vibration. Provided the vibration frequency f, length ℓ of the board and its density ρ are known, the dynamic modulus of elasticity E_{dyn} can be calculated, see equation (B5-1), featuring measurement of longitudinal oscillation as shown in Figure B5-9 (derivation in Annex 1). This measurement method is shown schematically in Figure B5-9, while Figure B5-10 shows the result of radiography with X-rays to measure knots.

$$E_{dyn} = 4 \cdot f^2 \cdot \ell^2 \cdot \rho \qquad \text{(B5-1)}$$

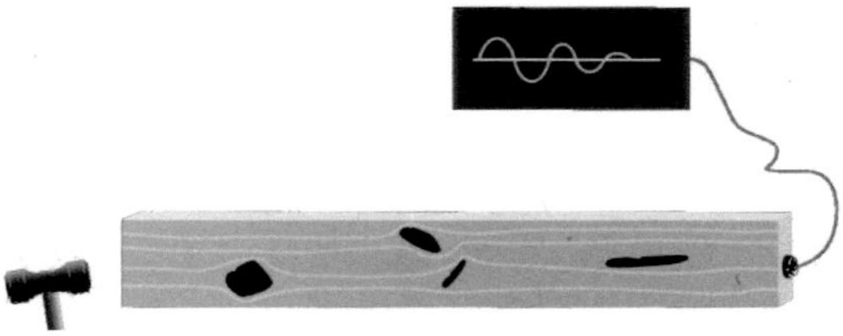

Figure B5-9 Oscillation measurement.

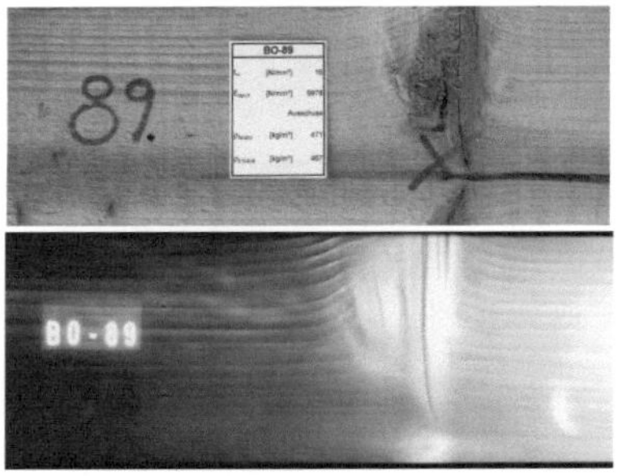

Figure B5-10 X-ray image (bottom) of a board with knot (top).

The best results are obtained by combining several grading parameters. Such a combination elicits improved correlation with strength and hence a better result when classifying into a strength class. The grading parameters for "dynamic modulus of elasticity" and "knots" are combined in virtually all modern grading machines, which may also involve using density obtained through weighing (or via X-rays) and the cross-sectional dimensions. Figure B5-11 shows a typical correlation between bending strength and a combination of grading parameters. The coefficient of determination is even higher compared to Figure B5-8, which means a better correlation between bending strength and grading parameters.

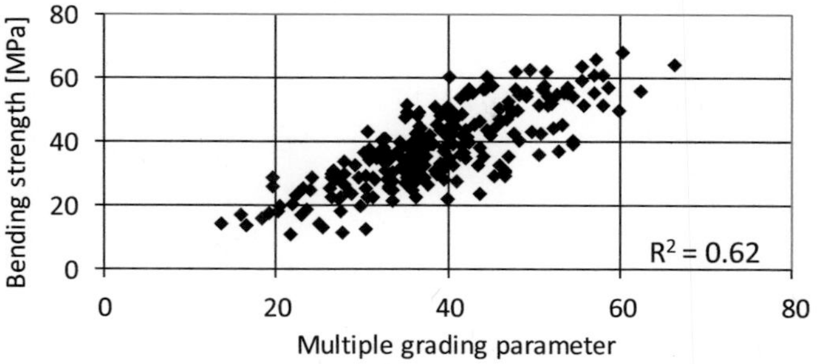

Figure B5-11 Correlation between bending strength and a combination of several grading parameters.

Four-point bending tests

The correlation between grading parameters and strengths is normally determined by so-called four-point bending tests (Figure B5-12). The strength derived from grading criteria is bending strength, whereupon all other strength properties of a strength class tend to be determined using regression equations derived from the bending strength (EN 338). The local modulus of elasticity between both force transmission points (see w_{local} in Figure B5-12) does not include any shear strain portion, given the absence of shear stresses in the relevant area. This means the local modulus of elasticity is determined based on the measured force and deformation portions excluding contribution of shear.

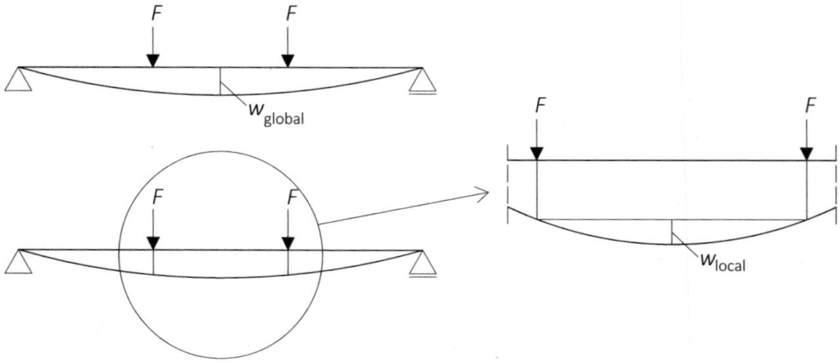

Figure B5-12 Four-point bending tests with deflection curve, w_{global} and w_{local}.

Calibration of grading machines

The correlation between grading criteria and bending strength must be set and checked. The "settings" (adjustment values) of a grading machine must also be calibrated, so that the machine is capable of assigning the tested sawn timbers to a strength class. This calibration can be implemented in two different ways, namely "machine-controlled" and "output-controlled".

Machine controlled:
In this case, numerous pieces of sawn timber are tested using four-point bending tests. Grading models are then derived from the results and used to determine the settings for the grading machines. During the production, no further sawn timber is tested.

Output controlled:
The initial settings of the grading machines are derived following a small number of tests, based on which sawn timber is regularly taken from graded wood and tested and checked to ensure it meets the selected criteria (characteristic values). If not, the settings are adapted accordingly.

Log grading

Visual log grading involves checking logs with the naked eye, as is done for sawn timber. Centring on the visibility of the characteristics for logs, the assessment encompasses similar characteristics as when grading sawn timber, all of which impact on strength. However, unlike sawn timber, logs are classified into quality classes.

B5.3 Literature

P. Glos, original Article A6, STEP 1995.

Kollmann F.F.P and Coté W.A. (1968). Principles of wood science and technology. Volume I, Solid wood. Springer Verlag, Berlin, 592 p.

B6 Wood products

Original articles: P. Glos, F. Colling, A. Ranta-Maunus, G. Steck, D. R. Griffiths, E. Raknes

At one time, the size of the tree in the forest used to dictate the dimensions of the structural timber which it was used to make. A century ago, square timbers, with a cross-sectional area of 150 mm x 450 mm and up to 20 m long were generally available. Nowadays, although timber with a cross-section exceeding 75 mm x 225 mm and more than 5 m long is increasingly rare and thus costlier than ever, when such larger dimensions are required, multiple pieces of wood can be combined to form a single member, e.g. of glued laminated timber. Wood is inevitably non-homogenous as a result of the natural growth process of trees. Meanwhile, knots, pitch pockets and other growth-related properties have a key influence on strength, which explains the considerable variance in strength properties within a member. If large pieces of wood are split up into smaller parts, then re-joined, any flaws within the materials are dispersed and the variance in material properties also declines. The larger load-bearing capacity of glued laminated timber compared to solid timber is not due to a higher average load-bearing capacity, but a reduced variance in strength properties, which elicits a higher characteristic strength. In general, the variance in strength of the wood products listed in Table B6-1 declines with increasing manufacturing cost and the homogenisation also increases. However, logs, which are very rarely processed, represent an exception. They are particularly strong, since the wood fibres remain uncut and the continuous fibres can convey the stresses around the knots.

Table B6-1 Wood construction products and their components.

Wood product		Components
Logs	Variance declines ↓	Stems
Sawn timber		Squared timber, planks, boards and battens
Glued laminated timber		Boards
Laminated veneer lumber		Veneers
Plywood		Veneers or sawn timber
Parallel strand lumber		Veneer strands
Particleboards		Particles (chips)
Fibreboards		Fibres

As a natural material, wood displays different properties in various directions. Parallel to the grain, namely in the longitudinal direction of the tree stem, wood is particularly strong, but conversely far lower at right angles to the grain, for example the tensile strength parallel to the grain is around 40 times higher than perpendicular to the grain. This is reflected in the ease of chopping wood along the fibres with an axe, as opposed to the far more difficult task of splitting a piece of wood at right angles to the grain. These significant differences in strength and stiffness properties in various directions no longer apply for most wood-based panels. Since the wood particles and fibres are randomly arranged in many types of panels, in panel planes, the loading direction is far less relevant to determine strength in the panel plane than in solid timber.

B6.1 Solid timber

The term solid timber is used for members completely made of wood in its natural structural form. This is opposed to so-called engineered wood products or wood-based panels, which are manufactured by disassembling and then reassembling parts.

Sawn timber

Sawn timber is produced by sawing logs longitudinally and refers collectively to products produced from logs in sawmills. The main (product) groups, according to their cross-sectional dimensions, are boards, planks, battens and square timbers. The cross-sectional areas are standardised (e.g. in DIN 4074-1), as preferred cross-sections e.g. determined for use as solid structural timber (KVH®) or separately determined depending on the purpose of use. The various process technologies (frame saw, band saw, chipper-canter) allow various yields of sawn timber and help achieve the required quantities of industrial residual wood. Figure B6-1 shows cutting options for square timbers, where the free-of-heart conversion is particularly relevant, as it helps ensure fewer shrinkage cracks.

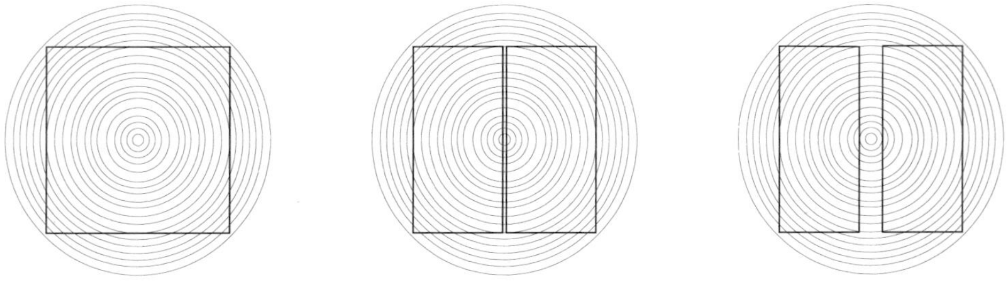

Figure B6-1 Cutting types. Left: Single-stem cut; Middle: Dual-stem split-heart cut; right: Dual-stem free-of-heart cut.

After the cut, the sawn timber is visually pre-graded, before being kiln-dried or dried naturally outdoors. Smaller sawmills in particular often lack any kiln-drying rooms, which imbues the sawn timber with excessive moisture. Once the wood has been dried to the desired moisture content, the sawn timber is visually or machine graded and assigned to a strength class, while wet wood is cut based on the subsequent measure of the dry wood. This involves cutting with a certain allowance for subsequent shrinkage of the wood. An example of a piece of squared timber with pith after drying is shown in Figure B6-2. All additional processing steps used for sawn timber are known as finishing and structural timber is often used rough-sawn. Glued solid timber comprising two or three bonded individual cross-sections is available, while sawn timber can also be finger-jointed, to extend its dimensions. For this purpose, the manufacturer in Germany requires proof of suitability for gluing load-bearing wooden members (DIN 1052-10).

Figure B6-2 Squared timber with pith, knots and drying cracks. (Studiengemeinschaft Holzleimbau, 1998)

Strength classes for solid timber

A system of strength classes has the advantage of aggregating the wide-ranging combinations of wood species, grades and origins available in the European internal market into a manageable number of groups, to simplify and improve the use of timber for load-bearing purposes. The system of strength classes as laid out in EN 338 is illustrated further in the following and comprises twelve classes for conifers (Table B6-2) and eight for deciduous species (Table B6-3). The scope goes from the lowest softwood class C14 up to the highest hardwood class D70 to cover all current strength classes in Europe. EN 338 includes details of the characteristic strength, stiffness and density parameters for each strength class.

It is possible to agree on strength classes, including a constant rate of strength and stiffness values in each case, because regardless of origin, practically all coniferous and deciduous species used for commercial applications have a similar ratio of strength and stiffness values. Existing test data revealed that an effective approximation of all key characteristic strength and stiffness values could be calculated from the bending strength, the dynamic modulus of elasticity (MOE) and the density (see Article B5). EN 1912 summarises which visual grades and wood species can be assigned to which strength classes in EN 338, while Table B6-4 sets out this assignment in EN 1912 for the strength class C30. Using this table, all the various national grades can be assigned to a strength class.

Table B6-2 Strength classes and characteristic values in accordance with EN 338:2009; softwood.

N/mm^2	C14	C16	C18	C20	C22	C24	C27	C30	C35	C40	C45	C50
$f_{m,k}$	14	16	18	20	22	24	27	30	35	40	45	50
$f_{t,0,k}$	8	10	11	12	13	14	16	18	21	24	27	30
$f_{t,90,k}$	0.4	0.4	0.4	0.4	0.4	0.4	0.4	0.4	0.4	0.4	0.4	0.4
$f_{c,0,k}$	16	17	18	19	20	21	22	23	25	26	27	29
$f_{c,90,k}$	2.0	2.2	2.2	2.3	2.4	2.5	2.6	2.7	2.8	2.9	3.1	3.2
$f_{v,k}$	3.0	3.2	3.4	3.6	3.8	4.0	4.0	4.0	4.0	4.0	4.0	4.0
kN/mm^2												
$E_{0,mean}$	7	8	9	9.5	10	11	11.5	12	13	14	15	16
$E_{0,05}$	4.7	5.4	6.0	6.4	6.7	7.4	7.7	8.0	8.7	9.4	10	10.7
$E_{90,mean}$	0.23	0.27	0.30	0.32	0.33	0.37	0.38	0.4	0.43	0.47	0.5	0.53
G_{mean}	0.44	0.50	0.56	0.59	0.63	0.69	0.72	0.75	0.81	0.88	0.94	1.00
kg/m^3												
ρ_k	290	310	320	330	340	350	370	380	400	420	440	460
ρ_{mean}	350	370	380	390	410	420	450	460	480	500	520	550

Table B6-3 Strength classes and characteristic values in accordance with EN 338:2009; hardwood.

N/mm^2	D18	D24	D30	D35	D40	D50	D60	D70
$f_{m,k}$	18	24	30	35	40	50	60	70
$f_{t,0,k}$	11	14	18	21	24	30	36	42
$f_{t,90,k}$	0.6	0.6	0.6	0.6	0.6	0.6	0.6	0.6
$f_{c,0,k}$	18	21	23	25	26	29	32	34
$f_{c,90,k}$	7.5	7.8	8.0	8.1	8.3	9.3	10.5	13.5
$f_{v,k}$	3.4	4.0	4.0	4.0	4.0	4.0	4.5	5.0
kN/mm^2								
$E_{0,mean}$	9.5	10	11	12	13	14	17	20
$E_{0,05}$	8	8.5	9.2	10.1	10.9	11.8	14.3	16.8
$E_{90,mean}$	0.63	0.67	0.73	0.8	0.86	0.93	1.13	1.33
G_{mean}	0.59	0.62	0.69	0.75	0.81	0.88	1.06	1.25
kg/m^3								
ρ_k	475	485	530	540	550	620	700	900
ρ_{mean}	570	580	640	650	660	750	840	1080

Table B6-4 Strength class C30; assignment of visual grades, species and origins in accordance with EN 1912:2013 (botanical identification is specified in EN 1912).

Strength class	Grading rule publishing country	Grade	Species commercial name	Origin	Botanical identification
C30	France	ST-I	Spruce, fir	France	1, 22
	Germany, Austria, Czech Republic	S13, S13K	Spruce, fir, pine, larch	CNE Europe	1, 15, 22, 47
	Italy	S1	Douglas fir	Italy	54
	Nordic countries	T3	Spruce, fir, pine, larch	NNE Europe	1, 15, 22, 47
	France	ST-I	Larch, pine	France	15, 39, 47
	Spain	ME1	Laricio pine	Spain	39
	Slovak Republic	S0	Spruce	Slovak Republic	22

B6.2 Glued laminated timber

Glued laminated timber (glulam) is a variety of laminated timber in which all components (boards in this case) are arranged parallel to the grain. In contrast, components are orthogonally arranged in plywood or cross laminated timber. One example of a glued laminated timber beam is shown in Figure B6-3. For production purposes, the individual components, so-called lamellae, are glued over the entire contact surface with adhesives, which results in rigid connections. The great advantage of using glued laminated timber is the way it allows homogenisation of wood as a construction material.

The development of waterproof and mildew-proof synthetic resin adhesives has made this an increasingly important area in the timber construction field and also prompted a notable upturn in the development of glued timber constructions in timber engineering. This field allows far larger cross-sections than with sawn timber, which is why glulam, like other load-bearing products such as cross-laminated timber, is classed as an "engineered wood product" and has revolutionised the timber construction world in recent decades.

The requirements for glued laminated timber are set out in EN 14080, which includes details of both production requirements and strength classes. Rules are also set out for use by each manufacturer, to establish its own strength class, which paves the way to combine specific lamella and finger-jointing qualities with each other.

Incidentally, it is also worth recapping that EN 14080 sets out a new generation of European standards and unifies many originally disparate standards in the process. This means EN 14080 replaces standards EN 385 to EN 387, EN 390 to EN 392 and EN 1194, which were previously used to regulate strength classes (EN 1194) or delamination tests of glue lines (EN 391) among others. This means that the regulations dispersed to date over wide-ranging of standards are now grouped together in a single standard.

Figure B6-3 Glued laminated timber. (Studiengemeinschaft Holzleimbau e. V., 1998)

Structure of glued laminated timber

The basic concept underpinning the development of wood-based products involves reducing the variance in properties and homogenising wood as a material. The basic structure of the tree is generally preserved when logs are used, which makes them stronger than sawn timber, in which many fibres are cut during the processing. Given the wide-ranging strength values for wood, the strength of a board is determined by its weakest point and the load-bearing capacity can only be influenced by grading the boards or logs. If you consider the average area of Table B6-1 (glulam to parallel strand lumber), additionally to grading, the wood characteristics can be boosted through homogenisation, whereby individual solid timber components are bonded with each other (usually using adhesive). If the engineered wood components are even smaller (particles or fibres), grading in accordance with the strength of each individual component is no longer feasible and this is when significant homogenisation is applied to the production process.

Within the compound structure of a glulam beam, weak points such as knots are less significant, since they no longer influence the overall beam cross-section. Areas with knots have a lower modulus of elasticity than the clear wood above and below the same, exert a load on the surrounding lamellae and are relieved. This systematic action involving lamellae, which are glued together under pressure, is known as the lamination effect. Dissecting and subsequently gluing the wood paves the way to homogenise the material, since the variance in mechanical properties in the subsequent material is lower than in individual lamellae. In addition, elements like larger knots, pitch pockets or bark ingrowths are removed. This means that cutting out a flawed area from a piece of timber, which must be classified in accordance with DIN 4074 in S10, means two timbers classified in S13 are formed (Figure B6-4). Figure B6-4 shows a schematic view of how an area with large knot size is removed, followed by finger-jointing to produce a long board. As standard, during glulam manufacturing, boards 3 to 6 m long are combined by finger-jointing to form an endless lamella. Cutting to length as required allows the finger-jointed structural timber in the glulam to be dispersed.

Figure B6-4 Removal of low-value sections and combining via finger-jointing. (Studiengemeinschaft Holzleimbau e.V., 1998)

Glulam has further, clear advantages compared to squared timber. Squared timber with a large cross-section tends to include pith and large shrinkage cracks (see Figure B6-5). As well as hindering many applications, they also provide easier access for moisture or wood pests, while the kiln-drying of large cross-sections can be particularly problematic when significant lengths are involved. For glulam conversely, drying of individual boards is problem-free and during the drying process, any deformations arising can be planed off in the production process (see Figure B6-6).

Figure B6-5 and Figure B6-6 reveal another glulam production principle. According to EN 14080, in glulam, the piths should generally be on the same side, but for glulam in service class 3, the surfaces must be always on the right sides (facing the pith), which results in a left-on-left side bond of both lower lamellae (away from the pith), as in Figure B6-5 on the right. Under adverse weather conditions, this helps prevent shrinkage cracks on the surface. Since the left side of the wood is subject to stronger tangential shrinkage, it is more prone to cracking. The shrinkage behaviour of lamellae or finished glulam-cross-sections and their subsequent processing is set out in Figure B6-6.

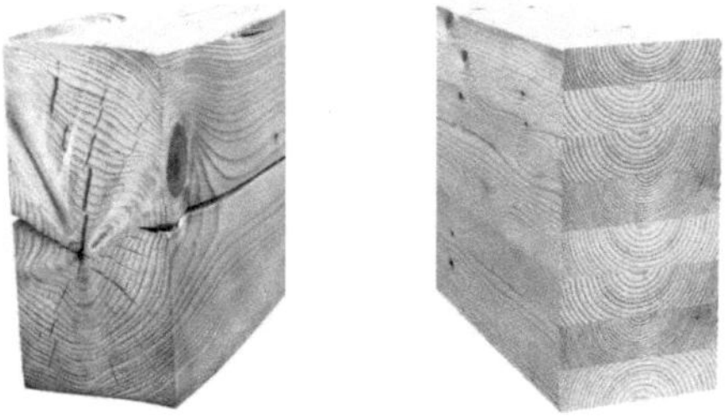

Figure B6-5 Squared timber cross-section with cracks compared to a glulam-cross-section. (Studiengemeinschaft Holzleimbau e.V., 1998)

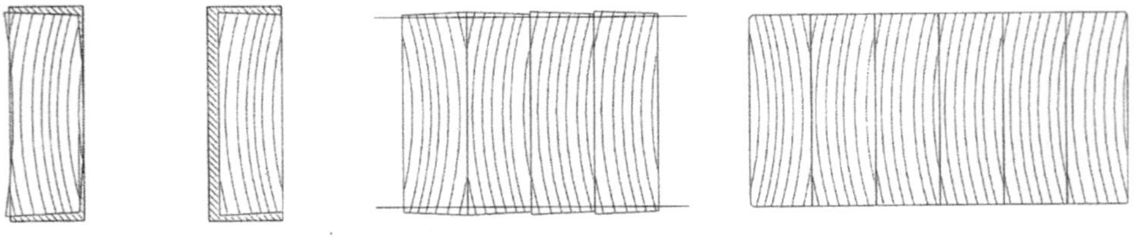

Figure B6-6 From left to right: Shrinkage deformation of a board cross-section in accordance with drying with respect to the fresh sawn board. Cross-section after planing the large sides with respect to the fresh sawn state with parallel areas for bonding. Glulam cross-section after gluing. Planed and chamfered glulam cross-section. Only the right sides are externally exposed. (Studiengemeinschaft Holzleimbau e.V., 1998)

The risk of cracking induced by changing humidity is greater for increasing lamellae thickness and it also becomes increasingly difficult to apply the required pressing power in the glue line. For these reasons, EN 14080 defines limit values for the cross-sections of lamellae. Glued laminated timber must be formed from lamellae between 6 and 35 mm thick and for straight members not exposed to extreme and alternating climatic stress, the maximum thickness may be increased to 45 mm. For members exposed to dramatically fluctuating climates, however, board thicknesses below 35 mm are recommended. To reduce the stresses generated within the boards due to changing humidity, the boards can be equipped with a relief groove in the longitudinal direction, which also helps prevent any tendency towards warping. The end grain of the boards should be protected regardless, since strongly fluctuating humidity may cause the glulam timber sections to crack. Moreover, during transport, storage and assembly, it is also important to control the humidity of the members e.g. caused by precipitation, ground moisture or drying. Reference can be made to EN 14080 for all additional rules and requirements.

Production

Glued laminated timber can be produced from various wood species, although in Germany, spruce is almost exclusively (up to 95%) the first choice. The use of pine, fir, larch, Douglas fir and beech is negligible, although glulam produced from larch and Douglas fir is suitable when increased durability requirements apply. Glulam is produced via the following steps:

- Pre-grading of the boards,
- Drying of the boards,
- Strength-grading of the boards,
- Finger jointing to generate "continuous lamellae",
- Capping of the continuous lamellae to the required length,
- Planing the lamellae,
- Applying adhesive,
- Inserting into a pressing jig and pressing,
- Planing to the final member size.

Experienced professionals using special equipment are required to produce glued load-bearing wood members. Requirements include covered and air-conditioned working spaces, an annex for kiln-drying, reliable measurement devices to determine wood moisture, machines for processing adhesive surfaces (e.g. planing machines; the wood surfaces to be bonded must have been freshly planed) and the board ends for which butt joints are envisaged (e.g. finger-jointing systems), glue application devices and devices to apply the required pressing power (presses). Plus, in Germany, the "Proof of suitability for gluing load-bearing wood members ("glue approval")" (DIN 1052-10) must have been issued.

The sawn timber used to produce glulam is stored and normally kiln-dried. To prevent subsequent shrinkage cracks, the moisture content for gluing should be established at a level commensurate with the subsequent average expected moisture content in construction. However, the moisture content should be set slightly below the subsequent average moisture content, since any compressive stresses perpendicular to the grain with swelling that follows are less damaging for glulam than tensile stresses perpendicular to the grain while shrinking. When used indoors, prior to processing, the boards are dried to a moisture content of 10 ± 2%. This is in the region of the equilibrium moisture content when installed, normally prevents any further damaging shrinking and helps minimise cracking. Unavoidable drying stresses which occur during conventional drying of sawn timber must be minimised, to reduce problems for further processing (e.g. the glue line warping and splitting open). The next step is to grade the boards visually or via machine grading, normally the latter, and the wood is pre-planed before grading. The quality control also involves weeding out any excessively dry or moist boards and sending them to be reconditioned. Glulam boards vary considerably in terms of their mechanical and optical properties. Classification into (visual and mechanical) grades and reducing unsightly natural wood characteristics are also vital when it comes to producing premium glulam beams. Following the finger-jointing and gluing (during which compliance with specific environmental conditions, temperature and relative humidity must be ensured), the final stage of production involves planing and chamfering (edge breaking), whereupon the desired cross-section is attained for the glulam member. This is followed by cosmetic repairs, cutting to length, packaging, delivery and erection.

Monitoring

Particular attention is paid to monitoring quality, which is established by factory production control (in-house monitoring) and external supervision to verify the self-monitoring process (external monitoring). The in-house monitoring mainly includes checking board quality, the finger joint strength and the quality of glue lines. A glue log is also maintained, which must include details of the production date and number, wood species, strength class, dimensions of the member, moisture content, time of initial glue application, start and end time of the pressing process, laminating pressure, resin and hardener, use of adhesive (g/m^2), calibration of the moisture analyser and the temperature and relative humidity of the various working spaces.

Types of glued laminated timber

EN 14080 differentiates horizontal laminated glued laminated timber with one or more adjacent lamellae (Figure B6-7). Alternatively, multiple narrow glued laminated timber members are bonded together to form a single wider member (block-glued glulam).

In load-bearing constructions, since glulam beams are particularly suited to accomodate bending stresses, the main forces exerted on outer lamellae are tensile or compressive forces. This explains the range of cross-section composition using various strength classes of lamellae, as shown in Figure B6-8. In this case, a distinction is established between homogenous and combined glued laminated timber (e.g. GL24h and GL24c respectively). In homogeneous glulam, all the lamellae comprise boards of a particular strength class. In combined glulam, the middle lamellae may have a lower strength class than the outer lamellae and this combination can be symmetrical or asymmetrical. For higher glulam beams, the outer lamellae comprise at least two lamellae, while for smaller cross-sections with up to ten lamellae, at least one lamella. A combination may encompass multiple wood species. This is appealing when the use of "lower quality" and hence more economical wood is considered. However, producing hybrid and even combined glulam makes production costlier, since the different boards have to be correctly arranged in the production process. For asymmetrical cross-sections, proper installation on the building site must also be guaranteed with a label.

This is particularly applicable for certain members made of glued laminated timber and shown in Figure B6-9. If, as is done for three-hinged frames and tapered beams, the external lamellae are cut, particular attention is required for combined glulam structures, since lamellae in the lower strength class tend to "shift" in the outer area.

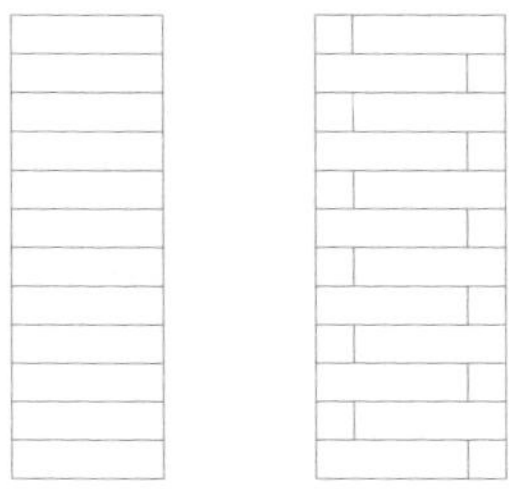

Figure B6-7 Horizontal laminated glulam with one (left) or multiple adjacent (right) lamellae.

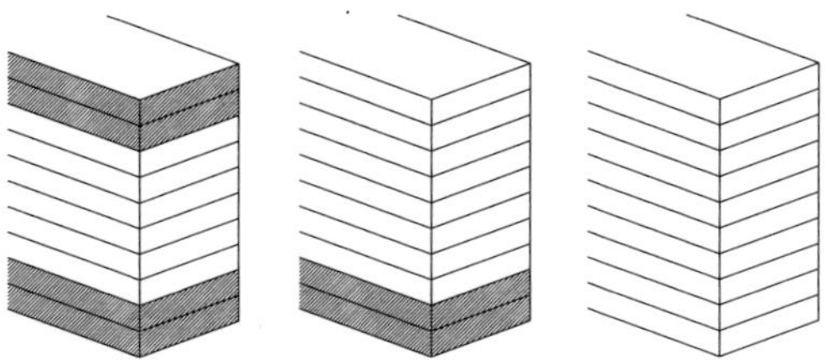

Figure B6-8 From left to right: Symmetrical and asymmetrical combined and homogenous glued laminated timber.

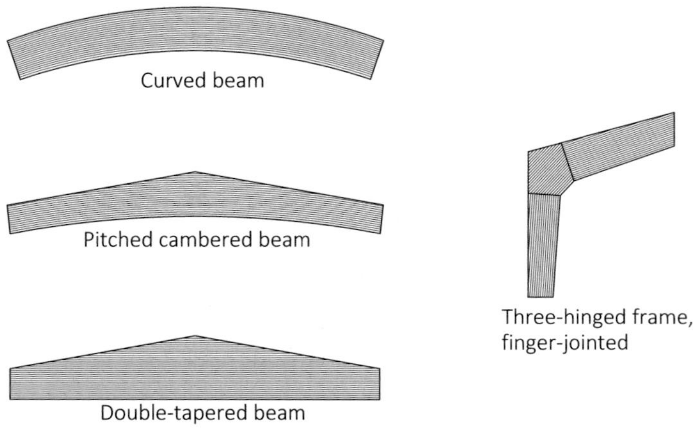

Figure B6-9 Special glued laminated timber members, examples.

Influencing parameters that determine strength

Glued laminated timber is mainly used in constructions exposed to bending stress, which is why this section focuses on the specific factors influencing the bending strength of glued laminated timber members. The additional strength properties will be discussed in the following section.

Systematic investigations (Colling, 1990; Frese, 2006) show that the strength of glued laminated timber members is a factor of both the strength of the boards as well as that of the finger joints. If a glued laminated timber beam is exposed to stress, the natural result is a move to relieve the stress as soon as possible. The breakage thus tends to occur at points where the ratio of stress over strength peaks and may be either a section of the board with knots or a finger joint. In glued laminated timber members including poor quality finger joints, the failure is usually due to the finger joints themselves, while glued laminated timber members containing substandard planks are prone to break in areas of knots. This means that efforts to attain premium glued laminated timber must consider all of components, boards and finger joints alike. There is little sense in seeking to unilaterally reinforce only one of these components, because the load-bearing capacity of the glued laminated timber beam will increasingly be determined by the other (weaker) component, making it impossible to exploit the targeted strength improvement. The following conclusions can be made following the cited investigations:

- Visual grading with stringent knot requirements for boards is not a suitable means of effectively boosting the load-bearing capacity of glued laminated timber members. This is because reducing the permitted degree of knots merely boosts the board strength itself, taking no account of finger joint strength. Accordingly, the failure of glued laminated timber members is increasingly triggered by the finger joints, which prevents higher board strength from being exploited.
- By using strength grading based on density and/or the modulus of elasticity of the wood, the strength of both the boards and finger joints can be controlled, which helps achieve clearly stronger glued laminated timber. The machine strength grading of the boards can thus be considered key for high-strength glued laminated timber.
- Since various finger-jointing profiles are used and a range of production conditions play a part (such as the age of the adhesives, pressing power applied or climatic conditions in the production spaces), it is imperative to monitor the quality of the finger-jointing to ensure sufficient strength can be guaranteed. This applies all the more because production-related factors interact and are often very difficult to control.

Background to the CEN provisions

EN 14080 specifies equations used to calculate the mechanical properties of homogenous glued laminated timber depending on the board and finger-jointing properties. Currently valid equations for certain important properties are specified in Table B6-5. Here, the specified equations apply for homogenous glued laminated timber, while for combined glued laminated timber, the equations apply to calculate the properties of the various cross-sectional parts. The equations, in turn, are based on members with a height/width of 600 mm for bending beams/tension members, or on a reference volume of 0.01 m^3 for beams exposed to tensile stresses perpendicular to the grain. This should be taken into consideration when deviating from these reference dimensions. In this case, reference is made to the so-called volume effect, which takes account of the fact that the strength of a brittle material declines with increasing member dimensions.

Table B6-5 Some mechanical properties of glued laminated timber.

Property	Equation in accordance with EN 14080:2013
Bending strength [N/mm²]	$f_{m,g,k} = -2.2 + 2.5 \cdot f_{t,0,l,k}^{0.75} + 1.5 \cdot \left(\frac{f_{m,j,k}}{1.4} - f_{t,0,l,k} + 6 \right)^{0.65}$ if $1.4 \cdot f_{t,0,l,k} \leq f_{m,j,k} \leq 1.4 \cdot f_{t,0,l,k} + 12$
Tensile strength [N/mm²]	
parallel to the grain	$f_{t,0,g,k} = 0.8 \cdot f_{m,g,k}$
perpendicular to the grain	$f_{t,90,g,k} = 0.5$
Compressive strength [N/mm²]	
parallel to the grain	$f_{c,0,g,k} = f_{m,g,k}$
perpendicular to the grain	$f_{c,90,g,k} = 2.5$
Shear strength [N/mm²]	$f_{v,g,k} = 3.5$
Rolling shear strength [N/mm²]	$f_{r,g,k} = 1.2$
MOE parallel to the grain [N/mm²]	$E_{0,g,mean} = 1.05 \cdot E_{t,0,l,mean}$
MOE perpendicular [N/mm²]	$E_{90,g,mean} = 300$
Shear modulus [N/mm²]	$G_{g,mean} = 650$
Rolling shear modulus [N/mm²]	$G_{r,g,mean} = 65$
Density [kg/m³]	$\rho_{g,k} = 1.1 \cdot \rho_{l,k}$

The manner in which the bending strength of the glued laminated timber depends on the tensile strength of lamellae and the bending strength of the finger joints is recorded via an empirical relationship, determined based on tests and numerical investigations. The higher bending strength of glulam in comparison to the tensile strength of the lamellae is attributable to various lamination effects, key examples of which are outlined below:

- The load-bearing behaviour of a board in a standard tensile test differs from that in a glued laminated timber member. The test procedure outlined in EN 408 to determine tensile strength envisages a specific minimum test length and does not provide any lateral support for the test pieces. Accordingly, eccentrically placed knots or areas with asymmetrical density distribution may result in lateral deformations and thus additional bending moments, which reduce the tensile strength of the boards. However, within a glued laminated timber beam, the boards are secured by bonded adjacent lamellae, meaning a board in a glued laminated timber beam has seemingly higher tensile strength than in a free tensile test.

- The bonding of the lamellae means areas with lower stiffness may transmit forces to more rigid neighbouring lamellae. This means, e.g. that the strain on board sections with knots can be relieved, which, in turn, equates to a seeming increase in tensile strength.

The way the structure of glued laminated timber includes individual lamellae results in a more homogenous material with reduced variance in associated density. Accordingly, the specified characteristic density values for glued laminated timber exceed those for individual lamellae.

Strength classes

EN 14080 specifies strength classes for homogenous glulam (see Table B6-6).

Table B6-6 Strength classes for homogenous glulam in accordance with EN 14080:2013.

N/mm^2	GL20h	GL22h	GL24h	GL26h	GL28h	GL30h	GL32h
$f_{m,g,k}$	20	22	24	26	28	30	32
$f_{t,0,g,k}$	16	17.6	19.2	20.8	22.3	24	25.6
$f_{t,90,g,k}$				0.5			
$f_{c,0,g,k}$	20	22	24	26	28	30	32
$f_{c,90,g,k}$				2.5			
$f_{v,g,k}$				3.5			
$f_{r,g,k}$				1.2			
$E_{0,g,mean}$	8400	10500	11500	12100	12600	13600	14200
$E_{0,g,05}$	7000	8800	9600	10100	10500	11300	11800
$E_{90,g,mean}$				300			
$E_{90,g,05}$				250			
$G_{g,mean}$				650			
$G_{g,05}$				540			
$G_{r,g,mean}$				65			
$G_{r,g,05}$				54			
kg/m^3							
$\rho_{g,k}$	340	370	385	405	425	430	440
$\rho_{g,mean}$	370	410	420	445	460	480	490

B6.3 Adhesives

Adhesives (glue) for load-bearing wooden members are used to join two or more timbers together in such a way that they behave as a single unit in structural terms. Here, the role of the adhesive is to fill the joint between the wooden parts and establish adhesion between the individual parts as strong and durable as the cohesion within the wood itself. In addition, the adhesive layer itself must have sufficient strength and durability, to remain effective in the service class during the expected service life of construction.

The binding forces between the adhesive and wood are, like the cohesive forces in wood, electrical forces of attraction between the molecules. The resulting binding forces are normally of the secondary type, namely hydrogen bonds and Van der Waals forces. In some adhesives, primary bonds are also likely to form (real chemical bonds), such as covalent forces. To ensure the effective contact required to establish these adhesive bonds, the adhesive must be in a liquid form at a specific stage of the setting process and the wood surfaces must have been freshly planed.

The binding process comprises two stages:

- Applying a liquid adhesive, which impregnates the surfaces of both the parts to be joined in such a way that forces of attraction between the adhesive and wood molecules exceed the surface boundaries,
- Transition of the liquid adhesives, which fill the gaps between the individual parts, into a solid state, remaining sufficiently strong and durable and with unchanged effectiveness during the construction service life.

The process above is known as curing and can be initiated in three ways:

- Via a physical process, such as the transformation of a solution or the solidification of a melt (thermoplastic adhesives like PVAc (polyvinyl acetate) and hot melts),
- Via a chemical process, which involves the glue molecules interreacting and polymer cross-linking (epoxy resin and polyurethane adhesives),
- Via a combination of solution transformation and chemical reaction (urea, melamine, phenol and resorcinol-formaldehyde resin).

Construction adhesives always involve a chemical reaction taking place. Purely physical hardening adhesives are thermoplasts and the excessive creep they allow renders them unsuitable for structural purposes.

Classification of adhesives

In EN 301, adhesives are sub-classified into:

- Adhesive type I for use in unlimited weathering and for temperatures exceeding 50°C,
- Adhesive type II for use in heated and ventilated buildings, when protected against external weathering, for short-term weathering and for maximum temperatures of 50°C. In Germany, since 2006, type II adhesives have no longer been used for glulam production.

Adhesive types

Adhesives are non-metallic and most often sub-classified into natural and synthetic groups. Synthetic adhesives are used for bonding wood, based on duroplastic resins:

- Urea formaldehyde resin UF,
- Urea melamine formaldehyde resin MUF,
- Melamine formaldehyde resin MF,
- Phenol formaldehyde resin PF,
- Resorcinol formaldehyde resin RF,
- Phenol resorcinol formaldehyde resin PRF,
- Polyurethane PUR.

Additional synthetic adhesives use thermoplastic resins (polyvinyl acetate PVAc – the most well-known white glue for DIY enthusiasts), although this is not used for bonding load-bearing members. For thick-layer constructive bonding (e.g. for glued-in threaded rods), epoxy resins (EP) are also used.

Formaldehyde-based adhesives harden when the formaldehyde reacts with the added resin. The environmental conditions required for this, including the pH level, temperature and curing duration depend on the resins used or a combination thereof. To start the curing process for UF, MF and RF adhesives, either salts have to be added or the pH content must be changed. Conversely, all alkaline-hardening PF-adhesives need is the addition of heat.

Urea formaldehyde resin UF

Given the good workability, fast curing, colourlessness, low price and usability with various wood species, UF adhesives are in very widespread use. However, they are not resistant to moisture and temperature, although adding the far more expensive melamine to MUF-adhesives improves the moisture resistance.

Melamin formaldehyde resin MF

Compared to UF adhesives, MF adhesives are far more expensive, but also far more resistant to moisture and temperature, which means they perform better in terms of the swelling and shrinking response.

Phenol formaldehyde resin PF

PF-adhesives are also moisture-resistant and mainly used as hot-curing PF-adhesives to produce particleboards and fibreboards, OSB, LVL and plywood. They possess good swelling and shrinking behaviour (low thickness swelling) and moreover, do not emit as much formaldehyde. PF-adhesives are economical and readily available.

Resorcinol formaldehyde resins RF and PRF

Adding resorcinol significantly increases the reactivity of the PF-adhesives and allows for cold curing, meaning heat is no longer needed for the adhesives to harden. They are highly resistant to moisture and temperature and this, together with the cold-curing property, makes RF- and PRF-adhesives popular conventional choices for glulam construction. The high costs of using resorcinol mean that in this case, as is done for MUF-adhesives, combinations of phenol- and resorcin adhesives (PRF) are often used. Phenol and resorcinol are dark in colour, which may be a disadvantage for visible applications.

All the adhesives described to this point contain formaldehyde. Recent years, however, have seen an increasing shift to the use of PUR-adhesives, which contain no formaldehyde. PUR-adhesives usually function with a dual-component approach, whereby the addition of isocyanate triggers the start of the curing process. PUR-adhesives often stand out for their "foamy" character. Since the curing and adhesion largely depend on the environmental conditions, temperature and moisture as well as the wood moisture content itself, these parameters are also regulated in EN 14080. The various curing times of adhesives (as specified by manufacturers) must also be taken into account, normally specified in terms of "open time" (time between mixing and the start of the reaction) and "pressing time".

B6.4 Cross-laminated timber

All types of cross-laminated timber (CLT) are classed as structural wood materials. Unlike plywood, CLT is always used for load-bearing members, which is why it merits a brief introduction in a dedicated section. The advantages of this plate-shaped CLT as opposed to solid timber are the approximated isotropy in-plane and lower variance in properties. In addition, the CLT elements are far more dimensionally stable. The effective stiffness values can be calculated using the "shear analogy method" (see Article D7).

Material

CLT comprises multiple cross-wise arranged board layers. The compound structure is usually formed via gluing, but hardwood dowels or aluminium nails may also be used as mechanical joining means. The individual layers are completely (Figure B6-10 on the left) or partially (Figure B6-10 on the right) filled with boards, with melamine resins or polyurethane (PUR) the most frequently used adhesives. All wood species used in practice to date have been conifers, although first producers try to use beech. In the vast majority of cases, CLT is made of spruce, but CLT panels are also made of fir, pine, larch or Douglas fir. These boards, most of which are between 15 and 40 mm thick, are kiln-dried before the gluing and strength-graded. Normally, the narrow sides of the boards are not systematically glued together, while the size of the board elements produced depends on the manufacturer. Finishing is performed in CNC facilities and there is also scope to manufacture bent elements.

The layer-by-layer structure means a very wide range of panel thicknesses (between around 60 to 400 mm) and compositions is possible. CLT is generally constructed symmetrically, although partially filled CLT may also show asymmetrical cross-sections (Figure B6-10 on the right). Numerous compositions are produced, the respective properties of which are adapted to specific usage applications.

Figure B6-10 CLT. Completely filled with boards (left), partially filled with gaps (right, photo: Lignotrend productions GmbH).

Types of loading acting on cross-laminated timber:

- Loading perpendicular to the plane: Bending parallel or perpendicular to the grain of the external layers,
- In-plane loading: Bending / compression / tension in plane and parallel or perpendicular to the grain of the external layers.

Since the modulus of elasticity of wood perpendicular to the grain only equates to around 3% of the values in the grain direction, the stiffness of a CLT element is determined in each case by the layers with the grain parallel to the stress direction. Based on the variation in thicknesses among the individual layers or the variance in layer composition (for example two longitudinal layers and only then another perpendicular layer), the elasto-mechanical properties can vary considerably, even among CLT elements with the same overall thickness. One additional key aspect is rolling shear, since rolling shear stiffness and strength are very low. When loads perpendicular to the plane generate a non-uniform bending moment, shear also acts in-plane, resulting in rolling shear in the layers perpendicular to the stress direction. In Figure B6-11 and Figure B6-12, the distribution of bending stress and the related distribution of shear stress are shown schematically.

To sum up, we can conclude that numerous dimensions, compositions and materials are available. The product standard for CLT is EN 16351 "Timber structures – cross-laminated timber – requirements" (currently not a harmonised version). Individual producers usually also have German national technical approvals or European technical assessments, in which the production requirements and material properties are specified. The approvals also often regulate characteristic values (particularly effective bending stiffnesses, bending and rolling shear strengths).

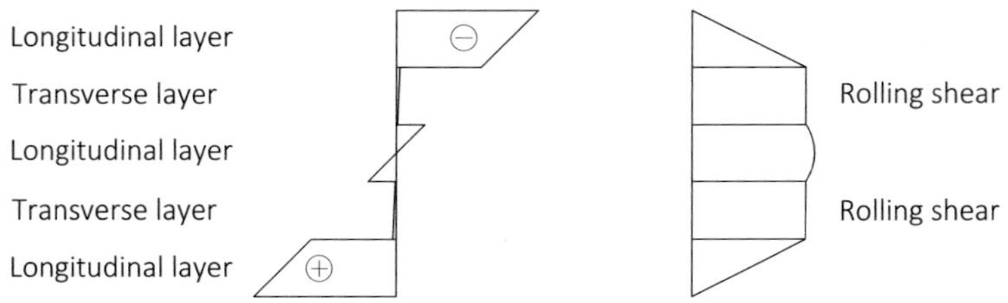

Figure B6-11 Distribution of bending stress (left) and distribution of shear stress (right) of a 5-layer CLT panel loaded perpendicular to the plane and parallel to the grain of the top layer.

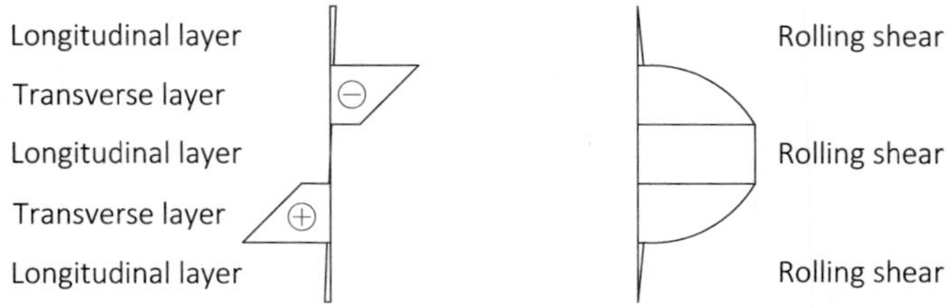

Figure B6-12 Distribution of bending stress (left) and distribution of shear stress (right) of a 5-layer CLT panel loaded perpendicular to the plane and perpendicular to the grain of the top layer.

Constructive design

The CLT construction method is quick, easy and also advantageous from a building physics perspective. The CLT elements are cut to size in the factory, thermal insulation and windows are often installed immediately, whereupon the completed elements are transported to the building site and installed. CLT are used as both wall and floor elements and typically joined using self-drilling screws and nails. Figure B6-13 to Figure B6-15 show typical buildings and joints.

Figure B6-13 Construction progress of a CLT building; after 2, 4 and 6 days respectively. (Rasom S.r.l., Pozza di Fassa)

Figure B6-14 Left: Corner connection with screws. Right: Connection floor-wall.

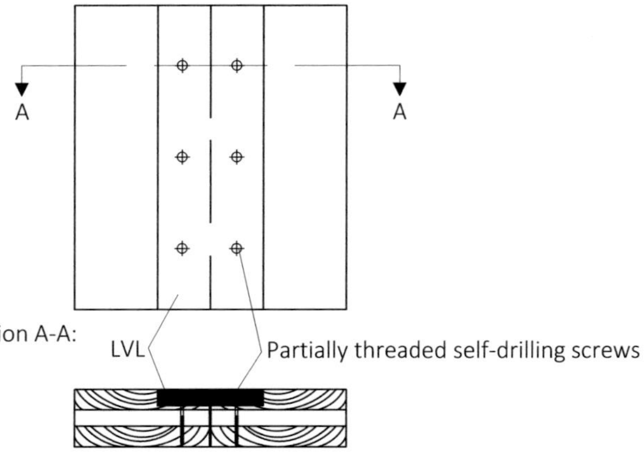

Figure B6-15 Possible in-plane joint between two wall panels.

B6.5 Wood-based panels

Wood-based panels are produced by disassembling and subsequently reassembling wood, mainly using suitable adhesives. The starting material of wood is broken up into sawn timber, veneers, particles, wood shavings or fibres. Raw materials for wood-based products include logs, wood residues (e.g. from the sawmill industry, such as wood chips and sawdust) and any recycled wood. An overview of the various types of dissasembly and reassembly used to produce wood-based panels is shown in Figure B6-16. The key types of wood-based panels are plywoods, particleboards and fibreboards and the advantages of these plate-shaped materials compared to solid timber include the approximated isotropy in plane and the low level of variance in properties. Numerous panel types are produced from various components and using a range of compositions, adapting their respective properties to the specific usage applications. Products solely produced by processing solid timber are not considered wood-based panels where the refining measures include steaming, drying, impregnating or densifying.

As a general rule, plate-shaped wood-based panels are either resin bonded or mineral (cement, gypsum) bonded and the key varieties are introduced in the following sections. No classification of the panels into a usage class is assumed here (see Article B4), also because the related standardisation has yet to be clarified. The umbrella standard for all wood-based panels is EN 13986, with additional details for individual wood-based panels contained in the individual product standards, reference to which is made in EN 13986.

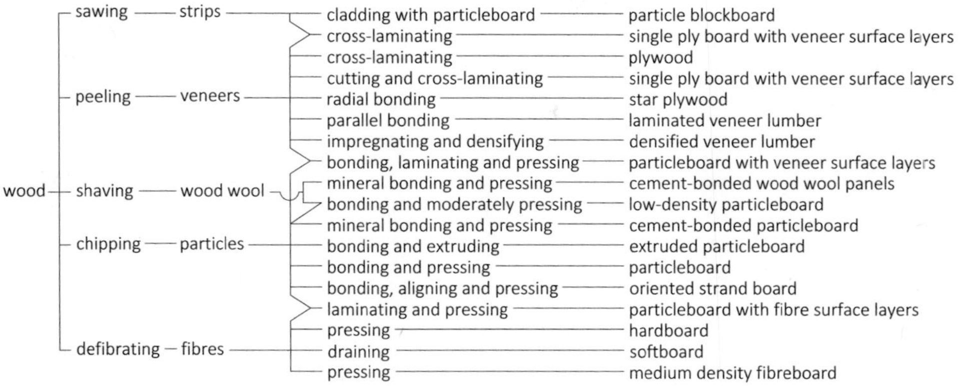

Figure B6-16 Schematic overview of the various wood-based panels and their production.

Solid wood panels (SWP), resin bonded

The panels comprise three or five board- or veneer layers of softwood, bonded together, where the grain directions between adjacent layers run at right angles. One example is shown in Figure B6-17. Modified melamine resins and phenol resins are used for gluing. The various thicknesses of the individual layers mean the elasto-mechanical properties may vary considerably, even among panels of the same overall thickness. The solid wood panels are basically CLT elements, which were already presented. However, given that they comprise thin board- or veneer layers, are not used as standalone elements and are only deemed to be load-bearing when used as sheathing for shear walls (like other wood-based panels in timber frame constructions), a brief explanation will be given here. Solid wood panels are mainly employed in furniture-making, divided into non load-bearing (acronym NS non-structural – SWP/1 NS) and load-bearing (acronym S structural – SWP/3 S) categories and regulated in EN 13986 and EN 13353. The technical class SWP/1 is usable in service class 1, SWP/2 in service classes 1 and 2 and SWP/3 in service classes 1, 2 and 3.

Figure B6-17 Triple-layered solid wood panel.

Veneer-based panels, resin bonded

Plywood

Plywood comprises at least three layers (plies) bonded to each other, which generally run at right angles, to ensure a barrier effect in-plane (dimensional stability) (for example see Figure B6-18). The central plies may consist of veneers (plywood), wooden strips (block board) or thin strips of wood (laminboard). In all three types, the top ply comprises veneers, which are thin sheets of wood, normally produced by peeling in a process that yields strips 0.5 to 6.0 mm thick with a width determined by the length of the peeled log portion. They are used to produce large-scale panels. Veneers used in timber construction are produced by rotary peeling (Figure B6-19). For this purpose, the wood used must be soft and as resistant to cracking as possible, which can be ensured by steaming the log portions or storing them in hot water. The log portions, still hot from the steaming process, are then conveyed to a centring station, which scans the log shape mechanically or visually to determine the optimal position for the spindles of the peeling machine; targeting maximum yield. A knife carrier and pressure strip advance by the thickness of a single veneer with each log rotation, while ensuring the adjustment angle of the knife conforms to the log contour and with the remaining rolls used as raw materials for the particleboard industry. After peeling, the veneers end up as a relatively regular platform, with many defective spots at the edges and holes and other defects on the transport line. This is where the veneers are scanned, a cutting device is used to eliminate defective portions and plies of equivalent width are cut. The high moisture from the steaming process is also reduced by 6 to 12%, whereupon adhesive is applied to the veneers, they are consolidated to form panels and hot-pressed together. Finally, the panels are climatised or conditioned, to ensure they cool down and moisture is uniformly distributed throughout the panel thickness. The final work process involves trimming the panels or cutting and sanding them to fixed dimensions. For certain applications, e.g. formwork panels, a coating process may follow. The production is shown schematically in Figure B6-20.

Figure B6-18 Plywood with longitudinal and perpendicular veneers of various thicknesses.

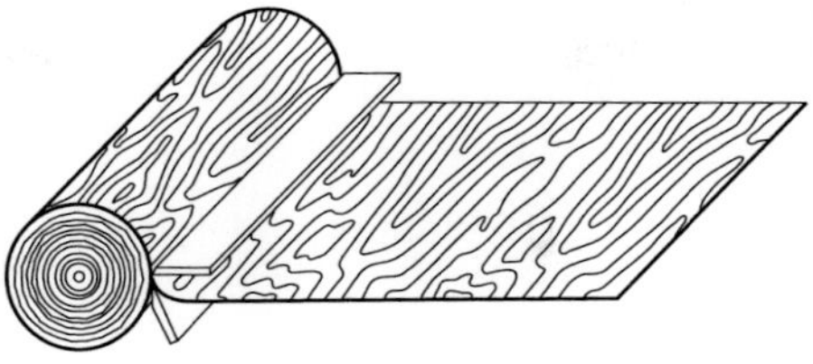

Figure B6-19 Rotary peeling (schematic illustration). (STEP 1995 Article A10)

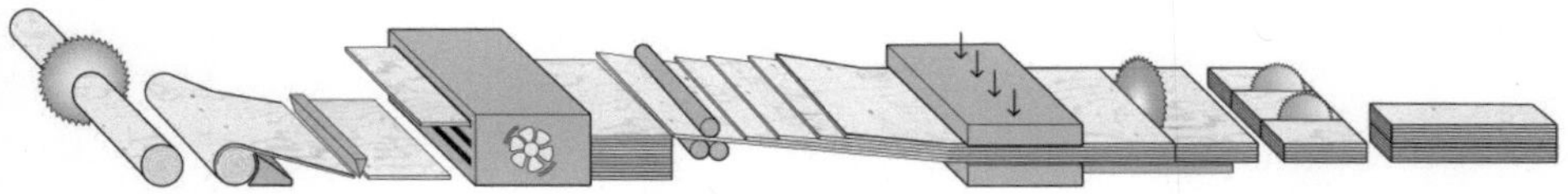

Figure B6-20 Production method for veneered materials. (Metsäwood, brochure „Ich bin Kerto", 2014 version)

Table B6-7 Technical plywood classes and usability.

Technical class	Use in service class
For internal use in dry conditions	1
For internal use in humid conditions	1, 2
For external use	1, 2, 3

Plywood is created by arranging and bonding veneers in alternative crosswise layers, whereby the veneers must be arranged symmetrically to the central plane. The adhesives used in this case are urea resins (only of the "dry" class), alkaline-hardening phenol resins, phenol resorcinol resins and resorcinol resins. Changing the number, thickness and arrangement of the individual plies opens up a whole host of options concerning panel composition, which means panels with targeted properties can be created. The type of veneer wood used is also a key criterion dictating the plywood properties. Plywoods can be constructed from various wood species, provided symmetry is maintained. Plywood is mainly used as a sheathing material in horizontal (floors, roofs) or vertical (shear walls) diaphragms. Since the modulus of elasticity of wood perpendicular to the grain only equates to around 3% of the value in the grain direction, the stiffness of a plywood panel, similarly to a cross-laminated timber panel, is respectively determined by the layers with grains running parallel to the stress direction. Accordingly, for plywood, a distinction is made between both the main axes running parallel and at right angles to the grain of the top plies.

Plywood is regulated in EN 13986 and in the EN 636 product standard. Besides in technical classes (Table B6-7), plywood is also classified in terms of bending strength and bending modulus. The technical classes govern the scope of use depending on the prevailing climate. The additional classification of plywood panels depends on the minimum required values per length or width of panels for bending strength, letter "F", or the modulus of elasticity with the letter "E", see Table B6-8. This means that a plywood panel classed as F 10/20 E 30/40 must have a minimum bending strength value along the length of $f_{m,0}$ = 15 N/mm² and widthwise, $f_{m,90}$ = 30 N/mm² and a minimum value for the bending modulus along the length of $E_{m,0}$ = 2700 N/mm² and widthwise, $E_{m,90}$ = 3600 N/mm².

Table B6-8 Bending strength and bending modulus classes for plywood, EN 636:2012.

Bending strength in N/mm²	Class	Minimum value
	F 3	5
	F 5	8
	F 10	15
	F 15	23
	F 20	30
	F 25	38
$f_{m,0}$	F 30	45
$f_{m,90}$	F 35	52
	F 40	60
	F 50	75
	F 60	90
	F 70	105
	F 80	120

Bending modulus in N/mm²	Class	Minimum value
	E 5	450
	E 10	900
	E 15	1350
	E 20	1800
	E 25	2250
	E 30	2700
	E 35	3150
$E_{m,0}$	E 40	3600
$E_{m,90}$	E 50	4500
	E 60	5400
	E 70	6300
	E 80	7200
	E 90	8100
	E 100	9000
	E 120	10800
	E 140	12600

Laminated Veneer Lumber (LVL)

Laminated veneer lumber is produced using rotary cut softwood or hardwood veneers, with phenol resin used for gluing. The fibre direction of the veneers is either generally parallel to the longitudinal direction of the laminated veneer lumber or predominantly parallel and slightly (up to around 25%) perpendicular to the longitudinal direction of the laminated veneer lumber (Example see Figure B6-21). The veneers within a layer are generally combined using a scarf or overlapping process. Laminated veneer lumber is regulated in product standards EN 14279 and EN 14374. Laminated veneer lumber is used as either load-bearing sheathing or, like glulam, as a bar-shaped member and classified in accordance with the three technical classes of EN 14279:

- LVL/1 for use in dry conditions (service class 1)
- LVL/2 for use in humid conditions (service class 2)
- LVL/3 for external use (service class 3)

Peeling cracks in veneers allow entire cross-sections of laminated veneer lumber to be impregnated, after which it can even be used in locations exposed to adverse weather conditions. The strength values required for design are determined in accordance with the method specified in EN 14374 and specified in the CE label for the individual products. Although LVL centres on the concept of increasing the homogeneity of this wood-based product over solid timber, it generally encompasses a range of densities throughout the board thickness, see Figure B6-22. This is often influenced by production factors (pressing power and heat), but may also be a desirable feature, particularly in wood-based panels like OSB and particleboards, to impact on the end panel properties.

Figure B6-21 Laminated veneer lumber made of beech veneers.

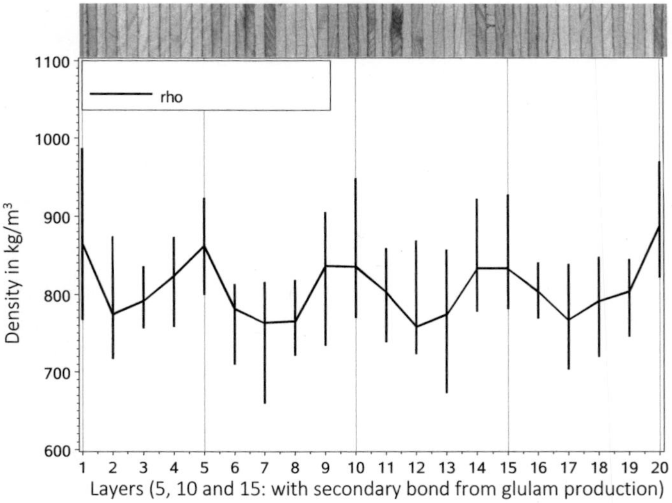

Figure B6-22 Density profile of glulam made of beech LVL. The chart shows the minimum, maximum and average values of 10 measurements on glulam, made of 4 laminated veneer lumber layers, each encompassing 5 veneer layers.

Parallel Strand Lumber (PSL)

PSL (Parallel Strand Lumber, see Figure B6-23) is a special invention, which was originally planned to facilitate the use of waste veneers from plywood production. For PSL, rotary cut veneers, normally from Douglas fir or Southern Yellow Pine and around 3 mm thick are used. The veneer sheets are cut into strips up to 2.50 m long and around 23 mm wide, whereupon the "strands" are waxed, positioned longitudinally and made to traverse a through-feed press in a process that can generate almost unlimited beam lengths. A subsequent quality control process sees the beams checked visually and in terms of weight before delivery. Parallel strand lumber is not regulated by product standards and requires general technical approval in Germany. PSL (trade name Parallam) stands out, thanks to its high strength values, increased stiffness and dimensional tolerance. The main PSL applications include beams, structural members, purlins, columns and trusses.

Figure B6-23 Parallel Strand Lumber, PSL.

Wood-based panels made of chips, resin bonded

Laminated Strand Lumber (LSL)

LSL is so-called laminated strand lumber and distributed under a range of trade names (including Intrallam), see Figure B6-24. LSL is generally technically approved in Germany. LSL comprises trim strands of poplar or aspen wood, bonded to each other and with dimensions of around 0.8 mm x 25 mm x 300 mm. The raw wood is debarked before the machining process and any excessively short chips are discarded. The relatively high densification and the type of adhesive allows high strengths to be attained, while there is also scope to manufacture a range of strength classes by changing the distance between the mat-forming machine to the mat surface. The greater the distance, the higher the proportion of trim strips in a perpendicular board direction, while for smaller distances, the trim strips tend to be arranged in parallel.

Figure B6-24 Laminated Strand Lumber, LSL.

Oriented Strand Board (OSB)

OSB is made using longitudinal strands, preferably parallel to the board surface, which are around 0.6 mm thick, 75 mm long and 35 mm wide, see Figure B6-25. The strands can be applied single or multiple-layered, with multilayer panels offering additional scope to vary the strand size in the individual layers, the degree of orientation, layer thicknesses, adhesives used and their proportion. Strands for top layers should preferably run parallel to the production direction, while those in the middle layer should be random or perpendicular to the production direction, which means OSB achieves different longitudinal and perpendicular properties. The bending strength in the longitudinal direction of the board significantly exceeds that in the perpendicular direction. Moreover, the density of OSB generally varies through the board thickness (similar to LVL, Figure B6-22). OSB is regulated in EN 13986 and in the EN 300 product standard, which includes four technical classes:

- OSB/1 non load-bearing boards for general purposes and interior fitments (including furniture) for use in dry conditions (service class 1)
- OSB/2 load-bearing boards for use in dry conditions (service class 1)
- OSB/3 load-bearing boards for use in humid conditions (service class 2)
- OSB/4 heavy-duty load-bearing boards for use in humid conditions (service class 2)

Figure B6-25 Oriented Strand Board, OSB.

Particleboard (chipboard)

Particleboards are formed by spreading out relatively small wood chips sprayed with adhesive on a level underlay and then pressing, as shown in Figure B6-26. The pressed boards are weighed, any defective portions are detected and discarded, using an ultrasound device and the thickness is finally measured. The boards then undergo the final production process, where they are conditioned, trimmed, sorted, labelled, sanded and stored. Particleboards are regulated in EN 13986 and in the EN 312 product standard. The technical classes and their usability are specified in Table B6-9, although technical classes for non load-bearing use (P1, P2, P3) are excluded.

Figure B6-26 Particleboard.

Table B6-9 Particleboard classes and usability for load-bearing purposes.

Technical class	Service class	Use
P4	1	Dry conditions
P6	1	Heavy-duty, dry conditions
P5	1, 2	Humid conditions
P7	1, 2	Heavy-duty, humid conditions

During the spreading process, the particles should preferably be oriented parallel to the panel plane, which gives particleboards their advantageous in-plane tensile and compressive strengths. They are generally formed of multiple layers or with a constant structural transition, although the loose union of particles in the middle layer weakens the tensile strength under out-of-plane loading. Bending properties, meanwhile, are improved thanks to a special board structure with an exceptional density profile, with density peaking in the external layers and an associated panelling effect. The boards are mainly used as structural floor or roof decking and structural wall sheathing.

Fibreboards, resin bonded

Classification of fibreboards

Fibreboards are quasi-isotropic wood-based panels, meaning their in-plane properties apply independently of any direction. They comprise individual or bundled fibres which, depending on the production method, contain a greater or lesser proportion of chemicals and adhesives. Fibreboards produced using wet production methods have almost no or minimal quantities of adhesive added, with mainly natural fibre bonding used. Conversely, the boards having undergone the dry production method contain larger quantities of adhesive, to ensure they attain the properties determined in the product standards. Fibreboards are regulated in EN 13986 and the EN 316 and EN 622 product standards.

Since the fibreboard designation is relatively unclear, all the relevant tables from EN 316 are shown at this point, with the first sub-grouping made according to the production method:

- Wet process: Fibreboard with moisture content exceeding 20% during the forming stage,
- Dry process: Fibreboard with moisture content of less than 20% in the forming stage.

A further distinction is applied to boards produced using the wet process based on their density:

- Hardboards HB: Density > 900 kg/m^3, Figure B6-27 on the right
- Low-density medium board MBL: 400 kg/m^3 to < 560 kg/m^3
- High-density medium board MBH: 560 kg/m^3 to < 900 kg/m^3
- Softboard SB: Density > 230 kg/m^3 to < 400 kg/m^3, Figure B6-27 on the left

Figure B6-27 Left: softboard, right: hardboard.

The dry-process fibreboards are abbreviated to the label "MDF" (medium-density fibreboards), while EN 622-2 explains the special technical classes. The additional classification is implemented depending on the application conditions and purpose of use, Table B6-10.

Table B6-10 Classification of fibreboards.

Application conditions, purpose of use, structural use	Abbreviation
Application conditions	
For internal use, dry conditions	No abbreviation
For internal use, humid conditions	H
For external use	E
Purpose of use	
General purpose	No abbreviation
Load-bearing purpose	L
For all load-duration classes	A
Only for instantaneous or short-term loading	S
Structural use	
General structural use	1
Heavy-duty	2

Examples:

MDF.HLS = dry-process medium density fibreboard for internal use as a structural component in humid conditions, only for for instantaneous or short-term loading
HB.HLA2 = hardboard for internal use as a heavy-duty structural component in humid conditions, for all load-duration classes

Wood-based panels, mineral bonded (cement, gypsum)

All mineral bonded boards, excluding wood wool boards, can be used as sheathing material in timber frame constructions. As a general rule, they are used for dry lining and fire protection in particular.

Cement-bonded particleboard

The cement-bonded particleboard, Figure B6-28, comprises treated spruce and fir wood particles, which function as reinforcement and cement acting as a binder. The boards are regulated in EN 13986 and the EN 634 product standard. There are two technical classes for cement-bonded particleboards, which only differ in terms of the requirement for the minimum bending modulus:

- Technical class 1: 4500 N/mm²
- Technical class 2: 4000 N/mm²

Figure B6-28 Cement-bonded particleboard.

Gypsum-bonded particleboard

Gypsum-bonded particleboards comprise gypsum and spruce or aspen particles, which function as reinforcement.

Gypsum fibreboard

Gypsum fibreboards (Figure B6-29) comprise gypsum and paper fibres, which are extracted in a recycling method and which function as reinforcement

Figure B6-29 Gypsum fibreboard.

Gypsum plasterboard

Gypsum plasterboards (Figure B6-30) comprise a gypsum core, which is surrounded by paper surfacing on the face, back and long edges commensurate with the purpose of use. The encased gypsum core may be porous and include additives intended to imbue it with specific properties. The key properties of the panels are a factor of the composite effect of the gypsum core and the paper encasing, whereby the paper encasing acts to reinforce the tensile area and provides the gypsum boards with the required strength and bending strength, when combined with the gypsum core.

Figure B6-30 Gypsum plasterboard.

Wood wool board

Wood wool boards are lightweight boards made out of wood wool and mineral binders (cement or magnesite). They are used in construction to insulate heat and sound and protect against fire.

B6.6 Literature

P. Glos, F. Colling, A. Ranta-Maunus, G. Steck, D.R. Griffiths, E. Raknes, original Articles A7 – A12, STEP 1995.

Studiengemeinschaft Holzleimbau e.V. (1998). Argumente für BS-Holz. Informationsblatt zum Brettschichtholz des Informationsdienstes Holz, Düsseldorf.

Colling F. (1990). Tragfähigkeit von Biegeträgern aus Brettschichtholz in Abhängigkeit von den festigkeitsrelevanten Einflußgrößen. Reports of the Versuchsanstalt für Stahl, Holz und Steine of Karlsruhe University, Volume 4, Issue 22, 205 pp.

Frese M. (2006). Die Biegefestigkeit von Brettschichtholz aus Buche. Experimentelle und numerische Untersuchungen zum Laminierungseffekt. Dissertation, Universität Karlsruhe (TH).

C

Principles for design

C1 Safety concept

Original article: H. J. Larsen

The Eurocode principles for safety and serviceability of structures and the basis of design common to all materials are explained. Moreover, specific rules for timber structures, required due to the impact of load duration, wood moisture content and the scatter of material properties, are specified. Requirements for normal design situations and special requirements for accidental design situations are also described.

C1.1 Calculation models

Prior to performing the actual calculations for a structure, the structure has to be analysed and a suitable caluclation model defined. When abstracting a structure in a structural model suitable for a calculation, there is usually a whole host of possible solutions. Structural engineers can decide between a simple solution, although this may well be too much on the safe side and use simplified calculations, or opt for a more complicated solution, which describes the actual construction response more effectively, but also includes a higher likelihood of errors and the risk of overlooking possible failure mechanisms. When selecting the structural model, the expected quality of workmanship must also be taken into consideration. Specifically, structures, within which the stress varies significantly with even small geometrical deviations, must be manufactured precisely in line with the assumptions made during the calculation and in them, the impact of initial deformations and deformations due to loading must be taken into consideration.

In joints, larger contact surfaces are often necessary, which may result in local eccentricities. These, in turn, may have a key impact on force transfer, at least to a localised extent. It is often possible to select the structural model freely, subject to compliance with the assumptions already made. Figure C1-1 shows an example: a possible assumption for the heel joint would be that the load is transferred in the shear plane. In this case, the rafter, tie beam and joints are loaded eccentrically at point *A* e.g. with moment $F \cdot e_A$. Another possibility would involve assuming the rafter is centrally loaded, whereby the tie beam and respective joint would be assessed for the moment $F \cdot (e_A + e_B)$.

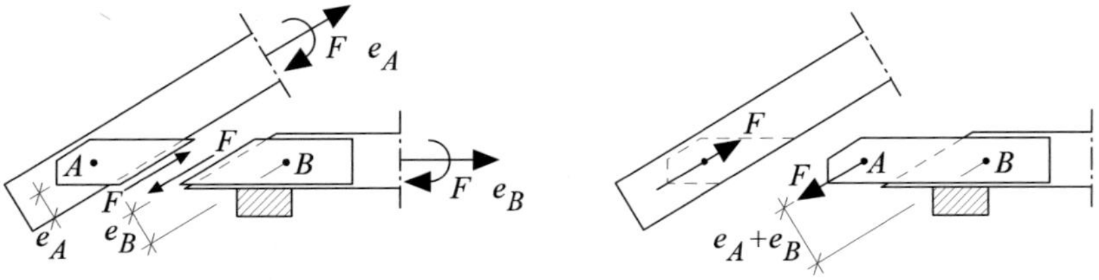

Figure C1-1 Heel joint. (STEP 1995 Article A2)

C1.2 Standards based on limit states

Eurocodes are standards based on limit states, which means that the requirements concerning structural reliability are subject to clearly defined limit states. Limit states are those which, when exceeded, render the structure no longer capable of meeting the adopted design requirements. Within the Eurocodes system, only two limit states are taken into consideration: the ultimate and the serviceability limit state.

The ultimate limit states are any states, which, due to any collapse or other failure affecting the structure, may put personal safety at risk. This may involve e.g. the loss of equilibrium of a structure or one of its parts, while failure can also be initiated by excessive deformations, rupture or loss of stability of a structural part.

The serviceablity limit states comprise deformations which would impact on the appearance or planned usage of a structure, vibrations, which cause unease or damage to construction or related equipment or which limit the functional scope of the construction and damage (including cracks), which has a long-term impact on the durability of the structure.

C1.3 Design in accordance with the partial factor method

Structural design in Eurocodes is based on the partial factor method. The key parameters are the actions F, the material properties X and the geometrical data a[1].

[1] The effects of actions E are a function of the actions F and the geometrical data a: $E = (F, a)$
The resistance R is a function of the material properties X and the geometrical data a: $R = (X, a)$

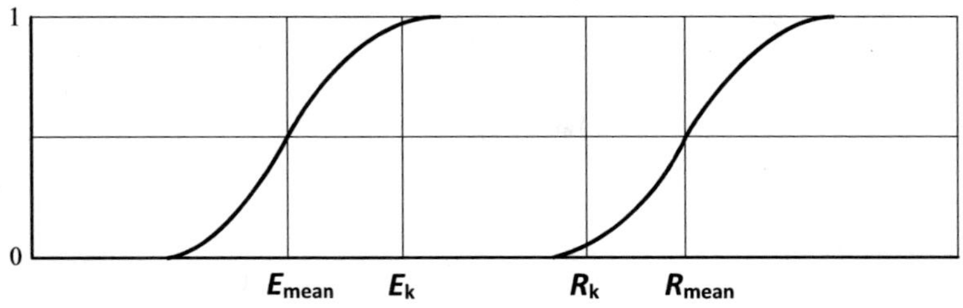

Figure C1-2 Idealised statistical distributions for the effect of actions *E* and resistance *R*. (STEP 1995 Article A2)

As a general rule, these variables are random with distribution functions, as shown in Figure C1-2 for the effect of actions *E* and the corresponding resistance *R*: e.g. bending stresses and bending strength or the normal force in a centrally loaded column and its buckling load. The corresponding distributions reveal the average values E_{mean} and R_{mean}, whereby the characteristic values E_k and R_k are determined as distribution quantiles. As regards the characteristic values of the effects of actions E_k, in this case, although an upper quantile value is usually applied (e.g. 98% quantile), applying a lower value is also often acceptable, e.g. when designing against uplift. A lower quantile value (e.g. 5% quantile) or average value is generally used for the characteristic resistance R_k. In exceptional cases, e.g. if a connection should fail earlier than the connected member, an upper quantile value may also be necessary. The goal of the design is to ensure a low probability of failure, namely minimising the possibility of the effect of action *E* exceeding the resistance *R*. The probability of failure is shown visually in Figure C1-3 and corresponds to the overlap area of the frequency curves. Using the method with partial safety factors allows a low probability of failure to be attained, whereby design values are used (Index "d" = "design"), which are determined for actions by multiplying the characteristic values (Index "k" = "characteristic") with partial safety factors and for the resistance by dividing the characteristic values by partial safety factors, see Figure C1-4.

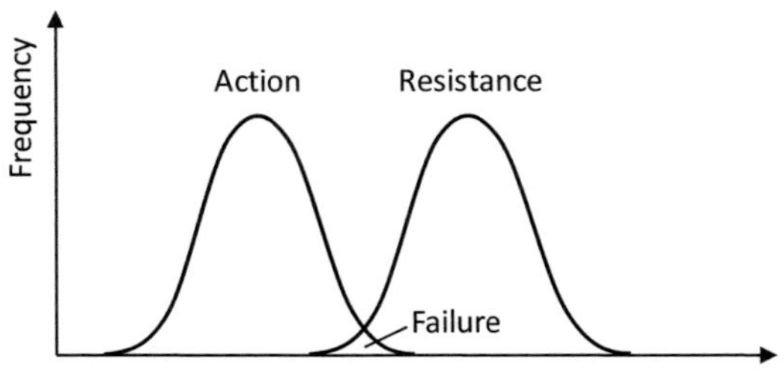

Figure C1-3 Frequency distributions for action and resistance.

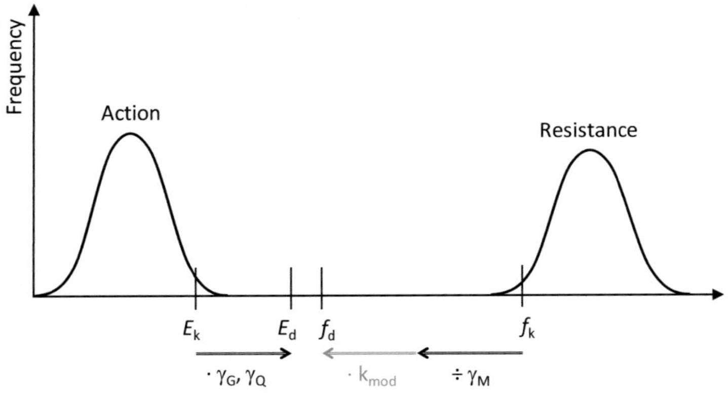

Figure C1-4 Reduction of the probability of failure through partial safety factors and k_{mod}.

In all key design situations, it is important to ensure that the limit states are not attained, when the design values of the actions, material properties and geometrical variables are used in the calculation model. In particular, it is important to ensure that:

- In the ultimate limit states, the design values of the action effects do not exceed the design values of the resistances and
- In the serviceability limit states, the design values of action effects do not exceed key limit values.

In a symbolic form (for the ultimate limit state corresponding to failure), it shall be verified that

$$E_d \leq R_d \qquad \text{(C1-1)}$$

For the serviceability limit state, it shall be verified that

$$E_d \leq C_d \qquad \text{(C1-2)}$$

Here

E_d Design value of the effect of action such as e.g. normal force, bending moment

R_d Design resistance

C_d Limit design value of the relevant serviceability criterion such as e.g. a limiting value for deflection

C1.4 Actions

Representative and characteristic values

This section only covers actions to the extent required to describe safety principles. Article C2 includes a detailed description of actions. Under normal design situations, a distinction is made between permanent actions *G* and variable actions *Q*. There are direct actions (e.g. external forces) and indirect actions (e.g. actions caused by shrinkage or uneven settlements), while actions, or parts of actions, can either be stationary or mobile. Mobile actions may be quantified as arbitrary values between predefined limits at any point of the structure. The variable actions are additionally sub-classified into 'leading variable action' and 'accompanying variable action'.

Material or product properties

The material properties correspond to either an average value or the 5% quantile and are determined by standardised tests under uniform test conditions: The test is conducted within five minutes at a temperature of 20°C and relative humidity of 65%. The average values are used for serviceability limit state verifications, while the ultimate limit state is verified by using the 5% quantiles of the material properties (the 5% quantile of stiffness is only taken into consideration when verifying the stability).

Geometrical data

The characteristic values of geometrical data, such as span, cross-sectional dimensions or initial deformation, generally correspond to the nominal values determined in the design.

Design values

The design values of the actions may cover a range of values, when verifying the various limit states. The first step is to determine the load cases, namely compatible load configurations or deformations and imperfections. A load configuration describes the position, size and direction of an action, whereupon the actions corresponding to the following symbolic expression are listed – in this case the fundamental combination of actions for persistent or transient design situations:

$$\sum \gamma_{G,j} \cdot G_{k,j} + \gamma_{Q,1} \cdot Q_{k,1} + \sum \gamma_{Q,i} \cdot \psi_{0,i} \cdot Q_{k,i} \qquad \text{(C1-3)}$$

The γ-values are partial safety factors for the corresponding action, which take into consideration the influence of possible unfavourable deviations and possible inaccuracies of actions and uncertainties when determining the final effects of actions. Certain partial safety factors for actions for verifications, which govern the strength of construction materials, are specified in Table C1-1 (other partial safety factors apply, for example, if the strength of the soil is the decisive element determining failure).

Table C1-1 Partial safety factor for ultimate limit states in buildings, EN 1990:2010.

Failure of the structure where the strength of construction materials governs		
Unfavourable permanent actions	$\gamma_{G,sup}$	1.35
Favourable permanent actions	$\gamma_{G,inf}$	1.00
Unfavourable variable actions	γ_{Q}	1.50

The first term in Equation (C1-3) expresses the design value of the permanent action $G_d = \gamma_G \cdot G_k$, while the second term is the design value of the leading variable action $Q_{d,1} = \gamma_{Q,1} \cdot Q_{k,1}$. The third term $Q_{d,i} = \gamma_{Q,i} \cdot \psi_{0,i} \cdot Q_{k,i}$ is the design combination value of the accompanying variable actions. The combination factor ψ_0 takes into consideration the fact that e.g. a simultaneous combination of maximum snow and wind loads is improbable, whereupon the lower value may be reduced by ψ_0. Table C1-2 specifies examples for ψ-values, while a complete overview is included in EN 1990 and in the NA to EN 1990.

Table C1-2 Some combination factors for buildings, NA to EN 1990:2010.

Action	ψ_0	ψ_1	ψ_2
Imposed loads	0.7 - 1.0	0.5 - 0.9	0.3 - 0.8
Snow loads at altitude < 1000 m a.s.l.	0.5	0.2	0.0
Wind loads	0.6	0.2	0.0

The combination of actions for persisting design situations specified in equation (C1-3) does not apply to accidental or seismic actions. The combination of actions for seismic design situations is as follows:

$$\sum G_{k,j} + A_{Ed} + \sum \psi_{2,i} \cdot Q_{k,i} \qquad \text{(C1-4)}$$

where A_{Ed} is the design value of the seismic action.

For such combinations of actions, no partial safety factors are applied and also lower combination factors ($\psi_0 \rightarrow \psi_2$). Similarly, no partial safety factors are applied to verify the serviceability limit state:

$$\sum G_{k,j} + Q_{k,1} + \sum \psi_{0,i} \cdot Q_{k,i} \tag{C1-5}$$

Finally, the effects of actions E are determined, e.g. internal forces, stresses, strains and deformations, based on the design values of the actions F (G_d, Q_d, etc.), the geometrical data a and, where applicable, the material or product properties X:

$$E_d = E\left(F_{1,d}, F_{2,d}, \ldots, a_{1,d}, a_{2,d}, \ldots, X_{1,d}, X_{2,d}, \ldots\right) \tag{C1-6}$$

C1.5 Resistance

The design value X_d of a material property with the characteristic value X_k is defined as follows in timber construction:

$$X_d = k_{mod} \cdot \frac{X_k}{\gamma_M} \tag{C1-7}$$

Where

γ_M Partial safety factor for the material property, see Table C1-3

k_{mod} Modification factor, which takes into consideration the impact of the load duration and moisture content on the strength properties. Examples for k_{mod} are specified in Table C1-4.

Table C1-3 Partial safety factors for the material properties, NA to EN 1995-1-1:2013.

Ultimate limit states:	
- Wood and wood-based products, joints	1.3
- Punched metal plate fasteners	1.25
Serviceability limit states	1.0

The factor k_{mod} depends on the service class of the structure and the load-duration class. There are three service classes 1, 2 and 3. The average equilibrium moisture content in most softwoods does not exceed 12% in service class 1 (internally) and 20% in service class 2. For service class 3 (externally), however, no limit value is determined for wood moisture content. The five various load-duration classes are distinguished based on the accumulated duration of the characteristic load (see Table C1-4). The coefficient k_{mod} thus takes into consideration the effect of various moisture contents and load durations on the mechanical properties of wood and wood-based products (see also Article B2).

In general, a structural model is used, which describes the relationship between resistance R and the strength values f, the stiffness values E (= material properties X) and the geometrical data a. This kind of model uses the corresponding design values to determine design resistance:

$$R_{\mathrm{d}} = R\left(f_{1,\mathrm{d}}, f_{2,\mathrm{d}}, \ldots, E_{1,\mathrm{d}}, E_{2,\mathrm{d}}, \ldots, a_{1,\mathrm{d}}, a_{2,\mathrm{d}}, \ldots\right) \tag{C1-8}$$

Table C1-4 Load-duration classes and factors k_{mod} for solid timber and glulam.

Load-duration class	Duration	Examples for loads	k_{mod} for service class	
			1 and 2	3
Permanent	More than 10 years	Self-weight	0.60	0.50
Long-term	From 6 months to 10 years	Storage	0.70	0.55
Medium-term	From 1 week to 6 months	Imposed floor load	0.80	0.65
Short-term	Less than one week	Snow, wind	0.90	0.70
Instantaneous		Accidental load	1.10	0.90

The design resistance R_d can also be directly determined from the specific characteristic value R_k based on tests:

$$R_{\mathrm{d}} = k_{\mathrm{mod}} \cdot \frac{R_{\mathrm{k}}}{\gamma_{\mathrm{M}}} \tag{C1-9}$$

Within a structure made of different structural materials, e.g. timber, steel and wood-based products, it may be difficult to select the correct factor k_{mod}. However, the selection of the lowest values for the construction materials used is always on the conservative side.

The design values of geometrical data generally correspond to the characteristic values, namely the nominal values determined in the design. Many cases involve the geometrical data being determined as follows (e.g. if deviations from the nominal value due to initial deformations must be taken into consideration → calculations in accordance with second order theory):

$$a_d = a_k + \Delta a \qquad \text{(C1-10)}$$

The values of Δa are specified in the corresponding sections of EC 5, e.g. in EC 5 Section 5.4.4 for plane frames and arches.

C1.6 Literature

H.J. Larsen, original Article A2, STEP 1995.

C2 Actions on structures

Original article: P. Racher

When structural engineers plan a building, they start by conceptually designing the structural system and the design depends on the type of construction and the construction material used. Based on this, the design gets underway by determining the actions on the selected construction. Both direct actions of external forces must be taken into consideration, as well as the indirect actions that result from imposed deformations (e.g. dimensional changes due to varying moisture or temperature levels or settlement of supports).

Regardless of which construction materials are used, the design must determine the range of actions which could occur during the planned service life of the structure. These actions depend on the form, construction type and how the construction is actually erected. The type of actions or the consequences of an action, which can be either static or dynamic, must be taken into account to achieve an accurate structural analysis. This means, e.g. that using a quasi-static assumption may be inapplicable in the following cases:

- Floors, which are subject to vibrations caused by people or machines.
- Flexible, plate-like constructions such as suspension bridge decks, which may be subject to flutter at critical wind speed.
- Constructions subject to stresses caused by the accelerations of seismic action.

In these cases, a dynamic analysis should be performed, to determine the actions of the force-time history and taking the stiffness, mass and damping behaviour of the structural members into consideration, whereby, however, the resonant component of these actions is low in most structures. Accordingly, a structural static calculation is frequently performed while applying an equivalent, dynamic magnification factor (vibration coefficient) for the static value of the action.

For these reasons, the current article only covers the direct actions and their combinations for static calculations. These calculations must also take into consideration the National Application Documents and the current provisions for countries within which the construction should be erected.

C2.1 General concepts

Structural classifications

The Eurocodes for design (EC 2 to EC 9) mainly centre on adapting successful traditional design methods, but even so, it is important to measure the criteria which underpin the reliability concept of EC 0. The required safety and serviceability depends on the service life and the design situation of the structures; taking into consideration any danger to people and economic losses (C.E.B., 1980).

The planned service life (= design working life) specified in Table C2-1 encompasses the specific period for which a structure may be used for its intended purpose, without requiring any significant maintenance (except planned maintenance measures). The design situations are based on incidents which could occur during the service life of the construction, whereupon the actions for key design situations, classified as follows, are determined:

- Persistent situations, which constitute normal conditions of use.
- Transient situations, which involve temporary conditions, such as during execution.
- Accidental situations related to exceptional conditions such as fire or impact.
- Seismic actions.

Table C2-1 Classification of design working life (planned service life).

Design working life category	Indicative design working life (years)	Examples
1	10	Temporary structures
2	10 to 25	Replaceable structural parts, e.g. gantry girders,bearings
3	15 to 30	Agricultural and similar structures
4	50	Building structures and other common structures
5	100	Monumental building structures, bridges and other civil engineering structures

Classification of actions

As well as the above-listed classifications, the actions must also be distinguished in terms of the way they vary over a spatial scope and over time. A normal design defines the actions or effects of actions as follows:

- Permanent actions (G), e.g. self-weight of the structure,
- Variable actions (Q), e.g. imposed loads, wind and snow loads.

Accidental (A) and seismic actions (A_E) are not covered in this article (for fire, see Article G1, earthquake see Articles G2 and G3).

The size of permanent actions remains virtually unchanged over time, except in the event of construction changes (see Figure C2-1). When variable actions apply, the changes can be modelled as a discrete process (e.g. snow or wind) or as a process from a permanently applied proportion Q_L and a proportion Q_T applied for only a short time (e.g. imposed loads) (Hendrickson et. al, 1987, Rackwitz, 1976).

For wood, where elapsed time plays a bigger role in determining strength than for other construction materials, special attention must be paid to the chronological change in actions as well as the duration of loads. Accordingly, the structural engineer must classify the variable actions in the specified load-duration classes (see also Table C1-4).

As regards their spatial variability, actions are considered as either stationary or free. Free actions can be dispersed arbitrarily over the structure or parts thereof in spatial terms, in which case a design taking into account unfavourable load configurations of free actions must be performed.

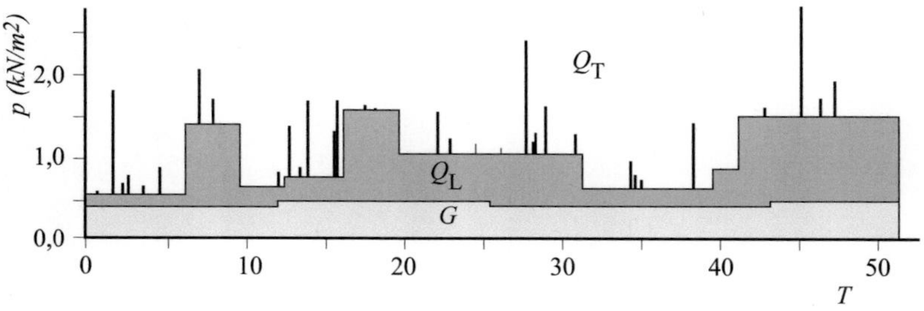

Figure C2-1 Chronological change in actions on a floor. (STEP 1995 Article A3)

Representative values of actions

The basic values of actions are the characteristic values, which are designated as G_k or Q_k and as a general rule, the permanent actions G_k correspond to the nominal values. If the construction responds sensitively to changes in G or if the coefficient of variation (COV, COV is defined as the standard deviation divided by the mean value) of G exceeds 10%, two characteristic values, one lower value $G_{k,inf}$ and one upper value $G_{k,sup}$ should be used, which correspond to the 5% quantile or the 95% quantile. Assuming a Gaussian distribution of G, these values are determined as:

$$G_{k,inf} = G_{mean} \cdot (1 - 1.645 \cdot COV) \qquad G_{k,sup} = G_{mean} \cdot (1 + 1.645 \cdot COV) \tag{C2-1}$$

The characteristic variable actions Q_k correspond to a given return period of N years, corresponding to the probability of exceedance of $p_N = 1/N$ per year. According to EC 0, the actions are defined as Q_k for N = 50 years or p_{50} = 0.02, which corresponds to a 98%-exceedance quantile.

In addition, the structural engineer has to take other representative values into consideration for the variable actions:

- The combination value ($\psi_0 \cdot Q_k$) (= characteristic combination), which is exceeded two percent of the time,
- The frequent value ($\psi_1 \cdot Q_k$), which is exceeded five percent of the time,
- The quasi-permanent[2] value ($\psi_2 \cdot Q_k$), which equates to the average value over a period of time.

In practice, the values G_k, Q_k, ($\psi_0 \cdot Q_k$) and ($\psi_1 \cdot Q_k$) are to be taken into consideration when verifying the ultimate limit state. However, when it comes to the serviceability limit state, these values are only relevant for calculating the short-term deformations. The long-term effects (e.g. creep) should be determined taking the values G_k and ($\psi_2 \cdot Q_k$) on the load side into consideration and the deformation factor k_{def} on the material side.

[2] The designation "quasi-permanent value" may briefly prove confusing, since ψ_2 is used for accidental design situations or earthquake actions, although these seldom occur. However, the designation actually relates to the fact that average values (= quasi-permanent values) of the variable actions are used in infrequent design situations, which are lower than their quantile values (ψ_0 corresponds to the 98% quantile, ψ_1 corresponds to the 95% quantile).

C2.2 Permanent actions

The permanent actions comprise the self-weight of the structural members and the weight of all components which must be permanently supported by the structural members, e.g. partition walls, insulations, panels and plasters. To determine permanent actions, insights into structural configurations and the construction building materials used are needed. The permanent actions are calculated from the nominal dimensions of components and the average weights of the construction materials used. If these are not specified e.g. in EC 1, the structural engineers should use the weights specified in accordance with the manufacturer's details. To simplify the calculations concerned, the self-weights of the bracings and lightweight partition walls should be applied as uniformly distributed loads, while a reasonable estimate can be obtained by referencing similar construction parts. The dead weight of floors (wooden joists with sheathing) or roofs (sheathing, rafter and purlins) is generally between 0.25 and 0.45 kN/m^2.

C2.3 Imposed actions

Imposed actions in buildings depend on their usage and correspond to the loads which move in their own right (e.g. persons, vehicles) and moveable loads (e.g. furniture, lightweight partition walls, stored goods), while the building areas exposed to load are differentiated in accordance with their intended use. In common buildings, three classes are taken into consideration:

- Office, residential and retail premises,
- Roofs and
- Storage and production facilities.

For storage and production facilities, a design is performed with the imposed loads arising during actual use of the building (category E). In other cases, the imposed loads take into account the density and type of public traffic, with the first class subdivided into four further categories A to D (Table C2-2). Roofs, meanwhile, are classified as non-accessible except for maintenance and repair (category H), or accessible (categories I and K). Categories F and G apply for roads and parking areas in buildings. The NA for EC 1 also defines categories T for stairs and Z for accesses and balconies. According to this classification, floors and roofs are designed with a uniformly distributed load q_k or a concentrated load Q_k. The concentrated load Q_k is exerted on a square area with 50 mm side length at the least favourable spot.

Table C2-2 Classification of floor areas in buildings.

Category	Type of use	Examples
A	Living space	Residences, hotel rooms
B	Office space	Classrooms, treatment rooms in hospitals
C	Meeting space	Meeting rooms, theatre, dining rooms
D	Sales space	Warehouses

In accordance with the load-duration classes under EC 5, for the imposed load q_k of categories A to D, medium-term load duration is generally assumed. This load is classified as long-term for category E and short-term for category H and the concentrated load Q_k corresponds to the short-term class. In addition to vertical loads from gravity, horizontal imposed loads on partition walls and parapets must also be taken into account. This includes e.g. short-term loads at the height of the hand rail (0.8 to 1.2 m). Depending on the type of use, the characteristic value of the horizontal imposed load q_k varies between 0.5 and 2.0 kN/m.

C2.4 Snow loads

Snow loads are based on snow depth measurements and snow density. Depending on the surrounding area and the prevailing weather, the specific density of the snow is between 0.1 (new snow) and 0.6 (old or wet snow). A statistical evaluation of these records is used to define the characteristic snow load s_k for a return period of 50 years. The values s_k are determined by national load standards, since they depend on geographical locations and altitude. In addition, the structural engineer should also take into consideration any local effect caused by modifying the value s_k. This means e.g. that the snow load exerted on a member may increase considerably when snow is transformed into ice, in the event of rain falling on the snow or areas of roof may heat up, causing localised melting of snow. During the static calculation, the engineer must take into consideration the load configurations on roofs as follows:

- Uniform distribution from uniform snowfall,
- Uneven loads due to snow drifts and snow slides.

After analysing the snowfall, the snow load is generally considered a variable load with short-term load duration (less than a week). For the horizontal projection of the area, the characteristic snow load is calculated as:

$$S_k = \mu_i \cdot s_k \qquad \text{(C2-2)}$$

The shape parameter μ_i takes the influence of roof geometry on the load configuration into consideration, e.g. whether the roof shape is more or less likely to cause snow slides or drifts. EC 1 defines two parameters μ_i depending on the roof pitch α (Figure C2-2). Both shape parameters apply, when the snow is allowed to slide unhindered, namely when no snow guards etc. are attached.

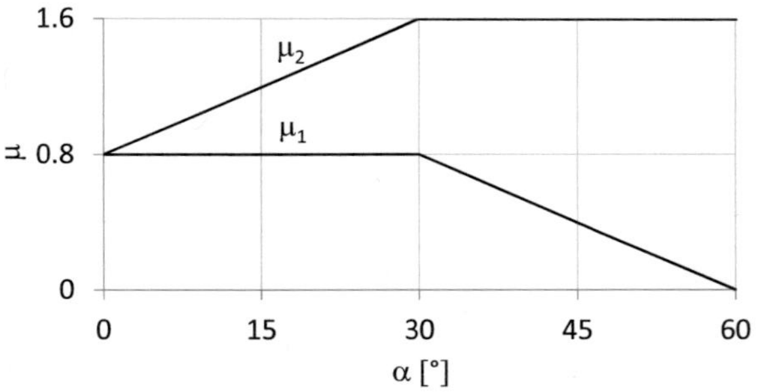

Figure C2-2 Shape parameters for snow on single pitch, double pitch and shed roofs. For barrel-shaped roofs, shape parameter μ_3 applies.

The shape parameter μ_1 applies to single and simple double pitch roofs, since no accumulation or pouching of snow is possible. Accordingly, with a roof pitch from 60° onward, no snow load can accumulate, since all the snow slides off. For double pitch roofs, a range of snow loads must be checked on both sides of the roof (albeit uniform), since snow may slide off on one side of the roof, but not the other. The shape parameter μ_2 is taken into consideration, if the formation of snow drifts is possible, namely in the event of lined-up double-pitch or shed roofs. In this case, the value for μ_2 peaks at a roof pitch of 60°, since this is when snow slides into the gusset and accumulates there.

C2.5 Wind loads

Wind loads change over time and are initially classified based on the short-term load-duration class. The action on the structure can be perceived as the combination of a quasi-static component and a resonance component and the latter component in particular may be significant for tall and slender structures, in which case more accurate wind verifications may be required. Despite this fact, the resonance component is less important for most constructions, which is why, in this article, wind loads are defined using the simplified method described in this article. Wind loads are represented in the form of static compressive forces on the construction surfaces or by global compressive forces and friction wind forces (E.C.C.S. 1987). The scope of this article is limited to wind loads exerted on constructions impervious to vibration, which can be represented via a static equivalent load. The wind load assumptions are based on a basic wind velocity v_b and a basic velocity pressure $q_b = ½ \cdot \rho \cdot v_b^2$ (where ρ = air density). Based on a return period of 50 years (= probability of occurrence of 2% per year), v_b is defined as the average ten-minute wind velocity at a height of 10 m over category II terrain (see Table C2-3).

Table C2-3 Terrain categories in accordance with EC 1.

Terrain category	
0	Lake or coastal areas exposed to the open sea
I	Lakes or areas with little vegetation and free of hindrances
II	Areas with little vegetation like grass and individual hindrances (trees, buildings)
III	Areas with uniform vegetation or development (e.g. villages, forest areas)
IV	Areas, in which at least 15% of the area is developed with buildings with an average height exceeding 15 m

Based on the basic wind velocity v_b and the resulting basic velocity pressure q_b, the peak velocity pressure $q_p(z)$ relevant for the design can be determined as a factor that depends on the construction height z. For this purpose, the ground roughness and topography in the vicinity of the building location (see also Table C2-3) as well as any turbulences have to be taken into consideration.

Pressure coefficients

The aerodynamic pressure coefficients $c_{pe/i}$ define the wind pressure, which impacts at right angles on building surfaces. The external and internal pressure coefficients (c_{pe} and c_{pi}) are positively defined, provided the wind pressure acts in the direction of the surface, whereas a negative value means suction on the member surfaces. The impact of wind direction θ can be differentiated using coefficients into two separate directions, to be examined with the gable (θ = 90°) or the longitudinal building (θ = 0 or 180°) on the windward side. The external pressure coefficients correlate strongly with the building shape, while wind tunnel tests have also shown higher wind suction forces in corner areas of the building (Lusch, 1964). These observations spawned the distributions shown in Figure C2-3 and Figure C2-4. As an example, Figure C2-3 specifies external pressure coefficients for a simple rectangular building. The values correspond to the upper value for all wind directions acting in a range of ± 45° from the normal to the surface under consideration. Figure C2-3 shows the coefficient $c_{pe,10}$ for wall surfaces exceeding 10 m² and building dimensions of $h/d(\theta = 0°)$ or $h/L(\theta = 90°) \leq 0.25$. These pressure distributions are based on the dimensional area facing the wind $e = \min(b, 2 \cdot h)$. Higher pressure coefficients apply for smaller wall areas.

As well as the forces on walls, special consideration of wind forces on roofs is also required, since uplift wind forces may have a decisive impact when it comes to designing joints and members. For flat roofs, Figure C2-4 shows the external pressure coefficients for wind directions of θ = 0 or 90°with a roof area of $A \geq 10$ m². The specified pressure coefficients must be increased if the roof areas are smaller. If sloping roofs facing the wind are involved, compressive or suction forces are exerted at roof pitches of α between 15° and 30° and the less favourable effects need to be taken into account during design.

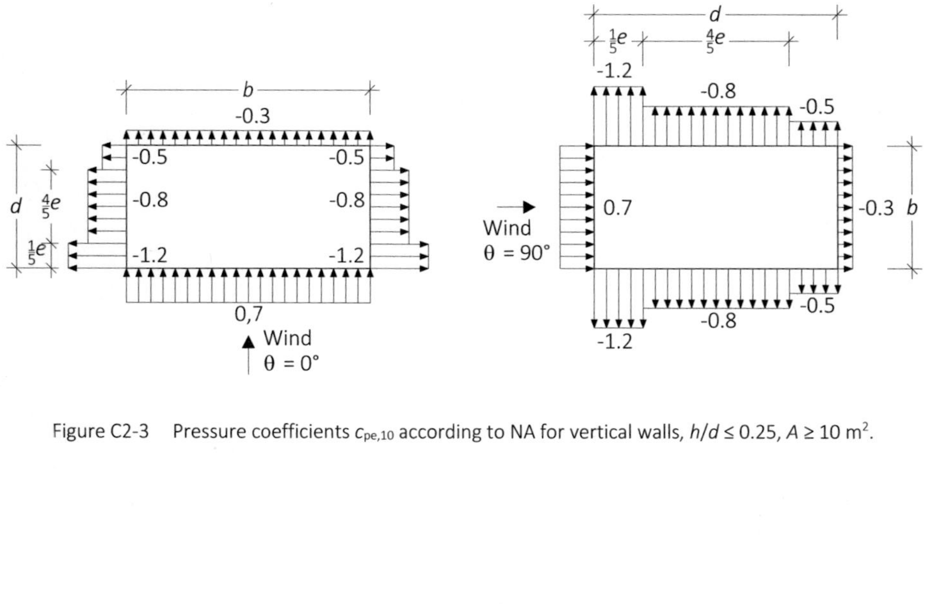

Figure C2-3 Pressure coefficients $c_{pe,10}$ according to NA for vertical walls, $h/d \leq 0.25$, $A \geq 10$ m².

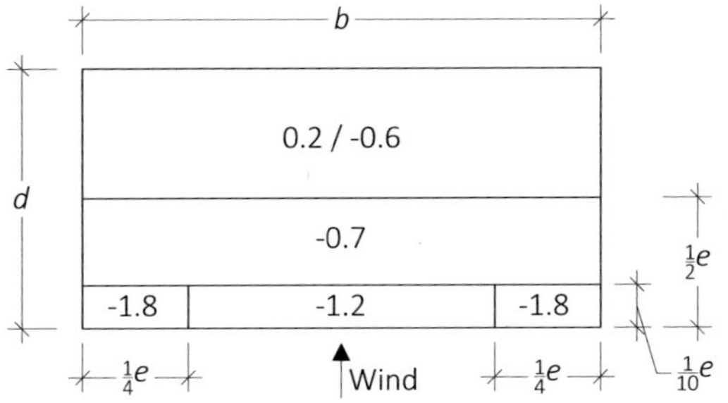

Figure C2-4 Pressure coefficients $c_{pe,10}$ according to NA for flat roofs with sharp-edged eaves area, $A \geq 10\ m^2$.

Design wind loads

When designing buildings, the effects of wind loads are generally determined using the wind pressure distribution on the surfaces, with the result determined as a combination of external (w_e) and internal (w_i) pressure as follows:

$$w_e = q_p(z_e) \cdot c_{pe} \qquad w_i = q_p(z_i) \cdot c_{pi} \tag{C2-3}$$

Since wind pressure rises with increasing height above ground, structural engineers have to consider the reference height z_e of the external surfaces. Depending on the building shape and dimensions of the width b on the windward side Figure C2-5 reveals reference heights for walls and roofs. Here, z_i is the reference height of the walls for closed buildings or the average height of the openings.

Alternatively, the wind pressure w can also be determined using aerodynamic force coefficients c_f instead of aerodynamic pressure coefficients $c_{pe/i}$. This is particularly important for constructions with linear structural members such as truss bridges, since only force and not pressure coefficients are given for trusses. For trusses, the fullness factor φ is important and defined as the ratio of the projected surface of the members relative to the overall area (product of length and width) of the truss. The fullness factor hence takes into consideration the fact that a truss has no closed surface, but is predominantly open, depending on the cross-sectional dimensions of the members (a completely closed wall has a fullness factor of 1). The force coefficient for trusses is determined based on said fullness factor. The wind force exerted on a structure can thus be determined by taking into consideration the tributary area or the projected reference surface of a building or of individual members. For larger areas exposed to the wind (e.g. free-standing roofs), a frictional force F_{fr} must also be taken into consideration.

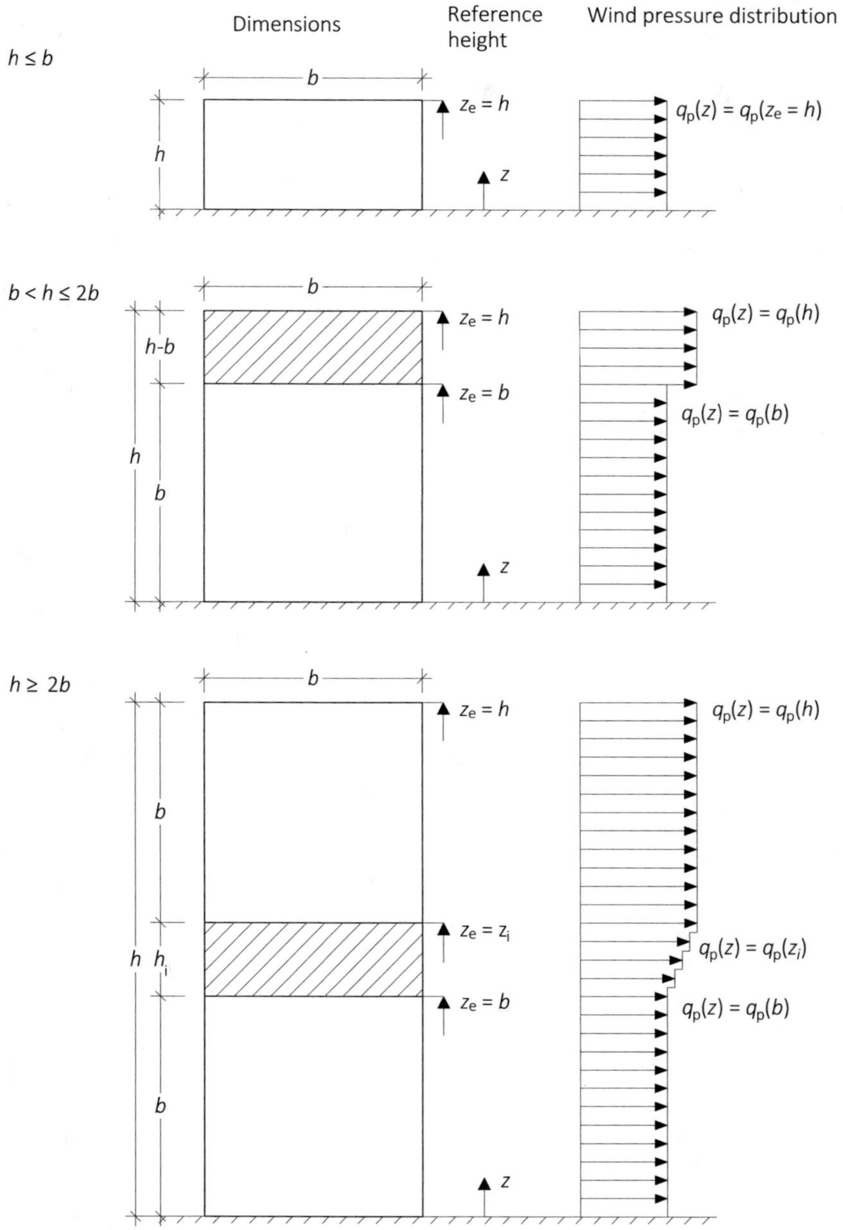

Figure C2-5 Reference height z_e depending on h and b and wind pressure distribution in accordance with EC 1.

C2.6 Combinations of actions

After the actions have been determined, the next design task is to analyse their effects, which also involves selecting realistic load configurations for which the structure or structural members are designed. The design values are then determined from the following load combinations. In the ultimate limit state, the fundamental combination for persistent and transient situations is:

$$\sum_{j\geq 1} \gamma_{G,j} \cdot G_{k,j} + 1.5 \cdot Q_{k,1} + \sum_{i>1} 1.5 \cdot \psi_{0,i} \cdot Q_{k,i} \qquad \text{(C2-4)}$$

$\gamma_{G,j}$ is the partial safety factor for permanent actions (see Article C1) and $Q_{k,1}$ is the leading variable action.

In the serviceability limit state, the combination depends on the action effects, meaning two combinations must be noted:

Characteristic combination

$$\sum_{j\geq 1} G_{k,j} + Q_{k,1} + \sum_{i>1} \psi_{0,i} \cdot Q_{k,i} \qquad \text{(C2-5)}$$

Quasi-permanent combination

$$\sum_{j\geq 1} G_{k,j} + \sum_{i\geq 1} \psi_{2,i} \cdot Q_{k,i} \qquad \text{(C2-6)}$$

Certain numeric values for the combination factors ψ for buildings are specified in Table C1-2.

In timber structures, structural engineers must focus particularly on determining the critical load cases, since they depend on the modification factors k_{mod} for the load duration and the moisture content and thus are values that are not independent of the strengths of construction materials. For all verifications in the ultimate limit state, the resistance must be determined taking into consideration the coefficient k_{mod}, whereby the selection of k_{mod} is made in accordance with the leading variable action $Q_{k,1}$. In the event of snow or wind as leading variable action, k_{mod} according to the short-term load-duration class must be used (see Table C1-4), while for other imposed loads the load-duration class is "medium-term", meaning k_{mod} is smaller. Conversely, for the load combination without variable actions considering only self-weight, k_{mod} must correspond to the "permanent" load-duration class to determine the design resistance.

While respecting the various limit states, the combination of actions for each critical load case is calculated, taking into consideration the various values for k_{mod}. During this, the structural engineer must also take into consideration any unfavourable load configurations. The uniformly distributed loads are generally governing when designing most structural members, although it is non-uniformly distributed loads that tend to have critical impacts on joints or bracing systems.

C2.7 Literature

P. Racher, original Article A3, STEP 1995.

C.E.B. (1980). Structural safety. Bulletins d'information Nos. 127 and 128, Brussels.

E.C.C.S. (1987). Recommendations for calculating the effects of wind on constructions. European Convention for Constructional Steelwork, Technical committee 12, Report No. 52, Brussels.

Hendrickson E.M., Ellingwood B. and Murphy J. (1987). Limit state probabilities for wood structural members. ASCE Journal of Structural Engineering, 113(1): 88-106.

Lusch G. (1964). Wind tunnel investigations on buildings with rectangular base and with flat and duo-pitched roofs. Bauforschung No. 41.

Rackwitz R. (1976). Practical probabilistic approach to design. C.E.B., Bulletin d'information No. 112, Brussels.

D

Structural members and systems

D1 Basic stresses

Original articles: B. Edlund, B. S. Choo, P. Aune

D1.1 Tension and compression

Wood is an anisotropic material, which means it demonstrates different properties when stress is applied in different directions, e.g. parallel or at right angles to the grain. The idealised form of a tree stem is deemed to be cylindrical and orthotropic, namely an orthogonal anisotropic structure, see Article B1. Axes *L*, *R* and *T* reveal the longitudinal, radial and tangential directions respectively, while the properties in *R* and *T* directions are often cumulatively expressed as properties perpendicular to the grain with the Index "90" (this is possible, since $E_L >> E_T \approx E_R$. Moreover, the type of cut and hence positioning of the *R* and *T* directions of the timber members is normally unknown). The longitudinal direction is specified with the Index "0". The following sections mainly cover strength and stiffness properties under short-term loading. Unless otherwise specified, the specified values apply for European softwood with a moisture content of 10 to 15%.

Timber is non-homogenous based on its growth irregularities and dimensions for load-bearing purposes. The material properties (density, strength, modulus of elasticity (MOE) etc.) vary widely, even within a single log cross-section, as well as throughout the log length. Properties also vary among individual trees of the same wood species and obviously among individual species. Even within a single annual ring, properties vary since early- and latewood have significantly different properties. This scatter is not covered in further detail in the current article.

Initially, the properties of clear wood with small dimensions are addressed, before moving on to structural timber with growth irregularities and structural timber dimensions, since only the latter information is relevant for timber structures. For timber which has been dried and strength graded, (Article B5) characteristic properties are established, with which the member resistances are determined. In the ultimate limit state, the partial safety coefficient γ_m and modification coefficient k_{mod} must be used to determine the design values of the member resistance, while in the serviceability limit state, creep must be taken into consideration applying the coefficient k_{def} (see Articles C1 and C2).

Clear wood exposed to tension and compression

Tensile stress

If small wood samples from material which is as flawless as possible are exposed to load, stress-strain curves can be derived, as shown in Figure D1-1. The tensile strength in the grain direction $f_{t,0}$ (Index "t" = tension) accordingly exceeds the compressive strength in the grain direction $f_{c,0}$ (Index "c" = compression). The stress-strain curve when tension is applied runs virtually up to the linear failure point, which occurs abruptly and is designated as brittle fracture. When compression is applied, instead, plastic behaviour is evident when failure occurs. Wood is at its weakest when tensile stress is applied at right angles to the grain. This strength $f_{t,90}$ is in the region of 1 to 2 N/mm^2 and depends very strongly on the volume involved, since tensile stress is strongly identified as a cause of brittle failure. This volume effect is less evident for other strength properties (for more on the volume effect, see Article D3). The tensile strength $f_{t,90}$ is significantly reduced when there are pre-existing cracks, particularly in earlywood. Likewise, the stiffness is also far lower with a modulus of elasticity of E_{90} = 400 to 500 N/mm^2 perpendicular to the grain than the same figure parallel to the grain at E_0 = 11000 to 15000 N/mm^2.

Timber constructions should generally be designed such as to ensure that tensile stresses perpendicular to the grain are avoided or minimised as far as possible. Knowing which areas of a construction are exposed to tensile stresses perpendicular to the grain is important, to instigate the required measures to reduce such stresses, such as reinforcements. Constructions exposed to such tensile stresses include curved beams, corner areas of portal frames, notched beam supports and beams with openings.

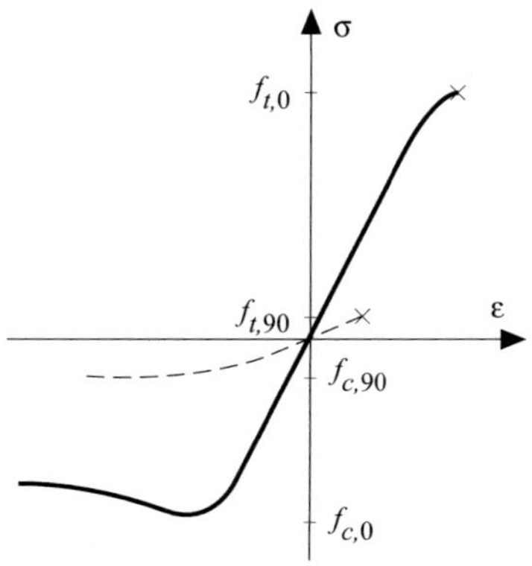

Figure D1-1 Stress-strain-curve of clear wood exposed to tensile (*t*) and compressive stresses (*c*) parallel to the grain (solid line) and perpendicular to the grain (dashed line) at constant strain increase. Typical strength values for softwood are: $f_{t,0}$ = 80 to 100 MPa, $f_{c,0}$ = 40 to 50 MPa, E_0 = 11000 to 15000 MPa, E_{90} = 400 to 500 MPa. (STEP 1995 arcticle B2)

Compressive stress

The curve progression in Figure D1-1 reveals that when compressive stress is applied in the grain direction, wood fibres will successively yield until the load peaks, whereupon the specimen fails, with localised buckling of wood fibres (see Figure D1-2). This is known as a local stability failure caused by shear along a sloping surface. As with tensile stress, the modulus of elasticity $E_{c,0}$ is between 11000 and 15000 N/mm^2. However, since the stress-strain curve levels off earlier than with tensile stress, the proportional limit must be noted.

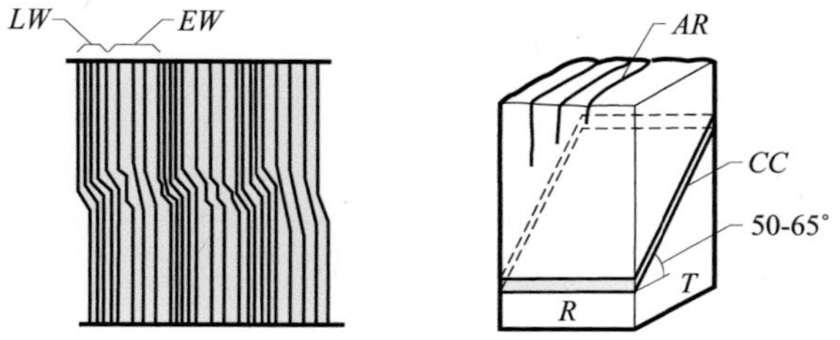

Figure D1-2 Compressive failure of compression test specimen with $f_{c,0}$ due to localised buckling of wood fibres, after Hoffmeyer (1990). LW latewood, EW earlywood, AR annual rings, CC compression creases. (STEP 1995 Article B2)

If the wood sample is completely exposed to compression perpendicular to the grain (as in case a in Figure D1-3), the wood fibres are practically squeezed, as in a bundle of tubes, until a type of squash load is attained, at which the tangent modulus (inclination of the tangent on the σ-ε-line) becomes very small. Under the maximum load, meanwhile, the strain levels become very high.

If only part of the area on the upper side of the wood sample is exposed to load, the stiffness value for the wood will be higher compared to case a in Figure D1-3 (cases b to e) and will only start declining at higher stresses than in case a. The transition from the initial stiffness to the plastic region is less pronounced in this case than in case a, because the load is also distributed on unloaded areas outside the part directly exposed to loads. In case b, the areas not directly exposed to loads are too small to ensure effective load distribution. The maximum load is attained immediately after traversing the transition area (from elastic to plastic). In cases c, d and e, meanwhile, the compressive load can still be absorbed, even at strain levels ε exceeding those shown in Figure D1-3, without any sign of failure occurring. Nevertheless, the deformation level rises considerably in this case, which means that for constructive reasons, it makes sense to limit the strain to a specific value, e.g. 1% of the cross-sectional height or 5 mm and to view the related stress as a type of strength (or "proof stress"). In this case, the figure determined is $f_{c,90}$ = 2 to 4 N/mm^2. However, the strengths defined in this manner depend on the position of the annual ring in the cross-section, see Figure D1-4.

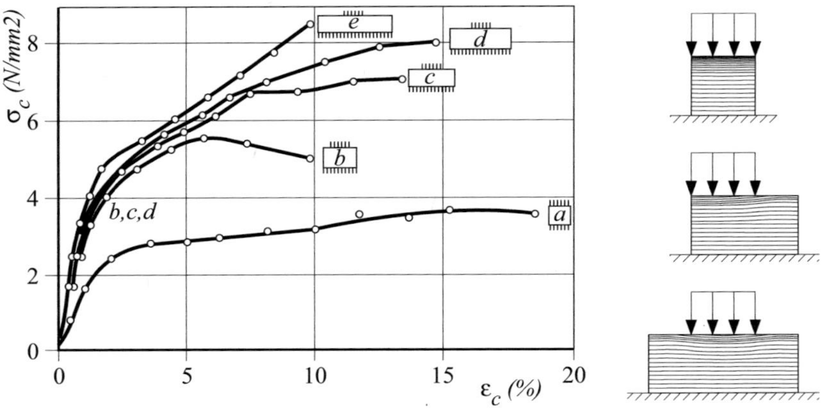

Figure D1-3 Compressive stresses σ_c acting on the upper side of the wood samples 150 mm x 150 mm at right angles to the grain depending on the strain levels. (according to Suenson, 1938, STEP 1995 Article B2) The right sketch shows the types of failure or the involvement of the wood fibres adjacent to the area exposed to load and the resulting increase in load-bearing capacity.

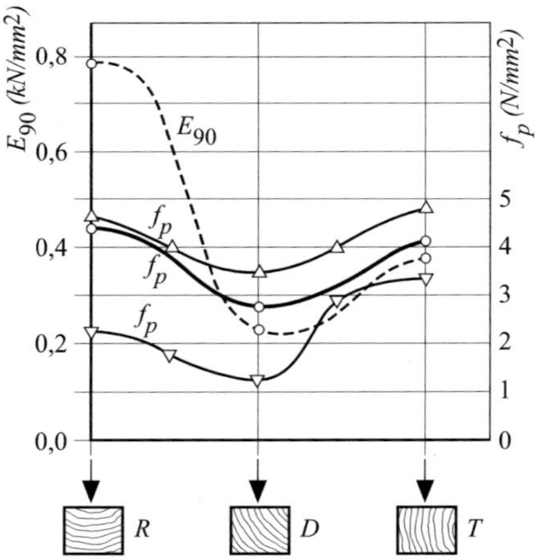

Figure D1-4 Modulus of elasticity and stress limit of wood under compression perpendicular to the grain. $f_{c,90}$ is defined in this case as stress at the proportional limit. (according to Siimes and Liiri, 1952, STEP 1995 Article B2)

However, both compressive and tensile stresses can also occur under an angle to the grain, whereby the angle between the load and grain directions is termed angle α. Figure D1-5 shows this kind of stress. Hankinson (1921) proposed the following equation (D1-1) (Figure D1-5 (b)) for the strength values $f_{c,\alpha}$ assuming linear interaction (see also Figure D1-7), which shows effective compliance with test results. (For the derivation of the Hankinson equation, see Annex 2.)

$$f_{c,\alpha} = \frac{f_{c,0} \cdot f_{c,90}}{f_{c,0} \cdot \sin^2 \alpha + f_{c,90} \cdot \cos^2 \alpha} \qquad \text{(D1-1)}$$

To determine the tension strength at an angle α to the grain, a corresponding expression is used, when replacing f_c with f_t, see Figure D1-5 (a).

For small angles α, the strength significantly depends on changes to the angle. Small changes in the slope of grain result in a considerable change in strength, particularly tensile strength. Conversely, in the area of $\alpha \approx 90°$, only slight influences on $f_{t,\alpha}$ and $f_{c,\alpha}$ are apparent (low gradient of the curves in Figure D1-5 (a) and (b)). For uniaxial tension under an angle α to the 1-axis, in accordance with Figure D1-5 (c) and (d), the following equilibrium conditions are valid (see also Annex 2, equations (II), (III) and (IV)):

$$\sigma_1 = \sigma_\alpha \cdot \cos^2 \alpha$$

$$\sigma_2 = \sigma_\alpha \cdot \sin^2 \alpha$$

$$\tau_{12} = \sigma_\alpha \cdot \sin\alpha \cdot \cos\alpha$$

For comparison, the limit curves for the three separate failure conditions $\sigma_1 \leq f_{c,0}$, $\sigma_2 \leq f_{c,90}$, $\tau_{12} \leq f_v$ are set out in Figure D1-5 (b) (dashed lines).

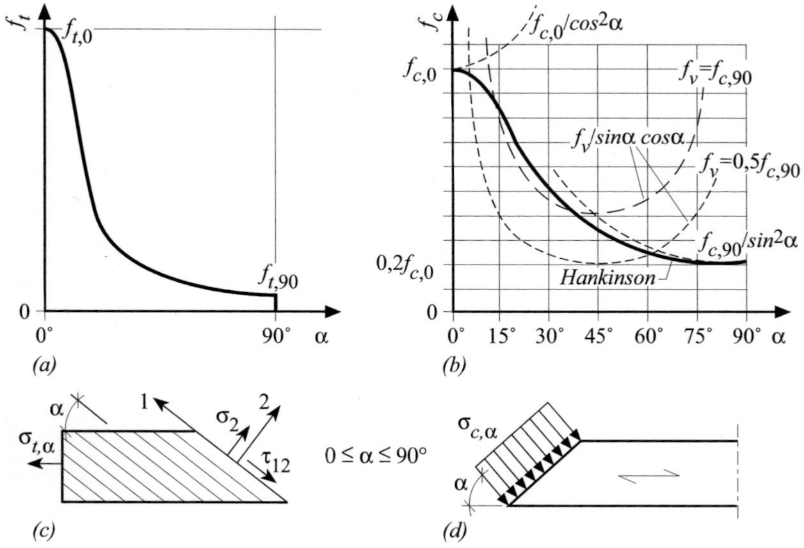

Figure D1-5 Tensile strengths (*a*) and compressive strengths (*b*) depending on the angle α. The Hankinson equation (D1-1) is shown as a solid line. The dashed lines indicate failure at stresses $f_{c,0}$, $f_{c,90}$ or f_v. (STEP 1995 Article B2)

Structural timber exposed to tension and compression

When using timber in dimensions used in load-bearing constructions, it is important to ensure the impact of various unavoidable growth irregularities such as knots and slope of grain are taken into consideration. A knot of "normal" size reduces the effective cross-section of a member and results in localised fibre disturbances, which often, in turn, trigger eccentricity of loads and high localised stresses. In addition, in uniaxially loaded sawn timber, stresses perpendicular to the grain are induced in the area where the fibres change directions around a knot. This is particularly significant for wood subject to tensile stress parallel to the grain, since this means that stresses perpendicular to the grain are also produced.

Although in clear wood, the tensile strength parallel to the grain far exceeds the compressive strength, the opposite applies for structural timber. This is partly due to the above-mentioned dependency on the slope of grain (see Figure D1-5 (a)) and partly due to the brittle fracture behaviour and volume effect (Article D3), which also encompasses all other influences. Given the significant roles played by the load duration (creep) and the moisture content, reference is made to Article B2.

Tensile stress

Inhomogeneities and other deviations from an ideal orthotropic material, as typically occur in structural timber, are often labelled as growth irregularities. As explained previously, these influences result in a clear reduction in tensile strength parallel to the grain. For Northern European softwood (spruce, fir), average values for $f_{t,0}$ are in the region of 10 to 35 N/mm^2, a significant reduction compared to the value cited in Figure D1-1 of $f_{t,0}$ from 80 to 100 N/mm^2 for clear wood. Various investigations showed the mean value of $f_{t,0}$ declining proportionally to the increasing diameter of the largest knot. However, the scatter is high and the correlation weak. The obtained values also depend on the test method, since failure may be induced by stress concentrations at the grip end devices. Moreover, the tensile strengths depend on the tested volume, which is why a height or width correction factor k_h is used in EC 5 (Article D3, Section D3.3).

Tension at an angle to the grain

EC 5 does not specify any means of verifying tensile capacity perpendicular to the grain ($\alpha = 90°$), except that the member size has to be taken into consideration. Tensile stresses perpendicular to the grain tend to occur in curved beams, notched beams or joints loaded perpendicular to the grain and are explained in the relevant articles (see Articles D4, D5, D8 and E11). EC 5 also excludes details on tension at an angle to the grain, although the NA proposes a means of verifying this type of stress. Stresses at an angle to the grain result in combined stress conditions, under which both tensile stresses parallel and perpendicular to the grain and shear stresses occur, which is why the Hankinson equation (D1-1) was expanded to encompass shear strength (derivation see Annex 2) and is then used to convert longitudinal tensile strength to tensile strength at an angle α to the grain:

$$\sigma_{t,\alpha,d} \leq k_{\alpha} \cdot f_{t,0,d} = \frac{1}{\frac{f_{t,0,d}}{f_{t,90,d}} \cdot \sin^2 \alpha + \frac{f_{t,0,d}}{f_{v,d}} \cdot \sin\alpha \cdot \cos\alpha + \cos^2 \alpha} \cdot f_{t,0,d} \tag{D1-2}$$

To verify tension in the grain direction, EC 5 includes a simple normal stress criterion where the design value of tensile stress $\sigma_{t,0,d}$ must be smaller than that of tensile strength $f_{t,0,d}$.

Compressive stress

Unlike tensile strength, compressive strength parallel to the grain is only moderately reduced by growth irregularities and averages from $f_{c,0}$ = 25 to 40 N/mm^2. Similarly to when verifying tensile strength in the grain direction, compression in the grain direction is checked via a simple normal stress criterion:

$$\sigma_{c,0,d} \leq f_{c,0,d} \tag{D1-3}$$

Compression perpendicular to the grain

When verifying compression perpendicular to the grain, the design value of strength is increased by a factor of $k_{c,90}$:

$$\sigma_{c,90,d} \leq k_{c,90} \cdot f_{c,90,d} \tag{D1-4}$$

$k_{c,90}$ takes into consideration the type of effect, the splitting risk and the extent of the deformation. Although the value for $k_{c,90}$ is generally 1.0, under specific support conditions, the design value for compression strength perpendicular to the grain $f_{c,90,d}$ may be increased (the reason for this increase in compression strength depending on the support type is clarified from Figure D1-3):

- $k_{c,90}$ = 1.25 for solid timber or 1.5 for glulam with continuous support if $\ell_1 \geq 2 \cdot h$, see Figure D1-6 (a) or
- $k_{c,90}$ = 1.5 for solid timber or 1.75 for glulam with discrete support, if $\ell_1 \geq 2 \cdot h$, see Figure D1-6 (b).

Moreover, the design value for compressive stress perpendicular to the grain $\sigma_{c,90,d}$ is determined with an effective contact surface A_{ef}, which takes a possible load distribution into consideration, see equation (D1-5). The actual contact length ℓ may be increased on both sides in the grain direction by 30 mm, but by a maximum of a, ℓ or $\ell_1/2$.

This increase in contact length to an effective value of ℓ_{ef} takes into consideration the fact that compressive stress perpendicular to the grain spreads out in the grain direction to extend to areas not directly subjected to stress (see also Figure D1-3).

$$\sigma_{c,90,d} = \frac{F_{c,90,d}}{A_{ef}} \tag{D1-5}$$

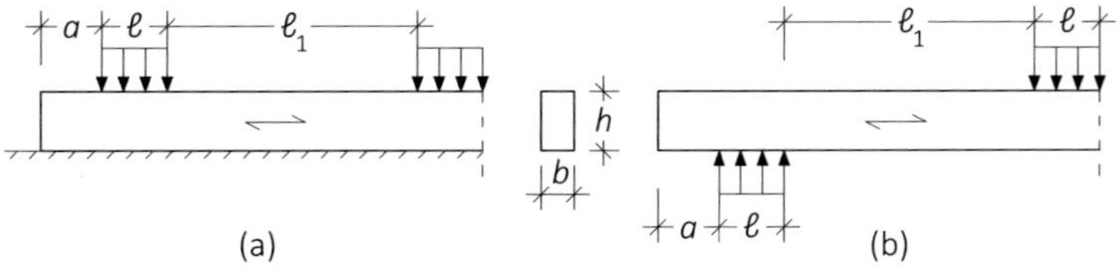

Figure D1-6 Member subject to (a) continuous support and (b) discrete support, distance ℓ_1 of continuous or discrete load from the support.

Compression at an angle to the grain

When verifying compression at an angle to the grain, equation (D1-6) applies, which corresponds to the Hankinson equation (D1-1), although in this case, coefficient $k_{c,90}$ is also considered:

$$\sigma_{c,\alpha,d} \leq \frac{f_{c,0,d}}{\frac{f_{c,0,d}}{k_{c,90} \cdot f_{c,90,d}} \cdot \sin^2 \alpha + \cos^2 \alpha} \tag{D1-6}$$

Shear strength $f_{v,d}$ is not used in equation (D1-6), but is used in equation (D1-2) to verify tensile stresses at an angle to the grain.

The verifications for compression stress cited here only apply to **members not at risk of buckling** (see Article D2).

D1.2 Bending

Beams often are horizontal structural members, which are supported on both sides and which transmit loads mainly through bending stresses. The bending moments in the beams are generated by loads acting perpendicular to the beam axis. When assessing a timber beam, it is important to ensure the following in particular:

- The design value of bending strength is not attained, the bending stresses do not result in any lateral torsional buckling of the beam and thus do not lead to a premature stability failure,
- The design value of shear strength is not attained,
- The design value of compressive strength perpendicular to the grain at the supports and under concentrated loads is not attained (equation (D1-4)),
- The deflection of the beam does not exceed the limit values recommended in EC 5,
- The vibration behaviour does not prove problematic.

This chapter focuses particularly on simple beams; namely straight beams of constant depth and without notches. The loss of strength in curved and tapered beams (Article D4) and the impact of notches (Article D5) are covered in additional articles. The bending stress is to be verified in the critical cross-section (e.g. square, T- or L-shaped), which, for simple beams in accordance with the above definitions, is usually at the point of maximum bending moment of the beam. EC 5 also requires that the influence of initial deformation, eccentricity and induced deflections be taken into consideration. The stability problem of lateral torsional buckling is covered in Article D2.

Beams not at risk of buckling

If the dimensions and support conditions of the beam are adequate to prevent instability, then the bending stresses according to elastic beam theory are as follows:

$$\sigma = \frac{M_y \cdot z}{I_y} \qquad \text{(D1-7)}$$

where

M_y Bending moment about the *y*-axis

I_y Moment of inertia (second moment of area) about the *y*-axis

z Distance from the neutral axis

σ Stress at distance z

This equation for bending stress in the beam is generally applicable, if the cross-section only needs to withstand bending about its minor principal axis or, when bent about its major principal axis, lateral support at narrow distances is provided so that the slenderness for bending is low.

Since EC 5 allows for timber structures to be designed while assuming elastic material behaviour, equation (D1-7) can be used for the design. The design value of bending strength $f_{m,d}$ is determined in this process by taking the partial safety coefficient γ_M and the modification coefficient k_{mod} (see also Article C1) into consideration. In addition, further factors, which impact on the bending strength, should be taken into consideration. The influence of member volume is considered e.g. by the coefficient k_h (see Section D1.1) and for beams of parallel structural systems including load distribution systems, the bending strength may be increased with coefficient k_{sys} (see Article D11). Accordingly, the verification of bending stresses is a simple normal stress criterion:

$$\sigma_{m,d} \leq f_{m,d} \qquad \text{(D1-8)}$$

Stress combinations

Equation (D1-8) only considers beams, which are only stressed about one principal axis by bending moments. In addition, stress combinations may also result due to

- Biaxial bending (bending about both principal axes),
- Uniaxial or biaxial bending and simultaneous axial tensile or compressive load.

Beams with biaxial bending must meet the following conditions:

$$\frac{\sigma_{m,y,d}}{f_{m,y,d}} + k_m \cdot \frac{\sigma_{m,z,d}}{f_{m,z,d}} \leq 1 \qquad \text{(D1-9)}$$

$$k_m \cdot \frac{\sigma_{m,y,d}}{f_{m,y,d}} + \frac{\sigma_{m,z,d}}{f_{m,z,d}} \leq 1 \qquad \text{(D1-10)}$$

where

$\sigma_{m,y(z),d}$ Design bending stress due to moment about the y(z)-axis

$f_{m,y(z),d}$ Design bending strength for bending about the y(z)-axis

k_m Coefficient, which takes into account that the bending capacity of the beam has not yet been attained, if, pursuant to the Euler-Bernoulli beam theory, the stresses calculated at a cross-sectional corner attain the bending strength. k_m is assumed for rectangular cross-sections made of solid timber, glulam and LVL to $k_m = 0.7$; otherwise $k_m = 1.0$ applies. It is advisable to only use k_m for compact cross-sections where $h/b \leq 4$.

The verifications for the combinations of bending and axial tension are:

$$\frac{\sigma_{\mathrm{t,0,d}}}{f_{\mathrm{t,0,d}}}+\frac{\sigma_{\mathrm{m,y,d}}}{f_{\mathrm{m,y,d}}}+k_{\mathrm{m}}\cdot\frac{\sigma_{\mathrm{m,z,d}}}{f_{\mathrm{m,z,d}}}\leq 1 \qquad \text{(D1-11)}$$

$$\frac{\sigma_{\mathrm{t,0,d}}}{f_{\mathrm{t,0,d}}}+k_{\mathrm{m}}\cdot\frac{\sigma_{\mathrm{m,y,d}}}{f_{\mathrm{m,y,d}}}+\frac{\sigma_{\mathrm{m,z,d}}}{f_{\mathrm{m,z,d}}}\leq 1 \qquad \text{(D1-12)}$$

For bending and axial compression:

$$\left(\frac{\sigma_{\mathrm{c,0,d}}}{f_{\mathrm{c,0,d}}}\right)^2+\frac{\sigma_{\mathrm{m,y,d}}}{f_{\mathrm{m,y,d}}}+k_{\mathrm{m}}\cdot\frac{\sigma_{\mathrm{m,z,d}}}{f_{\mathrm{m,z,d}}}\leq 1 \qquad \text{(D1-13)}$$

$$\left(\frac{\sigma_{\mathrm{c,0,d}}}{f_{\mathrm{c,0,d}}}\right)^2+k_{\mathrm{m}}\cdot\frac{\sigma_{\mathrm{m,y,d}}}{f_{\mathrm{m,y,d}}}+\frac{\sigma_{\mathrm{m,z,d}}}{f_{\mathrm{m,z,d}}}\leq 1 \qquad \text{(D1-14)}$$

The combinations of “biaxial bending” and “bending and tension” are linearly superposed, while the compressive normal stresses are quadratically included when designing for “bending and compression”. This is due to the plastic and thus favourable response of wood when exposed to compressive stresses. Moreover, the tension zone of a beam exposed to axial compressive stress attains lower resulting stresses, meaning it only attains bending strength at larger bending moments. Figure D1-7 clearly illustrates the significance of linear and quadratic stress superposition (see also Figure D2-5, which shows bending moment and normal force interacting). Whereas a linear superposition results in strengths declining linearly in a main direction, when a stress in the other main direction is applied, quadratic superposition allows a far slower reduction in strength for combined stresses. This is the case as explained, particularly for ductile types of failure.

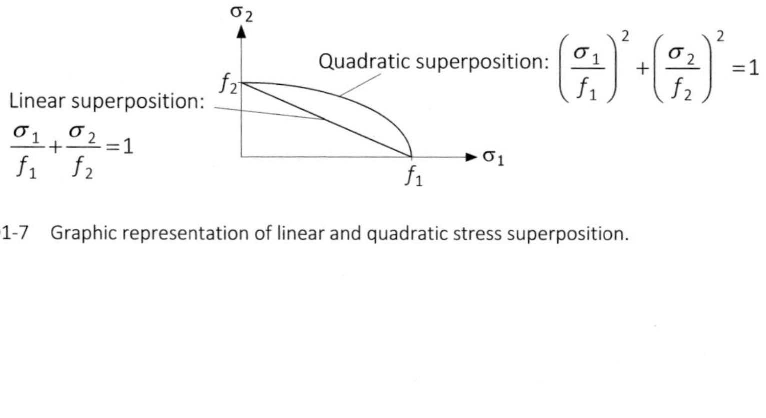

Figure D1-7 Graphic representation of linear and quadratic stress superposition.

D1.3 Shear and torsion

In accordance with the theory of elasticity, shear stresses are generated when bending is produced by loads perpendicular to the beam axis. Shear stresses perpendicular to beam axis and fibre direction always occur simultaneously with shear stresses of the same size, which act parallel to the beam axis and fibre direction. In solid timber and glued laminated timber, the shear strength parallel to the grain is considerably lower than that perpendicular to the grain, in which case fibres would have to be severed in the event of a shear fracture. Accordingly, shear parallel to the grain (longitudinal shear) is decisive for design. This article only covers longitudinal shear in solid timber and glulam members, while rolling shear, particularly relevant for cross-laminated timber members, is covered in Article D7.

Torsional stresses occur when a member is twisted by an external load. This can be caused by e.g. eccentric forces applied perpendicular to the member axis. One example is beams, which are curved due to geometric imperfections in the ground view and thus subject to torsional stresses.

Shear

In accordance with elastic beam theory (Euler-Bernoulli), shear stress τ can be specified at any point in the cross-section in generic form by:

$$\tau = \frac{V \cdot S}{I \cdot b} \tag{D1-15}$$

where V is the shear force, I the moment of inertia for the entire cross-section, b the cross-sectional width of the particular cross-sectional part at which τ is determined and S the first moment of area about the neutral axis of the considered cross-sectional part.

For a rectangular cross-section, the maximum value amounts to:

$$\tau = \frac{3}{2} \cdot \frac{V}{A} \tag{D1-16}$$

The shear stress distribution runs over the whole height of the cross-section, in a parabolic shape, with the figure peaking in the neutral fibre as shown in Figure D1-8.

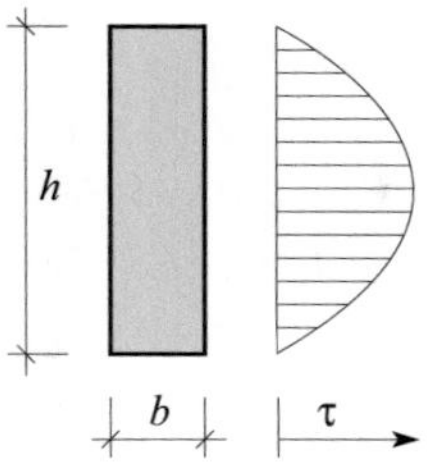

Figure D1-8 Shear stress distribution. (STEP 1995 Article B4)

In equation (D1-16), A indicates the cross-sectional area. The negative influence of cracks on the shear strength of beams must be taken into consideration, which is why area A is determined with an effective beam width b_{ef}:

$$b_{ef} = k_{cr} \cdot b \tag{D1-17}$$

The factor k_{cr} (cr = crack) is assumed in the NA for solid timber as $k_{cr} = 2.0/f_{v,k}$ and for glulam as $k_{cr} = 2.5/f_{v,k}$. The perpendicular layers included in cross-laminated timber mean that k_{cr} need not be taken into consideration and is assumed to constitute 1.0.

The factor k_{cr} was introduced since the characteristic shear strength values $f_{v,k}$, which can be taken from the corresponding applicable product standards (e.g. EN 338 for solid timber), were determined at uncracked cross-sections. However, from a practical construction perspective, this hardly ever happens, e.g. the formation of shrinkage cracks is inevitable, due to variable ambient moisture over the lifetime of a solid timber or glulam member. The specified values for $f_{v,k}$ were lower in earlier editions of product standards. For C24, $f_{v,k}$ in the EN 338 2003 edition amounted to 2.5 N/mm^2, while EN 338 in 2009 specified an $f_{v,k}$ value of 4.0 N/mm^2 and these changed nominal strength values are taken into account with k_{cr}. In accordance with the current regulation in EC 5, k_{cr} must always be taken into consideration for shear stresses τ of beams or if shear stresses are exerted in parallel to possible crack planes. As well as for beams, k_{cr} must also be applied e.g. to verify the loaded end distance of single step joints or block shear in joints. The question remains, however, how to handle other verifications, which despite requiring shear strength, do not stipulate the use of k_{cr}. Examples here include notches or verifications with stress interaction as in equation (D1-2).

Various researchers (e.g. Keenan, 1978) proved that shear stresses produced by concentrated loads near supports are smaller than values determined in accordance with elastic beam theory resp. that compressive stresses perpendicular to the grain generated by the reaction forces at supports actually increase shear strength, which is why reduced shear forces are used in EC 5. Beams which are supported at the lower edge and with the upper edge exposed to load, can be verified using a reduced shear force or the proportion of a concentrated load, applied near the support and on the upper side of the beam, on the total shear force may be disregarded. Accordingly, instead of increasing shear strength when compressive stress perpendicluar to the grain is present, as is done e.g. in the interaction equation for compressive stresses of tapers (equation (D4-7), see also Annex 2), the shear force is reduced in this case.

For biaxial bending, the shear stresses must be quadratically superposed according to the NA to EC 5. Here too, the factor k_{cr} is applied when determining shear stresses in parallel to possible crack planes:

$$\left(\frac{\tau_{y,d}}{f_{v,d}}\right)^2 + \left(\frac{\tau_{z,d}}{f_{v,d}}\right)^2 \leq 1 \qquad \text{(D1-18)}$$

Torsion

In accordance with the Euler-Bernoulli beam theory, the maximum torsional stresses τ_{tor} caused by a torsional moment M_T can be calculated for rectangular cross-sections as:

$$\tau_{tor} = \frac{M_T}{W_T} = \frac{M_T}{\alpha \cdot h \cdot b^2} \qquad \text{(D1-19)}$$

where $h \geq b$ and α is a factor depending on the ratio h/b. Timoshenko (1955) specifies the following values for α:

Table D1-1 Values for α.

h/b	1.00	1.50	1.75	2.00	2.50	3.00	4.00	6.00	8.00	10.00	∞
α	0.208	0.231	0.239	0.246	0.258	0.267	0.282	0.299	0.307	0.313	0.333

The distribution of torsional stresses along the principal axes of rectangular cross-sections is shown schematically in Figure D1-9, where the stress peaks in the middle of the longest side.

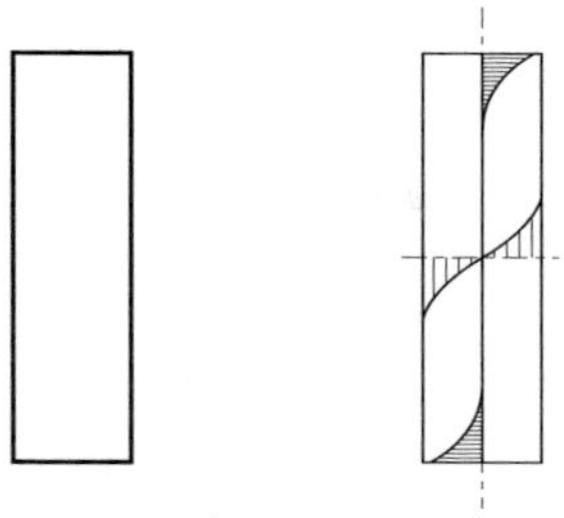

Figure D1-9 Distribution of torsional stresses. (STEP 1995 Article B4)

Torsional stress should meet the following condition:

$$\tau_{\text{tor,d}} \leq k_{\text{shape}} \cdot f_{\text{v,d}} \tag{D1-20}$$

In equation (D1-20), $f_{v,d}$ is the design shear strength. While the factor k_{shape} depends on the cross-sectional shape, EC 5 specifies values for circular and rectangular cross-sections (improved equations for k_{shape} in EC 5 Amendment A2). The use of k_{cr} is not stipulated here, although this is a point open to debate, since shear stresses are also exerted in parallel to possible crack planes when torsion applies.

Combination of shear and torsion

In some cases, both shear and torsion are exerted on a cross-section, although there have been few investigations into this phenomenon and knowledge in this area remains limited. EC 5 does not provide any pointers for this combined stress state, but the NA includes the following design equation:

$$\frac{\tau_{\text{tor,d}}}{k_{\text{shape}} \cdot f_{\text{v,d}}} + \left(\frac{\tau_{\text{y,d}}}{f_{\text{v,d}}}\right)^2 + \left(\frac{\tau_{\text{z,d}}}{f_{\text{v,d}}}\right)^2 \leq 1 \tag{D1-21}$$

Here, the factor k_{cr} is to be applied to determine the shear stresses $\tau_{y,d}$ and $\tau_{z,d}$.

D1.4 Literature

B. Edlund, B.S. Choo, P. Aune, original Articles B2, B3, B4, STEP 1995.

Hankinson R.L. (1921). Investigation of crushing strength of spruce at varying angles of grain. Air Service In-form Circular III, No. 259, US Air Service, Washington DC.

Hoffmeyer P. (1990). Failure of wood as influenced by moisture and duration of load. Dissertation, State University of New York.

Keenan F.J. (1978). The distribution of shear stresses in timber beams. Paper 9-10-1, CIB-W18 Meeting 9, Perth.

Siimes F. and Liiri O. (1952). Investigations of the strength properties of wood I. Tests on small clear specimens of Finnish Pine (Pinus Sylvestris). (In Finnish). Valtion Teknillinen Tutkimuslaitos, Tiedotus 103, Helsinki.

Suenson E. (1938). Zulässiger Druck auf Querholz. Holz als Roh- und Werkstoff 1(6): 213-216.

Timoshenko S. (1955). Strength of materials – Part 1. D. Van Nostrand, New Jersey, Third Edition.

D2 Stability

Original articles: B. S. Choo, H. J. Blass

This chapter shows stability analyses and their backgrounds, starting with the buckling of columns, before covering the way buckling lengths are derived. In addition to compressive and bending stress imposed on columns, compressive and bending stresses on beams also lead to stability problems. This is called lateral torsional buckling and is subsequently covered. **It is important to note that stability analyses are performed with the lower quantile values for stiffness and not the average values.**

D2.1 Columns

If a slender column is axially loaded, there is a risk that it deflects laterally which is called flexural buckling and is shown in Figure D2-1. The load-bearing capacity of a slender structural member when exposed to compressive stress thus depends not only on the strength of the construction material itself, but also particularly on its bending stiffness. The key material properties determining the load-bearing capacity of a wooden column are thus the compressive and bending strength and the modulus of elasticity (MOE). Since lateral deflection generates additional bending stresses in the column, these must also be taken into consideration in a stability design.

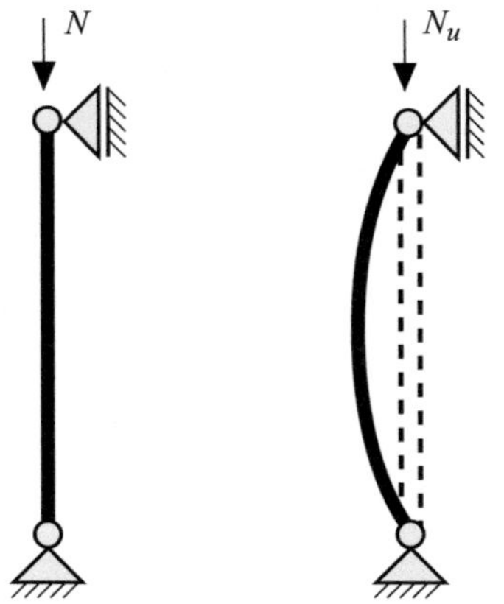

Figure D2-1 Column hinged at both ends, EI = const. (STEP 1995 Article B6)

Two different design methods are distinguished for columns: the first of which comprises a calculation of the column in accordance with second order theory, namely the equilibrium of moments and forces is determined on the deformed system. The second method uses buckling curves to take the lower load-bearing capacity of an actual column into consideration compared to a column of unlimited rigidity, which means stability design is performed via compressive stress analysis with reduced compressive strength. The extent to which the load-bearing capacity declines particularly depends on the slenderness of the corresponding member and was derived based on a column hinged at both ends (Figure D2-1). For members supported otherwise or structural systems such as portal frames, this method can be used, providing the buckling length has been determined beforehand (see Section D2.2) whereupon the analysis is performed as for a column hinged at both ends of the same length. In this context, only the column design based on buckling curves is covered.

Variables affecting the load-bearing capacity of columns

The key variables affecting the load-bearing capacity of wooden columns can be classified into two groups. The first group comprises the column dimensions, the support conditions and the construction material properties, which, in turn, are dictated by the selection of a strength class, the prevailing climate (→ service class) and the load duration class of the governing load case. The parameters included in this first group are either determined by the structural engineer or at least known to the same and the engineer can determine the load-bearing capacity of the column corresponding to the design requirements by changing these parameters. A second group of variables includes geometric and structural imperfections and their variations. Given the fact that actual structures are never perfect, these parameters must also be taken into consideration during the design. Since structural engineers generally lack any information about the extent of such imperfections for a given structure, their influence is already included in the design equations for columns in EC 5.

The key geometrical imperfections of wooden columns are initial curvature, inclination of members and tendency for actual cross-sectional values to deviate from nominal values. The extent of initial curvature for columns made of glulam or LVL is limited to 1/500, while the figure for columns made of solid timber is limited to 1/300 of the length (EC 5 Section 10.2). The buckling coefficients were determined within these limits, whereby in actual constructions, eccentricity was determined, with its statistic distribution used as input. Deviations of the nominal cross-sectional dimensions are limited for solid timber members to tolerance class 1 in EN 336 and for glulam members to the values in EN 390. Members with higher moisture content or within which moisture content varies (see Article B2, mechano-sorptive creep) show a clearly increased level of creep deformation (see also Figure B2-8) and hence larger deformations over time, hence for columns larger eccentricities *e*. Accordingly, the NA specifies that creep should also be taken into consideration in the ultimate limit state, if members in service classes 2 and 3 are subject to compression through high permanent loads.

Structural imperfections include growth irregularities and other properties like moisture content, which collectively determine the stress-strain behaviour of timber. As a general rule, the stress-strain curve of timber can be considered brittle when tensile and shear stresses are applied, while compressive stresses trigger considerable plastic deformations (see also Figure D1-1). For European conifers, the shape of the stress-strain curve depends particularly on the following properties (Glos, 1978): density, knot size, compression wood proportion and moisture content. Glos (1978) put forward mathematical relations between these properties and the shape of the stress-strain curve, for both laminations for glued laminated timber as well as solid timber cross-sections. Provided parameters such as density, knot size, compression wood proportion and moisture content are known, the expected shape of the stress-strain curve can be calculated via the cited relationships.

Background to the buckling curves in EC 5

In general, buckling curves reveal the impact of slenderness on the characteristic load-bearing capacity of columns hinged at both ends. In this context, each value of a buckling curve corresponds to the characteristic load-bearing capacity of columns of the corresponding slenderness ratio. The slenderness ratio is defined as the ratio of the buckling length to the radius of gyration and various approaches are available to determine the characteristic load-bearing capacity values of columns. One basic option here is to conduct load-bearing tests with representative selected columns, but the excessive number of tests required and resulting high cost makes this method unfeasible, which is why an alternative approach based on simulations of tests via computer, was used to derive the buckling curves in EC 5 (Blass, 1987; Blass, 1988a; Blass, 1988b). It involves modelling columns by assigning them material properties and geometric imperfections that are based on observations on real columns. This means that the strength and stiffness values and the extent of initial curvature or deviations of the nominal cross-sectional dimensions for a specific column are selected at random. Of course, the assigned properties must be realistic, which also means that the existing correlations between the various assigned properties must be taken into account in the simulation process. This also means, for example, that columns with a high modulus of elasticity also tend to show high strength values. Just like an actual column, the simulated column showcases a set of properties, which determine its load-bearing capacity.

The statistic distribution of the load-bearing capacity of columns with a specific slenderness ratio or strength class is determined by simulating numerous columns and calculating their load-bearing capacity. The load-bearing capacity values vary due to the scatter of strength and stiffness properties of the timber and the variance in geometrical imperfections. The statistic distribution of load-bearing capacity values for columns is used as the basis with which to determine the 5% quantile as a characteristic value. This characteristic value corresponds to a point of the buckling curve (see Figure D2-2). If corresponding simulations and load-bearing capacity calculations are also performed for other slenderness ratios, the resulting characteristic values reveal the course of the column load-bearing capacity depending on the slenderness ratio (buckling strength). Figure D2-3 shows an example of the specific course of characteristic values for buckling strength determined via simulations.

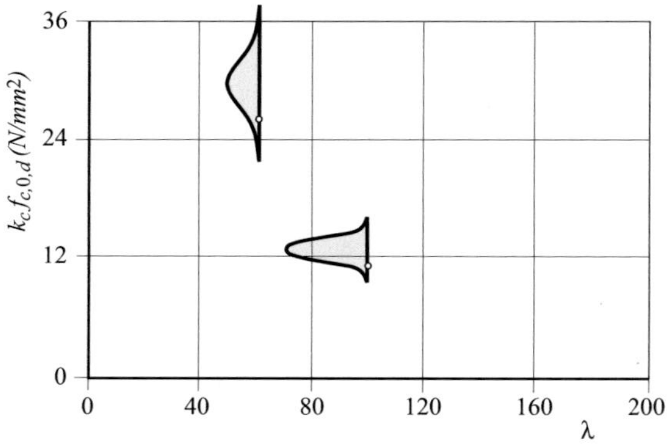

Figure D2-2 Distribution of the buckling strength and characteristic values for two different slenderness ratios λ. The 5% quantiles are specified. (STEP 1995 Article B6)

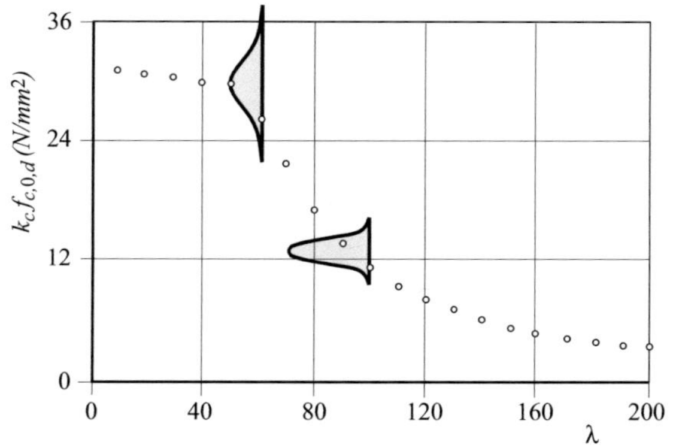

Figure D2-3 Course of characteristic buckling strength values. (STEP 1995 Article B6)

Since an equation is easier to handle in practical design terms than a diagram, approximation curves were adapted for the course of the characteristic load-bearing capacities, whereby the form of the design equations corresponds to that of the equations in EC 3 used to assess steel columns. Figure D2-4 shows one example of the course of characteristic load-bearing capacities with the corresponding approximation curve. The load-bearing capacity calculations for the simulated columns took place in accordance with second order plasticity theory; taking the plastic deformation potential of the wood when exposed to compressive stress into consideration. This approach spawned higher load-bearing capacities than when applying second order elasticity theory, where the load-bearing capacity is deemed already attained, when the compressive strength level of the wood is attained at the edge of the critical cross-section. However, when applying plasticity theory, the numerous iterations mean the calculations take longer to perform. The consideration of the plastic behaviour of the wood particularly increases the characteristic load-bearing capacity under combined compressive and bending stress.

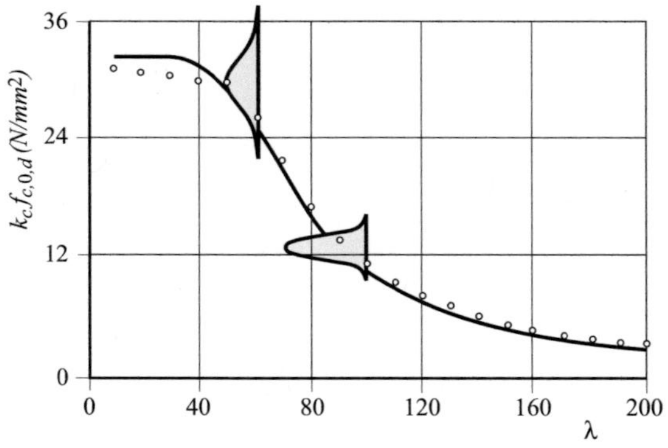

Figure D2-4 Approximation curve adapted to the course of characteristic values. (STEP 1995 Article B6)

Figure D2-5 portrays a bending moment-normal force interaction diagram for rectangular cross-sections. The straight line shows the load-bearing capacity assuming elastic and brittle behaviour, while the solid curve represents the characteristic load-bearing capacity, taking the plastic deformation capacity of the wood under compressive stress into consideration. The dashed line is the design rule in accordance with EC 5 for a combined stress comprising normal force and bending moment, when there is no risk of buckling or if the internal forces and moments have already been determined in accordance with second order theory (cf. equations (D1-13) and (D1-14)). For members at risk of buckling due to combined compressive and bending stress, the course of the interaction relationship (solid curve in Figure D2-5 for very compact members) gradually becomes a virtually linear relationship for very slender members. When assessing compact columns with slenderness ratios of up to around 20 (corresponding to $\lambda_{rel} = 0.3$), the dashed line in Figure D2-5 applies, whereas the simple linear interaction applies in all other cases.

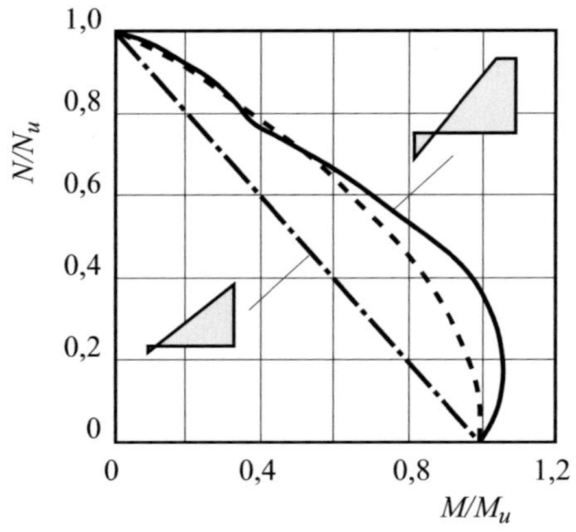

Figure D2-5 Bending moment-normal force interaction. (STEP 1995 Article B6)

Buckling curves in accordance with EC 5

The approximation curve shown in Figure D2-4 elicits the factors $k_{c,y}$ and $k_{c,z}$ required for the design, although they depend on slenderness, which is why limit values for the slenderness ratios must first be determined. The relative slenderness ratios applied are defined by:

$$\lambda_{rel,y} = \frac{\lambda_y}{\pi} \cdot \sqrt{\frac{f_{c,0,k}}{E_{0,05}}} \tag{D2-1}$$

$$\lambda_{rel,z} = \frac{\lambda_z}{\pi} \cdot \sqrt{\frac{f_{c,0,k}}{E_{0,05}}} \tag{D2-2}$$

λ_y and $\lambda_{rel,y}$ are to be used for bending about the *y*-axis (buckling in *z*-direction), λ_z and $\lambda_{rel,z}$ for bending about the *z*-axis (buckling in *y*-direction).

λ_y and λ_z are the slenderness ratios obtainable in accordance with the following equation for buckling about the *z*- or *y*-axis, whereby the effective buckling lengths ℓ_{ef} result from the Euler cases (see Section D2.2 and for example Figure D2-7), i = radius of gyration:

$$\lambda = \frac{\ell_{ef}}{i} \tag{D2-3}$$

If $\lambda_{rel,y} \leq 0.3$ and $\lambda_{rel,z} \leq 0.3$, the following conditions should be met, which correspond to the verifications under combined compressive and bending stress, see equations (D1-13) and (D1-14):

$$\left(\frac{\sigma_{c,0,d}}{f_{c,0,d}}\right)^2 + \frac{\sigma_{m,y,d}}{f_{m,y,d}} + k_m \cdot \frac{\sigma_{m,z,d}}{f_{m,z,d}} \leq 1 \tag{D2-4}$$

$$\left(\frac{\sigma_{c,0,d}}{f_{c,0,d}}\right)^2 + k_m \cdot \frac{\sigma_{m,y,d}}{f_{m,y,d}} + \frac{\sigma_{m,z,d}}{f_{m,z,d}} \leq 1 \tag{D2-5}$$

In all remaining cases, namely for larger slenderness ratios, stresses should meet the following conditions, whereby the reduced compressive strength now used is in accordance with Figure D2-4 and linear interaction is assumed:

$$\frac{\sigma_{c,0,d}}{k_{c,y} \cdot f_{c,0,d}} + \frac{\sigma_{m,y,d}}{f_{m,y,d}} + k_m \cdot \frac{\sigma_{m,z,d}}{f_{m,z,d}} \leq 1 \tag{D2-6}$$

$$\frac{\sigma_{c,0,d}}{k_{c,z} \cdot f_{c,0,d}} + k_m \cdot \frac{\sigma_{m,y,d}}{f_{m,y,d}} + \frac{\sigma_{m,z,d}}{f_{m,z,d}} \leq 1 \tag{D2-7}$$

where:

σ_m Bending stress in accordance with first order theory

$$k_{c,y/z} = \frac{1}{k_{y/z} + \sqrt{k_{y/z}^2 - \lambda_{rel,y/z}^2}}$$

$$k_{y/z} = 0.5 \cdot \left(1 + \beta_c \cdot \left(\lambda_{rel,y/z} - 0.3\right) + \lambda_{rel,y/z}^2\right)$$

β_c Coefficient for members, where the initial curvature does not exceed the above-specified limit values (columns made of glulam/LVL 1/500 and those made of solid timber 1/300 of the length):
For solid timber: $\beta_c = 0.2$
For glulam and LVL: $\beta_c = 0.1$

The difference between solid timber and glulam or LVL is mainly due to the smaller degrees of initial curvature in glulam and LVL members and the smaller deviations in cross-sectional dimensions from nominal dimensions. Moreover, in the case of columns made of glulam or LVL, both the average value as well as the coefficient of variation of moisture content are lower than with solid timber columns. Higher moisture content causes compressive strength to decline, which then reduces the load-bearing capacity of columns with low to medium slenderness. In contrast, the modulus of elasticity and hence the load-bearing capacity of very slender columns is only influenced to a very minor extent by higher levels of moisture content.

D2.2 Buckling lengths

Buckling curves used to design timber columns generally describe the course of the load-bearing capacity for columns with hinges at both ends depending on the slenderness. In the case of the column shown in Figure D2-1 the buckling length corresponds to the actual length. However, the support conditions of actual columns frequently differ from what is shown in Figure D2-1. To continue using the buckling curves specified in EC 5 for such cases, which involve deviating support conditions, the concept of an effective buckling length is used. Verifications performed in accordance with buckling curves are included in Section 6.3.2 of EC 5, while buckling length coefficients for cases frequently occurring in practice are specified in NCI NA.13.

Example members in which the actual buckling length differs include e.g. truss webs. The chords often have their outer edges braced to resist buckling. This means that the webs are not laterally secured at the centre of gravity of the chords and the buckling length of the webs approximately corresponds to the distance between the bracings of the upper or lower chord and hence exceeds the distance between the theoretical truss nodes.

The buckling length or effective length ℓ_{ef} of a member exposed to compressive loads is defined as the length of a column hinged at both ends with the same stiffness properties and elastic buckling load as the member in question. The buckling length can be shown as the distance between two adjacent inflexion points in the deflection curve of the buckled compression member (see Figure D2-6). For practical applications, an effective length factor $\beta = \ell_{ef}/L$ is often used, which indicates the ratio between buckling length ℓ_{ef} and the actual length L of the member. Figure D2-7 shows the four Euler buckling cases, whereby the buckling length ℓ_{ef} is specified for the various ideal support conditions. In Annex 3, the second Euler case (column hinged at both ends) is derived for example.

The NA, meanwhile, specifies approximate solutions for buckling lengths of various structural systems. In cases in which the applicable limits of approximate solutions are exceeded (e.g. for large initial curvatures), the corresponding structural system should be calculated in accordance with second order theory (and not in accordance with buckling curves). In other words, the equilibrium conditions should be met for the deformed system.

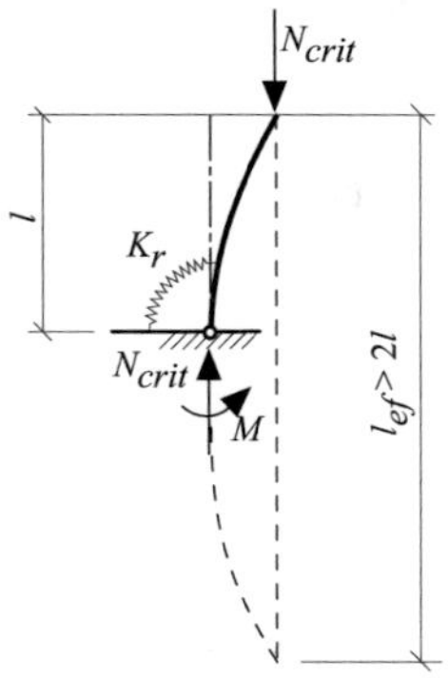

Figure D2-6 Buckling length ℓ_{ef} of a clamped column with a semi-rigid base connection. (STEP 1995 Article B7)

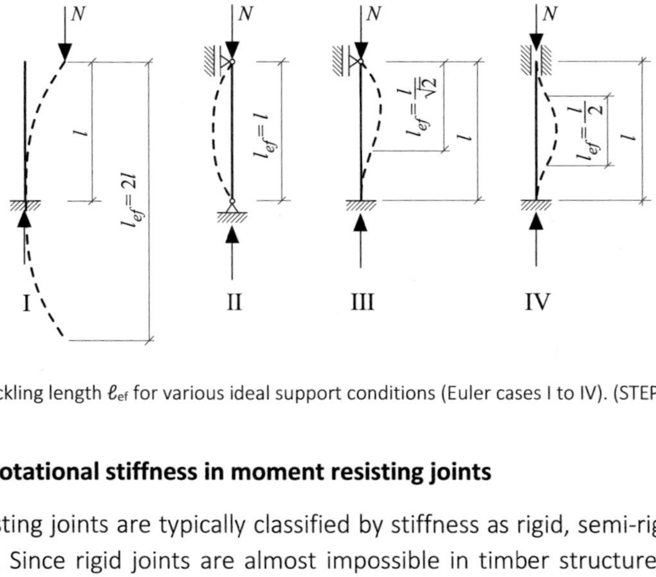

Figure D2-7 Buckling length ℓ_{ef} for various ideal support conditions (Euler cases I to IV). (STEP 1995 Article B7)

Influence of rotational stiffness in moment resisting joints

Moment resisting joints are typically classified by stiffness as rigid, semi-rigid and (nominally) pinned. Since rigid joints are almost impossible in timber structures, rotations in semi-rigid joints should be taken into account when determining buckling lengths. Semi-rigidity in moment resisting joints with mechanical connections means a greater buckling length compared to a system with rigid joints. Here, the rotational stiffness K_r of a semi-rigid joint is defined as the moment necessary to cause an angle of rotation of 1 rad in the joint. Using the slip modulus K_u of the fastener, the rotational stiffness K_r is calculated to:

$$K_{\mathrm{r}} = \sum_{i=1}^{n}\left(K_{\mathrm{u}} \cdot r_{\mathrm{i}}^{2}\right) \qquad \text{(D2-8)}$$

where r_i is the distance between the fastener i and the centre of gravity of the joint.

As an example, the buckling length of the column shown in Figure D2-8 is derived considering the influence of rotation in the semi-rigid joint at the base of the column. The approximate solutions for buckling lengths, which incorporate the impact of semi-rigidity in joints, remain valid, provided the semi-rigidity does not reduce the elastic buckling load by more than around 20%.

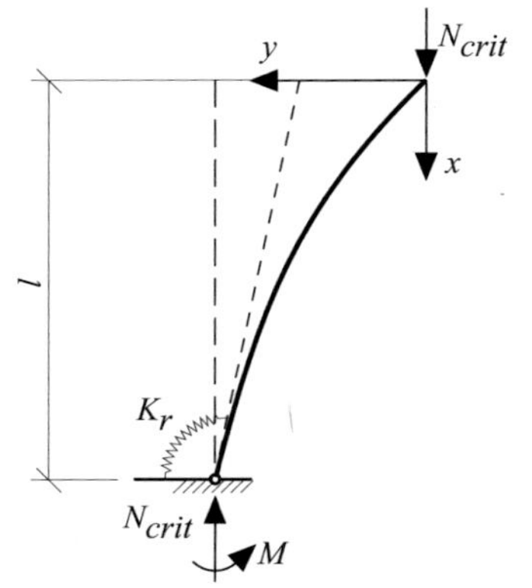

Figure D2-8 Deformed state of a column with a semi-rigid base joint. (STEP 1995 Article B7)

Using the notation of Figure D2-8, the bending moment M is determined as:

$$M(x) = N \cdot y(x) \tag{D2-9}$$

and the differential equation of the deflection curve as:

$$E \cdot I \cdot y'' = -M(x) \tag{D2-10}$$

which results in the differential equation

$$y'' + \frac{N}{E \cdot I} \cdot y = 0 \tag{D2-11}$$

with the solution:

$$y = A \cdot \sin(\mu \cdot x) \tag{D2-12}$$

Here

$$\mu = \sqrt{\frac{N}{E \cdot I}} \tag{D2-13}$$

With the boundary conditions

$$M(x=\ell)=N\cdot y(x=\ell)=K_r\cdot y'(x=\ell) \tag{D2-14}$$

the following buckling condition results:

$$(\mu\cdot\ell)\cdot\frac{E\cdot I}{\ell\cdot K_r}\cdot\tan(\mu\cdot\ell)=1 \tag{D2-15}$$

There is no analytical solution to equation (D2-15). However, for

$$-\frac{\pi}{2}<\mu\cdot\ell<\frac{\pi}{2} \tag{D2-16}$$

the following approximate solution can be established:

$$\tan(\mu\cdot\ell)\approx\frac{\mu\cdot\ell}{1-\frac{4\cdot\mu^2\cdot\ell^2}{\pi^2}} \tag{D2-17}$$

If the approximate solution to the equation (D2-17) is inserted in the buckling condition (D2-15), the resulting elastic buckling load can be calculated as:

$$N_{crit}=\frac{1}{\frac{4\cdot\ell^2}{\pi^2\cdot E\cdot I}+\frac{\ell}{K_r}} \tag{D2-18}$$

Compared with the elastic buckling load of the Euler buckling case II

$$N_{crit}=\frac{\pi\cdot E\cdot I}{\ell_{ef}^2} \tag{D2-19}$$

the resulting effective length factor β is

$$\beta=\frac{\ell_{ef}}{\ell}=\sqrt{4+\frac{\pi^2\cdot E\cdot I}{\ell\cdot K_r}} \tag{D2-20}$$

The same procedure can also be used to determine effective length factors for other structural systems such as interconnected columns, arches or frames. Some effective length factors for such systems are included in Annex 3 (without derivation).

D2.3 Lateral torsional buckling

Background

The first step when designing beams is to ensure sufficient load-bearing capacity and stiffness for bending about its major principal axis, usually on the vertical plane. This often results in cross-sectional forms in which the stiffness on the vertical plane may far exceed the value on a horizontal plane. For columns, meanwhile, it was shown that buckling failure could occur when axial compressive force is applied to a slender load-bearing element (loaded in its stiff plane), resulting in lateral deviation of the element (deflecting in its weaker planes). Figure D2-9 portrays the behaviour of a slender simply supported beam when subjected to bending stress on the vertical plane. This behaviour, which involves the beam laterally deviating and twisting in the process, is known as lateral torsional buckling. This type of instability is comparable to a column buckling when subject to axial (compression) force, since the stress imposed on the beam in its stiffer vertical plane causes failure due to buckling from this plane, in a less stiff direction.

The bending moment, under which the ideal beam is rendered unstable, is known as the critical moment, equations for which can be taken from textbooks, such as those from Timoshenko and Gere (1961). The critical moment used in EC 5 is derived in Annex 4, whereby, as a general rule, a beam of ideal elastic and isotropic material is assumed. Hooley and Madsen (1964) proved that this theory also applies to beams made of wood as an anisotropic material.

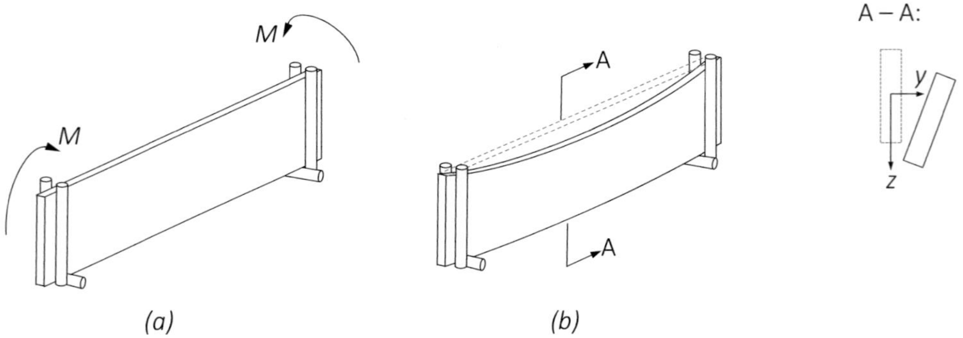

Figure D2-9 Lateral torsional buckling of a simply supported beam subject to constant moment. (*a*) simply supported beam, (*b*) buckled beam. (STEP 1995 Article B3)

The critical moment for the ideal beam in accordance with Figure D2-9 with $E \cdot I$ = const., simply supported in *y*- and *z*-directions and torsional restraints about the *x*-axis at both supports is as follows (derivation of M_{crit} shown in Annex 4):

$$M_{y,crit} = \frac{\pi}{\ell_{ef}} \cdot \sqrt{E \cdot I_z \cdot G \cdot I_{tor}} \tag{D2-21}$$

where

I_z	Moment of inertia (second moment of area) about the corresponding axis
E	Modulus of elasticity of the material
G	Shear modulus of the material
ℓ_{ef}	Effective length, in this case the distance of the torsional restraints
I_{tor}	Torsional second moment of the beam cross-sectional area

Since the lower quantile value of the critical moment remains to be determined, 5% quantiles should be used for the stiffness properties *E* and *G*.

For a timber beam with rectangular cross-section *b* x *h*, the critical stress can be calculated from the critical moment divided by the section modulus as

$$\sigma_{m,crit} = \frac{M_{y,crit}}{W_y} = \frac{\pi \cdot \sqrt{E_{0,05} \cdot I_z \cdot G_{0,05} \cdot I_{tor}}}{\ell_{ef} \cdot W_y} \tag{D2-22}$$

For isotropic material, there is only one value for the modulus of elasticity *E* and the shear modulus *G*, while the values for wood depend on the angle between the fibre and stress directions. In general, the modulus of elasticity parallel to the grain should be used and *G* is assumed to constitute $G = E/16$ for softwood. Accordingly, assuming rectangular and slender beams with high h/b ratio ($\alpha = 0.333$, see also Table D1-1), the critical stress for softwood is:

$$\sigma_{m,crit} = \frac{0,78 \cdot b^2}{h \cdot \ell_{ef}} \cdot E_{0,05} \tag{D2-23}$$

According to NCI NA 13.3, however, for simply supported beams with torsional restraints, the influences of a semi-rigid torsional restraint at the support, elastic foundation against lateral deformation and against twisting may be taken into consideration. However, these influences were not taken into consideration when deriving the equation for M_{crit} used in EC 5, see also Annex 4. The corresponding equations are specified in the NA. Moreover, due to the homogenisation of glulam beams, in the event of torsion and bending about the minor principal axis, the 5% quantile of the product of *E* and *G* may be increased by 40% for glulam beams in accordance with the NA.

Similar terms are also obtained for other load cases, load configurations and support conditions. For a simply supported beam with a concentrated load at the centre that is applied at the neutral axis of the beam, the equation for $\sigma_{m,crit}$ is very similar to equation (D2-22), merely replacing π with 4.24. The ratio $\pi/4.24$ is often referred to as the equivalent uniform moment or *m*-factor. It is a measure denoting the influence of individual moment diagrams compared to the basic case with a uniform moment diagram. The values of the *m*-factor for certain load cases are listed in Table D2-1.

In general, the more the moment diagram deviates from the uniform case, the greater the lateral stability. In EC 5, rounded *m*-factors are used; $m = 0.9$ for simply supported beams with a uniformly distributed load and 0.8 for a centrally placed concentrated load. Cantilever beams with a uniformly distributed load have an *m*-factor of m = 0.5, while for cantilever beams with a concentrated load at free end, $m = 0.8$. The NA cites certain additional factors for structural systems and load cases, with which effective lengths can be determined (Table NA.25). The location of the load in the beam height is also important, since loads on the upper edge of a slender beam have a destabilising effect, while those on the lower edge of the beam act to stabilise it. This is not taken into consideration in the existing verifications in Europe.

Of course, the support conditions are also important. Torsional restraints, which prevent lateral deviation and twisting of the beam, enhance its lateral torsional stability (or are required when deriving M_{crit}). Improving the lateral torsional stability through the support conditions is generally reflected in lower values for the *m*-factor. Lateral torsional stability of beams is a comprehensive topic, which goes beyond the scope of the current article. In textbooks however, e.g. by Timoshenko and Gere (1961), this topic is presented in detail. The following main influences on lateral torsional stability can be deduced from the comments to date:

- The distance of the lateral supports (i.e. braces, the distance between points at which lateral deflection is prevented),
- The lateral bending stiffness EI_z of the beam,
- The torsional stiffness GI_{tor} of the beam,
- The location of the load,
- The torsional restraint at the beam ends.

The load-bearing capacity of a beam at risk of lateral torsional buckling can be improved by installing bracing members. The main requirements are sufficient stiffness of the bracing to effectively prevent lateral deformations of the beam and sufficient load-bearing capacity of the bracing to accommodate forces transmitted by the beam (see also Article D9).

Table D2-1 Equivalent uniform moment factors according to Kirby and Nethercot (1979). The value of 0.74 for the load case „concentrated load on simply supported beam" corresponds to $\pi/4.24$. (STEP 1995 Article B3)

Load case	Actual moment diagram	m	Equivalent uniform moment diagram
	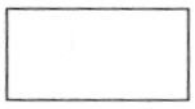	1.0	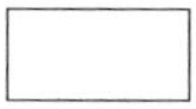
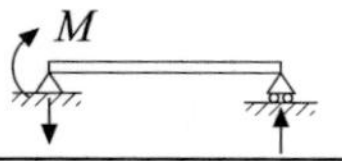		0.57	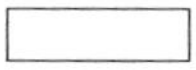
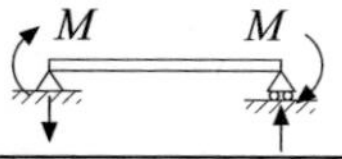	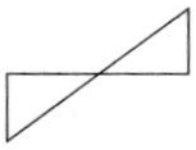	0.43	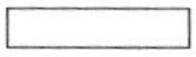
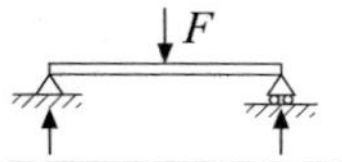		0.74	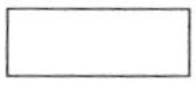
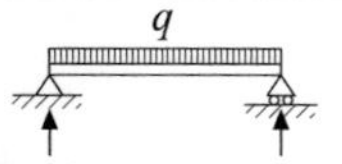		0.88	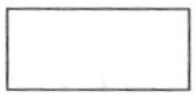
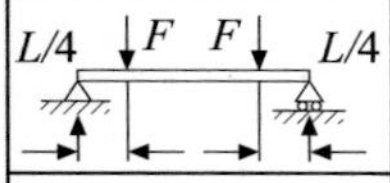		0.96	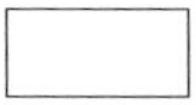
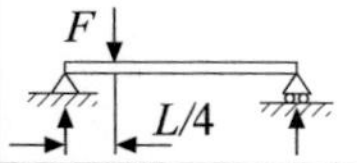		0.69	
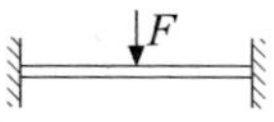		0.59	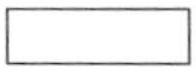
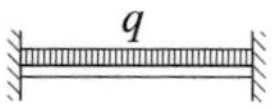		0.39	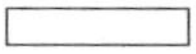

Lateral torsional stability in EC 5

Similar to the k_c values for columns, the strength of beams is also mathematically reduced by a coefficient k_{crit}, which is determined from the relative slenderness $\lambda_{rel,m}$ and the above-defined critical bending stress $\sigma_{m,crit}$. The coefficient k_{crit} was defined on a completely different basis than the coefficient k_c. The values for k_{crit} were derived for elastic systems, homogenous material and fixed eccentricity e of $\ell/500$ (see Annex 4), while to derive k_c, second order plastic theory assuming stochastically dispersed eccentricities, stiffness and strength properties of timber members was applied (see Section D2.1).

The verification thus reads as follows:

$$\sigma_{m,d} \leq k_{crit} \cdot f_{m,d} \tag{D2-24}$$

For $\lambda_{rel,m} \leq 0.75$, the beam is not at risk of lateral torsional buckling:

$$k_{crit} = 1 \tag{D2-25}$$

For $0.75 < \lambda_{rel,m} \leq 1.4$:

$$k_{crit} = 1.56 - 0.75 \cdot \lambda_{rel,m} \tag{D2-26}$$

For $1.4 < \lambda_{rel,m}$:

$$k_{crit} = \frac{1}{\lambda_{rel,m}^2} \tag{D2-27}$$

whereby the relative slenderness $\lambda_{rel,m}$ for bending is revealed by

$$\lambda_{rel,m} = \sqrt{\frac{f_{m,k}}{\sigma_{m,crit}}} \tag{D2-28}$$

The coefficient k_{crit} depending on $\lambda_{rel,m}$ is shown in Figure D2-10. Reference was already made to the similarity with the buckling strength-slenderness ratio curve for columns.

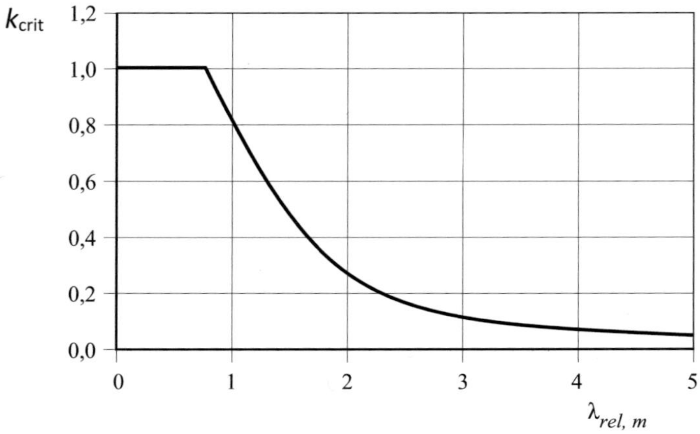

Figure D2-10 Coefficient k_{crit} depending on the relative slenderness $\lambda_{rel,m}$. (STEP 1995 Article B3)

Combined stresses

For beams exposed to a combined load of both a bending moment and a normal (compression) force, a combined verification (equations (D2-7) and (D2-29)) is provided:

$$\frac{\sigma_{c,0,d}}{k_{c,z} \cdot f_{c,0,d}} + \left(\frac{\sigma_{m,y,d}}{k_{crit} \cdot f_{m,y,d}} \right)^2 \leq 1 \qquad \text{(D2-29)}$$

With a simultaneous occurrence of bending moments in *y* and *z*-directions and aspect ratios of $h/b \leq 4$, the following equations apply according to the NA (the coefficient k_m from equations (D1-9) to (D1-14) is not applied in this case):

$$\frac{\sigma_{c,0,d}}{k_{c,z} \cdot f_{c,0,d}} + \left(\frac{\sigma_{m,y,d}}{k_{crit} \cdot f_{m,y,d}} \right)^2 + \frac{\sigma_{m,z,d}}{f_{m,z,d}} \leq 1 \qquad \text{(D2-30)}$$

$$\frac{\sigma_{c,0,d}}{k_{c,y} \cdot f_{c,0,d}} + \frac{\sigma_{m,y,d}}{k_{crit} \cdot f_{m,y,d}} + \left(\frac{\sigma_{m,z,d}}{f_{m,z,d}} \right)^2 \leq 1 \qquad \text{(D2-31)}$$

D2.4 Literature

B.S. Choo, H.J. Blass, original Articles B3, B6, B7, STEP 1995.

Blass H.J. (1987). Tragfähigkeit von Druckstäben aus Brettschichtholz unter Berücksichtigung streuender Einflußgrößen. Dissertation Universität Karlsruhe.

Blass H.J. (1988)a. Traglastberechnung von Druckstäben aus Brettschichtholz. Bauingenieur 63:245-251.

Blass H.J. (1988)b. Einfluß des Kriechens auf die Tragfähigkeit von Holzdruckstäben. Holz als Roh- und Werkstoff 46:405-411.

Glos P. (1978). Zur Bestimmung des Festigkeitsverhaltens von Brettschichtholz bei Druckbeanspruchung aus Werkstoff- und Einwirkungskenngrößen. Dissertation, Technische Universität München.

Hooley R.F. and Madsen B. (1964). Lateral stability of glued laminated beams. Journal of the Structural Division ASCE, 3:201-218.

Kirby P.A. and Nethercot D.A. (1979). Design for structural stability. Constrado Monographs, Crosby Lockwood Staples, Granada Publishing.

Timoshenko S. and Gere J.M. (1961). Theory of elastic stability. McGraw-Hill Book Co. Inc. New York, NY., 2nd Edition.

D3 Influence of volume and stress distribution

Original article: F. Rouger

Observations from numerous tests show that the strength of a material declines as the size of the test specimen increases. This so-called volume or size effect can easily be explained by the fact that as the size of a member increases, the number of defective points (knots, finger joints...) follows suit. Accordingly, the number of potential failure points rises and the point with the lowest ratio value in terms of strength against stress is where failure is likeliest. The volume effect only comes into play for brittle failure mechanisms (tension, shear), since load redistributions may occur as part of ductile (plastic) behaviour (compressive stress), which considerably mitigate the impact of the defective point on strength. Weibull (1939) put forward a theory describing the influence of the stressed member volume and stress distribution within this volume on the strength of homogeneous and isotropic materials with brittle failure. Although wood is neither homogenous nor isotropic, the Weibull theory can also be applied to wood (Colling, 1986). Denzler (2007) also cites further theories, concerning the shortcomings of Weibull theory for wood as a material. For timber engineers, for example, the volume effect has to be taken into consideration in each case of brittle failure, particularly tensile failure perpendicular to the grain. In practical terms, this means the volume effect is crucial in double-tapered beams, pitched cambered beams and in curved beams, since such beam types are prone to high tensile stresses perpendicular to the grain in the apex zone due to their geometry.

D3.1 Theory

The theory of the weakest link was developed by Pierce (1926), Tucker (1927) and Weibull (1939), who examined brittle material such as concrete. This theory states that "A chain exposed to tensile stress is as strong as its weakest link." To explain this theory, we consider a unit volume subject to tension, the failure probability P_f of which is given by:

$$P_f = F(\sigma) = \text{Probability}(\text{strength} \leq \sigma) \qquad \text{(D3-1)}$$

whereby $F(\sigma)$ is the distribution function of strength σ, Figure D3-1 and Figure D3-2 on the left.

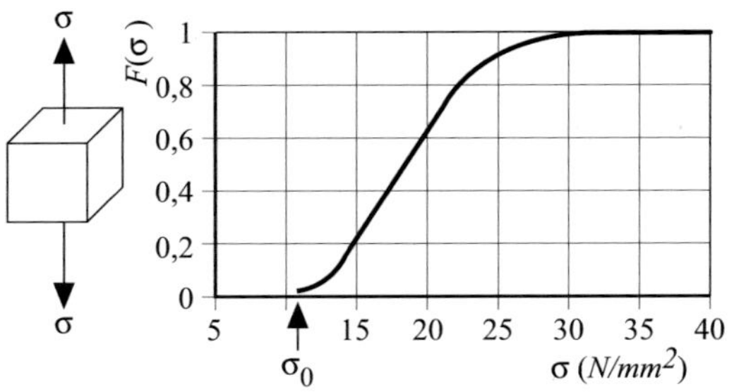

Figure D3-1 Distribution function for the failure of a unit volume. (STEP 1995 Article B1)

If we now consider a chain comprising N links, whereby each link represents a unit volume, this system will only be sustained, if each link survives, namely:

$$P_s = P_s(1) \cdot P_s(2) \cdot \ldots \cdot P_s(N) = [1 - P_f(1)] \cdot [1 - P_f(2)] \cdot \ldots \cdot [1 - P_f(N)] \tag{D3-2}$$

P_s is the probability of survival of the system and $P_s\,(i)$ is the probability of survival of an individual element i. Based on equation (D3-2) and assuming that the unit volumes show the same probabilities of failure and that the occurrences of failure are independent among all unit volumes (the power of which can be expressed with the exponent N), the resulting probability of failure of the system is:

$$P_f = 1 - P_s = 1 - [1 - F(\sigma)]^N = 1 - e^{N \cdot \ln(1 - F(\sigma))} \approx 1 - e^{-N \cdot F(\sigma)} \tag{D3-3}$$

The simplification applies, since $F(\sigma)$ is small.

We now assume that the lower tail of the strength distribution of a unit volume ($F(\sigma)$ from Figure D3-1) may be adapted to an exponential function with a lower limit value σ_0, namely

$$F(\sigma) = \left(\frac{\sigma - \sigma_0}{m}\right)^k \quad \text{mit} \quad \sigma \geq \sigma_0 \tag{D3-4}$$

The probability of failure of the system, namely a structural member comprising multiple infinitesimally small unit volumes dV, can then be expressed as follows:

$$P_f(\sigma) = 1 - e^{-N \cdot \left(\frac{\sigma - \sigma_0}{m}\right)^k} = 1 - e^{\int_V \left(\frac{\sigma - \sigma_0}{m}\right)^k \mathrm{d}V} \tag{D3-5}$$

If we now assume constant stress distribution in the member volume, equation (D3-5) can be simplified as follows:

$$P_f(\sigma)=1-e^{-V\cdot\left(\frac{\sigma-\sigma_0}{m}\right)^k} \tag{D3-6}$$

Equation (D3-6) is known as a three-parameter Weibull distribution. When $\sigma_0 = 0$, a two-parameter Weibull distribution results. Parameters m and k can be estimated from the average value of σ, $E(\sigma)$ and the coefficient of variation of σ, $COV(\sigma)$. Parameter k is also known as a shape parameter and determines the "appearance" of the Weibull distribution. When k = 1 for example, the exponential distribution is obtained and for k = 2 the Rayleigh distribution, while the parameter m is the scale parameter. However, the three-parameter Weibull distribution (D3-6) depends not only on the parameters k and m, but also on volume V. Figure D3-2 shows the distribution and density function of the two-parameter Weibull distribution.

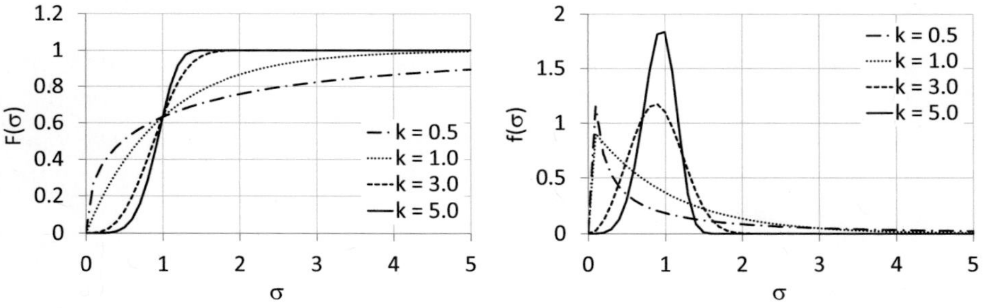

Figure D3-2 Distribution function (left) and density function (right) of a two-parameter Weibull distribution, $\sigma_0 = 0$, $V = m = 1.0$.

The Weibull distribution can hence be used e.g. to determine the dependency of tensile strength on stressed volume, assuming the equivalent probability of failures P_f of two test specimens. For the test specimen of volume V_1 and a given probability of failure $P_f(\sigma_1)$ under tensile stress σ_1 and a second test specimen with volume V_2 and $P_f(\sigma_2)$ under tensile stress σ_2, a comparison of the characteristic strength values of both volumes results in:

$$P_f(\sigma_1)=P_f(\sigma_1)\Rightarrow V_1\cdot\left(\frac{\sigma_1}{m}\right)^k=V_2\cdot\left(\frac{\sigma_2}{m}\right)^k\Rightarrow\frac{\sigma_2}{\sigma_1}=\left(\frac{V_1}{V_2}\right)^{\frac{1}{k}} \tag{D3-7}$$

This equation is the basis for explaining the size effect.

If there are no uniformly dispersed tensile stresses, these equations must be modified as follows for other stress distributions:

$$\sigma(x,y,z)=\sigma\cdot w(x,y,z) \tag{D3-8}$$

Here, σ is the maximum stress in volume V and $w(x,y,z)$ indicates the spatial distribution function of the stress.

The two-parameter Weibull distribution function ($\sigma_0 = 0$) can then be expressed as:

$$P_{\mathrm{f}}(\sigma)=1-\mathrm{e}^{-V^*\cdot\left(\frac{\sigma}{m}\right)^k} \tag{D3-9}$$

whereby V^* is defined as:

$$V^*=\int_V\left(w(x,y,z)\right)^k\mathrm{d}V \tag{D3-10}$$

This method used to calculate the influence of stress distribution was applied by Larsen (1986) and Colling (1986), to estimate the influence of volume and stress distribution on shear and tensile strengths perpendicular to the grain in curved and double-tapered beams. Larsen used the so-called “distribution coefficient”, to record the impact of stress distribution on strength:

$$k_{\mathrm{dis}}=\frac{V}{\left(V^*\right)^{\frac{1}{k}}} \tag{D3-11}$$

Accordingly, the coefficient k_{dis} is used to calculate design tension strengths perpendicular to the grain for different load configurations:

$$f_{\mathrm{t,90,d}}=k_{\mathrm{vol}}\cdot k_{\mathrm{dis}}\cdot f^*_{\mathrm{t,90,d}}\quad\text{with}\quad k_{\mathrm{vol}}=\left(\frac{V_0}{V}\right)^{\frac{1}{k}} \tag{D3-12}$$

whereby $f^*_{\mathrm{t,90,d}}$ equates to a reference volume V_0 under uniform stress (and thus corresponds to nominal strength, which was determined using test specimens with specific geometry and hence a specific volume V_0). Now, however, the other coefficients of the Weibull distribution also have to be determined, particularly the shape parameter k.

D3.2 Research results

Numerous published test data has emerged to explain the size effect for timber, but some of the results are contradictory (Barrett and Lam, 1992; Madsen, 1992), which could be for the following reasons:

- The size effect is explained using the brittle failure theory, which is applicable to tension parallel and perpendicular to the grain (Barrett, 1974; Colling, 1986) and to shear (Foschi and Barrett, 1976; Foschi, 1985; Colling, 1986). For compression and particularly for bending, whereby a combined failure mechanism comprising both compressive and tensile stresses is involved, this theory is debatable.
- The size effect assumes an equivalent probability of failure of the "unit volumes". This assumption may not apply to all wood species, however, particularlyif the knots are not dispersed at random within the wood volume.
- For visually graded timber in particular, the sizes of wood defects rise with increasing member sizes. In other words, the material itself changes in line with dimensions, meaning a pure size effect is distorted. In particular, the size effect can be superimposed when timber of various qualities is used by imposing a "grading effect". Denzler (2007) thus makes a distinction between direct size effects (dimensions) and indirect (knot area sizes). The combination of both effects as an effective and measurable size effect depends significantly on grading (grading influences knot area sizes) and testing conditions.
- If tests are conducted for constant span to depth ratios in bending, the size effect is a combination of length and depth effect (Barrett and Fewell, 1990). Both effects cannot be identified separately.

Above all, to take the strong anisotropy of the wood into consideration, the size effect according to Weibull was subdivided into respective length and depth effects. In Table D3-1, the following parameters are listed for the size effect when bending stress is applied.

The length factor S_L (namely members of the same height $h_1 = h_2$) results from:

$$\frac{\sigma_2}{\sigma_1} = \left(\frac{L_1}{L_2}\right)^{\frac{1}{k_L}} = \left(\frac{L_1}{L_2}\right)^{S_L} \qquad \text{(D3-13)}$$

The depth factor S_h (namely members of the same span $L_1 = L_2$) results from:

$$\frac{\sigma_2}{\sigma_1} = \left(\frac{h_1}{h_2}\right)^{\frac{1}{k_h}} = \left(\frac{h_1}{h_2}\right)^{S_h} \qquad \text{(D3-14)}$$

The size factor S_R (namely members with a constant span to depth ratio $L_1/h_1 = L_2/h_2$) results from:

$$\frac{\sigma_2}{\sigma_1} = \left(\frac{L_1}{L_2}\right)^{S_L} \cdot \left(\frac{h_1}{h_2}\right)^{S_h} = \left(\frac{h_1}{h_2}\right)^{S_L} \cdot \left(\frac{h_1}{h_2}\right)^{S_h} = \left(\frac{h_1}{h_2}\right)^{S_L + S_h} = \left(\frac{h_1}{h_2}\right)^{S_R} \qquad \text{(D3-15)}$$

Additional test results for glued laminated timber with a far lower sample size were published by Ehlbeck and Colling (1990). From simulations on the bending strength of glued laminated timber, a value S_R of 0.12 (Frese and Blass, 2009) was determined. The size effect for glued laminated timber is smaller than for solid timber, since ultimately, the so-called "lamination effect" increases strength (lamination effect see Section B6.2).

The results for tension differ slightly from those for bending, since a brittle behaviour prevails in the first case, see also Table D3-2. Based on simulations, in turn, conducted on glued laminated timber subject to tension, the resulting length factor is $S_L = 0.11$. (Frese et al., 2010).

Table D3-1 Size factors for bending.

	S_L	S_h	S_R
Barrett and Lam (1992), solid timber	0.17	0.23	0.40
Madsen (1992), solid timber	0.20	0.00	0.20
Ehlbeck and Colling (1990), glulam	0.15	0.15	0.30

Table D3-2 Size factors for tension, solid timber.

	S_L	S_h	S_R
Barrett and Lam (1992)	0.17	0.23	0.40
Madsen (1992)	0.20	0.10	0.30

Table D3-1, Table D3-2 and the simulation results of Frese show that size factors of solid timber have a value of around 0.2 (corresponding to a shape parameter k of 5) and are somewhat lower for glulam, around 0.1. Both these values are echoed in the height factor k_h, which was already addressed in Section D1.1 and which is applied when establishing strength values for tension parallel to the grain and bending.

Although the results differ, they still prove the size effect (→ k_h, k_{vol}) for many types of stress with an additional stress distribution factor (→ k_{dis}), which may be in the same order of magnitude. For the purpose of practical application, these approximations, particularly influences from stress distribution, were further simplified.

D3.3 Size and stress distribution effects in accordance with EC 5

Size effects are taken into consideration by modifying the characteristic strength values determined in EN 338. The characteristic values for bending and tensile strength are based on a reference height of 150 mm for solid timber and 600 mm for glued laminated timber[3]. For depths less than these reference values, strength values are multiplied by a size factor, which is limited by an upper value. This means that the size effect need only be used in one direction (Figure D3-3) as was already explained in the previous section and in Section D1.1.

For solid timber:

$$k_h = \min\begin{cases} \left(\dfrac{150}{h}\right)^{0.2} \\ 1.3 \end{cases} \tag{D3-16}$$

For glued laminated timber:

$$k_h = \min\begin{cases} \left(\dfrac{600}{h}\right)^{0.1} \\ 1.1 \end{cases} \tag{D3-17}$$

where h represents the beam depth.

[3] Namely, the test specimens, which were used to determine the characteristic strengths, were 150 mm or 600 mm in height.

Since the size effect depends significantly on the grading and test conditions, equations (D3-16) and (D3-17) remain under discussion (in DIN 1052:2008, for example, there was no k_h for solid timber). Moreover, no purely brittle behaviour can be observed for bending members. Accordingly, it is very difficult to determine a pure size effect in isolation. Based on the type of grading and classification of solid timber used in Europe, the scope of the size effect is already partially included in the characteristic values determined.

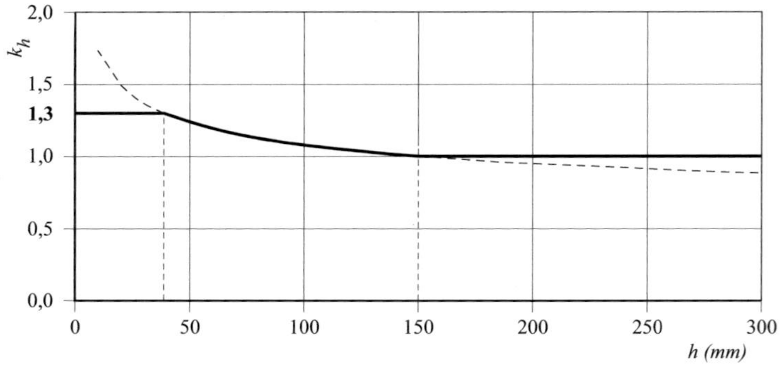

Figure D3-3 Size factor for solid timber in bending or tension in accordance with EC 5 (solid line) and with theory (dashed line). (STEP 1995 Article B1)

When designing double-tapered and curved beams above all, it is indisputably crucial to take a volume and stress distribution effect in the apex zone into consideration, since this is where, based on geometry, high tensile stresses perpendicular to the grain frequently occur with a simultaneous large volume. The following requirement must be met in the apex zone for beams made of glulam or LVL (see also equations (D3-12) and (D4-15)):

$$\sigma_{t,90,d} \le k_{dis} \cdot k_{vol} \cdot f_{t,90,d} = k_{dis} \cdot \left(\frac{V_0}{V}\right)^{0.2} \cdot f_{t,90,d} \qquad \text{(D3-18)}$$

where

k_{dis} Stress distribution factor (see also equation (D3-11)), which is determined as follows:
k_{dis} = 1.4 for double-tapered and curved beams
k_{dis} = 1.7 for pitched cambered beams

k_{vol} Volume factor, k_{vol} = 1.0 for solid timber

V_0 Reference volume of 0.01 m^3

V Volume subject to tension perpendicular to the grain in the apex zone, for definition see Figure 6.9 in EC 5

Another interesting point here is the fact that the verification in equation (D3-18) adopted a slightly different format in DIN 1052:

$$\sigma_{t,90,d} \leq k_{dis} \cdot \left(\frac{h_0}{h_{ap}}\right)^{0.3} \cdot f_{t,90,d} \qquad \text{(D3-19)}$$

In DIN 1052 therefore, with $(h_0/h_{ap})^{0.3}$ a depth effect had been taken into consideration and not a volume effect as in EC 5 with $(V_0/V)^{0.2}$, which also explains the different shape parameter $k = 10/3$ in DIN 1052. In DIN 1052, the coefficient for the distribution of tensile stresses perpendicular to the grain k_{dis}, the reference depth h_0, the exponent 0.3 and the design value of tensile strength perpendicular to the grain $f_{t,90,d}$ were coordinated with each other to ensure any probability of failure would be sufficiently small by comparison with the maximum value $\sigma_{t,90,d}$. The reasons why a depth effect was taken into consideration rather than a volume effect can be attributed to the easier handling, since the beam depth h_{ap} in the apex zone is easy to establish, whereas the volume V has to be determined first.

D3.4 Literature

F. Rouger, original Article B1, STEP 1995.

Barrett J.D. and Fewell A.R. (1990). Size factors for the bending and tension strength of structural lumber. Paper 23-10-3, CIB-W18 Meeting 23, Lisbon.

Barrett J.D. and Lam F. (1992). Size effects in visually graded softwood structural lumber. Paper 25-6-5, CIB-W18 Meeting 25, Ahus.

Barrett J.D. (1974). Effect of size on tension perpendicular to grain strength of Douglas fir. Wood and Fiber 6(2):126-143.

Colling F. (1986). Influence of volume and stress distribution on the shear strength and tensile strength perpendicular to grain. Paper 19-12-3, CIB-W18 Meeting 19, Florence.

Denzler J.K. (2007). Modellierung des Größeneffektes bei biegebeanspruchtem Fichtenschnittholz. Dissertation, Technische Universität München.

Ehlbeck J. and Colling F. (1990). Bending strength of glulam beams, a design proposal. Paper 23-12-1, CIB-W18 Meeting 23, Lisbon.

Foschi R.O. (1985). Longitudinal shear design of glued laminated beams. Paper 18-10-2, CIB-W18, Beit Oren.

Foschi R.O. and Barrett J.D. (1975). Longitudinal shear strength of Douglas Fir. Canadian Journal of Civil Engineering, 3(2):198-208.

Frese M. and Blass H.J. (2009). Bending strength of spruce glulam. European Journal of Wood and Wood Products 67:277-286.

Frese M., Chen Y. and Blass H.J. (2010). Tensile strength of spruce glulam. European Journal of Wood and Wood Products 68:257-265.

Larsen H.J. (1986). EC 5 and CIB structural timber design code. Paper 19-102-2, CIB-W18 Meeting 19, Florence.

Madsen B. (1992). Structural behaviour of timber. Timber Engineering Ltd., Vancouver.

Pierce F.T. (1926). Tension tests for cotton yarn. Journal of the Textile Institute, S. T155-T368.

Tucker J. (1927). A study of compressive strength dispersion of material with applications. Journal of the Franklin Institute 204:751-781.

Weibull W. (1939). A statistical theory of the strength of materials. Royal Swedish Institute for Engineering Research N. 141, 45p.

D4 Members with variable cross-sections or curved shapes

Original articles: J. Ehlbeck, J. Kürth, H. J. Larsen

Glued laminated timber beams are often realised with varying depths and with or without curvature, in line with architectural perspectives, due to gabled roofs or the need to maximise interior space or reduce the height of exterior walls at supports. The most frequently used beam shapes are single and double-tapered beams, curved beams of constant depth, fish-bellied beams and pitched cambered beams (see Figure D4-1) and here, fish-bellied, double-tapered and pitched cambered beams optimise the use of materials, since their depth follows the moment diagram. Many beam shapes also offer the benefit of allowing secondary components or roof diaphragms to be directly applied while eliminating the need for woodblocks to make a minimum roof slope. The incline should not generally exceed 10° even though the limit value specified in the NA is unrealistically high at 24°.

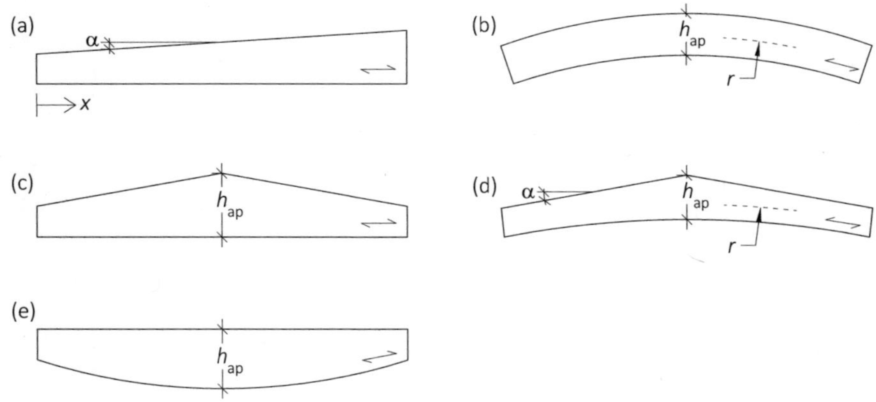

Figure D4-1 (a) Single-tapered beam, (b) curved beam of constant depth, (c) double-tapered beam, (d) pitched cambered beam, (e) fish-bellied beam.

Single-tapered beams

The distribution of bending stresses in areas with variable beam depth $h(x)$ is non-linear already due to the beam geometry, since the slope of the beam edge results in additional shear stresses τ_{zx} and stresses perpendicular to the grain σ_Z, see Figure D4-2, which trigger an increase in bending stress at the edge parallel to the grain and thus an overall non-linear distribution as is also shown in Figure D4-3. For small angles of inclination, α up to around $\alpha \leq 10°$, however, this non-linearity is small (for derivations, see Petersen (1993), Section 26.7). Accordingly, the bending stresses in the outer fibres can be determined more simplistically in accordance with the Euler-Bernoulli beam theory.

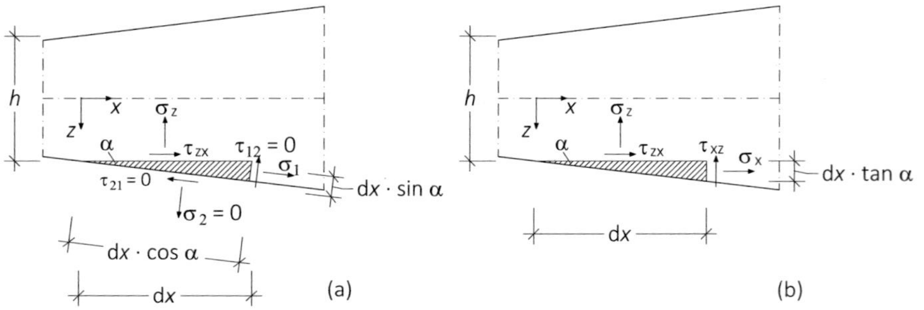

Figure D4-2 Stresses at tapered edge; (a) cut normal to edge and horizontal, (b) cut vertical and horizontal (according to Petersen 1993, Figure 57)

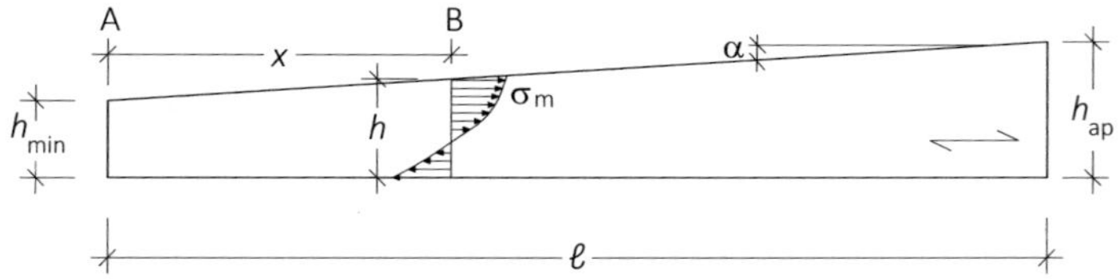

Figure D4-3 Single-tapered beam with beam width b. (STEP 1995 Article E4)

In addition to the geometry, the anisotropy of the wood also has an impact on stress distribution, although this impact cannot be properly recorded with the Euler-Bernoulli beam theory. Nevertheless, an approximate value can still be calculated for the same based on anisotropic plate theory and taking into consideration the ratios E_0/E_{90} and E_0/G and Poisson's ratio (Blumer, 1979).

The third parameter is the fact that areas with cut wood fibres (at the tapered edge) exist due to the beam shape and production methods. Since the area of the tapered edge is, additionally to the axial stresses, also subject to tranverse and shear stresses σ_Z and τ_{zx} depending on the taper angle α (Figure D4-2), this means the activation of very low stiffness and strength values at the tapered edge when normal stress perpendicular to the grain and shear are applied, see Figure D4-4. Tensile stresses perpendicular to the grain form at a tapered tensile edge and compressive stresses perpendicular to the grain at a tapered compressive edge. As a general rule, cut wood fibres should preferably be arranged in the compression zone of beams, so that the lamellae run parallel to the beam edge in the tensile zone. During design, this stress combination is taken into consideration by reducing the design values for bending strength f_m by a factor $k_{m,\alpha}$. Moreover, for beams with variable cross-sectional depth, it is important to identify the critical cross-section, in which bending stresses peak. The strength of the beam in Figure D4-3 must hence be verified in two cross-sections: in cross-section A, in which the shear stresses peak and cross-section B, in which the bending stresses peak.

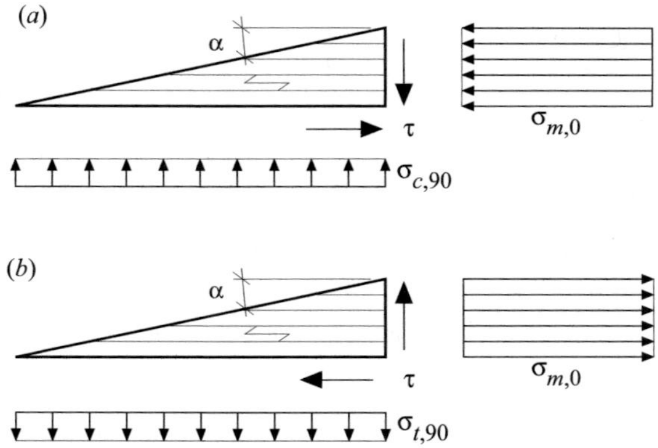

Figure D4-4 Axial stresses (parallel to the grain, longitudinal),stresses perpendicular to the grain and shear stresses at the tapered edge with cut wood fibres with (*a*) compressive bending stresses and (*b*) tensile bending stresses. (STEP 1995 Article B8)

Double-tapered beams

Double-tapered beams (Figure D4-5) represent an expansion of single-tapered beams, but include an apex, at which normal stresses must be deflected. This results in high tensile stresses perpendicular to the grain, which cannot be ignored. For double-tapered beams, this means that as for single-tapered beams, the strength for two cross-sections must be checked (cross-sections A and B, see Figure D4-5). In addition, the tensile stress perpendicular to the grain and bending stress in the apex cross-section must also be checked. This bending stress depends on the taper angle α, as calculations using anisotropic plate theory have shown (Blumer, 1979). Both additional stress components must also be verified for curved beams; the formation of which is explained for both beam types in the following section.

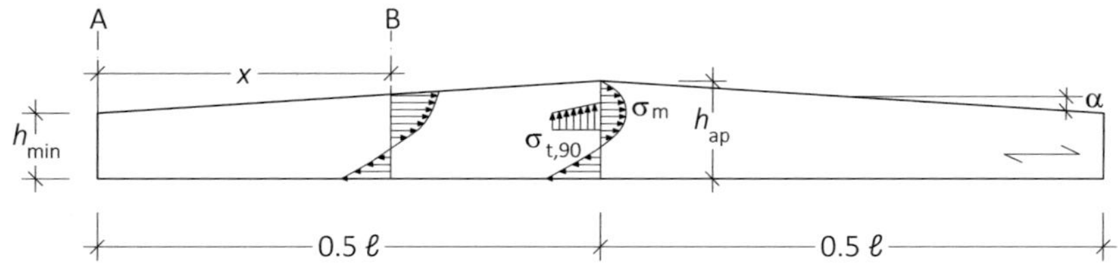

Figure D4-5 Double-tapered beam with beam width *b*. (STEP 1995 Article E4)

Curved beams

Unlike single and double-tapered beams, curved beams with a constant cross-section have no tapered edge (and thus no cut wood fibres) and no variable depth. However, here also the bending stresses are non-linear, regardless of the material used and, in turn, ultimately based on the beam shape. For clarification, Figure D4-6 shows the distribution of bending stress at a section of a curved beam. Since the internal fibres are considerably shorter than the external fibres, the strains at the edges are as follows, assuming the Bernoulli hypothesis that plane sections remain plane applies and that the neutral axis is going through the centre of gravity:

$$\varepsilon_{\mathrm{i}} = \frac{\Delta d \cdot \ell_{\mathrm{i}}}{d \cdot \ell_{\mathrm{i}}} > \frac{\Delta d \cdot \ell_{\mathrm{o}}}{d \cdot \ell_{\mathrm{o}}} = \varepsilon_{\mathrm{o}} \qquad \text{(D4-1)}$$

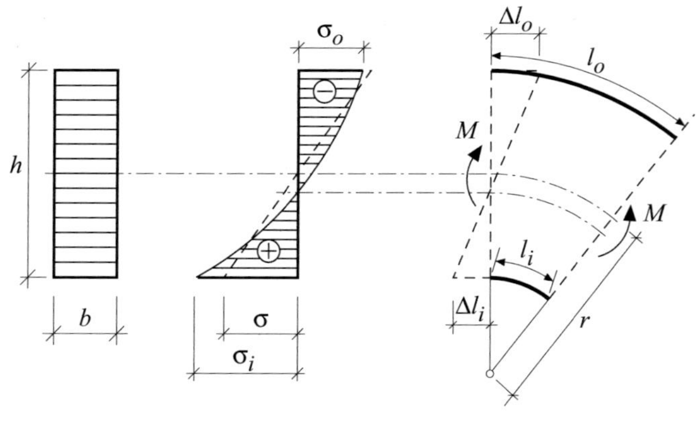

Figure D4-6 Distribution of bending stresses in a curved beam. (STEP 1995 Article B8)

Consequently, Hooke's law states that the maximum bending stress $|\sigma_i|$ at the inner edge exceeds the same at the outer edge $|\sigma_o|$. To maintain the equilibrium of internal forces in the cross-section, the neutral axis must shift from the centre of gravity position towards the inner edge resulting in a non-linear, hyperbolic stress distribution. The maximum bending stress in a symmetrical single-span girder under a uniformly distributed load occurs at the inner edge of the apex cross-section, while the extent and distribution of the maximum bending stress depends on the curvature (Blumer, 1979). As with beams of varying depth, the distribution of bending stresses is also influenced by the anisotropy of the wood, which means the bending stresses increase at the inner edge (Blumer, 1979). For curved beams, the rule is that the bending stresses in the apex cross-section must be checked, the size of which varies depending on the radius of curvature. The bending stress in the apex can be approximately determined (Blumer, 1975 and 1979) from M_{ap}/W_{ap} („ap" = apex) multiplied by a coefficient k_ℓ ($k_\ell > 1$, see Figure D4-7 on the left), which depends on the ratio of the apex height h_{ap} to the radius of curvature r and for beams where the depth is not constant (double-tapered beams), additionally on the taper angle α. For curved beams of constant depth, $\alpha = 0$.

However, bending moments in curved and apex areas of double-tapered beams also generate stresses perpendicular to the grain. Figure D4-8 shows the apex area of a curved beam subject to a constant moment and specifying the tensile stresses perpendicular to the grain forming. These tensile stresses also form due to climate-related stresses, for example changes in moisture content caused by variable environmental conditions.

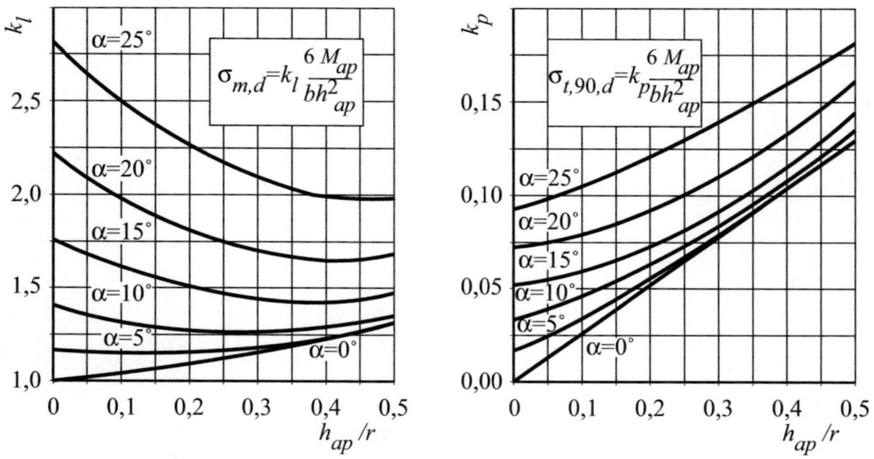

Figure D4-7 Coefficients k_ℓ (influence of curvature or taper angle on bending stress, left) and k_p (influence of curvature or taper angle on tension perpendicular to the grain, right) for different curvatures h_{ap}/r and taper angles α. (STEP 1995 Article B8)

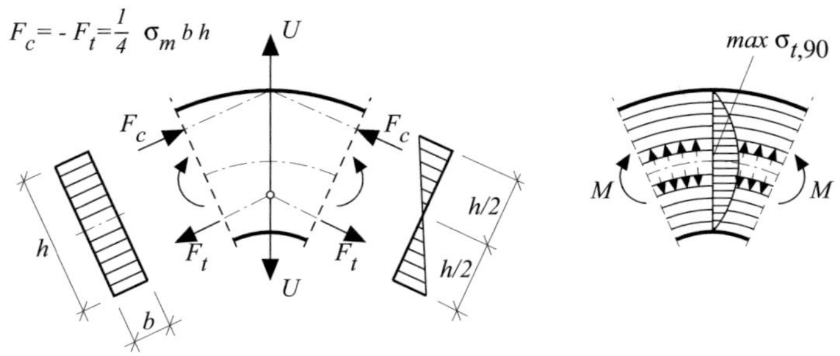

Figure D4-8 Tension perpendicular to the grain caused by a constant moment. (STEP 1995 Article B8)

Taking the simplified assumption of a linear longitudinal stress distribution, it is easy to show that the resulting tensile and compressive forces, F_t and F_c, generate a deviation force U in a radial direction. If the bending moment has a tendency to increase the radius of curvature, this causes tensile stresses perpendicular to the grain to form. The extent of these tensile stresses depends on the radius of curvature or the taper angle of the upper chord and the stresses involved cannot be ignored. Given the generally large apex volume in curved and double-tapered beams and the fact that tensile stresses perpendicular to the grain cause brittle failure, the influence of volume and stress distribution on strength must be taken into consideration, see Article D3. The maximum tensile stress perpendicular to the grain max $\sigma_{t,90}$ in the apex cross-section, can, like bending stress, be approximately calculated from M_{ap}/W_{ap} multiplied by a coefficient k_p ($k_p < 1$, see Figure D4-7 on the right). In addition to the stresses from external loads, for curved beams with small radii of curvature, the stresses generated prior to bonding from the bending of individual lamellae must be taken into consideration. In a board of thickness t, curvature $r_{in} / t = 240$ and a modulus of elasticity of $E_0 = 10000$ N/mm^2, the theoretical bending stress would equate to:

$$\sigma_m = \frac{E_i \cdot t}{2 \cdot r_{in}} = \frac{10000}{2 \cdot 240} = 20.8\,\frac{\mathrm{N}}{\mathrm{mm}^2} \tag{D4-2}$$

However, since such stresses are partially dissipated by plastic processes and relaxation, they only need to be taken into consideration for small curvature radii. This is done by reducing the design values of bending strength by k_r.

Pitched cambered beams

This beam shape combines all the properties discussed to date. As well as a variable depth and tapered edge with cut wood fibres, pitched cambered beams also include a curvature of the bottom chord. All the verifications performed to date must therefore be applied.

Conclusion

The distribution of bending stress in beams with variable cross-sections or a curved shape is non-linear; one of the reasons for which is the beam geometry, which is intensified by the anisotropic properties of the wood. Moreover, stress interactions must also be taken into consideration for tapered edges with cut wood fibres, since this is where additional shear stresses and stresses perpendicular to the grain form and the respectively assigned stiffnesses and strengths of wood are low. The following verifications must be made and are shown in the following sections:

- Verification in the critical cross-section with reduction $k_{m,\alpha}$ for the tapered edge with cut wood fibres:
  For beams with variable cross-section; namely single and double-tapered beams and pitched cambered beams.
- Verification of bending stress in the apex cross-section:
  For double-tapered beams with reduction k_ℓ depending on the taper angle α of the upper chord, for curved beams with reductions k_ℓ and k_r depending on the curvature, pitched cambered beams like curved beams, although k_ℓ additionally depends on the taper angle α of the upper chord.
- Verification of tensile stresses perpendicular to the grain in the apex cross-section:
  For double-tapered beams with reduction k_p depending on the taper angle α of the upper chord, for curved beams with reduction k_p depending on the curvature, for pitched cambered beams with reduction k_p depending on the taper angle α of the upper chord and curvature. In addition, when verifying tensile stresses perpendicular to the grain, the volume effect must be taken into consideration via k_{dis} and k_{vol}, see also equation (D3-18).

D4.1 Single-tapered beams

Single-tapered beams, like that shown in Figure D4-9, must be verified at the cross-section where bending stress peaks, while the stresses perpendicular to the grain and shear stresses which occur at a tapered edge are taken into consideration using a reduction coefficient $k_{m,\alpha}$.

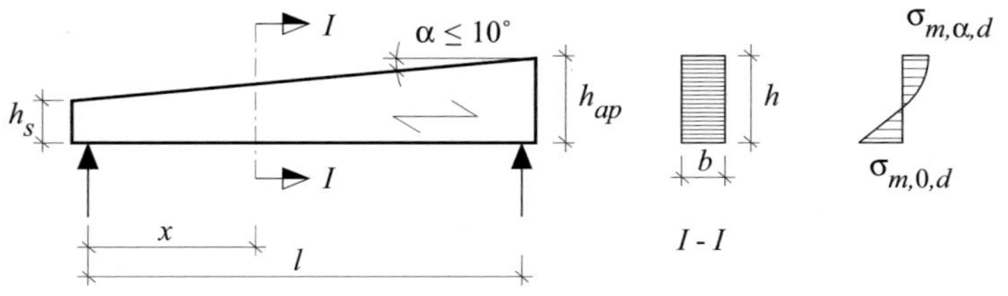

Figure D4-9 Single-tapered beam with critical cross-section I-I. (STEP 1995 Article B8)

Verification of bending stresses in the critical cross-section

The maximum stress acts at a point x, at which $\partial\sigma/\partial x = 0$. In the case of a uniformly distributed load, the following applies for x:

$$x = \frac{\ell}{1 + {}^{h_{ap}}\!/_{h_s}} \tag{D4-3}$$

If other load configurations apply or if the characteristic bending strength value is increased by the factor k_h for beam depths under 600 mm, the maximum value is determined by calculating the stresses in various cross-sections, by proceeding step by step along the beam, from the point of maximum bending moment in the direction in which the cross-sectional area declines.

The stresses in cross-section I-I (Figure D4-9) are determined as follows:

$$\sigma_{m,\alpha,d} = \sigma_{m,0,d} = \frac{6 \cdot M_d}{b \cdot h^2} \tag{D4-4}$$

Equation (D4-4) shows that EC 5 equates the bending stresses in the outer fibres at the straight and tapered edges as M/W simplistically. As was explained in the previous section, this simplified verification is still permissible for the inclination angles encountered in timber structures.

The following condition must be met for the edge with the cut wood fibres:

$$\sigma_{m,\alpha,d} \leq k_{m,\alpha} \cdot f_{m,\alpha,d} \quad \text{(D4-5)}$$

The stress combinations at the edge with cut wood fibres are taken into consideration using the factor $k_{m,\alpha}$ (see also the Hankinson equation (D1-1) and equation (6) in Annex 2). For tensile stresses along the edge with cut wood fibres, the following applies for $k_{m,\alpha}$:

$$k_{m,\alpha} = \frac{1}{\sqrt{1+\left(\frac{f_{m,d}}{f_{t,90,d}}\right)^2 \cdot \tan^4\alpha + \left(\frac{f_{m,d}}{0.75 \cdot f_{v,d}}\right)^2 \cdot \tan^2\alpha}} \quad \text{(D4-6)}$$

For compressive stresses along the edge with cut wood fibres, the following applies ($f_{t,0,d}$ is replaced by $f_{c,0,d}$ and the shear strength is increased due to the simultaneous, favourably acting compression perpendicular to the grain):

$$k_{m,\alpha} = \frac{1}{\sqrt{1+\left(\frac{f_{m,d}}{f_{c,90,d}}\right)^2 \cdot \tan^4\alpha + \left(\frac{f_{m,d}}{1.5 \cdot f_{v,d}}\right)^2 \cdot \tan^2\alpha}} \quad \text{(D4-7)}$$

In addition, shear stress and compression perpendicular to the grain at the supports must be verified in cross-section A (see Figure D4-3) (see Article D1).

D4.2 Curved, double-tapered, pitched cambered and fish-bellied beams

The verifications shown for single-tapered beams, namely shear stress and compression perpendicular to the grain at the supports and bending stresses in the critical cross-section, must also be conducted for double-tapered, pitched cambered, fish-bellied and curved beams, whereby the depth of curved beams tends not to vary, meaning the critical cross-section for verifying bending stress is usually at the point where the bending moment peaks. As well as verifying bending stress at the critical cross-section, for double-tapered and pitched cambered beams, the bending strength and tensile strength perpendicular to the grain must also be verified in the apex cross-section. The reason for the tensile stress verification was already shown in Figure D4-8, for which the volume effect must also be taken into consideration. Figure D4-10 sets out important geometric details, which are used in the following.

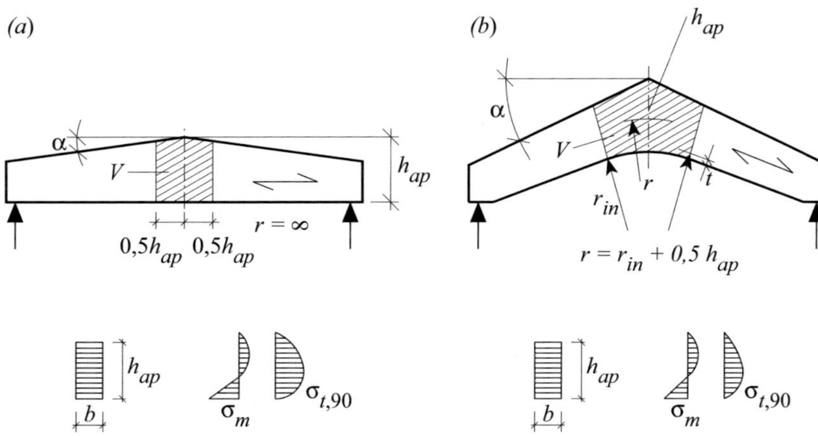

Figure D4-10 Geometric details and stress distribution at apex for (*a*) double-tapered beams and (*b*) pitched cambered beams. (STEP 1995 Article B8)

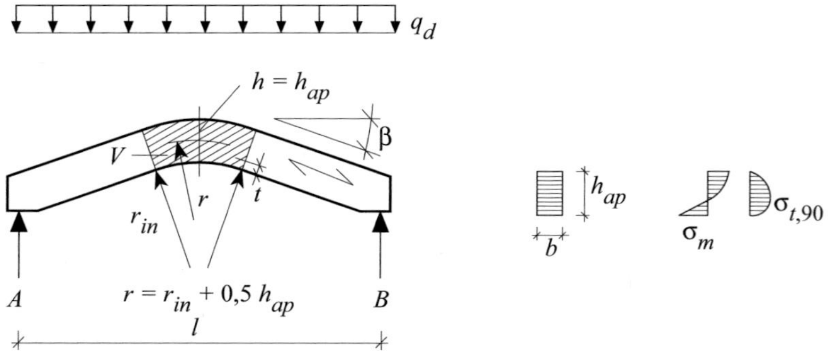

Figure D4-11 Geometric details and stress distribution at apex for curved beams of constant cross-section subject to uniformly distributed load. (STEP 1995 Article B8)

Verification of bending stresses at apex

In the apex cross-section, the design value of the maximum bending stress can be calculated as per:

$$\sigma_{\mathrm{m,d}} = k_\ell \cdot \frac{M_{\mathrm{ap,d}}}{W_{\mathrm{ap}}} = k_\ell \cdot \frac{6 \cdot M_{\mathrm{ap,d}}}{b \cdot h_{\mathrm{ap}}^2} \tag{D4-8}$$

Coefficient k_ℓ is (see also Figure D4-7 on the left)

$$k_\ell = k_1 + k_2 \cdot \left(\frac{h_{ap}}{r}\right) + k_3 \cdot \left(\frac{h_{ap}}{r}\right)^2 + k_4 \cdot \left(\frac{h_{ap}}{r}\right)^3 \quad \text{(D4-9)}$$

where

$k_1 = 1 + 1.4 \cdot \tan\alpha_{ap} + 5.4 \cdot \tan^2 \alpha_{ap}$

$k_2 = 0.35 - 8 \cdot \tan\alpha_{ap}$

$k_3 = 0.6 + 8.3 \cdot \tan\alpha_{ap} - 7.8 \cdot \tan^2 \alpha_{ap}$

$k_4 = 6 \cdot \tan^2 \alpha_{ap}$

$r = r_{in} + 0.5 \cdot h_{ap}$

The coefficient k_ℓ is reduced for double-tapered beams to the factor k_1, which is solely dependent of the taper angle α of the upper chord. All factors were determined using regression equations based on the work of Blumer (1979).

The verification is established as follows:

$$\sigma_{m,d} \le k_r \cdot f_{m,d} \quad \text{(D4-10)}$$

The coefficient k_r required in the verification has already been addressed and takes into consideration the stresses generated prior to bonding due to the bending of individual lamellae for curved beams with small radii of curvature. It is determined as follows, where t = lamellae thickness:

For $r_{in}/t \ge 240$:

$$k_r = 1 \quad \text{(D4-11)}$$

For $150 \le r_{in}/t < 240$:

$$k_r = 0.76 + 0.001 \cdot \frac{r_{in}}{t} \quad \text{(D4-12)}$$

Verification of tension perpendicular to the grain at apex

In the apex cross-section, the design value of the maximum tensile stress perpendicular to the grain can be calculated as per:

$$\sigma_{t,90,d} = k_p \cdot \frac{M_{ap,d}}{W_{ap}} = k_p \cdot \frac{6 \cdot M_{ap,d}}{b \cdot h_{ap}^2} \tag{D4-13}$$

Coefficient k_p is (see also Figure D4-7 on the right):

$$k_p = k_5 + k_6 \cdot \left(\frac{h_{ap}}{r} \right) + k_7 \cdot \left(\frac{h_{ap}}{r} \right)^2 \tag{D4-14}$$

where

$k_5 = 0.2 \cdot \tan\alpha_{ap}$

$k_6 = 0.25 - 1.5 \cdot \tan\alpha_{ap} + 2.6 \cdot \tan^2 \alpha_{ap}$

$k_7 = 2.1 \cdot \tan\alpha_{ap} - 4 \cdot \tan^2 \alpha_{ap}$

As with coefficient k_ℓ, the equation for double-tapered beams is reduced to the factor k_5, while the relevant factors, in turn, are determined by regression equations based on the work of Blumer (1979).

The verification is established as follows (see also equation (D3-18)), whereby the factors k_{dis} and k_{vol} to be used here take the influence of stress distribution (dis = distribution) and volume into consideration, see Article D3:

$$\sigma_{t,90,d} \le k_{dis} \cdot k_{vol} \cdot f_{t,90,d} \tag{D4-15}$$

The factor k_{dis} takes into consideration the distribution of tensile stresses perpendicular to the grain in the stressed volume. The factor k_{dis} is assumed to be 1.4 for double-tapered and curved beams and 1.7 for pitched cambered beams.

The factor k_{vol} is the quotient from the reference volume $V_0 = 0.01$ m^3 and volume V subject to tensile stress perpendicular to the grain and takes into consideration the influence of volume on tensile strength perpendicular to the grain. However, no more than 2/3 of the entire beam volume V_b must be mathematically applied. k_{vol} for solid timber:

$$k_{vol} = 1 \tag{D4-16}$$

For glulam and LVL:

$$k_{\text{vol}} = \left(\frac{V_0}{V}\right)^{0.2} \tag{D4-17}$$

If shear and tensile stress perpendicular to the grain apply simultaneously, a combined verification is performed with linear stress interaction:

$$\frac{\tau_{\text{d}}}{f_{\text{v,d}}} + \frac{\sigma_{\text{t,90,d}}}{k_{\text{dis}} \cdot k_{\text{vol}} \cdot f_{\text{t,90,d}}} \leq 1 \tag{D4-18}$$

The verifications for tensile stresses perpendicular to the grain in the apex only apply to non-reinforced beams, while reinforced beams are covered in Article D8.

D4.3 Climate action on curved, double-tapered and pitched cambered beams

The specified tensile stress verifications perpendicular to the grain (equations (D4-15) and (D4-18)) for the apex area of curved, double-tapered and pitched cambered beams only apply for non-reinforced beams. All verifications are performed with design loads; stresses due to swelling and shrinking processes, which, in turn, are triggered by climatic fluctuations are not taken into consideration. However, the beam shapes shown are very vulnerable to tensile stresses perpendicular to the grain, which may also be triggered by climate change and which may exert additional strain on the beams, additionally to the pre-existing external forces. Accordingly, the NA recommends that such beams should **always** be reinforced to take account of additional climate-related tensile stresses perpendicular to the grain. For reinforced beams, the verifications for reinforced members, Article D8, apply. Finally, according to the NA, double-tapered beams should be reinforced from a utilisation factor of 80% in equations (D4-15) and (D4-18).

D4.4 Literature

J. Ehlbeck, J. Kürth, H.J. Larsen, Original Articles B8, E4, STEP 1995.

Blumer H. (1975). Spannungsberechnung an Brettschichtholz mit gekrümmter Längsachse und veränderlicher Trägerhöhe. Holzbau (Zürich), (6):158-161; (7):191-194; (8):235-237.

Blumer H. (1979). Spannungsberechnung an anisotropen Kreisbogenscheiben und Satteldachträgern konstanter Dicke. Veröffentlichung des Lehrstuhls für Ingenieurholzbau und Baukonstruktionen, Universität Karlsruhe.

Petersen C. (1993). Stahlbau: Grundlagen der Berechnung und baulichen Ausbildung von Stahlbauten. 3. überarbeitete und erweiterte Auflage. Vieweg Verlag Braunschweig Wiesbaden.

D5 Non-reinforced notches and holes

Original article: P. J. Gustafsson

Figure D5-1 shows beams with various notches and holes. Notches or holes can considerably reduce the load-bearing capacity of a member and should therefore be avoided in the construction as early as the design stage. One way this can be done is using truss structures instead of solid girders with large depths, although it is not always possible to eliminate notches. For example, when timber beam flooring has to be brought to the desired level, a clearance height or a fit between members has to be ensured. Old timber constructions in particular tend to use many different notches and recesses for installing joints, while large holes in glued laminated timber members are required, e.g. to accommodate ventilation pipes.

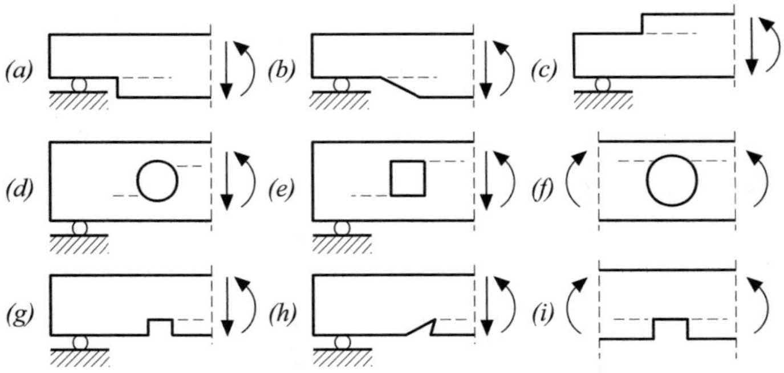

Figure D5-1 Notched beams and beams with holes, the dashed lines indicate probable crack propagation paths. (STEP 1995 Article B5)

A crack may propagate from a notch or a hole along the dashed lines and parallel to the grain (Figure D5-1). Such failure is brittle and happens suddenly, without any large deformations or visible signs beforehand. Depending on the beam geometry, a crack which progresses swiftly along the beam axis may trigger the complete failure of the beam in question.

The cracks are generally initiated by a combination of shear and tensile stresses perpendicular to the grain, which may become excessive at the notch tip. According to the theory of elasticity, the stresses at the tip of a sharp notch are even infinite (Figure D5-2). In this case, the extent of the stress remains undefined at the point of singularity. In reality, however, infinite stress is not possible, since the strength of all material is finite. The localised failure of the material results in a stress distribution at the time of crack growth as shown schematically in Figure D5-2 by the dashed line.

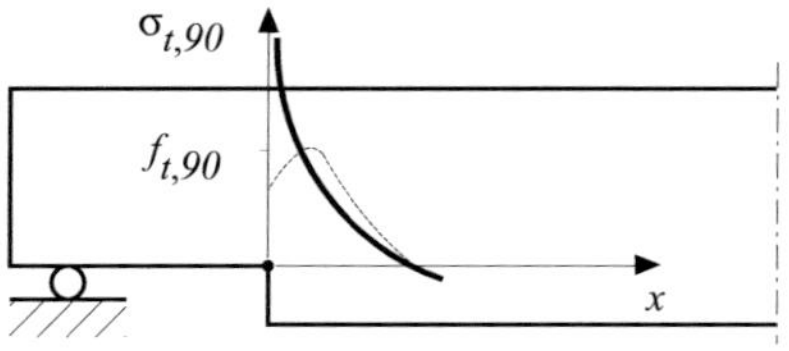

Figure D5-2 Stress distribution at the notch tip in accordance with the (linear) theory of elasticity (solid line) and assumed actual stress distribution (dashed line). (STEP 1995 Article B5)

Moisture content changes can generate considerable additional stresses, compounding the pre-existing localised high stresses imposed by external forces, which may already trigger the propagation of a crack in a notch or hole. The fact that the end grain surface is exposed in a notch or hole can accelerate the rate of localised moisture changes considerably. To reduce the risk of drying-related failure, these end grain surfaces can be coated, to prevent rapid moisture exchange. The general recommendation to avoid notches and holes is particularly relevant under fluctuating climatic conditions.

Currently, EC 5 only regulates beams with notches, rules for beams with holes are specified in the NA and the backgrounds vary for both verifications. Whereas the verifications for notched members are based on concepts from fracture mechanics, stress-based verifications are performed for members with holes in accordance with the NA. The following sections initially explain the fundamentals of fracture mechanics as a design format for notched members, whereupon the design format for members with holes is also explained. In cross-sectional areas where systematic tensile stresses perpendicular to the grain are involved, non-reinforced notches and holes should generally be avoided.

D5.1 Fundamentals of fracture mechanics

Background

Since excessive stresses tend to be focused on a very small area, it is difficult – and also irrelevant given the theoretically infinite stresses – to determine the load-bearing capacity of a member with a notch or a hole using a conventional stress criterion. Such a criterion would involve comparing the extent of stress at the critical cross-section with fracture stress, which equates to the strength of the material itself. However, the theoretically infinite stresses at the point where said stress peaks make this impossible. With this in mind, determining the load-bearing capacity involves either a combination of analytical approaches and test results, or the use of fracture mechanics instead of a conventional stress criterion.

Fracture mechanics – general

Fracture mechanics is a branch of the science of material strength. When exposed to high levels of stress, a solid body responds with significant deformation or fracture. The incidence of fracture, involving separation and consequential loss of contact between two parts of a body, is relevant to fracture mechanics. From an engineering perspective, conversely, it is more important to calculate the extent of the load, which causes the fracture. As a general rule, fracture mechanics differentiates three different kinds of crack opening, see Figure D5-3.

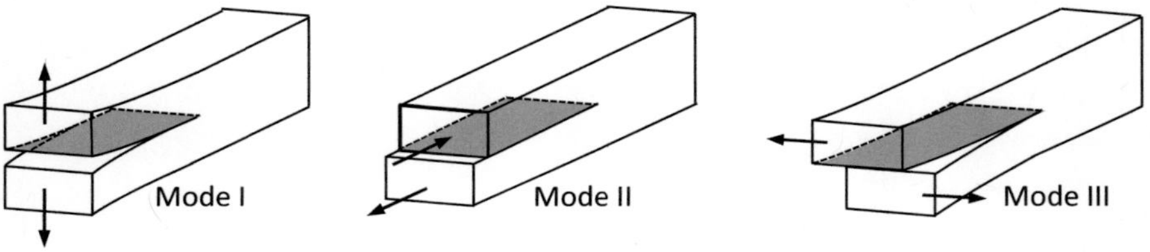

Figure D5-3 Modes I, II and III.

If there are no or only minimal stress concentrations, e.g. when uniform tensile or bending stress is exerted on a member, the fracture load can be calculated using a conventional stress criterion. For high stress concentrations, meanwhile, for example at the tip of a sharp notch or crack, another approach is required. The scope of linear-elastic theory allows appropriate calculation of the fracture load, either by evaluating the stress intensity at the tip of the notch or evaluating the energy release rate as the crack propagates. Although formal distinctions are established between both approaches, their basic prerequisites coincide and at this point, we focus solely on the second method. Analysing cracks as part of linear-elastic theory is often referred to as linear-elastic fracture mechanics. Other models attempt to take the non-linear behaviour of the material in the vicinity of the crack tip into consideration, particularly the fracture process and crack growth, which take place in the fracture process region immediately behind the crack already having opened up. In linear-elastic fracture mechanics, it is assumed that this area of energy dissipating fracture process region is very small compared to the scale of the actual structural details and thus mathematically a point, i.e. a region of zero size.

Energy release analysis – an example

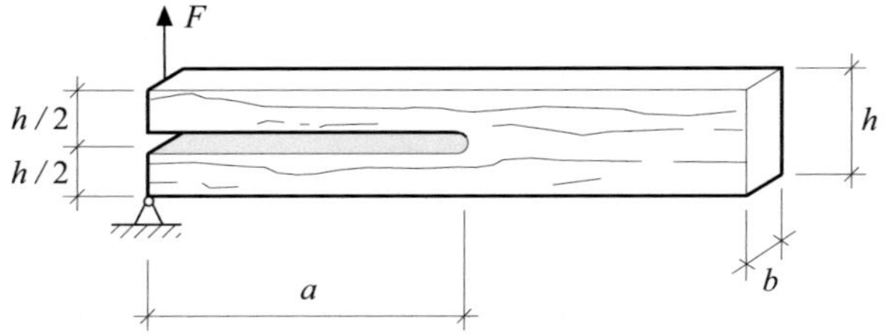

Figure D5-4 Test specimen with an end crack. (STEP 1995 Article B5)

A beam with a large longitudinal crack is considered; exposed to load in accordance with Figure D5-4. It is assumed that for $F = 0$, no stresses and strains occur within the beam. In accordance with linear-elastic theory, the potential energy of the existing system, a combination of beams and load, equates to:

$$W = -\frac{1}{2} \cdot F \cdot u \qquad \text{(D5-1)}$$

whereby u represents the deformation at the point of loading.

The area of the longitudinal crack can now be considered a system of two cantilevers of height $h/2$ and length a. In accordance with Euler-Bernoulli beam theory, the total displacement u for both cantilevers amounts to:

$$u = 2 \cdot \frac{F \cdot a^3}{3 \cdot E \cdot I} \qquad \text{(D5-2)}$$

where E is the modulus of elasticity and $I = b \cdot (h/2)^3/12$ the second moment of area of each cantilever.

Taking u from equation (D5-2) and inserting it into equation (D5-1) reveals:

$$W = -\frac{F^2 \cdot a^3}{3 \cdot E \cdot I} \qquad \text{(D5-3)}$$

The change in potential energy dW due to a small crack growth da is obtained from the derivative:

$$\mathrm{d}W = -\frac{F^2 \cdot a^2}{E \cdot I} \cdot \mathrm{d}a \qquad \text{(D5-4)}$$

This reduction in potential energy corresponds to a positive release of energy -dW and a simultaneous increase in the fractured surface of $b \cdot da$. The released energy -dW applied to the fractured surface $b \cdot da$ is normally identified with G (after A. A. Griffith, who established the principles of fracture mechanics back in the 1920s):

$$G = \frac{-\mathrm{d}W}{b \cdot \mathrm{d}a} = \frac{F^2 \cdot a^2}{b \cdot E \cdot I} \tag{D5-5}$$

If the force F is so large that the crack begins to propagate, this signals that G has reached its critical value G_c and this value corresponds to the potential of the material to dissipate energy. For European conifers, the value G_c, depending on the density, is around 150 to 600 J/m^2 for mode I, tension perpendicular to the grain as shown in Figure D5-3 (Larsen and Gustafsson, 1990). Accordingly, the fracture criterion (the moment at which the material fails) equates to

$$G = G_c \tag{D5-6}$$

which elicits the fracture load F_c with G in accordance with equation (D5-5) (where $I = b \cdot (h/2)^3/12$):

$$F_c = \frac{\sqrt{G_c \cdot b \cdot E \cdot I}}{a} = \frac{\sqrt{G_c \cdot E}}{\sqrt{h}} \cdot \frac{b \cdot h^2}{\sqrt{96} \cdot a} \tag{D5-7}$$

This equation includes two fundamentally important general results:

- The material properties which dictate the resistance to crack propagation are present in the modulus of elasticity E and fracture energy G_c. The tension strength perpendicular to the grain of the material is not required to determine F_c.
- The load-bearing capacity is heavily dependent on the dimensions, namely the computed strengths (e.g. $F_c/(b \cdot h)$) decline with increasing absolute size of the specimen.

In this example, in applying equation (D5-2), the implicit prerequisite was a slender sample, namely the fact that the ratio h/a is small and the deformation u is solely attributable to bending stress, while additional deformations caused by shear deformations and actual semi-rigid clamping of the cantilever were not taken into consideration. However, this method can also be applied to other crack geometries and fracture mechanisms (see Figure D5-3). In this case, naturally, equation (D5-2) must be replaced with an equation which applies to the actual geometric conditions.

D5.2 End-notched members, theoretical and experimental results

Using the above described method shown as an example, the force F_c during crack growth can be calculated for a loaded member and with an end crack in accordance with Figure D5-5 (a) (Gustafsson, 1988):

$$F_c = \frac{\alpha \cdot h \cdot b \cdot \sqrt{G_c / h}}{\sqrt{0.6 \cdot \left(\alpha - \alpha^2\right) / G_v} + \beta \cdot \sqrt{6 \cdot \left(1/\alpha - \alpha^2\right) / E_0}} \tag{D5-8}$$

α and β are geometric conditions, as defined in Figure D5-5. G_v and E_0 indicate the shear modulus and modulus of elasticity parallel to the grain.

Equation (D5-8) also applies to rectangular notches (Figure D5-5 (b)) and various notch types in accordance with Figure D5-1 (g, h and i), but not Figure D5-5 (c). Compared to equation (D5-7), when deriving equation (D5-8), various other geometries were also taken into consideration via the factors α and β and the deformation u was determined, not only due to bending stress (equation (D5-2)), but also shear deformations and a semi-rigid clamping of the cantilever. Here, it was assumed that the individual deformation components can be superposed, namely $u_{total} = u_{bending} + u_{shear} + u_{rotation}$. For small notches, namely for α against 1.0, the resistance against failure is considerable. In this case, the possibility of shear and bending failure of the net cross-section $\alpha \cdot h \cdot b$ must also be considered.

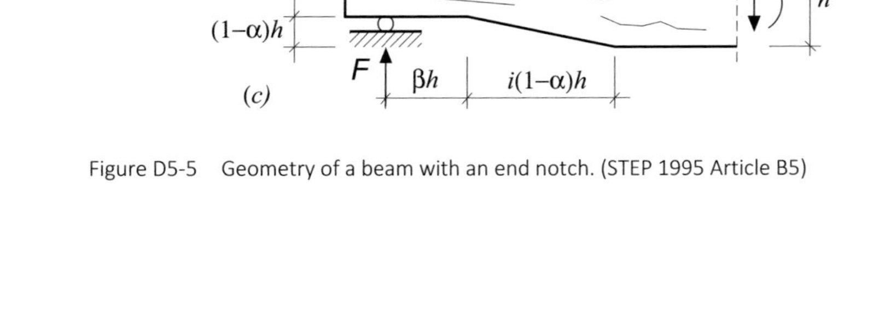

Figure D5-5 Geometry of a beam with an end notch. (STEP 1995 Article B5)

Table D5-1 lists the results of some short-term tests (Gustafsson, 1988; Riberholt et al., 1991). The values indicated are mean values; determined using dry wood with uniform moisture content. The coefficient of variation of a test series is around 20. Table D5-1 clarifies the actual low strengths of notched beams, even under favourable conditions. Moreover, also clarified is the significant extent to which strengths are influenced by dimensions, as could already theoretically be derived from equation (D5-7). It is also important to note that the mean values for F_c for specimens with knots in the vicinity of the notch exceed those for specimens without knots (Larsen and Riberholt, 1972; Möhler and Mistler, 1978).

Table D5-1 Test results. Strength (mean values) for notched beams.

Material	h [mm]	b [mm]	α	β	i	$\tau = 3/2 \cdot F_c/(\alpha \cdot h \cdot b)$ [N/mm²]
Pine	50 200	44	0.75	0.5	0	4.04 1.91
Glulam	300 567	90 160	0.50	0.15	0	2.16 1.41
Spruce	120	32	0.83 0.75 0.50 0.33	0.25	0	2.90 2.52 2.39 2.22
Glulam	600	100	0.92 0.75 0.50	0.42	0	3.00 1.32 1.13
Spruce	95	45	0.75	0.33 0.66	0	3.33 2.94
Glulam	305	79	0.70	2.5 5.5	0	0.69 0.36
Spruce	95	45	0.75	0.33	0 1 3	3.33 3.44 4.71
Glulam	300	90	0.50	0.15	0 2 8	2.16 2.76 4.16

D5.3 Background to EC 5 equations for notched members

Equation (D5-8) was modified to develop a simplified design equation for EC 5 (Larsen, 1992). The ratio E_0/G_v was set to 16. In addition, the introduction of the "new" material parameter G_c was avoided by assuming that $\sqrt{E_0 \cdot G_c}$ is proportional to the shear strength f_v of the material. The corresponding proportionality constant k_n was determined by testing. A range of constants emerged for solid timber, glued laminated timber and LVL and the final simplified form of the equation (D5-8) elicited the factor k_v shown in equation (D5-10). In addition, via tests (Riberholt et al., 1991) a function f(i) was determined, which allowed the influence of the chamfer i (Figure D5-5 (c)) to be taken into consideration. Through these considerations, in EC 5, the risk of a crack growth in a notched member can be taken into consideration through a formal reduction by a factor k_v of the design shear strength $f_{v,d}$ of the net cross-section $\alpha \cdot h \cdot b_{ef}$ (with $b_{ef} = k_{cr} \cdot b$ according to equation (D1-17)):

$$\tau_d = \frac{3}{2} \cdot \frac{V_d}{\alpha \cdot h \cdot b_{ef}} \leq k_v \cdot f_{v,d} \qquad \text{(D5-9)}$$

The reduction factor k_v (≤ 1.0) is calculated in accordance with:

$$k_v = \frac{k_n \cdot \left(1 + 1.1 \cdot i^{1.5} / \sqrt{h}\right)}{\sqrt{h} \cdot \left(\sqrt{\alpha - \alpha^2} + 0.8 \cdot \beta \cdot \sqrt{\left(1/\alpha - \alpha^2\right)}\right)} \qquad \text{(D5-10)}$$

For solid timber, we can assume $k_n = 5.0$, for glued laminated timber $k_n = 6.5$ and for LVL $k_n = 4.5$. It is also important to note that the beam depth h must be established in mm. The actual fracture mechanics background is no longer visible in equations (D5-9) and (D5-10), the verification was formally converted in a shear stress verification.

To ensure no shear failure occurs in the remaining cross-section, k_v in equation (D5-9) must not exceed 1.0. For beams notched at the opposite side to the support (Figure D5-1 (c)), $k_v = 1.0$ may be used. Figure D5-6 shows k_v depending on α and the dependency of k_v on β, i and h is also clear.

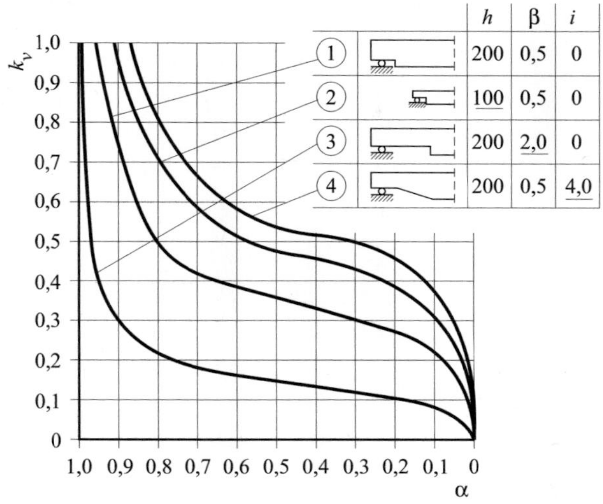

Figure D5-6 Factor k_v depending on α for solid timber with various h, β and i. (STEP 1995 Article B5)

D5.4 Climate-related tension perpendicular to the grain in notched members

The NA regulations stipulate that non-reinforced, notched beams may only be used in service classes 1 and 2 and that notches in service class 3 must be reinforced (see Article D8). The background to this rule corresponds to the regulation for curved, double-tapered and pitched cambered beams. In addition to external forces (e.g. dead weight, wind or snow), climate fluctuations intensify tension perpendicular to the grain in the area of the notch. These tensile stresses perpendicular to the grain, generated by swelling and shrinking processes, must either be taken into account by applying reinforcements or, as stipulated in the NA, the use of notched beams must be limited to service classes 1 and 2.

D5.5 Holes in glued laminated timber members

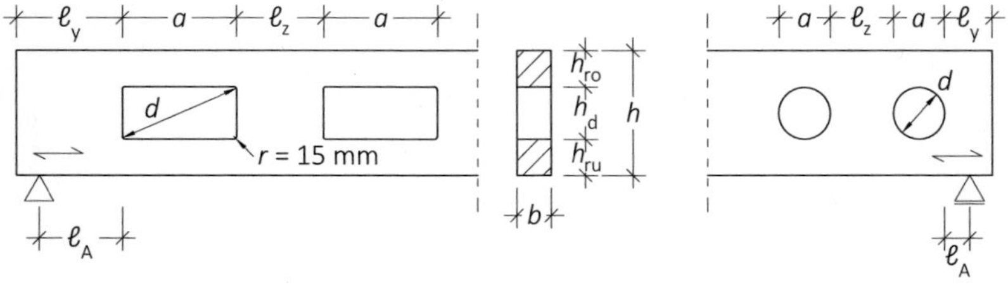

Figure D5-7 Non-reinforced holes.

EC 5 does not specify any design equations for beams with holes (holes see Figure D5-7), the only regulations for which are included in the NA. The regulations only apply to non-reinforced holes in glulam and LVL, which must not be subject to systematic tension perpendicular to the grain in the hole area. Similar to the regulations for notched beams, non-reinforced holes must only be used in service classes 1 and 2. Reinforcements of holes are covered in Article D8.

When reference is made to holes in beams of height h, this refers to holes of $d > 50$ mm, h_d is the height of the hole, h_{ro} is the remaining height above the hole and h_{ru} that underneath the hole (Figure D5-7). Unlike the fracture mechanics design format for notched beams regulated in EC 5 (which was formally converted in a shear stress verification there), verifications for beams with holes are based on the Euler-Bernoulli beam theory, extended with a volume effect. Comprehensive works have shown that holes in beams under bending stress show crack propagation as illustrated in Figure D5-1 (d) to (f), which means tensile stresses perpendicular to the grain and therefore forces $F_{t,90}$ occur at the corners of the holes, the extent of which can be calculated from a shear force component $F_{t,V}$ and moment component $F_{t,M}$ (Kolb and Epple, 1985):

$$F_{t,90} = F_{t,V} + F_{t,M} \tag{D5-11}$$

The shape of the hole also influences failure and must be taken into consideration in terms of geometric factors (Kolb and Epple, 1985). Kolb and Epple (1985) were able to use FE simulations to confirm the validity of the approach shown in equation (D5-11) and explained in the following.

Shear force component

The shear stresses occurring in a rectangular beam without a hole are shown in Figure D5-8 schematically. Via the crack propagation path, a portion of the shear force, $F_{t,V}$, corresponding to the hatched area shown in Figure D5-8 in the shear stress diagram, must be transmitted to the remaining cross-section above or underneath the hole.

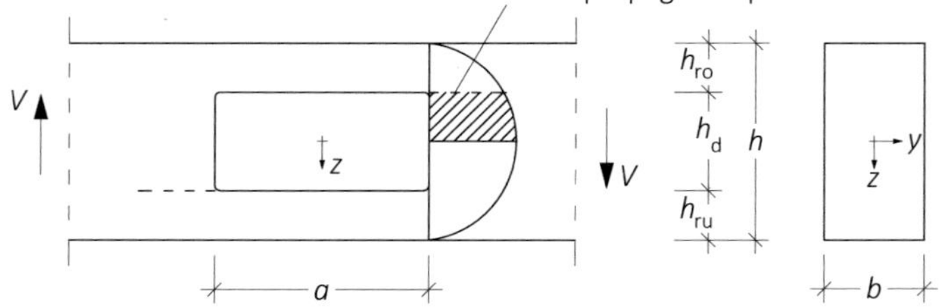

Figure D5-8 Shear stress diagram in a beam with rectangular hole. (according to Blass et al., 2005, Figure 11/12)

This component $F_{t,V}$ can be determined via the equations of the Euler-Bernoulli beam theory (geometric definitions see Figure D5-8). The shear stress diagram of a rectangular cross-section can be calculated as follows:

$$\tau(z)=\frac{V\cdot S(z)}{I\cdot b}=\frac{3}{2}\cdot\frac{V}{b\cdot h}\cdot\left(1-4\cdot\frac{z^2}{h^2}\right) \tag{D5-12}$$

The integral over the hatched area reveals the shear force component $F_{t,V}$:

$$F_{t,V}=b\cdot\int_{-\frac{h_d}{2}}^{0}\tau(z)\cdot dz=\frac{3}{2}\cdot\frac{V}{h}\int_{-\frac{h_d}{2}}^{0}\left(1-4\frac{z^2}{h^2}\right)\cdot dz=\frac{1}{4}\cdot V\cdot\frac{h_d}{h}\cdot\left(3-\frac{h_d^2}{h^2}\right) \tag{D5-13}$$

The equation specified in the NA corresponds precisely to equation (D5-13), $F_{t,V,d}$ is the design value component of the shear force V_d:

$$F_{t,V,d}=\frac{V_d\cdot h_d}{4\cdot h}\cdot\left(3-\frac{h_d^2}{h^2}\right) \tag{D5-14}$$

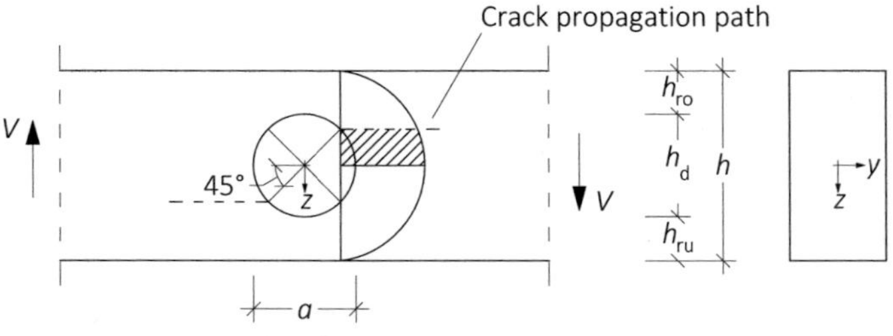

Figure D5-9 Shear stress diagram in a beam with circular hole. (according to Blass et al., 2005, Figure 11/12)

In circular holes, the crack appears approximately at a point on the circumference, where the diameter, at an angle of 45°, intersects the hole edge (Figure D5-9) and the integral for the shear stresses declines in this case below the figure for rectangular holes. In this case, instead of h_d, the value $0.7 \cdot h_d$ is used.

However, the shear force component derived in this manner assumed the hole was precisely centrally positioned, which renders it **invalid for asymmetrically configured holes**.

Moment component

Kolb and Epple (1985) could show with test results on beams with holes, that such beams were prone to tension failure perpendicular to the grain in hole corners also in areas not exposed to internal shear forces. Accordingly, they could confirm that the tensile force perpendicular to the grain $F_{t,90}$ in equation (D5-11) also had to include a moment component. This empirical moment component was iteratively adapted to the test results and determined as follows:

$$F_{t,M} = \xi \cdot \frac{M}{h_r} = 0.008 \cdot \frac{M}{h_r} \tag{D5-15}$$

The format shown in equation (D5-15) indicates that the moment component is determined from the ratio of the bending moment M at the critical cross-section to the edge distance of the crack to be expected; the coefficient ξ was iteratively determined as 0.008. In the NA, in turn, the design value M_d of the critical bending moment is used, to determine the design value of the moment component $F_{t,M,d}$.

For rectangular holes therefore, the required height h_r is the smallest value of the remaining height underneath or above the hole, $h = \min(h_{ro};\ h_{ru})$, since this is where the tensile stresses perpendicular to the grain peak and hence where cracks emerge. This regulation is excessively conservative for circular holes, since the maximum tensile stresses perpendicular to the grain only occur at the point, at which the circle diameter is at an angle of 45°. The following applies in this case:

$$h_r = \min\left(h_{ro} + 0.15 \cdot h_d;\ h_{ru} + 0.15 \cdot h_d\right) \tag{D5-16}$$

Using the design value for tensile force perpendicular to the grain $F_{t,90,d}$, determined in accordance with equation (D5-11), the tensile strength perpendicular to the grain can thus be verified:

$$F_{t,90,d} = F_{t,V,d} + F_{t,M,d} \leq 0.5 \cdot \ell_{t,90} \cdot b \cdot k_{t,90} \cdot f_{t,90,d} \tag{D5-17}$$

where

b Beam width at hole

$f_{t,90,d}$ Design value of tensile strength perpendicular to the grain of the glulam or LVL

$k_{t,90}$ = $\min(1; (450/h)^{0.5})$, h in mm

$\ell_{t,90}$ = $0.5 \cdot (h_d + h)$ for rectangular holes

$\ell_{t,90}$ = $0.353 \cdot h_d + 0.5 \cdot h$ for circular holes

$b \cdot \ell_{t,90}$ in this case corresponds to the area subject to tensile stress perpendicular to the grain, whereby $\ell_{t,90}$ is the length starting from the hole corner. However, the stresses are not uniformly dispersed over the area $b \cdot \ell_{t,90}$, but the tensile stresses perpendicular to the grain peak in the hole corners, similar to a notch corner, see Figure D5-2. These stress peaks are taken into consideration, whereby in equation (D5-17) the tensile force perpendicular to the grain $F_{t,90,d}$ is dispersed in a triangular stress, which explains the 0.5 factor. The coefficient $k_{t,90}$ takes into consideration the impact of height, which makes it a coefficient in accordance with Article D3 and thus takes the impact of volume on strength into consideration.

It is important to note the fact at this point that rounded corners in holes are stipulated, to prevent areas of singularity and reduce the stress peaks forming in the corners, see also details of the minimum radius in Figure D5-7.

D5.6 Literature

P.J. Gustafsson, original Article B5, STEP 1995.

Blass H.J., Ehlbeck J., Kreuzinger H. and Steck G. (2005). Erläuterungen zur DIN 1052:08-2004. Herausgeber: Deutsche Gesellschaft für Holzforschung. Bruder-Verlag, Karlsruhe.

Gustafsson P.J. (1988). A study of strength of notched beams. Paper 21-10-1, CIB-W18, Parksville.

Kolb H. and Epple A. (1985).Verstärkung von durchbrochenen Brettschichtbindern. Forschungsvorhaben
I.4-34810 der MPA Stuttgart.

Larsen H.J. and Riberholt H. (1972). Tests with not classified structural timber. Rapport nr R 31 (in Danish), Technical University of Denmark.

Larsen H.J. and Gustafsson P.J. (1990). The fracture energy of wood in tension perpendicular to the grain – results from a joint testing project. Paper 23-19-2, CIB-W18 Meeting 23, Lisbon.

Larsen H.J. (1992). Latest development of EC 5. Paper 25-102-1a, CIB-W18 Meeting 25, Åhus.

Möhler K. and Mistler H.-L. (1978). Untersuchungen über den Einfluß von Ausklinkungen im Auflagerbereich von Holzbiegeträgern auf die Tragfestigkeit. Forschungsbericht, Lehrstuhl für Ingenieurholzbau und Baukonstruktionen, Universität Karlsruhe.

Riberholt H., Enquist B., Gustafsson P.J. and Jensen R.B. (1991). Timber beams notched at the support. Report TVSM-7071, Lund University.

D6 Glued composite members

Original articles: K. H. Solli, H. J. Blass, J. G. M. Raadschelders

This article explains the design of glued composite components regulated in EC 5 with thin webs or thin flanges. At this point, determining internal forces is crucial and explained in detail. Glued components can be used in very wide-ranging applications, frequent examples of which include glued thin-webbed beams (I-joists or box girders, Figure D6-1 on the left) and glued thin-flanged beams (stressed skin panels, Figure D6-1 on the right).

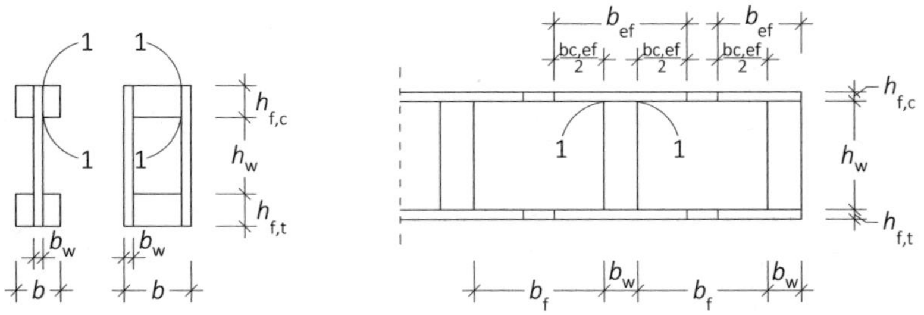

Figure D6-1 Thin-webbed beams (left) and thin-flanged beam (right). 1-1 are glue lines.

While the webs are relatively thin compared to flanges in I-joists and box girders, stressed skin panels are frequently seen in the form of beams with thick webs and thin flanges. For both component types, thin-webbed and thin-flanged beams, linear strain distribution over beam depth is assumed, although shear deformations in the flanges of stressed skin panels mean normal stresses in the centre plane of the flanges are non-linearly distributed. A separate explanation of both component types is given in the following sections accordingly.

Glued components often comprise various materials. Since various stiffnesses affect the stress distribution, the stiffness of individual components must also be taken into consideration during the design. Meanwhile, if the materials used also exhibit differing creep behaviour (various creep coefficients k_{def}, see also Article B2), this should also be taken into consideration during the design. In such cases, the stress distribution for the instantaneous and final states must also be determined and for the instantaneous state, the mean values of stiffness properties are normally used. The stress distribution in the final state, meanwhile, can be calculated in accordance with EC 5 using the final mean values of stiffness properties. Equation (D6-1) shows how final values are determined, citing the example of the modulus of elasticity.

$$E_{\text{mean,fin}} = \frac{E_{\text{mean}}}{1+\psi_2 \cdot k_{\text{def}}} \quad \text{(D6-1)}$$

The rule adopted is that $\psi_2 = 1.0$ for permanent loads, otherwise the combination coefficient should be used for the quasi-permanent load component. For load combinations, ψ_2 may be proportionally weighted.

D6.1 Glued thin-webbed beams

A glued thin-webbed beam (examples Figure D6-1 on the left) comprises three main components:

- Flanges,
- Web,
- Glued joints between flanges and web.

Although the flanges are frequently made of finger-jointed solid timber, they can also be manufactured from other materials such as glued laminated timber or laminated veneer lumber (LVL). Their main function is to absorb stresses generated by bending moments and axial forces. Given the generally small dimensions of the flanges, it is imperative to ensure that the material has few and minimal defects.

The web can be made of various types of wood materials, such as plywood, particleboards, fibreboards or OSB. There are also beam shapes including truss-like webs, particularly formwork girders. The web primarily absorbs the stresses from shear forces, while for longer beams, joints in the webs may be required. If the web splices are in an area subject to lower shear forces, they can be configured as butt joints, while in other cases, the web splice should be force-locking. There is also often a need to reinforce the web in the support area, which can be done via additional wood-based panels, which are nailed or preferably glued onto the web. The reinforced web at the load input position or support must be configured such as to ensure that the shear forces exerted can be effectively transmitted.

Production

Glued, thin-webbed beams are normally manufactured on an industrial scale and it is crucial to ensure the correct gluing temperature to optimise the joint between the web and flanges. It is equally important to ensure that the surfaces of the glued joints are planed or milled immediately before the glue is applied and that the moisture content of the flange and web material is checked.

Use of glued thin-webbed beams

Glued thin-webbed beams, in relation to their dead weight, have exceptional load-bearing capacity and stiffness, while their low weight also facilitates transport and assembly. Moreover, using tools, they can be easily processed by hand. As a general rule, these beams can be used instead of solid timber beams. Glued thin-webbed beams are often used to construct floor structures, when it is difficult to procure sufficiently large cross-sections made of solid timber or if the cost of using glued laminated timber (from 5 to 8 m long) would be prohibitive. When used in floor and wall structures, beam depths of 300 to 500 mm are possible, which helps pave the way to install a range of technical building equipment. The construction depth also guarantees sufficient room for any insulation material that may be required. In countries subject to cold winters, the required insulation thickness for the cross-sectional dimensions of studs is a crucial parameter and here, if a glued thin-webbed cross-section is used, the proportion of insulation material can be maximised. In service class 3, however, the use of glued thin-webbed beams is infeasible due to the restrictions on using web materials in this service class.

Special production and transport aspects

Stiffness about the *z*-axis is very low compared to that about the *y*-axis (see Figure D6-2 for axis definition), which must be noted during production and transport from the manufacturing company to the building site. Web materials are also prone to damage during the transport and assembly processes and beams must be protected against moisture during the construction phase. If the moisture content in the web soars, this considerably increases the risk of permanent deformations in the finished construction.

Lateral stability

Flanges carrying compressive stresses must be supported, to avoid lateral deflection and buckling. For beams used as simply supported beams in floors, meanwhile, the joint between the compression flange and floor panel generally suffices to prevent lateral deflection. It is also important to ensure that in the lower chord, compressive stresses can be generated e.g. at the intermediate supports of continuous beams.

Effective cross-sectional values

For subsequent calculations, it is assumed that flanges and web are glued together such as to form a single constructive unit. It is also assumed that any strains will run linearly over the beam depth. According to Hooke's law for linear-elastic stress-strain behaviour, stresses at a particular point can be calculated from the product of strain and the modulus of elasticity. Since a beam may be made of materials including various moduli of elasticity, this means that stresses over beam depth are non-linearly distributed. Figure D6-2 shows an example of how the stresses due to bending moment vary in the beam cross-section.

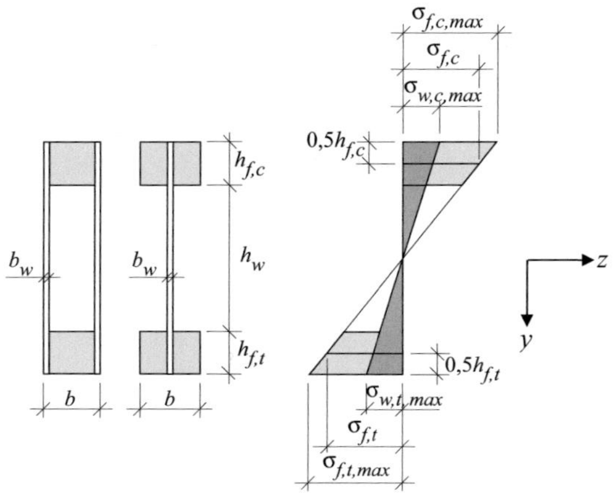

Figure D6-2 Example of the stress distribution in glued I-joists and box girders, index *w* = web, Index *f* = flange. (STEP 1995 Article B9)

Given the fact that the moduli of elasticity in the cross-section vary, as a general rule, effective cross-sectional values are calculated. In this case for example, the entire profile can be considered a cross-section of homogenous material with properties equivalent to the flange material. When adopting such an approach, the web width must be reduced by the ratio of the average moduli of elasticity. All specified equations apply for permanent loads with $\psi_2 = 1.0$; while for other loads the rule is $(1 + \psi_2 \cdot k_{def})$ instead of $(1 + k_{def})$, see equation (D6-1).

The effective cross-sectional area (index f = flange, index w = web) is thus determined using a weighted web surface:

$$A_{ef} = A_f + \frac{E_{mean,w}}{E_{mean,f}} \cdot \frac{1 + k_{def,f}}{1 + k_{def,w}} \cdot A_w \quad \text{(D6-2)}$$

The effective second moment of area likewise:

$$I_{ef} = I_f + \frac{E_{mean,w}}{E_{mean,f}} \cdot \frac{1+k_{def,f}}{1+k_{def,w}} \cdot I_w \tag{D6-3}$$

The approach used here is the so-called **composite theory**, for which all known equations on stress and deformations derived via Euler-Bernoulli beam theory apply. The only difference here is that the moduli of elasticity are no longer constant. Given that fact, to now take account of a full cross-section and a single modulus of elasticity, the cross-sectional values have to be adapted, to take the differences in stiffness and hence the different deformation behaviour, into consideration. As a reference value E_0 for the modulus of elasticity, the modulus of elasticity of the key material, the flanges in this case, is often used. The individual areas can then be converted into effective areas using the ratio of the moduli of elasticity E_i/E_0: $A_{ef,i} = E_i/E_0 \cdot A_i$. The effective second moments of area thus result in $I_{ef,i,y} = \int z^2 \cdot E_i/E_0 \cdot dA$ and the internal forces and moments in $M_{i,y} = E_0 \cdot I_{ef,i,y} \cdot \psi'$, $M_{i,z} = E_0 \cdot I_{ef,i,z} \cdot \psi'$, $N_i = E_0 \cdot A_{ef,i} \cdot \varepsilon$ (where ψ' = derivative of the angle).

Verification of stresses in the flanges

Within a beam cross-section under bending stress, the flanges are particularly prone to axial compressive and tensile stresses, amid a small bending component. With a symmetrical cross-section, which is only required to absorb a bending moment, the absolute stresses in the compression and tension flanges are equal. If a beam is subject to additional axial stresses, the maximum stresses in the outer fibres of the compression flange can be calculated as the sum of stresses due to the bending moment M_d and axial compressive force F_d, where y_{max} equates to the distance from the outer fibres to the neutral axis, tension flange analogous:

$$\sigma_{f,c,max,d} = \frac{M_d}{I_{ef}} \cdot y_{max} + \frac{F_d}{A_{ef}} \tag{D6-4}$$

Equation (D6-4) clearly shows how cross-sectional values are replaced by effective values in accordance with composite theory, while all other details remain as to date, based on Euler-Bernoulli beam theory. Conversely, to calculate axial stresses in the centre of gravity of the compression flange, equation (D6-4) must take into account the distance y_c between the centre of gravity and the neutral axis of the beam (tension flange analogous):

$$\sigma_{f,c,d} = \frac{M_d}{I_{ef}} \cdot y_c + \frac{F_d}{A_{ef}} \tag{D6-5}$$

When calculating the stresses, the position of the neutral axis thus always has to be determined. For symmetrical cross-sections, this is trivial and $y_{max} = h/2$ with the beam depth h (see equations (D6-23) and (D6-24) for an exemplary derivation of the position of the neutral axis).

The calculated stresses must now be compared with the design strength values of the compression flange:

$$\sigma_{f,c,max,d} \leq f_{f,m,d} \quad \text{and} \quad \sigma_{f,c,d} \leq k_c \cdot f_{f,c,0,d} \tag{D6-6}$$

In Equation (D6-6) the verification of the compression flange also takes possible buckling into consideration, which involves the use of the coefficient k_c already familiar from Article D2 (there coefficient $k_{c,z}$ in equation (D2-7)). This coefficient k_c may be conservatively determined (particularly conservative for box girders) in accordance with EC 5 with the following slenderness ratio λ_z (steps: $\lambda_z \rightarrow \lambda_{rel} \rightarrow k_z \rightarrow k_c$):

$$\lambda_z = \sqrt{12} \cdot \frac{\ell_c}{b} \quad \left(\lambda_z = \frac{\ell_{ef}}{i} = \frac{\ell_{ef}}{\sqrt{I/A}} = \frac{\ell_{ef}}{\sqrt{b^2/12}} \right) \tag{D6-7}$$

ℓ_c is the distance between the lateral supports of the compression flange, while b is the flange width (see Figure D6-2). The compression flange is considered a column without taking the lateral support from the web into account. The stresses in the tension flange are calculated correspondingly, but without the buckling coefficient k_c.

Verification of axial stresses in the web

The web must particularly absorb shear stresses generated by shear forces, although proportions of the bending moment and axial forces generate additional normal stresses in the web, which is why the load-bearing capacity of the web to accommodate these stresses must also be checked. Given the assumption of linear strain distribution over the beam depth, the axial stresses σ_w in the web can be calculated with composite theory in the following general form from the stresses in the flange σ_f:

$$\sigma_w = \sigma_f \cdot \frac{E_{mean,w}}{E_{mean,f}} \tag{D6-8}$$

Respectively, taking into consideration the load duration and service class:

$$\sigma_{\mathrm{w}} = \sigma_{\mathrm{f}} \cdot \frac{E_{\mathrm{mean,w}}}{E_{\mathrm{mean,f}}} \cdot \frac{1+k_{\mathrm{def,f}}}{1+k_{\mathrm{def,w}}} \tag{D6-9}$$

To calculate the axial stresses σ_{w} in the web, the first task is to determine the flange stresses σ_{f} at the point at which σ_{w} should be calculated. This is performed similarly to the equations (D6-4) and (D6-5):

$$\sigma_{\mathrm{f,d}} = \frac{M_{\mathrm{d}}}{I_{\mathrm{ef}}} \cdot y_1 + \frac{F_{\mathrm{d}}}{A_{\mathrm{ef}}} \tag{D6-10}$$

The maximum stress in the compression zone of the web $\sigma_{\mathrm{w,c,max,d}}$ can then be calculated with equations (D6-10) and (D6-9) as:

$$\sigma_{\mathrm{w,c,max,d}} = \left(\frac{M_{\mathrm{d}}}{I_{\mathrm{ef}}} \cdot y_{\mathrm{w,c}} + \frac{F_{\mathrm{d}}}{A_{\mathrm{ef}}} \right) \cdot \frac{E_{\mathrm{mean,w}}}{E_{\mathrm{mean,f}}} \cdot \frac{1+k_{\mathrm{def,f}}}{1+k_{\mathrm{def,w}}} \tag{D6-11}$$

$y_{\mathrm{w,c}}$ here is the distance of the neutral axis of the beam from the point where the stress value is calculated.

This stress must meet the following condition:

$$\sigma_{\mathrm{w,c,max,d}} \leq f_{\mathrm{w,c,d}} \tag{D6-12}$$

The maximum stress in the tension zone of the web is correspondingly calculated as:

$$\sigma_{\mathrm{w,t,max,d}} = \left(\frac{M_{\mathrm{d}}}{I_{\mathrm{ef}}} \cdot y_{\mathrm{w,t}} + \frac{F_{\mathrm{d}}}{A_{\mathrm{ef}}} \right) \cdot \frac{E_{\mathrm{mean,w}}}{E_{\mathrm{mean,f}}} \cdot \frac{1+k_{\mathrm{def,f}}}{1+k_{\mathrm{def,w}}} \tag{D6-13}$$

$y_{\mathrm{w,t}}$ is the distance of the neutral axis of the beam from the point where the stress value is calculated.

This stress must meet the following condition:

$$\sigma_{\mathrm{w,t,max,d}} \leq f_{\mathrm{w,t,d}} \tag{D6-14}$$

$f_{\mathrm{w,c,d}}$ and $f_{\mathrm{w,t,d}}$ are the design values for compressive and tensile bending strengths of the web. In the absence of any other values, the design values for in-plane tensile and compressive strength of the web should be used.

Verification of shear stresses in the web

If equation (D6-15) is satisfied, no more accurate shear buckling analysis is required:

$$h_w \leq 70 \cdot b_w \quad \text{(D6-15)}$$

In this case, the design shear force in each web $F_{w,v,Ed}$ must maintain the following condition (for geometric details, see Figure D6-2):

$$F_{w,v,Ed} \leq b_w \cdot h_w \cdot \left(1 + \frac{0.5 \cdot (h_{f,t} + h_{f,c})}{h_w}\right) \cdot f_{v,0,d} \quad \text{for } h_w \leq 35 \cdot b_w$$
$$F_{w,v,Ed} \leq 35 \cdot b_w^2 \cdot \left(1 + \frac{0.5 \cdot (h_{f,t} + h_{f,c})}{h_w}\right) \cdot f_{v,0,d} \quad \text{for } 35 \cdot b_w \leq h_w \leq 70 \cdot b_w \quad \text{(D6-16)}$$

$f_{v,0,d}$ is the design in-plane shear strength. As Figure D6-2 shows, since the load-bearing depth of the web exceeds the clearance depth h_w, the web depth is determined as the distance between the centres of gravity of the flanges. The slenderer the web, the larger h_w is in relation to b_w and the greater the risk of the web buckling. Limiting the effective height h_w to $35 \cdot b_w$ (lower equation) thus constitutes an indirect buckling verification.

Verification of the shear stresses in the glue line between flanges and web

The strength of a correctly glued joint between web and flanges exceeds the shear strength of the flange and web materials and the weakest element in such a joint is normally the rolling shear strength of the web $f_{v,90}$. It is assumed that the design shear stress $\tau_{mean,d}$ in the cross-section under consideration (namely in both glued joints between the web and compression or tension flange, see Figure D6-2) is uniformly distributed if flange heights $h_f \leq 4 \cdot b_{ef}$, which is why the shear stress can be determined using Euler-Bernoulli beam theory:

$$\tau_{mean,d} = \frac{V_d \cdot S_f}{I_{ef} \cdot \ell_g} \quad \text{(D6-17)}$$

where

S_f First moment of area of a flange, related to the neutral axis of the beam

$\ell_g = 2 \cdot h_{f,c(t)}$ For cross-sections in accordance with Figure D6-2

The shear stress should meet the following condition:

$$\tau_{\text{mean,d}} \leq f_{\text{v,90,d}} \qquad \text{for } h_{\text{f}} \leq 4 \cdot b_{\text{ef}}$$
$$\tau_{\text{mean,d}} \leq f_{\text{v,90,d}} \cdot \left(\frac{4 \cdot b_{\text{ef}}}{h_{\text{f}}} \right)^{0.8} \qquad \text{for } h_{\text{f}} > 4 \cdot b_{\text{ef}} \tag{D6-18}$$

where

$f_{\text{v,90,d}}$	Design rolling shear strength of the web
h_{f}	Either $h_{\text{f,c}}$ or $h_{\text{f,t}}$, depending on whether the glued joints in the compression or tension flange are verified
$b_{\text{ef}} = b_{\text{w}}$	For box girders
$b_{\text{ef}} = b_{\text{w}}/2$	For I-joists

Equation (D6-18) must be allocated depending on flange height h_{f}, since the shear stresses in the glued joint between the flange and web are not actually uniformly distributed. While this uneven distribution can be ignored up to a glue line depth h_{f} of $4 \cdot b_{\text{ef}}$, at greater flange depths, either the shear stress has to be increased or, as shown in equation (D6-18), it must be reduced to take account of this non-uniform distribution with higher maximum values. However, the equations shown to date to determine normal and shear stresses in the individual components can also be formulated more generically, as will be shown at the end of this article.

D6.2 Glued thin-flanged beams (stressed skin panels)

The previous section focused on glued components with thin webs and thick flanges. However, glued components with thin flanges also exist, for which a potential stability failure of the flanges (buckling) as well as their shear deformations, both due to their significant slenderness, must be taken into consideration. (Glued) stressed skin panels (Figure D6-3) can be considered thin-flanged beams when considering effective tributary areas. The glued joints between the webs and flanges (sheathing) are deemed rigid in the assessment. Consequently and also at this point, a linear strain distribution is assumed over the depth of the composite component, Figure D6-4 (b). The joints between flanges and web are also frequently implemented with mechanical fasteners, whereupon the slip between flanges and web must be taken into consideration when stresses are determined, as in Figure B6-4 (d). Although such mechanically jointed elements do not constitute part of the current article, they can be designed using the methods set out in Article D7.

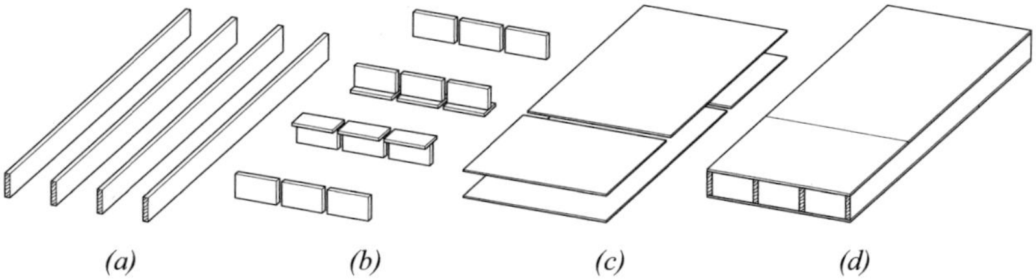

Figure D6-3 Construction of a stressed skin panel. (*a*) webs, (*b*) blocks and flange splices, (*c*) sheathing (flanges), (*d*) stressed skin panel. (STEP 1995 Article B10)

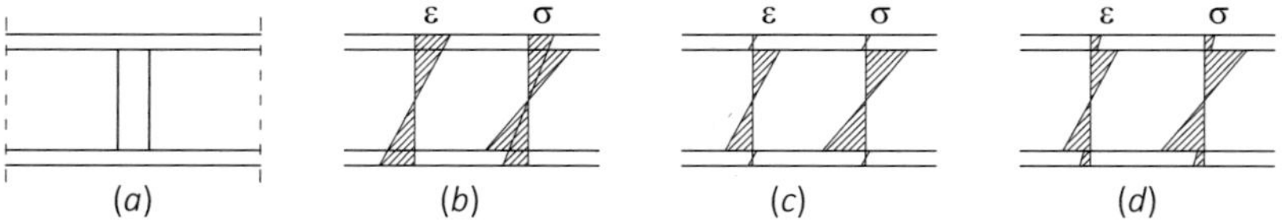

Figure D6-4 Strain and stress distribution in the cross-section. (*a*) cross-section (with variable modulus of elasticity of flange and web), (*b*) rigid connection, (*c*) no connection, (*d*) semi-rigid connection.

Effective flange width

Shear deformations in the plane of the flanges mean the normal stresses in the centre plane of the flanges are not uniformly distributed (see Figure D6-5). The extent to which flanges contribute to bending stiffness or bending capacity of the composite component thus declines with increasing lateral distance from the webs. The extent of this stress reduction particularly depends on the ratio of the b_f/ℓ and E/G values and here, b_f represents the clear distance between webs, ℓ the span, E the modulus of elasticity of the flange in the direction of the webs and G the shear modulus of the flange, while the mean values for in-plane loading are used for both moduli. The effective flange width declines with increasing E/G and b_f/ℓ values.

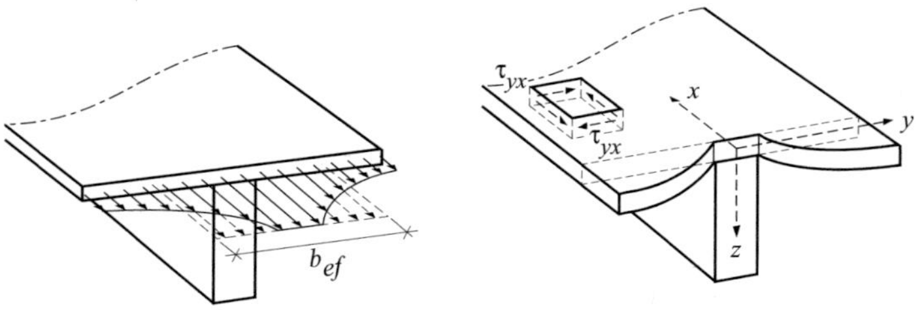

Figure D6-5 Stress distribution in the flange. (STEP 1995 Article B10)

The derivation of the effective flange width taking into consideration the shear deformations of flanges is specified in Möhler et al. (1963) and the resulting ratio obtained between effective and actual flange widths b_{ef}/b_f for simply supported beams under a uniformly distributed load is:

$$\frac{b_{ef}}{b_f}=\frac{\left(\lambda_1\cdot\tanh\alpha_1-\lambda_2\cdot\tanh\alpha_2\right)}{\pi\cdot\left(\lambda_1^2-\lambda_2^2\right)}\cdot\frac{2\cdot\ell}{b_f} \tag{D6-19}$$

where

$$\alpha_1=\frac{\lambda_1\cdot\pi\cdot b_f}{2\cdot\ell} \quad\text{and}\quad \alpha_2=\frac{\lambda_2\cdot\pi\cdot b_f}{2\cdot\ell}$$

$$\lambda_1=\sqrt{a+\sqrt{a^2-c}} \quad\text{and}\quad \lambda_2=\sqrt{a-\sqrt{a^2-c}}$$

$$a=\frac{E_y}{2\cdot G}-\mu_{xy} \quad\text{and}\quad c=\frac{E_y}{E_x}$$

μ_{xy} is the Poisson's ratio of the flanges.

To allow the Euler-Bernoulli beam theory to be used when designing thin-flanged beams, the concept of effective flange width is used. Here, the effective flange width b_{ef} is defined as the width of a fictitious flange, in which the normal stresses in the centre of gravity of the flange calculated with the Euler-Bernoulli beam theory correspond to the related maximum value in accordance with the correct theory (Möhler et al., 1963). The entire normal force in the flange as well as the section modulus of the overall cross-section remains identical in both cases (Beam theory and correct theory). In EC 5, the approximation values for the effective flange width b_{ef} for internal webs are specified as follows (for notations, see Figure D6-6):

$$b_{ef}=b_{c,ef}+b_w \quad\text{or}\quad b_{ef}=b_{t,ef}+b_w \tag{D6-20}$$

The corresponding approximation equation for edge webs is:

$$b_{ef}=0.5\cdot b_{c,ef}+b_w \quad\text{or}\quad b_{ef}=0.5\cdot b_{t,ef}+b_w \tag{D6-21}$$

Table D6-1 from EC 5 provides maximum values for the effective flange widths $b_{c,ef}$ and $b_{t,ef}$, whereby the buckling failure is only relevant for the compression flange (determining $b_{c,ef}$). Accordingly, the effective widths $b_{c,ef}$ or $b_{t,ef}$ should not be assumed to exceed the maximum value calculated in accordance with equation (D6-19) which takes shear deformation into consideration. Moreover, $b_{c,ef}$ should not be assumed to exceed the maximum value in accordance with Table D6-1 which takes buckling of the flanges into consideration. Table D6-1 in EC 5 deviates considerably from the table given in DIN 1052 (Table 5). Not only do the given maximum values differ slightly, but the information provided is also far less detailed. The NA also stipulates that laminated veneer lumber (LVL) with perpendicular layers can be treated like plywood and cross-laminated timber like OSB.

Figure D6-7 shows the course of effective flange width in accordance with equation (D6-19) and the corresponding approximation in accordance with EC 5 for shear deformations. In the most practical cases, the ratio b_f/ℓ is below 0.3.

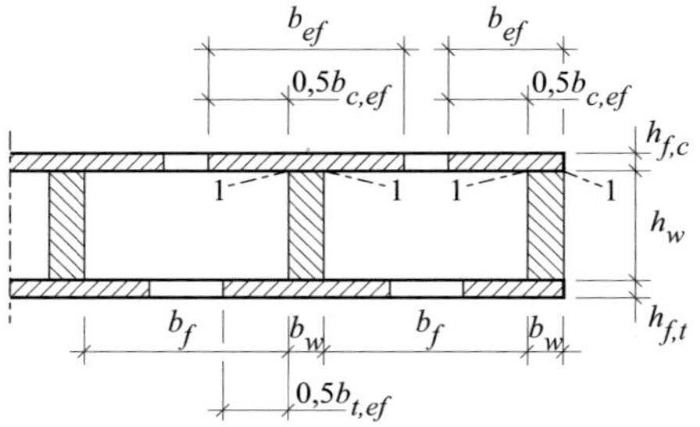

Figure D6-6 Stressed skin panel (thin-flanged beam). (STEP 1995 Article B10)

Table D6-1 Maximum values of the effective flange width taking shear deformation ($b_{c,ef}$ and $b_{t,ef}$) and buckling ($b_{c,ef}$) into consideration.

Flanges	Shear deformation	Buckling
Plywood with the fibre direction of the outer plies:		
- Parallel to the webs	$0.1 \cdot \ell$	$25 \cdot h_f$
- Perpendicular to the webs	$0.1 \cdot \ell$	$20 \cdot h_f$
OSB	$0.15 \cdot \ell$	$25 \cdot h_f$
Particleboards of fibreboards with random fibre orientation	$0.2 \cdot \ell$	$30 \cdot h_f$

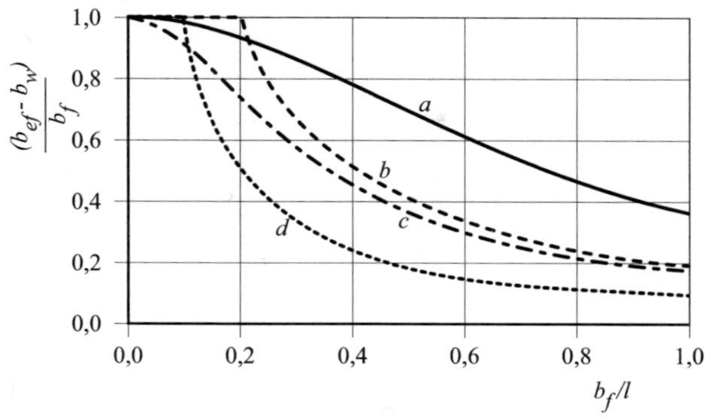

Figure D6-7 Effective flange width. *a* particleboard equation (D6-19), *b* particleboard EC 5, *c* plywood equation (D6-19), *d* plywood EC 5. (STEP 1995 Article B10)

Flanges subject to compression are at risk of buckling and a more accurate verification can be conducted e.g. in accordance with the method of Halász and Cziesielski (1966). In the absence of any more accurate buckling verification, the clear distance between webs b_f should be assumed as not exceeding double the effective flange widths $b_{c,ef}$ and $b_{t,ef}$ determined while taking buckling into consideration (Table D6-1). For thin-flanged beams with nailed or stapled joints between flanges and webs, the withdrawal capacity of the fasteners must be sufficient to prevent any buckling of the flanges.

Verification

Thin-flanged beams (stressed skin panels) are computed assuming rigid joints between web and flange (sheathing). The procedure resembles that for thin-webbed beams (Section D6.1), only differing in terms of the shear verification of (in this case thin sheathing) flanges (in Section D6.1 equation (D6-18)). Here too, a uniform distribution of shear stress $\tau_{mean,d}$ in the glued joint in question (Section 1-1 in Figure D6-6) is assumed and the determined shear stress should meet the following condition:

$$\tau_{mean,d} \le f_{v,90,d} \qquad \text{for } b_w \le 8 \cdot h_f$$

$$\tau_{mean,d} \le f_{v,90,d} \cdot \left(\frac{8 \cdot h_f}{b_w}\right)^{0.8} \qquad \text{for } b_w > 8 \cdot h_f \qquad \text{(D6-22)}$$

where

$f_{v,90,d}$ Design rolling shear strength of flange (sheathing)

h_f Either $h_{f,c}$ or $h_{f,t}$

Also relevant is the fact that $8 \cdot h_f$ should be replaced by $4 \cdot h_f$, if U-shaped cross-sections are considered.

Generalisation

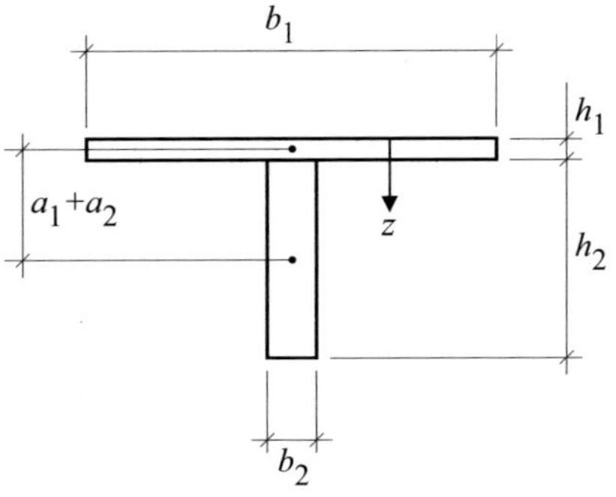

Figure D6-8 Cross-section of a glued stressed skin panel. 1 designates the sheathing and 2 the web. (STEP 1995 Article B10)

In the following section, using the example of a glued stressed skin panel with one-sided sheathing, it is shown how bending stiffnesses and stresses in the different components can be calculated while taking a range of material stiffnesses into consideration. All geometric details are taken from Figure D6-8. The first task is to determine the position of the neutral axis and likewise here, applying the composite theory, the influence of different moduli of elasticity has to be taken into consideration:

$$a_1 = \frac{\sum_{i=1}^{2} z_i \cdot E_i \cdot A_i}{\sum_{i=1}^{2} E_i \cdot A_i} = \frac{h_1/2 \cdot E_1 \cdot A_1 + \left(h_1 + h_2/2\right) \cdot E_2 \cdot A_2}{E_1 \cdot A_1 + E_2 \cdot A_2} - \frac{h_1}{2} = \frac{E_2 \cdot A_2 \cdot \left(h_1 + h_2\right)}{2 \cdot \left(E_1 \cdot A_1 + E_2 \cdot A_2\right)} \tag{D6-23}$$

$$a_2 = \frac{h_1 + h_2}{2} - a_1 \tag{D6-24}$$

The effective bending stiffness is:

$$\left(E \cdot I\right)_{ef} = \sum_{i=1}^{2} \left(E_i \cdot I_i + E_i \cdot A_i \cdot a_i^2\right) \tag{D6-25}$$

The compressive stress in the centre of gravity of the flange is:

$$\sigma_{1,c} = \frac{M}{\left(E \cdot I\right)_{ef}} \cdot E_1 \cdot a_1 \tag{D6-26}$$

The compressive stress in the outer upper fibre of the flange is:

$$\sigma_{1,c,max} = \frac{M}{(E \cdot I)_{ef}} \cdot E_1 \cdot \left(a_1 + \frac{h_1}{2} \right) \tag{D6-27}$$

The tensile stress in the centre of gravity of the web is:

$$\sigma_{2,t} = \frac{M}{(E \cdot I)_{ef}} \cdot E_2 \cdot a_2 \tag{D6-28}$$

The bending stress in the outer lower fibre of the web is:

$$\sigma_{2,t,max} = \frac{M}{(E \cdot I)_{ef}} \cdot E_2 \cdot \left(a_2 + \frac{h_2}{2} \right) \tag{D6-29}$$

The shear stress in the joint between flange and web is (required to verify equation (D6-22)):

$$\tau_{max} = \frac{V \cdot S_1}{(E \cdot I)_{ef} \cdot b_2} \cdot E_1 = \frac{V \cdot A_1 \cdot a_1}{(E \cdot I)_{ef} \cdot b_2} \cdot E_1 \tag{D6-30}$$

S_1 is the first moment of area of the flange, which is why E_1 must be selected as the modulus of elasticity, b_2 is the width of the glue line.

Equations (D6-26) to (D6-30) correspond to equations (D6-4) or (D6-17) already given and are only formulated generally, without being based on a modulus of elasticity E_0, so recalculations such as in equation (D6-11) are no longer needed when the stresses in the components should be calculated by a parameter other than the reference modulus of elasticity E_0. The design stresses must not exceed the corresponding design strengths. The stresses determined here correspond to those in the instantaneous state, while for the stresses in the final state, the moduli of elasticity must be re-divided by $(1 + \psi_2 \cdot k_{def})$ (equation (D6-1)). Composite beams with mechanical fasteners between the web and flange can be calculated with the γ-method presented in Article D7.

D6.3 Literature

K.H. Solli, H.J. Blass, J.G.M. Raadschelders, original Articles B9, B10, STEP 1995.

Von Halász R. and Cziesielski E. (1966). Berechnung und Konstruktion geleimter Träger mit Stegen aus Furnierplatten. Berichte aus der Bauforschung Heft 47, pp 75-118.

Möhler K., Abdel-Sayed G. and Ehlbeck J. (1963). Zur Berechnung doppelschaliger, geleimter Tafelelemente. Holz als Roh- und Werkstoff 21:328-333.

D7 Mechanically jointed members and CLT elements

Original article: H. Kreuzinger

Mechanically jointed beams comprise at least two individual components joined either by mechanical fasteners or interim layers with comparatively low shear stiffness. Shear forces are applied and generate relative displacements between the components, due to semi-rigidity of the joints with mechanical fasteners or shear deformations in the interim layers. As a result of these relative displacements, the Bernoulli hypothesis that plane sections remain plane no longer holds and the Euler-Bernoulli beam theory cannot be applied to the overall cross-section of the mechanically jointed beam. Longitudinally, the components of mechanically jointed beams are generally without abutment. The joints between the individual components mainly serve to accommodate shear forces, while the mechanical fasteners used include nails, screws, bolts, connectors or punched metal plate fasteners.

While glued joints are deemed rigid (see Article D6), interim layers with low shear stiffness, such as perpendicular layers in cross-laminated timber subject to rolling shear, can cause similar relative displacements between longitudinal layers to joints with mechanical fasteners. The load-bearing capacity and stiffness of mechanically jointed beams lie between the corresponding values for composite beams whose individual components are not jointed and composite beams with rigid (glued) joints. Amid rising composite stiffness between components, the load-bearing capacity and stiffness of the overall beam follow suit. Figure D7-1 clarifies the effect of a semi-rigid joint in terms of overall deformation and the course of bending stresses over the beam depth for a full cross-section (A), a cross-section comprising three loosely superimposed individual cross-sections (C) and a cross-section comprising three individual cross-sections connected via semi-rigid joints (B).

Assuming constant moduli of elasticity for the wood in beams A, B and C, as well as identical loads and span, the maximum value of bending stress in the outer fibre σ_C in beam C in Figure D7-1 is three times as large as the corresponding value of bending stress in the outer fibre σ_A for the full cross-section of beam A. For a simply supported beam with uniformly distributed load, this results in a ratio between the maximum deflection u_C for beam C and the maximum deflection u_A for beam A of nine. If the individual components are mechanically jointed, beam B, the bending stress in the outer fibre and deflection can be lowered compared to beam C. Assuming constant moduli of elasticity and dimensions of the beams, $\sigma_A \leq \sigma_B \leq \sigma_C = 3 \cdot \sigma_A$ and $u_A \leq u_B \leq u_C = 9 \cdot u_A$ applies for the system shown in Figure D7-1.

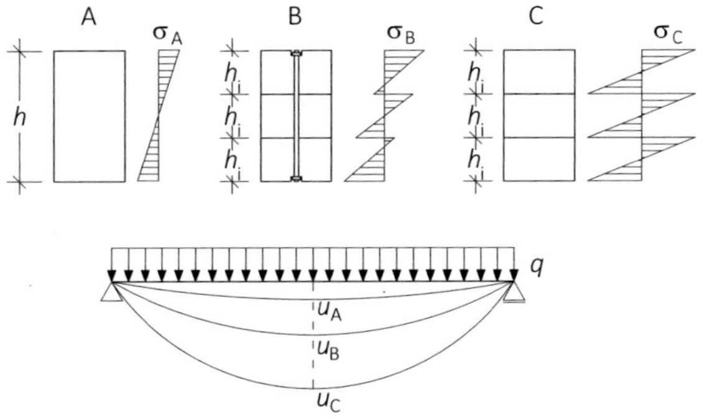

Figure D7-1 Deflection and bending stress distribution of a full cross-section (A), a cross-section comprising three individual cross-sections connected via semi-rigid joints (B) and a cross-section comprising three loosely superimposed individual cross-sections (C).

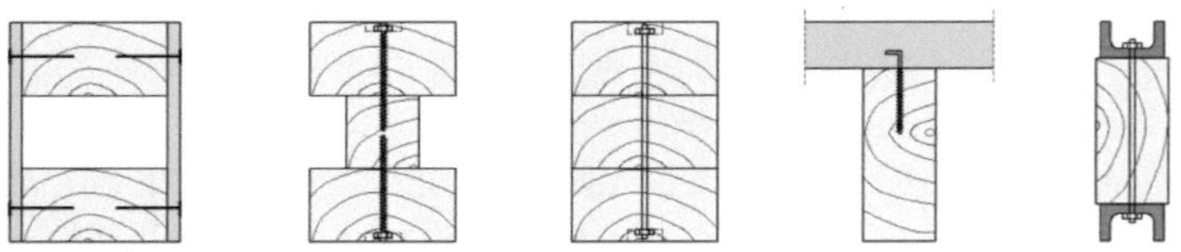

Figure D7-2 Selection of possible cross-sectional types for mechanically jointed beams.

Composite beams may comprise solely wood components or other construction materials too. Using suitable mechanical fasteners means that, in principle, any desired construction materials can be combined to form a composite beam. Examples of possible cross-sectional types for mechanically jointed beams are shown in Figure D7-2 (timber-concrete-composite see also Article D10). The key calculation methods used for mechanically jointed beams are the γ-method and the shear analogy method. The γ-method for mechanically jointed beams and columns with up to three single components is optimal for calculation by hand. Conversely, the shear analogy method is valid for any cross-sectional types also containing more than three individual components and allows freely variable component cross-sections, load or connection configurations along the beam axis. However, the shear analogy method requires the use of computer software (plane frame analysis programs), which can consider shear deformations in bar elements.

The "γ-method" (Möhler, 1956) is suitable for designing beams and columns with up to three mechanically jointed components and which are not longitudinally abutted. For mechanically jointed beams with more than three components, Schelling (1982) proposed a solution, albeit one which requires solving a system of equations, rendering it impractical to perform manually. The coefficient γ denotes a value between 0 and 1, by which the Steiner parts ($E \cdot A \cdot a^2$) of the bending stiffness contributions of the individual components within the total cross-section are reduced. If $\gamma = 1$, no shear deformations occur, while the Euler-Bernoulli beam theory applies for the overall beam. For beams with unconnected individual components, $\gamma = 0$ applies. The joint efficiency coefficient γ is therefore a measure of composite stiffness. The γ-method has limited applicability to simply supported beams with approximately uniformly distributed load, while for continuous beams, it can be used with approximation and under specific boundary conditions.

For beams with any number of mechanically jointed components, random amount of spans and a random configuration of concentrated or distributed loads, the shear analogy method (Kreuzinger, 2001; Kreuzinger and Scholz, 2003; Scholz, 2004) is used, although the shear analogy method requires the use of computer programs. Both methods are explained in the following sections whereby the shear analogy method is crucial in assessing cross-laminated timber elements.

D7.1 γ-method

The γ-method described in Annex B of EC 5 and used to calculate simple mechanically jointed beams is a vital part of timber engineering. Using coefficients γ, effective bending stiffnesses can be calculated for mechanically jointed beams from up to three individual cross-sections, via equation (D7-1). This equation corresponds to equation (D6-25) from Article D6 used to calculate rigidly glued composite components, whereby coefficients γ_i were included in equation (D7-1), which are always between 0 and 1 thereby reducing the Steiner parts ($E_i \cdot A_i \cdot a_i^2$) of the bending stiffness (where a_i is the distance between the centres of gravity of the individual cross-sections to the neutral axis):

$$(E \cdot I)_{ef} = \sum_{i=1}^{3} \left(E_i \cdot I_i + \gamma_i \cdot E_i \cdot A_i \cdot a_i^2 \right) \qquad \text{(D7-1)}$$

The joint efficiency coefficients γ_i take into consideration shear deformations in the semi-rigid joint connecting two individual components, beam B in Figure D7-1. When $\gamma = 0$, the Steiner part $E \cdot A \cdot a^2$ vanishes. In this case, the overall effective bending stiffness solely comprises the bending stiffnesses of the individual components, beam C in Figure D7-1. When $\gamma = 1$, the effective bending stiffness peaks, corresponding to that of a composite beam with rigid joints or with a geometrically identical full cross-section, beam A in Figure D7-1. Using the effective bending stiffness $(E \cdot I)_{ef}$, the γ-method facilitates deformation calculation and stress distribution of the composite cross-section whilst using the Euler-Bernoulli beam theory. The γ-method is also applicable for composite columns.

The reduction coefficient γ was derived by Möhler (1956). By studying the deformation relationships of composite cross-sections with two or three individual components without butt joints within the entire beam length and that have been continuously connected via mechanical fasteners leading to semi-rigid joints, he presented coupled differential equations of beam deflections and joint slips, which he then solved for a simply supported beam under uniformly distributed loads and a centred concentrated load, for a column centrally loaded and for two-span beams under a uniformly distributed load. Each static system revealed its own reduction coefficients γ. The γ-method disregards the shear deformation of the individual components and thus only applies to beams with a span so large that the shear deformations of the individual components remain negligible. The stiffnesses of the joints with continuous mechanical fasteners are considered with a joint stiffness assumed to be constant and dispersed over the entire beam length. Strictly speaking, the γ-value from equation (D7-2) only applies to beams with sinusoidal distributed loads or those centrally loaded, since these have a sinusoidal flexural buckling curve (to derive the γ-values, see Annex 4). The mechanical reason for this is that "...the coefficient γ and hence $(EI)_{ef}$ is only constant in the case of a sinusoidal distributed load over the beam length" (Scholz, 2004). However, given the minimal difference between simply supported beams with a uniformly or sinusoidally distributed load, the γ-method also applies to simply supported beams with a uniformly distributed load. To ensure a more comprehensive overview, the conditions to validate the γ-methods are listed here in note form. The γ-method is valid for:

- Simply supported beams with a uniformly distributed load and centrally loaded columns
- A maximum of three individual components non-abutted over the member length
- Continuous fasteners with constant stiffness over the member length
- Shear deformations of the individual components are negligible (large spans)

According to Möhler (1956), the coefficient γ_i for the individual components i = 1 and i = 3 is calculated as follows (see also Annex 5):

$$\gamma_i = \frac{1}{1+\frac{\pi^2 \cdot E_i \cdot A_i \cdot s_i}{K_i \cdot \ell^2}} \quad \text{for i=1 and i=3} \tag{D7-2}$$

For the reduction coefficient γ_2 of component 2, the following applies:

$$\gamma_2 = 1 \tag{D7-3}$$

The last, still lacking information is the position of the neutral axis which, as shown in Article D6, is determined as follows (for geometric definitions, see Figure D7-3):

$$a_2 = \frac{\sum_{i=1}^{3} x_i \cdot E_i \cdot A_i}{\sum_{i=1}^{3} E_i \cdot A_i} = \frac{-h_3/2 \cdot E_3 \cdot A_3 + h_2/2 \cdot E_2 \cdot A_2 + \left(h_2 + h_1/2\right) \cdot E_1 \cdot A_1}{E_1 \cdot A_1 + E_2 \cdot A_2 + E_3 \cdot A_3} - \frac{h_2}{2} = \frac{E_1 \cdot A_1 \cdot \left(h_1 + h_2\right) - E_3 \cdot A_3 \cdot \left(h_2 + h_3\right)}{2 \cdot \left(E_1 \cdot A_1 + E_2 \cdot A_2 + E_3 \cdot A_3\right)} \tag{D7-4}$$

This results in the following distances between the centre of gravity of the individual components 1 and 3 to the neutral axis:

$$a_1 = -\frac{h_1 + h_2}{2} + a_2 \quad \text{and} \quad a_3 = \frac{h_2 + h_3}{2} + a_2 \tag{D7-5}$$

where

E_i	Modulus of elasticity
A_i	Cross-sectional area
I_i	Second moment of area
s_i	Fastener spacing in joint between individual components
k_i	Slip modulus of fasteners
ℓ	Member length
h_i	Depth of individual component

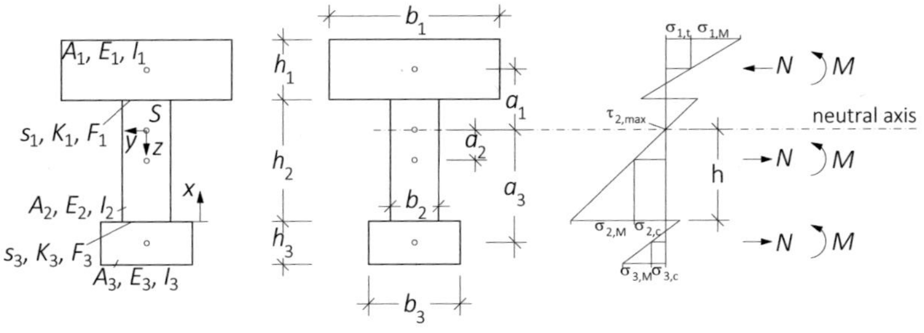

Figure D7-3 Example of a composite beam with three components.

Stress analysis

Now, all parameters required to calculate normal and shear stresses in the individual components are known, Annex 5 includes accurate information on stress analysis and the influence of bending stiffness on stress distribution. The bending stresses in the outer fibres of the individual components *i* are determined from the external moment *M* (cf. equation (D6-29)):

$$\sigma_{\mathrm{i,m}} = \frac{M}{(E \cdot I)_{\mathrm{ef}}} \cdot E_{\mathrm{i}} \cdot \frac{h_{\mathrm{i}}}{2} \tag{D7-6}$$

When calculating the normal stresses in the centre of gravity of the individual components, however, unlike glued composite beams, the coefficients γ_i are taken into consideration (for an explanation, see Annex 5):

$$\sigma_{\mathrm{i,t(c)}} = \frac{M}{(E \cdot I)_{\mathrm{ef}}} \cdot \gamma_{\mathrm{i}} \cdot E_{\mathrm{i}} \cdot a_{\mathrm{i}} \tag{D7-7}$$

The overall stress of the beam is determined by superposing the normal and bending stress in the outer fibre.

The shear stress peaks in the neutral axis (here in component 2) and amounts to (derivation see Annex 5):

$$\tau_{\max} = \frac{V_{\max} \cdot \left(\gamma_3 \cdot E_3 \cdot A_3 \cdot a_3 + 0{,}5 \cdot E_2 \cdot b_2 \cdot h^2\right)}{(E \cdot I)_{\mathrm{ef}} \cdot b_2} \tag{D7-8}$$

The load on the fasteners in the joint between the individual components amounts to (derivation see Annex 5):

$$F_{1(3)} = \frac{V_{max} \cdot \gamma_{1(3)} \cdot E_{1(3)} \cdot A_{1(3)} \cdot a_{1(3)} \cdot s_{1(3)}}{(E \cdot I)_{ef}} \qquad \text{(D7-9)}$$

Here, there is still a need to divide by the number of rows of fasteners, if multiple such rows are installed adjacent.

Strictly speaking, the γ-method only applies to simply supported beams. EC 5 Annex B, however, means the γ-method can also be used for the approximate calculation of continuous and cantilevered beams. In this case, the actual length of a continuous or cantilevered beam is translated into a fictitious length, which for cantilevered beams, equates to double the length of the cantilever ($\ell = 2 \cdot \ell_K$). For continuous beams, the fictitious length assumed should be 80% of the length of the span under examination ($\ell = 0{,}8 \cdot \ell_F$). When verifying a continuous beam at the intermediate supports, the smaller length of both adjacent spans is the key parameter. In addition, EC 5 also allows the fastener spacing in the joint to be adjusted to the course of the shear force diagram between s_{min} and s_{max} ($\leq 4 \cdot s_{min}$). In this case, the γ-method may be applied with an effective distance s_{ef} of the fasteners:

$$s_{ef} = 0.75 \cdot s_{min} + 0.25 \cdot s_{max} \qquad \text{(D7-10)}$$

Increased stresses imposed by reduced cross-sections can also be taken into consideration. For this purpose, the normal stresses in the centre of gravity of the individual components in accordance with equation (D7-7) are to be multiplied by $A_i/A_{i,net}$ and the bending stresses in the outer fibres of the individual components in accordance with equation (D7-6) by $I_i/I_{i,net}$.

Time-dependent deformations of composite beams are generally only to be taken into consideration in the ultimate limit state. However, in the event of composite beams made from various construction materials having variable creep behaviour, creep must also be taken into consideration for the ultimate limit state (EC 5 Section 2.2.2 (1)P). Time-dependent deformations are taken into consideration by reducing the moduli of elasticity E_i of the individual components and the slip moduli K_i of the fasteners in the joint with k_{def} (see also equation (D6-1)).

D7.2 Shear analogy method

Unlike the γ-method, the shear analogy method by Kreuzinger (1999) can be used to calculate any composite members under any load configuration more accurately. Details are included in the NA to EC 5. The concept of the shear analogy centres on transforming a real composite beam comprising multiple mechanically jointed components into a single fictitious and homogenised beam. In this case, two components A and B of a fictitious beam are defined, as in Figure D7-4. The fictitious component A is given the sum of the bending stiffnesses B^A along the individual neutral axis of the real components. The fictitious component B is allocated the Steiner parts ($E_i \cdot A_i \cdot z_i^2$) of the bending stiffness B^B of the real components and a shear stiffness S^B. This shear stiffness S^B includes the shear deformations of the individual components themselves and the joint slip due to the semi-rigidity of the joint between the individual components. Fictitious component A thus represents the sum of the bending stiffnesses ($E_i \cdot I_i$) and fictitious component B the interaction of the individual components through Steiner's parts ($E_i \cdot A_i \cdot z_i^2$) of the bending stiffness, shear deformation and joint slip of the overall composite beam. The following section focuses on the shear analogy method for mechanically jointed beams. This approach is also particularly applicable to uniaxially loaded cross-laminated timber elements, since a strip of any CLT element can be considered a composite beam and thus transformed into a plane frame model comprising fictitious beams A and B. However, twodimensional structural systems comprising mechanically jointed layers can also be designed by applying the shear analogy method (as described in the NA). For calculations within the framework of second order theory, see Scholz (2004).

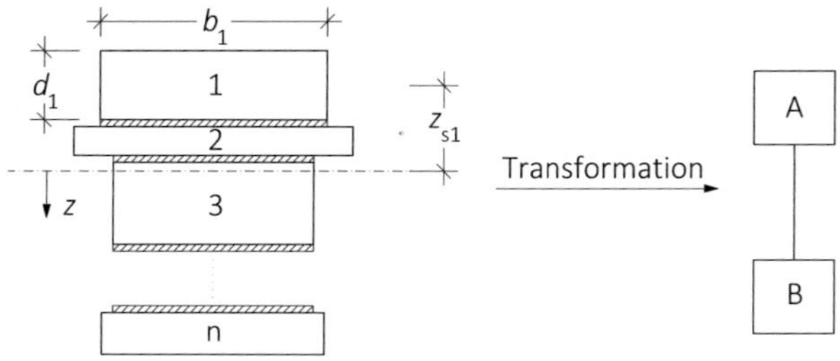

Figure D7-4 Transformation of a composite cross-section into a fictitious cross-section. (Scholz, 2004)

The bending stiffness B^A of the fictitious component A is the sum of bending stiffnesses ($E_i \cdot I_i$) of the individual components i:

$$B^A = \sum_{i=1}^{n}(E_i \cdot I_i) = \sum_{i=1}^{n}\left(E_i \cdot \frac{b_i \cdot d_i^3}{12}\right) \qquad \text{(D7-11)}$$

The bending stiffness B^B of the fictitious component B can be calculated from the sum of Steiner's parts ($E_i \cdot A_i \cdot z_i^2$) of the bending stiffnesses of the individual components *i* to:

$$B^B = \sum_{i=1}^{n}\left(E_i \cdot A_i \cdot z_{si}^2\right) = \sum_{i=1}^{n}\left(E_i \cdot b_i \cdot d_i \cdot z_{si}^2\right) \qquad \text{(D7-12)}$$

where z_{si} are the distances between the centres of gravity of the individual components *i* from the centre of gravity *S* of the entire beam.

Since the fictitious component A does not allow for shear deformations, all the shear deformations of the composite beam are allocated to the fictitious component B. The shear stiffness S^B hence includes all shear deformations of the individual components as well as the relative displacement of adjacent individual components due to the semi-rigidity in the joint, Figure D7-5.

The shear stiffness due to the shear deformation of individual components is recorded via the related shear modulus G_i and the thickness d_i of the individual components. The shear stiffness due to semi-rigidity in the joint is recorded via the slip modulus $k_{i,i+1}$ of the fasteners in the joint between the individual components *i* and *i*+1. The following applies to the relative displacement of the individual components:

$$u_i = \frac{\tau}{G_i} \cdot d_i = \frac{t}{G_i \cdot b_i} \cdot d_i \, ; \qquad u_{i,i+1} = \frac{\tau}{k_{i,i+1}} \approx \frac{t}{k_{i,i+1} \cdot b_i} \qquad \text{(D7-13)}$$

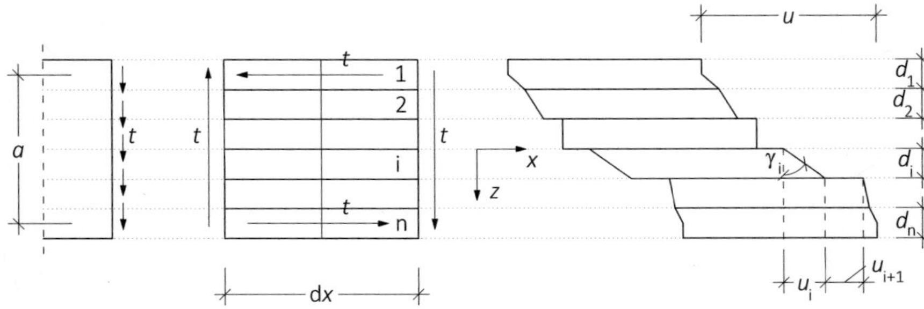

Figure D7-5 Deformation of a composite member (layered structure) subject to constant shear flow *t*. (Scholz, 2004)

The shear stiffness S^B allocated to fictitious component B is formulated such that a homogeneous beam of stiffness S^B subject to shear deforms in just the same way as the overall composite beam (the 1/2 factor in the second and fourth summand reflects the fact that both external layers 1 and n are only half applied, since the shear deformation between the centres of gravity of both external layers is taken into consideration; see also Figure D7-5):

$$\frac{1}{S^B}=\frac{1}{a^2}\cdot\left[\sum_{i=1}^{n-1}\frac{1}{k_{i,i+1}}+\frac{d_1}{2\cdot G_1\cdot b_1}+\sum_{i=2}^{n-1}\frac{d_i}{G_i\cdot b_i}+\frac{d_n}{2\cdot G_n\cdot b_n}\right] \quad \text{(D7-14)}$$

If only the semi-rigidity of the joint is considered and the individual components themselves do not allow for shear deformations, equation (D7-14) can be simplified as follows:

$$\frac{1}{S^B}=\frac{1}{a^2}\cdot\sum_{i=1}^{n-1}\frac{1}{k_{i,i+1}} \quad \text{(D7-15)}$$

To calculate the internal forces and moments in fictitious beams A and B, these are coupled to each other via infinitely rigid web members, to ensure they undergo the same deformation along their axis, Figure D7-6.

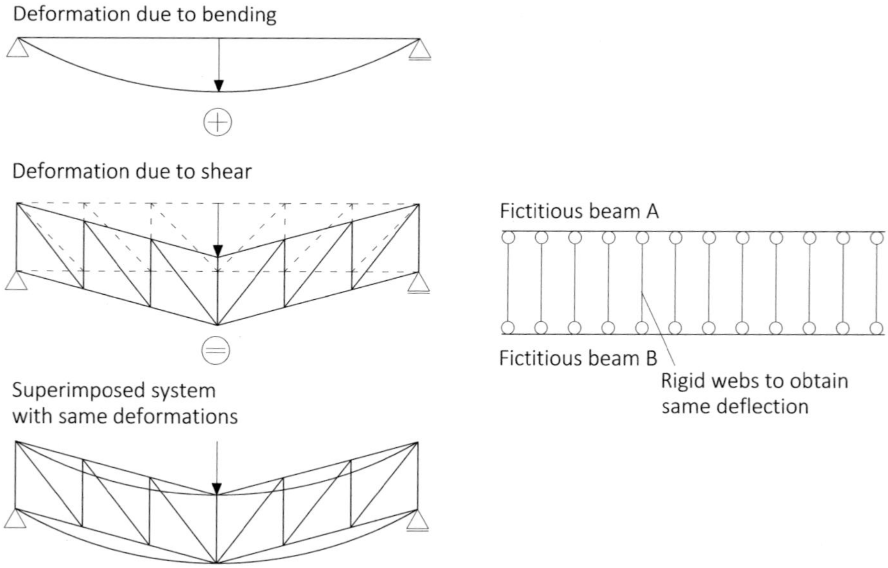

Figure D7-6 Fictitious beam with coupled fictitious components A and B.

While analytical solutions are available for this fictitious beam model to determine internal forces and moments, there is also the option of using software to determine the internal forces. The shear analogy method is designed in such a way that in practice, the use of conventional plane frame analysis programs, which take shear deformations of beams into consideration, is possible. Once the internal forces for the fictitious beams have been successfully determined, this is followed by a retransformation into the real composite system. The displacements of the fictitious beam already correspond to those of the composite beam, with only the internal forces having to be retransformed. The retransformation of the internal forces and moments into real internal forces and moments is performed in proportion to the stiffnesses. Plane frame analysis programs are hence used to determine the bending moments M^A and M^B and the shear forces V^A and V^B for the fictitious components A and B (consideration of shear deformations must be activated). The internal forces and moments of beam A, M^A and V^A, deliver the stress components along the own neutral axis on the overall stress (Figure D7-7 left) and the internal forces and moments of beam B, M^B and V^B, add those stresses to the overall stress which are due to the interaction of the individual components (Steiner parts, shear deformations of individual components and slip due to semi-rigidity of joints or perpendicular layers of CLT). Figure D7-7 shows the individual stress components and stress distribution determined by addition on the entire cross-section for a layered structure (mechanically jointed beam or CLT).

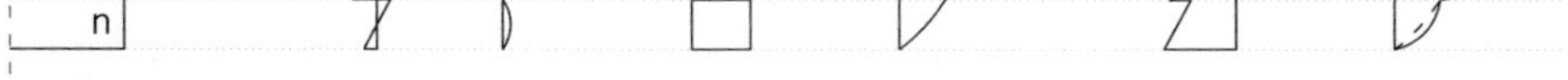

Figure D7-7 Schematic presentation of the stress analysis using the example of a layered structure.

The bending stresses in the outer fibres of the real components *i* are calculated using the bending moment M^A of the fictitious component A:

$$M_i^A = \frac{E_i \cdot I_i}{B^A} \cdot M^A \Rightarrow \sigma_{m,i}^A(\pm \frac{d_i}{2}) = \pm \frac{M_i^A}{I_i} \cdot \frac{d_i}{2} = \pm \frac{M^A}{B^A} \cdot E_i \cdot \frac{d_i}{2} \qquad \text{(D7-16)}$$

The bending moments of the fictitious component B, namely Steiner's parts on the overall stress, in turn, elicit the normal stresses, which are deemed constant over the individual components (z_{si} are the distances between the centres of gravity of components i from the centre of gravity S of the entire beam):

$$\sigma_{t/c,i}^{B}(\pm z_{si}) = \frac{M^{B}}{B^{B}} \cdot E_{i} \cdot z_{si} \tag{D7-17}$$

The position z_{0s} of the centre of gravity S of the entire beam, in relation to a reference axis 0, is determined similarly to equation (D7-4), with z_{0i} the distances between the centres of gravity of the individual components i to the reference axis 0:

$$z_{0s} = \frac{\sum_{i=1}^{n} z_{0i} \cdot E_{i} \cdot A_{i}}{\sum_{i=1}^{n} E_{i} \cdot A_{i}} \tag{D7-18}$$

From the equilibrium of forces at an infinitesimally small element come the shear stresses τ, which can be recorded from the integration, area by area, over individual component cross-sections using the shear forces of the fictitious component A:

$$\tau_{i}^{A}(z_{i}) = -\int_{-\frac{d_i}{2}}^{z_i} \frac{d\sigma^{A}}{dx} \cdot dz = -\int_{-\frac{d_i}{2}}^{z_i} \frac{dM^{A}}{dx} \cdot \frac{E_{i}}{B^{A}} \cdot z_{i} \cdot dz = -V^{A} \cdot \frac{E_{i}}{B^{A}} \cdot \left(\frac{z_{i}^{2}}{2} - \frac{d_{i}^{2}}{8} \right) \tag{D7-19}$$

and using shear forces of the fictitious component B:

$$\tau_{i}^{B}(z_{si}, z_{i}) = -\int_{-\frac{d_i}{2}}^{z_i} \frac{d\sigma^{B}}{dx} \cdot dz = -\int_{-\frac{d_i}{2}}^{z_i} \frac{dM^{B}}{dx} \cdot \frac{E_{i}}{B^{B}} \cdot z_{si} \cdot dz = -V^{B} \cdot \frac{E_{i}}{B^{B}} \cdot z_{si} \cdot \left(z_{i} + \frac{d_{i}}{2} \right) \tag{D7-20}$$

The maximum shear stresses in the centres of gravity of the individual components generated by the shear forces V^{A} can be determined using the well-known equation for rectangular cross-sections:

$$\tau_{i}^{A} = 1.5 \cdot \frac{E_{i} \cdot I_{i}}{B^{A}} \cdot \frac{V^{A}}{d_{i} \cdot b_{i}} \tag{D7-21}$$

The stress curves along the real composite beam are obtained by superposing individual stresses. Figure D7-7 clarifies the stress analysis via retransformation on the example of a cross-section comprising multiple layers.

The γ-method and the shear analogy method are suitable for calculating mechanically jointed beams, while the latter shear analogy method in particular is also useful for calculating twodimensional load-bearing structures, i.e. solid wood panels and cross-laminated timber elements. Although solid wood panels and cross-laminated timber elements with perpendicular layers lack any semi-rigid joints between single layers, such products can be considered like mechanically jointed beams and calculated likewise. The individual longitudinal layers arranged in the principal direction of the panel correspond to the individual components of a mechanically jointed beam, while the semi-rigid joint between individual components is mapped by the perpendicular layers arranged perpendicular to the principal panel direction. If solid wood panels and cross-laminated timber elements are exposed to out-of-plane bending stress, the longitudinal layers shift relatively to their individual positions, which ultimately leaves the perpendicular layers distorted due to shear. This shear stress in the tangential-radial plane is also referred to as rolling shear (Figure D7-8).

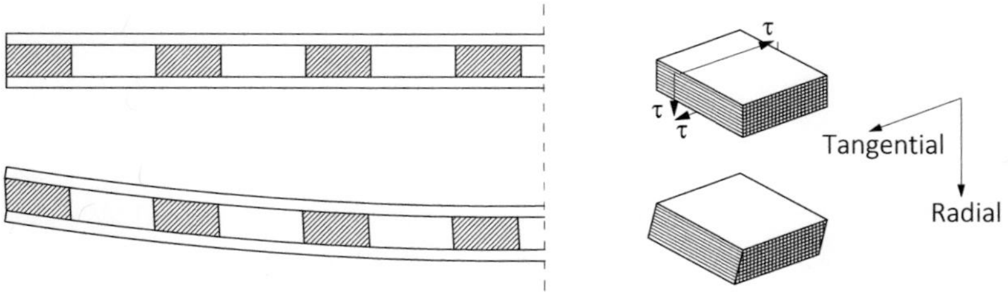

Figure D7-8 Left: Deflected three-layered panel. Right: Rolling shear stress of the middle layer.

The rolling shear stiffness and strength of wood are clearly lower than the corresponding stiffness and strength values for longitudinal shear in the tangential-longitudinal or radial-longitudinal plane. The rolling shear modulus comprises around 10% of the shear modulus of the wood, while the rolling shear strength is around 40% of its (longitudinal) shear strength. The shear analogy method can be directly applied to calculate twodimensional structural members with cross-wise arranged individual layers. In this case, the rolling shear in the middle layers is taken into consideration when shear stiffness s^B in accordance with equation (D7-14), i.e. $G_i = G_{R,i}$ for perpendicular layers. When applying the γ-method, the "weak" perpendicular layers loaded in rolling shear are considered to constitute a semi-rigid joint between the longitudinal layers arranged in the principal beam direction. This semi-rigidity is recorded in the γ-coefficients, whereby instead of the stiffness K/s, the rolling shear stiffness $G_R \cdot b/d$ of the perpendicular layer is the basis. In this case, b is the width and d the thickness of the perpendicular layer:

$$\gamma_i = \frac{1}{1 + \frac{\pi^2 \cdot E_i \cdot A_i \cdot d_i}{G_{R,i} \cdot b_i \cdot \ell^2}} \quad \text{for i=1 and i=3} \qquad \text{(D7-22)}$$

D7.3 Literature

H. Kreuzinger, original Article B11, STEP 1995.

Kreuzinger H. (2001). Verbundkonstruktionen. Holzbaukalender 2002, pp. 598-621. Bruderverlag Karlsruhe.

Kreuzinger H. and Scholz A. (2003). Flächentragwerke – Berechnung und Konstruktion. Schlussbericht AIF Forschungsvorhaben, Technische Universität München.

Möhler K. (1956). Über das Tragverhalten von Biegeträgern und Druckstäben mit zusammengesetzten Querschnitten und nachgiebigen Verbindungsmitteln. Habilitation, Technische Universität Karlsruhe.

Schelling W. (1982). Zur Berechnung nachgiebig zusammengesetzter Biegeträger aus beliebig vielen Einzelquerschnitten. In: „Ingenieurholzbau in Forschung und Praxis", Herausgeber J. Ehlbeck und G. Steck; Bruderverlag, Karlsruhe. ISBN 3-87104-049-5.

Scholz A. (2004). Ein Beitrag zur Berechnung von Flächentragwerken aus Holz. Dissertation, Technische Universität München.

D8 Reinforcements

Original article: H. J. Larsen

In a structural timber context, the term 'reinforcements' refers to measures used to increase the ability of structural members to withstand tension or compression perpendicular to the grain or shear. The members can be boosted using internal or external reinforcements, where cross-sectional weakening caused by internal reinforcements must be taken into account:

Internal reinforcements:

- Glued-in threaded rods,
- Glued-in rebars,
- Fully threaded wood screws.

External reinforcements:

- Glued-on plywood,
- Glued-on LVL,
- Glued-on boards,
- Punched metal plate fasteners.

In areas where notches, holes and joints loaded perpendicular to the grain are present, as well as in double-tapered, pitched cambered and curved beams, climate fluctuations trigger swelling and shrinking processes, which can cause tension perpendicular to the grain exerted by external loads to intensify significantly. This is why, in accordance with NA, reinforcements perpendicular to the grain are required in service class 3.

This article sets out verifications for reinforcements in notches, holes and apex zones of double-tapered, pitched cambered and curved beams, while joints loaded perpendicular to the grain are covered in Article E11. All verifications are given in the NA; EC 5 does not include any design rules for reinforcements. As a general rule, any reinforcement to increase resistance against tension perpendicular to the grain should be, conservatively, capable of withstanding the entire tensile force perpendicular to the grain; this equates to a state in which the wood has failed perpendicular to the grain.

Other types of reinforcements include those of supports with fully threaded screws to enhance compression behaviour perpendicular to the grain, where the screws are inserted perpendicular to the support area. This type of reinforcement is covered in Article E5.

D8.1 Effects of fluctuations in moisture content

Tensile stresses perpendicular to the grain are generated not only by external loads, but also by changing levels of moisture content, where the cross-sectional moisture gradient is of particular significance. For members in service class 1, which have been dried to an appropriate level of moisture content before assembly, the general assumption is that the impact of any additional load imposed by moisture changes will be sufficiently covered by characteristic values and partial safety coefficients. Under certain circumstances, however, the impact of moisture changes may have to be assessed.

For example, assume the moisture content in the external area of a cross-section (respectively 1/6) declined by 3%. This would correspond to unhindered strain of $\Delta\varepsilon = 3 \cdot 0.002 = 0.006$ (with a shrinkage coefficient of 0.2% per 1% change in moisture content below the fibre saturation point (see Article B2)). With $E_{90} = 300$ N/mm^2, it follows that $\Delta\sigma = 2/3 \cdot 300 \cdot 0.006 = 1.2$ N/mm^2, namely the stresses generated by hindered shrinking are in the same order of magnitude as the short-term strength and the risk of failure exists. In practice, however, while such stresses are reduced by relaxation, they should not be ignored.

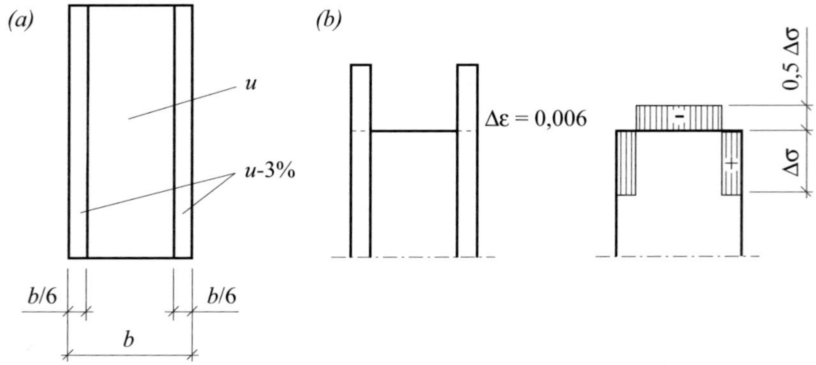

Figure D8-1 Influence of a decrease in moisture content in the (*a*) external fibres of a rectangular cross-section, (*b*) unhindered strain, (*c*) resulting internal stresses. (STEP 1995 Article E5)

One effect of an increase in moisture content is an increase in beam depth from h to $h \cdot (1 + \varepsilon)$, where ε corresponds to the strain resulting from this increase in humidity (for a beam 600 mm deep, this would equate to an increase of 4 mm).

The influence of moisture content parallel to the grain is generally negligible. However, in curved beam areas, the angle $d\varphi$ declines to $d\varphi'$ and the radius of curvature increases from r to r', Figure D8-2 (a):

$$d\varphi' = \frac{d\varphi}{1+\varepsilon} \approx d\varphi \cdot (1-\varepsilon) \tag{D8-1}$$

$$r' = r \cdot (1+\varepsilon) \tag{D8-2}$$

These changes indicate an increase in the distance between supports and a reduction in curvature of (see Figure D8-2 (b)):

$$\Delta v = r \cdot \varepsilon \cdot (1-\cos\varphi) + \frac{a \cdot \varepsilon \cdot \varphi}{2} \tag{D8-3}$$

The reduction in curvature seen in this example results in a lower level of tensile stress perpendicular to the grain.

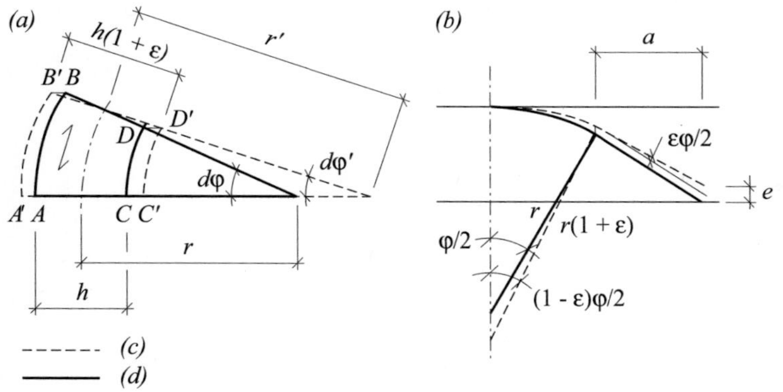

Figure D8-2 Influence of an increase in moisture content on the curvature of a beam. (*a*) geometry after the increase in moisture content, (*b*) undeformed state. *e* is the entire deflection at centre span. (STEP 1995 Article E5)

D8.2 Reinforcement of notched beams

Reinforcements of beams notched on the same side as the support are designed for a tensile force $F_{t,90,d}$ perpendicular to the grain, which corresponds to the tensile stress perpendicular to the grain of area 1 as specified in Figure D8-3. This tensile force can subsequently be transferred by glued-in rods, glued-on reinforcement plates or fully threaded screws, Figure D8-4. The first step involves calculating the tensile component perpendicular to the grain generated from an external load in the critical area 1. The derivation for reinforcements of notches corresponds to the derivation for joints loaded perpendicular to the grain (Ehlbeck and Görlacher, 1983) and is described in detail in Article E11. At this point, only the required changes to the design equations compared to those in Article E11 are shown.

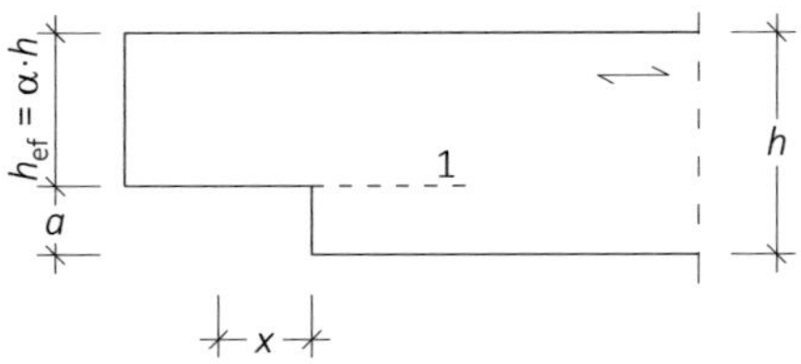

Figure D8-3 Rectangular notch on the same side as the support with critical area 1.

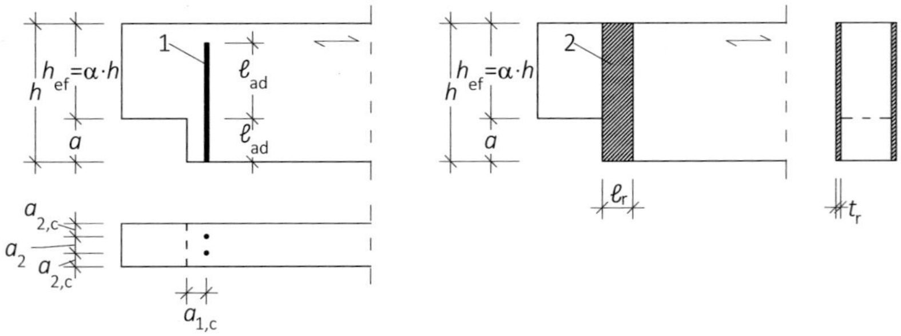

Figure D8-4 Reinforcement of notched beams with geometric details, 1: Reinforcement with fully threaded screw / glued-in rod, 2: Reinforcement with glued-on plates.

The equation specified in NA and used to determine the force component $F_{t,90,d}$ generating tension perpendicular to the grain in critical area 1 is as follows:

$$F_{t,90,d} = 1.3 \cdot V_d \cdot \left(3 \cdot (1-\alpha)^2 - 2 \cdot (1-\alpha)^3\right) = 1.3 \cdot V_d \cdot \eta \tag{D8-4}$$

The factor η used in equation (D8-4) for notches corresponds to the factor η specified in equation (E11-15) for joints loaded perpendicular to the grain if α is replaced with $a/h = h_{ef}$ and the term is converted, whereby a represents the distance from the loaded edge (for notches, this is the distance between the notch corner and the upper edge, see Figure D8-3):

$$\eta = 3 \cdot (1-\alpha)^2 - 2 \cdot (1-\alpha)^3 = 1 - 3 \cdot \left(\frac{a}{h}\right)^2 + 2 \cdot \left(\frac{a}{h}\right)^3 \tag{D8-5}$$

However, the stresses in the notch corners are not uniformly distributed (see also Figure D5-2), which results in peak stresses, which decline with increasing distance from the notch corner. FE simulations by Henrici (1984) resulted in the factor 1.3 used in equation (D8-4), which involves the force component $F_{t,90,d}$ increasing by 30%.

If the reinforcement is done with **glued-in rods or fully threaded screws** (number 1 in Figure D8-4), on the one hand, the means of reinforcement used must be capable of absorbing the tensile force $F_{t,90,d}$ and on the other, the bond between the wood and reinforcement must be guaranteed. This can be verified for fully threaded screws by designing for stress in the direction of the screw axis (withdrawal capacity $F_{ax,Rd}$ and tensile capacity $F_{t,Rd}$, see Article E5; the smaller value must exceed $F_{t,90,d}$ from equation ((D8-4)). For glued-in rods, the bond line must be checked, in which case a uniformly distributed bond line stress is assumed:

$$\tau_{ef,d} \leq f_{k1,d} \quad \text{with} \quad \tau_{ef,d} = \frac{F_{t,90,d}}{n \cdot d_r \cdot \pi \cdot \ell_{ad}} \tag{D8-6}$$

The characteristic value for the bond line strength $f_{k1,k}$ is specified in the NA (see also Table D8-1). The effective bond line stress $\tau_{ef,d}$ is calculated from the tensile force $F_{t,90,d}$ in accordance with equation (D8-4) divided by the bond line area, which is calculated from the effective anchorage length ℓ_{ad}, the external rod diameter d_r and the number n of glued rods used. The effective anchorage length ℓ_{ad} corresponds to the height of the notched area ($h - h_{ef}$) and is shown in Figure D8-4. The constraint is that in longitudinal member axis direction, only one rod can be taken into account, which must have a minimum length of $2 \cdot \ell_{ad}$ and which must not exceed the external diameter $d_r = 20$ mm. Figure D8-4 shows, in turn, the minimum distances of the rods. The spacings a_2 have to be a minimum of $3 \cdot d_r$, while the minimum figure for edge distances $a_{2,c}$ and for end distances $a_{1,c}$ is at least $2.5 \cdot d_r$.

Table D8-1 Characteristic bond line strength in reinforcements, DIN EN 1995-1-1/NA:2013.

	Characteristic strength [N/mm²]	Effective anchorage length ℓ_{ad} of the rod [mm]		
		≤ 250	$250 \leq \ell_{ad} \leq 500$	$500 \leq \ell_{ad} \leq 1000$
Bond line between rod and borehole wall	$f_{k1,k}$	4.0	$5.25 - 0.005 \cdot \ell_{ad}$	$3.5 - 0.0015 \cdot \ell_{ad}$
Bond line between beam surface and reinforcement plate	$f_{k2,k}$	0.75		
Bond line between beam surface and reinforcement plate if shear stresses are uniformly introduced	$f_{k3,k}$	1.50		

Laterally glued-on reinforcement plates are accounted for correspondingly; the bond line stress, which is assumed to be uniformly distributed, is calculated from the tensile force $F_{t,90,d}$ divided by the bond line area. However, the design value of bond line strength $f_{k2,d}$ equates approximately to the design rolling shear strength of the member to be reinforced and also takes into account a non-uniform shear stress distribution (characteristic value see Table D8-1):

$$\tau_{ef,d} \leq f_{k2,d} \quad \text{with} \quad \tau_{ef,d} = \frac{F_{t,90,d}}{2 \cdot (h - h_{ef}) \cdot \ell_r} \tag{D8-7}$$

As with glued-in rods, the effective bond line area taken into account here is the area below critical area 1: $2 \cdot \ell_r \cdot (h - h_{ef})$, with factor 2 due to reinforcement on both sides, see Figure D8-4.

The tensile force $F_{t,90,d}$ determined in accordance with equation (D8-4) must be transmitted via the bond line to the reinforcement plates, which must, in turn, be capable of transferring the tensile stresses formed, whereby a triangular stress distribution is taken into consideration in this case:

$$k_k \cdot \sigma_{t,d} \leq f_{t,d} \qquad \text{with} \qquad \sigma_{t,d} = \frac{F_{t,90,d}}{2 \cdot t_r \cdot \ell_r} \tag{D8-8}$$

where

t_r Thickness of a reinforcement plate

ℓ_r Width of a reinforcement plate

$f_{t,d}$ Design tension strength of the plate material

Coefficient k_k takes the non-uniformly distributed tensile stress into consideration, which peaks at the notch corner and can be assumed as $k_k = 2.0$ without further verification.

The effective area of the glued-on reinforcement plates must include a certain ratio between height $(h - h_{ef})$ and width ℓ_r. The minimum value should prevent any crack development in critical area 1; the maximum value should ensure that the only area of the reinforcement plate taken into account is that located in the notch area subject to tension perpendicular to the grain:

$$0.25 \leq \frac{\ell_r}{h - h_{ef}} \leq 0.5 \tag{D8-9}$$

D8.3 Reinforcement of beams with holes

The procedures used to verify reinforcements of holes are similar to those used for notches, but the tensile force component $F_{t,90,d}$ perpendicular to the grain and the effective bond line surfaces are calculated differently. Verifications for reinforcements of areas of rectangular and circular holes subjected to tension perpendicular to the grain in accordance with Figure D8-5 are shown; possible reinforcements see Figure D8-6. In addition, the holes to be reinforced must comply with the minimum and maximum dimensions specified in Table D8-2. If the geometric boundary conditions are fulfilled, the tensile force to be applied may be calculated in accordance with equation (D5-11). Similar to the approach when designing reinforced notches, this tensile force $F_{t,90,d}$ is used to verify bond lines of glued-in rods, bond lines and tensile stresses of glued-on plated and fully threaded screws.

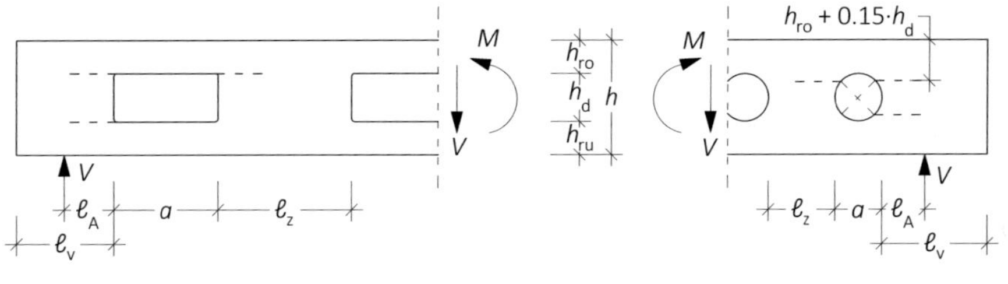

Figure D8-5 Beam with rectangular and circular hole.

Table D8-2 Minimum and maximum dimensions for reinforced holes, DIN EN 1995-1-1/NA:2013.

$\ell_v \geq h$	$\ell_z{}^c \geq h$, but at least 300 mm	$\ell_A \geq h/2$	$h_{ro(ru)} \geq 0.25 \cdot h$	$a \leq h$	$h_d \leq 0.3 \cdot h^a$
				$a/h_d \leq 2.5$	$h_d \leq 0.4 \cdot h^b$

[a] for internal reinforcement, [b] for external reinforcement,
[c] ℓ_z = distance between two adjacent holes, see also Figure D5-7.

When using **fully threaded screws**, they are verified for a tensile force $F_{t,90,d}$ in the direction of the screw axis; in turn, with the withdrawal capacity $F_{ax,Rd}$ and tensile capacity $F_{t,Rd}$ from Article E5, where the smaller value must exceed $F_{t,90,d}$ from equation (D5-11).

Verification of the bond line stress for **glued-in rods**:

$$\tau_{ef,d} \leq f_{k1,d} \quad \text{where} \quad \tau_{ef,d} = \frac{F_{t,90,d}}{n \cdot d_r \cdot \pi \cdot \ell_{ad}} \tag{D8-10}$$

Compared to equation (D8-6), only the determination of the effective anchorage length ℓ_{ad} changes and all other rules remain the same. In addition, however, for holes with internal reinforcements, increased shear stresses in the hole area must be verified (see below).

Effective anchorage length for rectangular holes (Figure D8-5):

$$\ell_{ad} = h_{ru} \quad \text{or} \quad \ell_{ad} = h_{ro} \tag{D8-11}$$

For circular holes (Figure D8-5):

$$\ell_{ad} = h_{ru} + 0.15 \cdot h_d \quad \text{or} \quad \ell_{ad} = h_{ro} + 0.15 \cdot h_d \tag{D8-12}$$

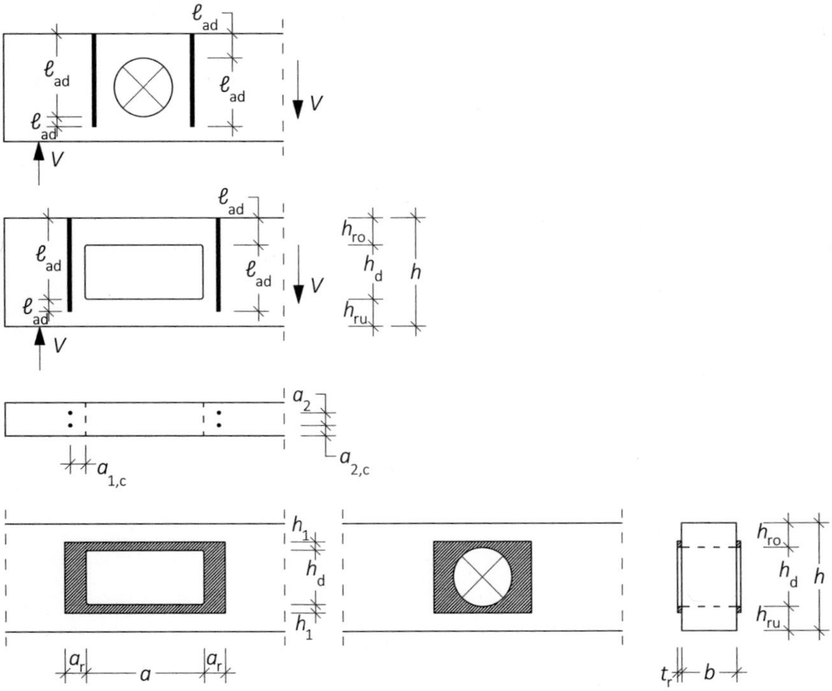

Figure D8-6 Reinforcement of holes with geometric details.

Verification of bond line stress for **glued-on plates**:

$$\tau_{\text{ef,d}} \leq f_{\text{k2,d}} \qquad \text{with} \qquad \tau_{\text{ef,d}} = \frac{F_{\text{t,90,d}}}{2 \cdot a_{\text{r}} \cdot h_{\text{ad}}} \tag{D8-13}$$

For rectangular holes (Figure D8-6):

$$h_{\text{ad}} = h_1 \tag{D8-14}$$

For circular holes (Figure D8-6):

$$h_{\text{ad}} = h_1 + 0.15 \cdot h_{\text{d}} \tag{D8-15}$$

Verification of tensile stresses in glued-on reinforcement plates, which likewise here, correspond to verifications of those of notches:

$$k_k \cdot \sigma_{t,d} \le f_{t,d} \quad \text{with} \quad \sigma_{t,d} = \frac{F_{t,90,d}}{2 \cdot a_r \cdot t_r} \tag{D8-16}$$

where

t_r Thickness of a reinforcement plate

a_r Width of the reinforcement plate to be taken into consideration, see Figure D8-6

$f_{t,d}$ Design tension strength of the plate material

The coefficient k_k takes into consideration the non-uniformly distributed stresses and also in this case, can be assumed to be $k_k = 2.0$ without further evidence required.

As with reinforced notches, a certain ratio between height and width must be maintained for the effective area of the glued-on reinforcement plates:

$$\begin{aligned} &0.25 \cdot a \le a_r \le 0.6 \cdot \ell_{t,90} \quad \text{with} \quad \ell_{t,90} = 0.5 \cdot (h_d + h) \\ &h_1 \ge 0.25 \cdot a \end{aligned} \tag{D8-17}$$

Verification of the **increased shear stress in the hole area** for internal reinforcements is not specified in the NA but can be carried out using the approach by Blass and Bejtka (2004), as well as the "Erläuterungen zur DIN 1052" (Blass et al., 2005, Section E11.4.4). Accordingly, the key shear stresses peak at the following value:

$$\tau_{max} \le \kappa_{max} \cdot \frac{1.5 \cdot V_d}{b \cdot (h - h_d)} \tag{D8-18}$$

where

$$\kappa_{max} = 1.84 \cdot \left(1 + \frac{a}{h}\right) \cdot \left(\frac{h_d}{h}\right)^{0.2} \tag{D8-19}$$

and $0.1 \le a/h \le 1.0$ and $0.1 \le h_d/h \le 0.4$.

D8.4 Reinforcement of double-tapered, pitched cambered and curved beams

As already explained in Article D4, tensile stresses perpendicular to the grain are generated in double-tapered, pitched cambered and curved beams, due to the action of external forces. Changes in moisture content trigger additional tensile stresses perpendicular to the grain. As a general rule, verifications of the external forces acting on such beam shapes are carried out and climate-related tensile stresses perpendicular to the grain are not taken into consideration. Accordingly, as specified in Article D4, the NA includes rules to take climate-related tensile stresses perpendicular to the grain into account. Reinforcements in the apex zone subject to tensile stresses perpendicular to the grain are frequently employed for double-tapered, pitched cambered and curved beams, which are exposed to changes in moisture content or where the calculated level of tensile stresses perpendicular to the grain is excessive. The NA regulates two different measures for reinforcement:

- Reinforcements intended to absorb additional climate-related stresses perpendicular to the grain (→ service classes 1 and 2) and
- Reinforcements to completely absorb tensile stresses perpendicular to the grain (external forces and climate loads), (→ service classes 1 and 2, mandatory for use in service class 3).

As with reinforcements for notches and holes, both measures involve determining the tensile force, which is generated from the tensile stresses perpendicular to the grain. In turn, this tensile force must then be capable of being transferred by the bond line between the wood and glued-in rods, by the anchorage area of screws or the bond line between wood and glued-on reinforcement plates. In addition, the tensile stresses in the glued-in rods, screws or reinforcement boards must be verified. The two reinforcement measures outlined in the NA are discussed in the following section and understanding of the current article presumes previous understanding of Article D4.

Figure D8-7 shows a schematical view of the course of tensile stresses perpendicular to the grain within the apex zone along the axis of a curved beam. In the design, a simplified (stepped) trajectory of tensile stresses perpendicular to the grain is considered, rather than the actual non-linear course, which allows the values of the forces $F_{t,90,d}$ which the reinforcements have to transfe to be calculated. The tensile stresses perpendicular to the grain decline based on stress and geometry with increasing distance from the apex and two separate areas of tensile stresses perpendicular to the grain can be differentiated, as shown in Figure D8-7: an internal area, the internal half or the two internal quarters of the apex cross-section and an external area with lower tensile stresses perpendicular to the grain, which encompasses both external quarters of the apex cross-section.

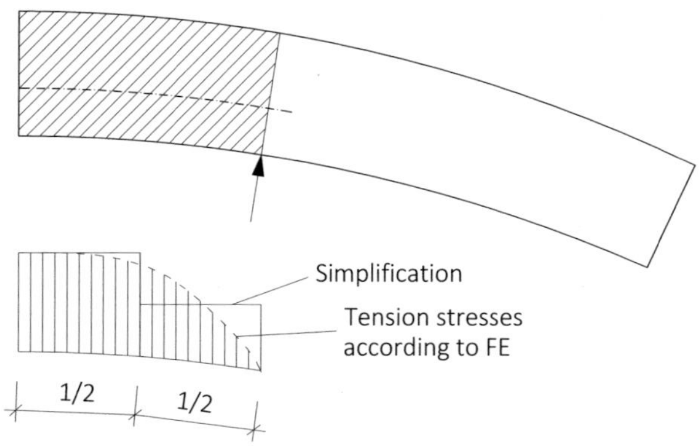

Figure D8-7 Tensile stresses perpendicular to the grain in the apex zone of a curved beam, shown qualitatively.

Reinforcement to absorb additional climate-related tensile stresses perpendicular to the grain

For reinforced double-tapered, pitched cambered and curved beams in service classes 1 and 2, the rule according to NA is that conditions in accordance with equations (D4-15) and (D4-18) can be overlooked, if it is possible to prove that the combination of tensile stresses perpendicular to the grain and shear stresses in the apex zone matches equation (D8-20):

$$\left(\frac{\tau_d}{f_{v,d}}\right)^2 + \frac{\sigma_{t,90,d}}{k_{dis} \cdot \left(\frac{h_0}{h_{ap}}\right)^{0.3} \cdot f_{t,90,d}} \leq 1 \qquad \text{(D8-20)}$$

where

k_{dis} = 1.3 for double-tapered and pitched cambered beams
= 1.15 for curved beams

h_0 Reference height of 600 mm

h_{ap} Beam height at apex, see Figure D4-10 and Figure D4-11

$\sigma_{t,90,d}$ Tensile stresses perpendicular to the grain at apex, calculated with equation (D4-13)

Equation (D8-20) is very similar to equation (D4-18), which was used to verify combined tensile perpendicular to the grain and shear stresses. However, equation (D8-20) specified in the NA is less conservative due to the semi-quadratic interaction between tensile and shear stresses. In addition, the volume factor k_{vol} from equations (D4-16) and (D4-17) is replaced with $(h_0/h_{ap})^{0.3}$. Although this elicits values which change somewhat, the function is the same as with k_{vol}; namely to take into consideration the volume effect on the tensile strength perpendicular to the grain of wood. The coefficient k_{dis}, that considers the stress distribution in the apex zone, is assigned values other than those specified in EC 5 (k_{vol} and k_{dis} see also Article D3).

The reinforcement elements to be deployed in this case must be capable of absorbing a portion of the additional climate-related tensile stresses perpendicular to the grain, while when determining the additional tensile forces perpendicular to the grain, the strain in wood and reinforcement elements were assumed to be equal (Brünninghof et al., 1993). Therefore, interaction between the wood and reinforcement is presumed, meaning that in this case, the reinforcement is not configured to absorb the full extent of the tensile forces perpendicular to the grain. Brünninghof et al. (1993) showed that the strains in the wood and reinforcement element were only approximately equal, when around a quarter of the tensile forces perpendicular to the grain $F_{t,90,d}$ generated from external loads was absorbed by the reinforcement elements. Moreover, most investigations centring on the derivation of $F_{t,90,d}$ assumed the use of beams $b = 160$ mm wide. However, given the intensifying impact of climate as beams widen, a factor of $b/160$ was introduced, to take account of the harsher situation applicable for such wider beams. Accordingly, the reinforcements applied to such beams to absorb climate-related tensile stresses perpendicular to the grain should be designed such to withstand the tensile force $F_{t,90,d}$, which can be calculated according to equation (D4-13) from the tensile stresses perpendicular to the grain. $F_{t,90,d}$ is the tensile force per reinforcement (where $\sigma_{t,90,d} \cdot b \cdot a_1$ equates to the entire tensile force exerted over the length a_1, a ¼ of which should be absorbed by the reinforcements):

$$F_{t,90,d} = \frac{1}{n} \cdot \frac{1}{4} \cdot \sigma_{t,90,d} \cdot b \cdot a_1 \cdot \frac{b}{160} = \frac{\sigma_{t,90,d} \cdot b \cdot a_1}{n} \cdot \frac{b}{640} \qquad \text{(D8-21)}$$

where

- b Beam width in mm
- a_1 Spacing of reinforcements in axial beam direction = tributary width to determine the tensile force perpendicular to the grain
- n Number of reinforcements within length a_1, namely the number of reinforcements per row

It must be possible to accommodate this tensile force by both glued-in rods or glued-on plates as well as by the shear stresses in the joints between the reinforcement elements and timber. The characteristic values specifying the strength of the bond line $f_{k1,k}$ or $f_{k2,k}$ and required to verify the bond line stress are specified in Table D8-1.

Bond line stress in **glued-in or screwed-in threaded rods**, respectively, is verified similarly to the verification on notches and holes, as in equations (D8-6) and (D8-10), whereby the tensile force $F_{t,90,d}$ has already been calculated for an individual rod and factor 2 in the numerator takes into consideration the fact that in what are now generally longer reinforcement rods, the shear stress is no longer uniformly distributed over the rod length. This means that the stresses involved are doubled, similar to coefficient k_k in equations (D8-8) and (D8-16) and a triangular distribution of shear stress is assumed over the effective anchoring length ℓ_{ad} above and underneath the beam axis:

$$\tau_{ef,d} \le f_{k1,d} \quad \text{with} \quad \tau_{ef,d} = \frac{2 \cdot F_{t,90,d}}{d_r \cdot \pi \cdot \ell_{ad}} \qquad \text{(D8-22)}$$

The withdrawal capacity of threaded rods with threads similar to screws is assessed similarly to the verifications of fully threaded screws exposed to axial loads (Article E5). Here, reference can also be made to the European Technical approvals (ETA) of these rods to determine their characteristic withdrawal parameter.

The verification of the bond line stress of a **glued-on reinforcement plate**, in turn, is determined similarly to the verifications for notches and openings, as in equations (D8-7) and (D8-13), but assuming higher values for the bond line strength ($f_{k3,d}$ instead of $f_{k2,d}$), since an uneven distribution of shear stresses has already been taken into consideration in $\tau_{ef,d}$. ℓ_r is the length of the glued-on reinforcement plate and ℓ_{ad} is its height, above or underneath the beam axis:

$$\tau_{ef,d} \le f_{k3,d} \quad \text{with} \quad \tau_{ef,d} = \frac{2 \cdot F_{t,90,d}}{\ell_{ad} \cdot \ell_r} \qquad \text{(D8-23)}$$

The tensile stresses in a reinforcement plate of thickness t_r, length ℓ_r and tensile strength $f_{t,d}$ are verified as follows:

$$\sigma_{t,d} \le f_{t,d} \quad \text{with} \quad \sigma_{t,d} = \frac{F_{t,90,d}}{t_r \cdot \ell_r} \qquad \text{(D8-24)}$$

Here, compared to equations (D8-8) and (D8-16), coefficient k_k is missing, since uneven stress distribution was already taken into consideration when determining $F_{t,90,d}$. Moreover, $F_{t,90,d}$ was already determined per reinforcement element.

Reinforcements to completely absorb tensile stresses perpendicular to the grain

Reinforcements are required according to NA when using double-tapered, pitched cambered and curved beams in service class 3, but may also be used in service classes 1 and 2. This then eliminates the need for tensile stress perpendicular to the grain verifications in accordance with equations (D4-15) and (D4-18), since all tensile stresses perpendicular to the grain are absorbed by reinforcement elements in this case. Here, the tensile forces $F_{t,90,d}$ exerted on the reinforcements must be determined more accurately, before the aforementioned verifications can be conducted via equations (D8-22), (D8-23) and (D8-24). This is followed, in turn, by verifications of tensile stresses in the cross-section of glued-in rods or screwed-in threaded rods. In such cases, the tensile forces $F_{t,90,d}$ to be applied are determined for both internal and external areas of the beam, subject to tensile stresses perpendicular to the grain, whereby for simplicity, the tensile forces in the external area are assumed to be 2/3 of the values of the internal area. Likewise in this case, the decisive tensile stress perpendicular to the grain $\sigma_{t,90,d}$ is deemed to be the value $\sigma_{t,90,d}$ calculated with equation (D4-13).

The tensile force in a reinforcement element in the internal area (both internal quarters), corresponds to equation (D8-21) without factor $b/640$:

$$F_{t,90,d} = \frac{\sigma_{t,90,d} \cdot b \cdot a_1}{n} \tag{D8-25}$$

Tensile force in a reinforcement element in the external area (both external quarters):

$$F_{t,90,d} = \frac{2}{3} \cdot \frac{\sigma_{t,90,d} \cdot b \cdot a_1}{n} \tag{D8-26}$$

D8.5 Literature

H.J. Larsen, original Article E5, STEP 1995.

Blass H.J. and Bejtka I. (2004). Selbstbohrende Holzschrauben und ihre Anwendungsmöglichkeiten. Holzbaukalender 2004, pp. 516-541. Bruderverlag Karlsruhe.

Blass H.J., Ehlbeck J., Kreuzinger H. and Steck G. (2005). Erläuterungen zur DIN 1052:08-2004. Herausgeber: Deutsche Gesellschaft für Holzforschung. Bruder-Verlag, Karlsruhe.

Brünninghof H., Schmidt K. and Wiegand T. (1993). Praxisnahe Empfehlungen zur Reduzierung von Querzugrissen. Bauen mit Holz 11:928-937.

Ehlbeck J. and Görlacher R. (1989). Tragverhalten von Queranschlüssen mittels Stahlformteilen, insbesondere Balkenschuhen, im Holzbau. Forschungsbericht Universität Karlsruhe.

Henrici D. (1984). Beitrag zur Spannungsermittlung in ausgeklinkten Biegeträgern aus Holz. Dissertation, Technische Universität München.

D9 Diaphragms and bracings

Original articles: T. Alsmarker, H. Brüninghoff, S. Winter

A building is not only exposed to vertical forces such as permanent loads and imposed loads, but horizontal forces like wind or earthquake loads must also be absorbed and wind affects a building in various ways. Its direct impact translates into pressure on one or multiple external surfaces and suction on the remaining external surfaces. Figure D9-1 shows the main wind load distribution in a building where the wind is blowing in a direction perpendicular to the longitudinal wall. The wind in the direction shown in Figure D9-1 exerts pressure on the facing wall and half the roof and suction on the area away from the wind. When the roof pitch is low, suction is also exerted on the wind-facing half of the roof (in Figure D9-1: „alt“ = „or“). Notable here is the fact that the suction effect on the gable walls acts perpendicular to the wind direction. In addition to the main wind loads, the wind can also exert pressure or generate suction on the interior surface of the building.

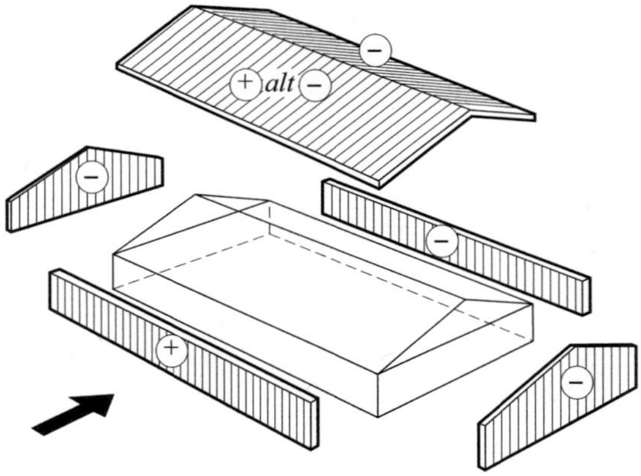

Figure D9-1 Wind load distribution with the wind direction perpendicular to the longitudinal wall. The arrow indicates the wind direction. (STEP 1995 Article B13)

In addition to wind loads, horizontal bracing loads also often have to be accommodated and the latter occur when force is applied to slender beams, deflecting them laterally (lateral torsional buckling). This is prevented using bracing with significant stiffness perpendicular to the principal load-bearing direction of the slender beams. The bracing is not subject to any stresses if the beams to be braced are completely straight and the forces are transmitted centrally and vertically i.e. when deflection occurs only in the principal plane and no stress components perpendicular to the principal load-bearing beam directions are developing. However, in reality, inevitable production and assembly irregularities (imperfections) lead to deviations from the ideal position, while additional changes in form due to wind loads or otherwise, from externally imposed horizontal forces exacerbate any eccentricity due to imperfections. The basic load-bearing behaviour when exposed to wind perpendicular to the longitudinal wall is shown in Figure D9-2 for a very simple building, which uses floor and wall diaphragms. In this case, the walls are only connected to the foundation and the (horizontal) roof, which means half of the entire wind force is directed towards the horizontal roof diaphragm, which acts as a deep beam. The roof diaphragm rests on the side walls, which transmit the forces into the foundations via their in-plane shear action. The bracing system comprises multiple elements, which must be correctly connected to ensure a complete load path of shear forces. In addition to other measures, this includes proper fastening of the horizontal diaphragm to the vertical diaphragms (shear walls) and hold-down anchors of the walls to the foundation.

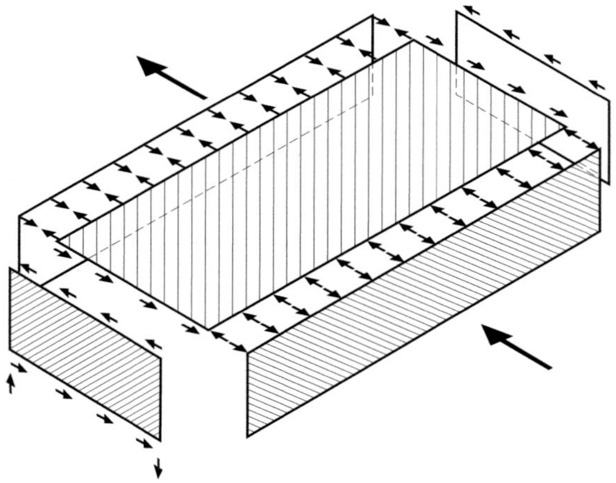

Figure D9-2 In-plane load-bearing in a simple building, where the roof functions as a horizontal diaphragm and the side walls function as vertical diaphragms = shear walls. (STEP 1995 Article B13)

D9.1 Possibilities, arrangement and design of bracing

Load distribution

Lateral load resisting systems designed to distribute horizontal forces can comprise vertical elements or combined vertical and horizontal elements, where the following minimum criteria apply:

- For existing floor diaphragms, at least three shear walls must be present, which do not intersect with each other at any point and which are not all parallel.
- In the absence of a floor diaphragm, at least four shear walls must be present, of which no more than two walls respectively may intersect at any point.

There is also scope to add additional structural members to this base unit, which are braced sufficiently to transfer all or part of the horizontal forces to which they are exposed, via the base unit into the foundation.

Bracing elements within a structural system should be arranged such as to allow a symmetrical load transfer. In all other cases, the additional forces generated by the eccentricity between the centre of gravity of the introduced load and that of the bracing must also be taken into consideration. The support forces for continuous floor diaphragms must be determined in a manner similar to that applied to the series of simply supported beams.

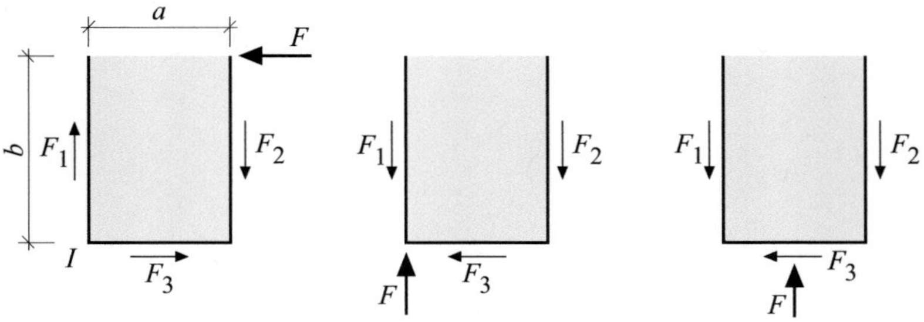

$\Sigma M = 0$
$F \cdot b - F_2 \cdot a = 0$
$F_2 = F \cdot b/a = F_1$
$F_3 = F_1$

$F_1 = F$
$F_2 = F_3 = 0$

$F_1 = F_2 = F/2$
$F_3 = 0$

Figure D9-3 Forces in shear walls, statically determinate system. (STEP 1995 Article E14)

For each load direction, the bracing elements used should have the same stiffnesses, since otherwise additional forces will be generated due to eccentricity. The following approximated calculation method for statically indeterminate systems (Figure D9-4) **only applies provided the stiffnesses of all shear walls are the same**. Where a more accurate consideration of the stiffness of the wall panels is required, as Steinmetz (1992) remarks, the bending deformation, shear deformation and the semi-rigidity of all fasteners must be taken into consideration.

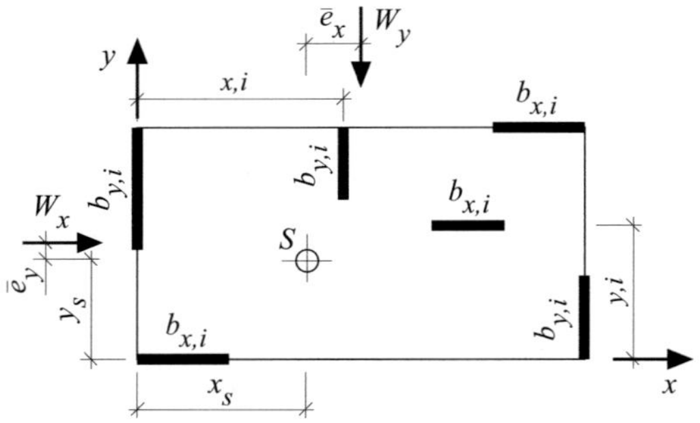

Figure D9-4 Shear walls, statically indeterminate system. (STEP 1995 Article E14)

The forces in the statically indeterminate system in accordance with Figure D9-4 can be calculated corresponding to the following equations. Eccentricities $\bar{e}_x$ and $\bar{e}_y$ are the lever arms of the load application point to the centre of gravity of the bracing.

Centre of gravity S (x_s, y_s) of the bracing:

$$x_s = \frac{\Sigma b_{yi} \cdot x_i}{\Sigma b_{yi}} \quad \text{and} \quad y_s = \frac{\Sigma b_{xi} \cdot y_i}{\Sigma b_{xi}} \tag{D9-1}$$

The eccentricities of individual panels *i* from the centre of gravity *S* of the bracing are:

$$s_{xi} = \left(x_i - x_s\right) \quad \text{and} \quad s_{yi} = \left(y_i - y_s\right) \tag{D9-2}$$

The forces on wall panels *i* with wind direction W_x are determined as the sum (i) of the portion of W_x per length b_{xi} of wall panels and (ii) the force from the additional moment ΔM_x caused by the eccentricity $\bar{e}_y$ of the load application point (note algebraic signs!):

$$H_{xi} = \frac{b_{xi}}{\Sigma b_{xi}} \cdot W_x + \Delta M_x \cdot \frac{s_{yi} \cdot b_{xi}}{I_p} = \frac{b_{xi}}{\Sigma b_{xi}} \cdot W_x + W_x \cdot \bar{e}_y \frac{s_{yi} \cdot b_{xi}}{\Sigma\left(b_{xi} \cdot s_{yi}^2\right) + \Sigma\left(b_{yi} \cdot s_{xi}^2\right)} \tag{D9-3}$$

I_p is the polar moment of inertia of the wall panels about the centre of rotation = centre of gravity *S*.

The additional moment ΔM_x also generates horizontal loads in a *y*-direction with wind direction W_x:

$$H_{yi} = \Delta M_x \cdot \frac{s_{xi} \cdot b_{yi}}{I_p} = W_x \cdot \bar{e}_y \frac{s_{xi} \cdot b_{yi}}{\sum\left(b_{xi} \cdot s_{yi}^2\right) + \sum\left(b_{yi} \cdot s_{xi}^2\right)} \tag{D9-4}$$

The forces on the wall panels *i* with wind direction W_y are similarly determined:

$$H_{yi} = \frac{b_{yi}}{\sum b_{yi}} \cdot W_y + \Delta M_y \cdot \frac{s_{xi} \cdot b_{yi}}{I_p} = \frac{b_{yi}}{\sum b_{yi}} \cdot W_y + W_y \cdot \bar{e}_x \frac{s_{xi} \cdot b_{yi}}{\sum\left(b_{xi} \cdot s_{yi}^2\right) + \sum\left(b_{yi} \cdot s_{xi}^2\right)} \tag{D9-5}$$

$$H_{xi} = \Delta M_y \cdot \frac{s_{yi} \cdot b_{xi}}{I_p} = W_y \cdot \bar{e}_x \frac{s_{yi} \cdot b_{xi}}{\sum\left(b_{xi} \cdot s_{yi}^2\right) + \sum\left(b_{yi} \cdot s_{xi}^2\right)} \tag{D9-6}$$

Examples for vertical bracing systems

Portal frames with moment-resisting connections in their corners can absorb both vertical and horizontal forces and are widely used in single-storey hall structures in particular. Dowel-type fasteners may be used to produce such moment-resisting connections (e.g. a dowel circle) or finger joints. As a rule, for horizontal loading, portal frames are only used to resist wind loads and when additional horizontal forces are exerted, e.g. horizontal braking forces of crane systems, the cross-sectional sizes would rapidly increase mainly due to the need to accommodate more fasteners. One further downside is the limited clear height of the construction and these reasons often spawn what are uneconomical solutions.

Cantilever columns can be used, particularly for hall structures involving crane operation and made of reinforced concrete or steel members (e.g. reinforced concrete cantilever column with sleeve foundations). For lower horizontal forces, meanwhile, timber poles or glued laminated timber columns (Heimeshoff, 1983) can also be used, e.g. for agricultural constructions or smaller hall structures. When clamping glued laminated timber columns directly in sleeve foundations, a national German technical approval Z-9.1-136 applies. However, given the high risk of fungal attack on the moulded column footing due to moisture, the clamped end must be very carefully protected.

A **diagonal bracing** may include either elements resistant to both compression and tension, e.g. diagonals in trusses, or solely tensile braces. Normally, the set-up includes cross-arrayed components, e.g. round steel bars with turnbuckles. Diagonal bracings are either used within wall constructions or as visible elements, which often fulfil a creative function at the same time.

Strip braces are usually used for bracing of roof structures and, in smaller buildings, also for bracing walls or floors. They are normally nailed to their respective components. The design load-bearing capacity is determined either by the load-bearing capacity of the joint or the load-bearing capacity of the strip brace in the net cross-section. Strip braces must be tensioned at the time of assembly and their advantages include fast and simple assembly and low costs, although the assembly is often defective in practical terms. Further downsides include the low load-bearing capacity and the sensitivity of the material to temperature. If a strip brace is assembled amid very low temperatures, significant deformation will occur during the summer months due to the steel expanding in response to heat. The system must withstand significant displacement before being able to accommodate forces, which may lead to deformations of the overall structural system.

Round steel bars with turnbuckles are particularly suitable for bracing structural systems and tend to be designed on an individual basis. Numerous options apply to the joint design, certain examples of which are specified in Figure D9-5. Bracings with round steel bars can resist large forces and also have the advantage of being adjustable during assembly to align the construction. Provided corresponding detailing applies, bracings with round steel bars can boost the overall visual impression of the construction.

Struts made of solid timber or glued laminated timber are designed taking buckling lengths into consideration and they can resist compressive and tensile forces. The joints can be established e.g. by nails, bolts, wood-based panels, steel plates (Figure D9-5 (a)) and this type of bracing may be more economical than using round steel bars. One downside, however, is the more complex assembly. Once the struts have been installed, there is no further scope to align the construction, hence the need for ultra-precise execution. The ability to dissipate compressive and tensile forces results in more rigid constructions with fewer deformations.

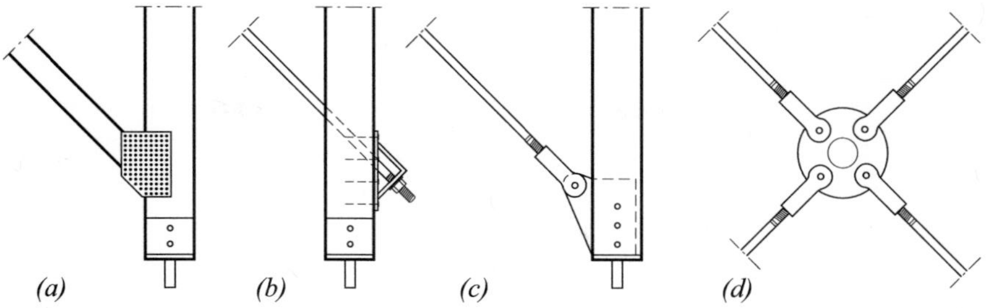

Figure D9-5 Various diagonal bracing systems. (*a*) timber diagonal with nailed steel plate, (*b*) round steel bar diagonal, (*c*), (*d*) round steel bar diagonal with cast iron connector. (STEP 1995 Article E14)

Vertical diaphragms (**shear walls**) are often executed in the form of timber frame constructions, comprising solid timber members (top and bottom plate, studs) and wood-based panel materials as sheathing. Such shear walls, made of wood-based panels and timber frames, are used very frequently when constructing single- and multi-storey homes. The multiple functionality of the wall, i.e. the ability to resist to horizontal forces as well as protect against fire, heat and sound, make this type of bracing the most economical option overall. When using wood-based panels, moreover, there is great scope for prefabrication. Suitable strength properties mean that particleboards, OSB and plywood are particularly good choices when applied as sheathing. The shear forces are transferred via numerous slender fasteners (nails or staples) along the panel edges in the timber frame. The joints of the wall elements must each be separately verified and the load-bearing capacity of the shear walls against compressive and tensile stresses must take into consideration the combination of vertical and horizontal forces. The ends of shear walls must be prevented from uplifting, while the compressive forces acting on the respective opposite side must also be taken into consideration during the design (compression perpendicular to the grain on the bottom plate). Massive shear walls made from cross-laminated timber are an increasingly popular alternative to timber frame walls.

Examples for horizontal bracing systems

Every floor of a building and its roof must be braced with horizontal elements, while bracing elements may include either horizontal diaphragms or diagonals respectively. For sloping roofs, meanwhile, the bracing is normally located within the plane of the roof, while in floors, wood-based panels are used to accommodate vertical and horizontal loads as well as absorbing shear forces in the horizontal plane. Edge beams and floor beams carry the flexural forces of the horizontal diaphragm. The floor functions like a beam, which transfers the horizontal forces into the vertical bracing. Horizontal diagonals or roof diagonals are parts of a truss, which transfer the horizontal forces into the adjacent vertical bracings. Accordingly, all joints and the compression and tension members of this truss have to be verified. Roof-level bracings may include round steel bars or strip braces in combination with purlins or small trusses. Trusses with punched metal plate fasteners always have to be laterally braced by additional trusses at roof level. For trusses or glued laminated timber beams, the bracing at roof level is not only to resist the horizontal forces generated, but also to protect against lateral buckling.

Procedure for conceptual and structural design

- Defining the building geometry and the lateral load resisting walls and building parts intended to accommodate horizontal forces. In areas at risk of earthquakes, as well as defining a system to resist wind loads, a bracing system to resist earthquake loads should also be determined. The ductility of the vertical bracing systems influences the earthquake design load (Article G2). Under certain circumstances, brittle bracing elements or joints can absorb higher wind loads, but when designing for earthquake loads, this leads to uneconomical results.
- Determination of horizontal forces. The current systems for accommodating wind load, e.g. continuous columns in gable walls of halls up to roof level, must be taken into consideration, since determining forces via tributary areas alone may distort results.
- Calculation of the resulting forces in the bracing elements. For symmetrically braced buildings, this is done in accordance with simple rules to determine the reaction forces in statically determinate structures. In a building in which the bracing elements are unevenly distributed or with uneven stiffness distribution of the bracing, e.g. due to large openings in the walls, the resulting torsional forces in the building must be taken into consideration (cf. Figure D9-4). The greater the distance between the bracing elements, the greater the torsional resistance of the structural system.
- Verification: $E_d \leq R_d$
- Verification of joints of the bracing elements. The flow of forces from the load application point to the foundation must be verified, as must the following joints:
  - Introduction of the forces of immediately stressed components such as vertical studs or rafters in horizontal elements, e.g. floor diaphragms, particularly tensile forces due to wind suction.
  - Joints within horizontal bracing, e.g. joints of floor or wall diaphragms, joints of horizontal trusses, joints of purlins, etc.
  - Joints between horizontal and vertical elements, e.g. connecting floor diaphragms to wall diaphragms or connecting chords of horizontal trusses to vertical diaphragms or diagonals.

  - Joints within vertical bracing, e.g. nailing of wood-based panels onto timber frames or connecting of prefabricated wall elements.
  - Joints between vertical elements and foundations, e.g. the end anchoring of shear walls or anchoring of steel diagonals to the bottom of columns.

- Any form of joint which may lead to stress being exerted perpendicular to the grain in the timber member should be avoided since the low strength may trigger sudden failure of the joints in question.
- Verification of deformations. The permissible scope of deformations depends on the building type and verification is particularly important for members simultaneously stressed by high vertical forces. In this case, high eccentricity of the forces will generate significant additional forces. However, also with a view to avoiding cracks in cladding, deformations of structural systems should be limited.
- Verification of conditions during erection. It is important to guarantee the structural stability of the building while erecting the structural system. However, when constructing houses with prefabricated walls, a short-term makeshift form of bracing, e.g. via adjustable inclined steel columns is sufficient and there is generally no need for separate verification. For hall structures, however, separate verifications may be required. In the normal case, the construction progress involves the vertical members being erected first and subsequently connected to the horizontal bracing elements.

In the following section, floor and wall diaphragms which constitute key bracing elements for timber structures will be explained in more detail, before discussing the design and arrangement of bracing.

D9.2 Diaphragms

Horizontal diaphragms

Floors and roofs can be used to transmit horizontal forces to the supporting walls. In timber frame buildings, these horizontal structures generally comprise wooden joists, which are sheathed with a range of wood-based panels. The double top plate of the walls is often used as a chord of the horizontal diaphragm, while the wood-based panels have a staggered arrangement and are joined using nails or bolts. Alternatively, a continuous header or trimmer joist may be used as the chord. The remainder of the article focuses solely on horizontal diaphragms in accordance with Figure D9-6, which are exposed to a horizontal and uniformly distributed load.

For this type of floor diaphragm, it is safe to assume that the load-bearing behaviour resembles that of a deep I-beam with a depth of b. According to EC 5, this is applicable, provided the span ℓ is between $2 \cdot b$ and $6 \cdot b$. The sheathing functions as a web resisting shear forces, while the chords act like flanges, which transmit the bending moment exerted; see Figure D9-7, forces F_c and F_t with lever arm b.

According to NA, the simplified method may also be used for diaphragms where the span ℓ is smaller than $2 \cdot b$, if joists, which are continuous in the load direction and over the diaphragm width b, transmit the loads uniformly into the diaphragm or if the diaphragm width b is only calculated as equating to half the span of the diaphragm. (NCI for 9.2.3.2 (NA.5))

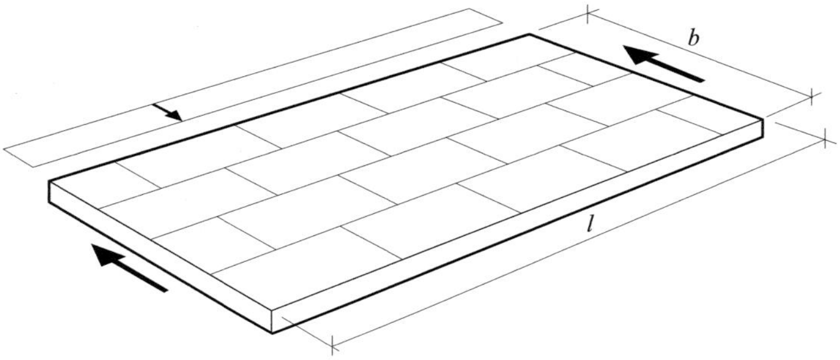

Figure D9-6 Horizontal (floor) diaphragm. (STEP 1995 Article B13)

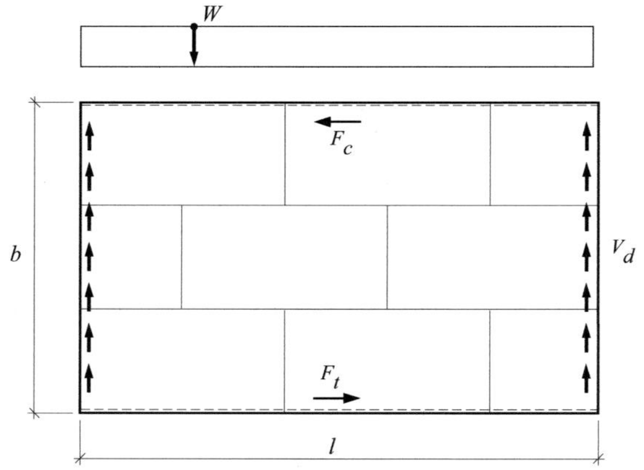

Figure D9-7 Principle behaviour of floor diaphragms. (STEP 1995 Article B13)

The simplified verification presumes that the entire moment is absorbed by the chords. Accordingly, the chords must be designed to resist tensile or compressive forces of

$$F_{t,d} = F_{c,d} = \frac{M_{max,d}}{b} \quad \text{(D9-7)}$$

where

$M_{max,d}$ Design value of maximum bending moment

b Diaphragm width

The sheathing must be designed for a shear flow v_d (see Figure D9-7) of

$$v_d = \frac{F_{v,d}}{b} \quad \text{(D9-8)}$$

where

$F_{v,d}$ Design value of maximum shear force

b Diaphragm width

In this case, v_d is assumed to be uniformly distributed over the diaphragm width b.

Finally, the spacing s of the fasteners which is used to connect the sheathing to the joists is determined as follows:

$$s = \frac{F_{v,Rd}}{v_d} \quad \text{(D9-9)}$$

where

$F_{v,Rd}$ Design load-bearing capacity of a joint with laterally loaded fasteners

v_d Shear flow from equation (D9-8)

For simply supported diaphragms in accordance with Figure D9-7, the shear force is transmitted from the horizontal diaphragm to the shear walls via the edge members at the end of the diaphragm. Here, the shear force is assumed to be uniformly distributed along the diaphragm edge. Edge members and chords must also be correctly connected to the top plate, to transmit the shear forces to the shear wall underneath. It is important to ensure that a flow of forces is possible via other structural elements, if the sheathing is not directly connected to these load-bearing elements.

In the proposed calculation model, the sheathing is assumed to function as one, which is why the individual sheathing panels must be properly connected. The optimal floor diaphragm design involves a staggered arrangement of sheathing panels, Figure D9-7. However, since the diaphragm action must be present in both directions, staggering should therefore be oriented in the most unfavourable loading direction. When large openings are present in the floor diaphragm, it is important to ensure that the forces in the area of the openings are reliably transferred, which can be done for compressive and tensile forces via trimmer beams and steel straps. Effective transmission of the shear forces is ensured if the sheathing is appropriately nailed or screwed to the trimmer beams and joists around the opening.

Vertical diaphragms (shear walls)

Timber frame walls generally comprise vertical studs installed at regular intervals, which then form a frame, together with the top and bottom plates. This frame is generally sheathed on one or both sides with different types of wood-based panels which are nailed or stapled onto the frame. Structurally, the wall can be considered a cantilevered diaphragm, the top plate of which is exposed to a concentrated horizontal load. Since the sheathing is used for bracing, this load can be effectively transferred to the foundations. Figure D9-8 sets out the load-bearing behaviour schematically.

The studs are connected to the bottom and top plate with nails or other metal fasteners, while from a structural perspective, the frame joints are deemed hinges, Figure D9-8 (b). The sheathing and fasteners fixing the sheathing to the frame must prevent any displacement of the timber frame and meanwhile, the fasteners are most highly stressed at the corners, since this is where the largest slip between the frame and sheathing takes place. The fasteners in the top left and bottom right corners will have force directions towards the free edge of the sheathing. In Figure D9-8, it was assumed that the **studs are anchored to the foundation**. The key criterion dictating the load-bearing capacity of a timber frame shear wall is whether or not the uplift of the studs can be prevented. In addition to verification of uplift, verification of the ability of the studs to withstand concentrated compressive load is required. Both the load-bearing capacity of joints as well as the shear strength of the sheathing are additional important factors impacting on the load-bearing capacity of shear walls.

The load-bearing capacity of a wall comprising multiple wall elements is calculated as the collective load-bearing capacities of the individual elements, even when wall elements include a range of different sheathing material and various fasteners. The same approximation can be used when different sheathing material and fasteners are used on both sides of the frame, but according to EC 5, in this case, the specified load-bearing capacity of the weaker side should be halved. Within a wall with window openings, when calculating the overall load-bearing capacity for the wall, wall element lengths with these openings should be disregarded.

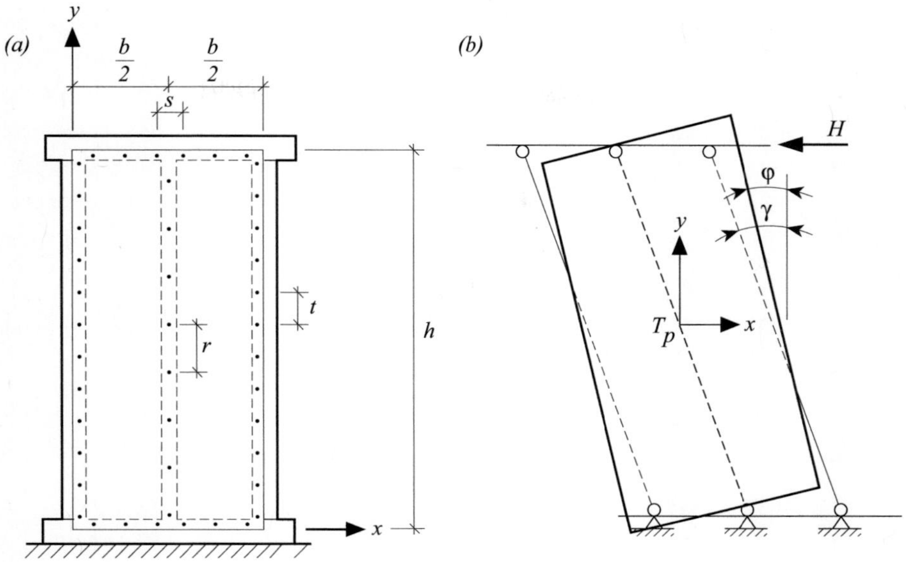

Figure D9-8 (*a*) Typical wall element, (*b*) principal structural behaviour. (STEP 1995 Article B13)

According to method A of EC 5, the design racking resistance $F_{i,v,Rd}$ of each wall panel *i* can be calculated as

$$F_{i,v,Rd} = \frac{F_{f,Rd} \cdot b_i \cdot c_i}{s} \tag{D9-10}$$

whereby $F_{f,Rd}$ is the design joint capacity and *s* the fastener spacing, while coefficient c_i considers the ratio of the wall panel height to the wall panel.

The tension studs, hold-down anchors and compression studs must resist the force

$$F_{i,c,Rd} = F_{i,t,Rd} = \frac{F_{i,v,Ed} \cdot h}{b_i} \tag{D9-11}$$

where $F_{i,v,Ed}$ equates to the design horizontal force acting on wall panel *i* of width b_i and height *h*.

Any vertical load may be taken into consideration when determining the vertical tensile force and must be considered when determining the vertical compressive force. To apply method A, the end studs of the wall panel must be vertically secured, either by the vertical load or by using hold-down anchors, while the bottom plate must also be sufficiently anchored to the foundation, to allow transmission of the horizontal forces. However, other design models are available in the event that the end studs are not vertically secured. In multi-storey buildings, the wall panels must be interconnected such as to allow transmission of the tensile forces from one floor to another.

In conclusion, the prerequisites for the validity of the simple calculation model presented in this document should be cited (Colling, 2011):

- All edges of the sheathing are supported by studs and continuously fastened to the same.
- The forces are uniformly distributed and continuously transferred via the fasteners between the sheathing and top plate (concentrated loads cannot be considered).
- The sheathing does not buckle.
- The studs and sheathing are rigid in comparison to the joints, plastic hinges are developing in the fasteners and thus the load-bearing capacity of the wall panel is determined by the load-bearing capacity of the joints.

Internal walls

The distribution of the horizontal load to the internal walls depends on the ratio of stiffness of the floor or roof diaphragm in relation to that of the wall. One assumed limit case involves a rigid horizontal diaphragm, supported by flexible walls, while the other involves a flexible horizontal diaphragm, supported by rigid walls. The first case involves the horizontal load being distributed to the wall panels based on the stiffness of the walls. If there is a rigid floor diaphragm on three walls, each of equivalent stiffness, each wall absorbs one third of the overall load. Here, it should be noted that when the walls are arranged asymmetrically, the resulting torsional moment is to be taken into consideration. If we assume a flexible floor diaphragm supported by rigid walls, it depends on whether the floor diaphragm is configured as a horizontal continuous beam or as simply supported beams. A safe assumption would be that the walls at diaphragm ends are considered supports of a simply supported beam and the internal walls as intermediate supports of a continuous beam. When timber floor diaphragms are used on timber shear walls, this is between both the above-mentioned limit cases and the assumption of a rigid floor diaphragm should be made cautiously. Rigid floor diaphragms should only be assumed, if the ratio of the diaphragm width b to the span ℓ between the internal walls is around one.

D9.3 Bracing

Background to the design equations in accordance with EC 5

Variables influencing actions on bracing members

Regarding actions on bracing, a distinction has to be made between compression and bending members. Also relevant is whether the scope of consideration includes a highly stressed individual support or a series of supports, which e.g. form a bracing structure in the shape of a truss. The actions on the bracing members particularly depend on the cross-sectional and longitudinal dimensions of the structure to be braced, the support conditions and the construction material properties, which, in turn, are determined by the selection of a strength class, service class and the load duration class of the governing load case. The stiffnesses of members and joints play a key role, not only as attributed of the structure to be braced, but particularly for the bracing structure itself. Verification in accordance with the second order theory moreover has to take geometric and structural imperfections into consideration, e.g. in the form of initial deformations.

Individual supports of compression members

Compression members of length ℓ, which are supported (= braced) at regular intervals by flexible (elastic) supports, generate very considerable support (= bracing) forces, assuming the deformation patterns in accordance with Figure D9-9 (b) and (c).

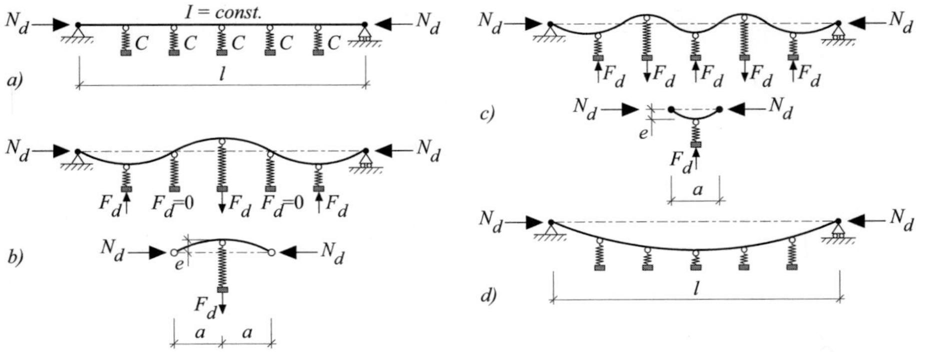

Figure D9-9 Static system and deformations of braced compression members. (STEP 1995 Article B15)

Möhler and Schelling (1960) showed that the minimum spring stiffness *C* of the supports should have the following value:

$$C = k_s \cdot \pi^2 \cdot \frac{E \cdot I}{a^3} \tag{D9-12}$$

where

$$k_s = 2 \cdot \left(1 + \cos\frac{\pi}{m}\right) \tag{D9-13}$$

and where a is the length and m the number of waves, so that with $\ell = m \cdot a$ and with hinged supports at both ends, a wave-shaped deformation pattern is attained with one wave for $k_s = 2$ or a limitless number of waves for $k_s = 4$ (in EC 5, $k_s = 4$ is recommended).

If we now use the (elastic) critical load N_d of the Euler buckling case II (equation (D2-19)) in equation (D9-12), the following equation given in EC 5 to determine the minimum spring stiffness follows:

$$C = \frac{k_s}{a} \cdot \frac{\pi^2 \cdot E \cdot I}{a^2} = \frac{k_s}{a} \cdot N_d \tag{D9-14}$$

The spring force F_d (see Figure D9-10) can be calculated conservatively and in accordance with second order theory at:

$$F_d \leq 5.2 \cdot N_d \cdot \frac{e}{2 \cdot a} \tag{D9-15}$$

where e equals the initial deformation of the member axis.

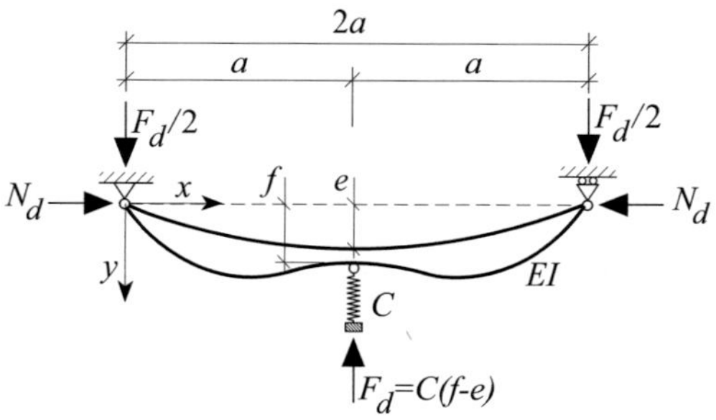

Figure D9-10 Deformation of an elastically supported beam. (STEP 1995 Article B15)

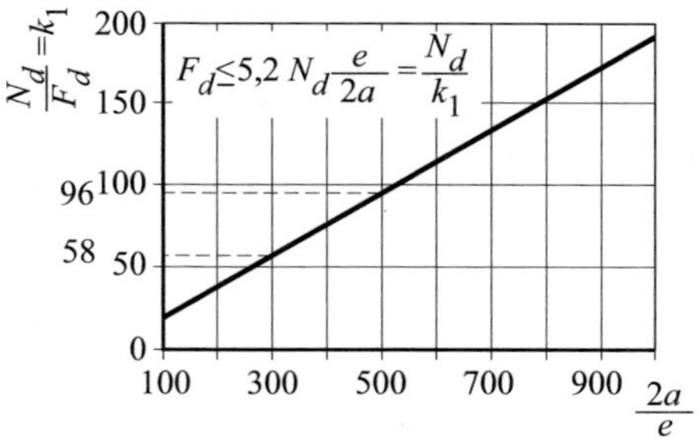

Figure D9-11 Coefficient of bracing force as a function of initial deformation. (STEP 1995 Article B15)

Figure D9-11 shows the emergence of support forces in the springs of $F_d = N_d/58$ and $F_d = N_d/96$, assuming initial deformations of $\ell/300$ and $\ell/500$ for members made of solid timber or glued laminated timber. These values were rounded up in EC 5 to the recommended values of $N_d/50$ and $N_d/80$.

Individual support of bending members

Burgess (1989) proposed specifying the design value of mean compressive force N_d in the compression zone of a bending member as follows:

$$N_d = M_d \cdot \frac{N_{crit}}{M_{crit}} \tag{D9-16}$$

whereby N_{crit} and M_{crit} signify the elastic buckling load and elastic buckling moment in accordance with classical theory of stability (Article D2). EC 5 specifies the following approximation:

$$N_d = \left(1 - k_{crit}\right) \cdot \frac{M_d}{h} \tag{D9-17}$$

This involves calculating the coefficient k_{crit} for the member without bracing (lateral torsional buckling of beams, Article D2), in which the torsional stiffness of a beam is also taken into consideration. Where $k_{crit} = 1$, no bracing is required. The method is only applicable for bending members braced along their compression zone.

Bracing of beam or truss systems

To determine the load in the bracing, it is assumed that the compression or bending members to be braced by a bracing structure of stiffness $(E \cdot I)_{ef}$ have an initial deformation in the shape of a half sine wave (see Figure D9-9 (d)).

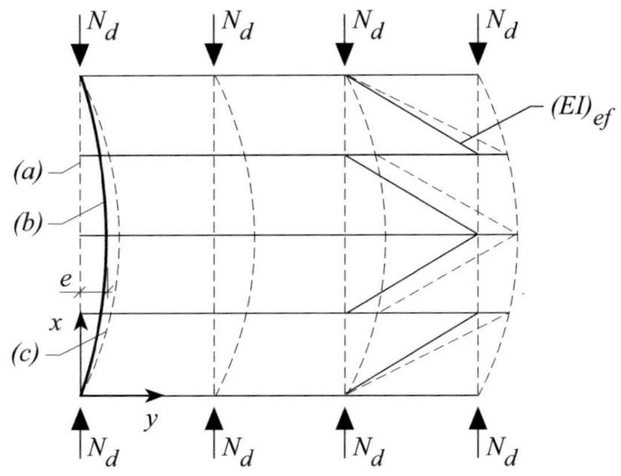

Figure D9-12 Deformation of braced compression members. (a) Straight line, (b) axis with initial deformation, (c) deformed axis. $(E \cdot I)_{ef}$ is the stiffness of the bracing structure. (STEP 1995 Article B15)

The compressive forces N_d generate a moment (for all definitions see Figure D9-12):

$$M_d = n \cdot N_d \cdot y \tag{D9-18}$$

Disregarding the bending stiffness $E \cdot I_z$ of the slender members to be braced, the differential equation of the elastic curve is obtained, taking the initial deformation y_0 into consideration:

$$\frac{M_d}{(E \cdot I)_{ef}} = -\frac{d^2 \cdot (y - y_0)}{d^2} = -(y - y_0)'' \tag{D9-19}$$

Replacing the moment M_d from equation (D9-18) in equation (D9-19) we have:

$$y'' + \frac{n \cdot N_d}{(E \cdot I)_{ef}} \cdot y = -e \cdot \left(\frac{\pi}{\ell}\right)^2 \cdot \sin\left(\frac{\pi}{\ell} \cdot x\right) \tag{D9-20}$$

with the initial deformation function

$$y_0 = e \cdot \sin\left(\frac{\pi}{\ell} \cdot x\right) \tag{D9-21}$$

The solution of the differential equation results in

$$q_d = \frac{e \cdot \left(\frac{\pi}{\ell}\right)^2 \cdot n \cdot N_d}{1 - \frac{n \cdot N_d}{\left(\frac{\pi}{\ell}\right)^2 \cdot (E \cdot I)_{ef}}} \cdot \sin\left(\frac{\pi}{\ell} \cdot x\right) \tag{D9-22}$$

see also Brüninghoff (1983).

Applying equation (D9-22) requires that the stiffness of the bracing structure $(E \cdot I)_{ef}$, including joint slip, be used as an input parameter. To simplify the design for frequently occurring practical cases, EC 5 limits the deflection caused by q_d and additional external actions of the bracing structure to $\ell/500$:

$$\max y = q_d \cdot \frac{\ell^4}{\pi^4} \cdot \frac{1}{(E \cdot I)_{ef}} \leq \frac{\ell}{500} \tag{D9-23}$$

If we cancel $(E \cdot I)_{ef}$ from equations (D9-22) and (D9-23) and convert the sinusoidal load into a uniformly distributed load, the result is

$$q_d = k_\ell \cdot \frac{n \cdot N_d}{30 \cdot \ell} \tag{D9-24}$$

where $k_\ell = 1$.

For spans exceeding 15 m, the expected level of execution accuracy means that the initial deformations will no longer increase proportionally with the span, whereupon the initial deformation can be reduced in accordance with equation (D9-21) by a factor of

$$k_\ell = \sqrt{\frac{15}{\ell}} \tag{D9-25}$$

where ℓ is the span in [m].

The structural engineers should verify the deflections of the bracing structure, if it is thought likely that the above-specified deflection limits will be exceeded. If bending members are to be braced rather than compression members, the side exposed to compression should also be braced, to ensure equations in accordance with EC 5 are applicable. The torsional stiffness of the bending members can be taken into consideration by calculating the compressive force in accordance with equation (D9-17).

Uses of bracing

Bracing is designed to resist external forces, which do not arise from the structure itself and which thus have to be transferred to the foundations:

- Wind loads,
- Horizontal loads, e.g. braking forces of a crane, seismic loads.

Bracing structures are also capable of absorbing internal forces, generated by existing or forced deviations from the planned position of the structural members. These forces can be balanced within the correctly detailed structure itself and therefore do not generally need to be transmitted to the foundations:

- Forces due to lateral displacements of frames and columns,
- Forces arising from bracing beams and compression chords of trusses; both of which are liable to lateral torsional buckling,
- Forces on intermediate supports of compression members,
- Forces on the lateral supports of buckled tension chords.

Figure D9-13 to Figure D9-18 show examples of bracing structures.

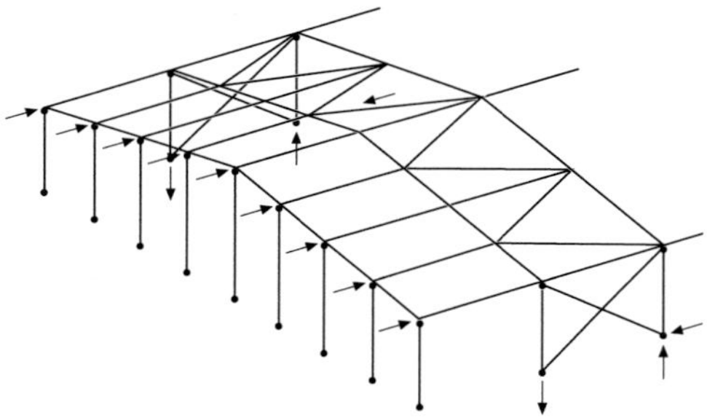

Figure D9-13 Transfer of wind loads from gable columns through roof and wall bracing. (STEP 1995 Article D9)

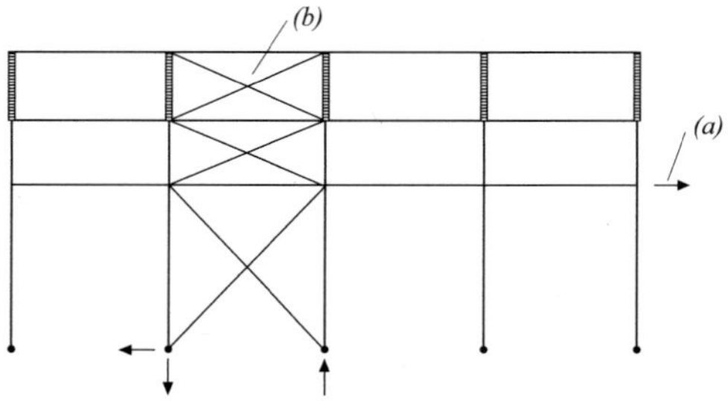

Figure D9-14 Transfer of external loads through bracing into the foundation. (a) Braking forces of a crane, (b) bracing to provide torsional resistance at beam ends. (STEP 1995 Article D9)

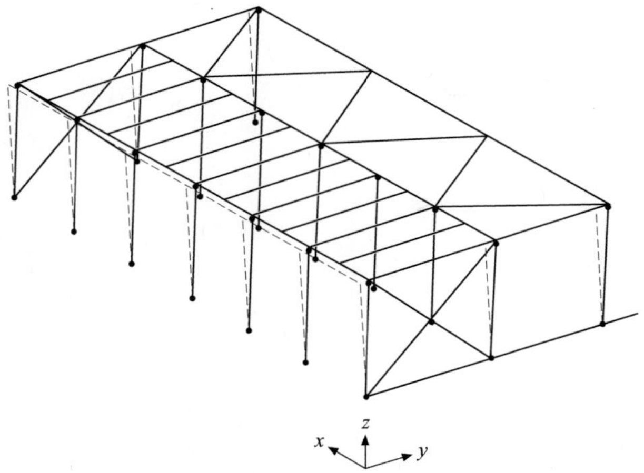

Figure D9-15 Transfer of P-Δ forces from inclined columns though roof and wall bracing. (STEP 1995 Article D9)

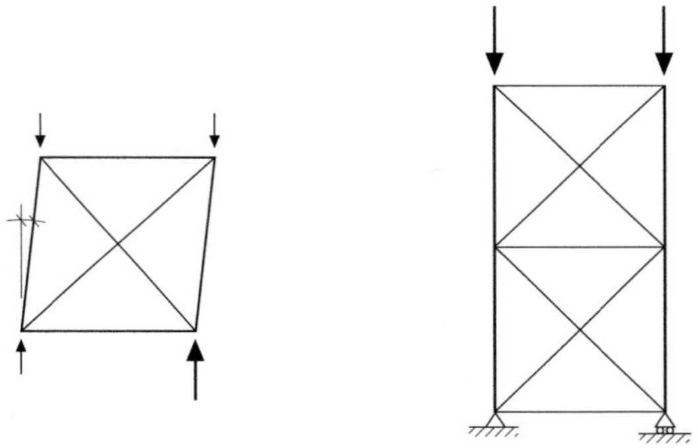

Figure D9-16 Support reactions for inclined structures with vertical actions (left) and lateral intermediate support to decrease column lengths (right). (STEP 1995 Article D9)

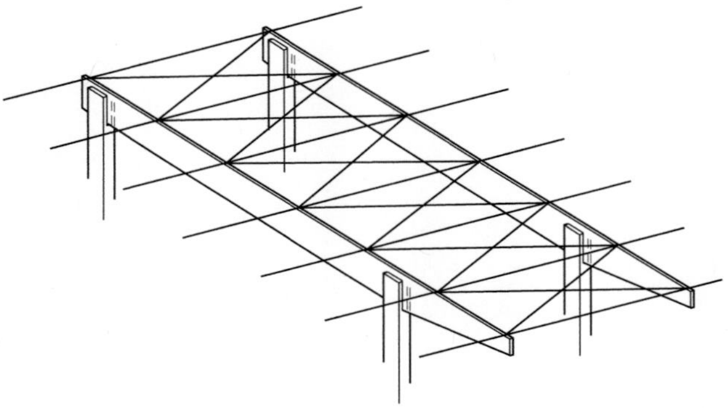

Figure D9-17 Lateral bracing of beams at risk of buckling. (STEP 1995 Article D9)

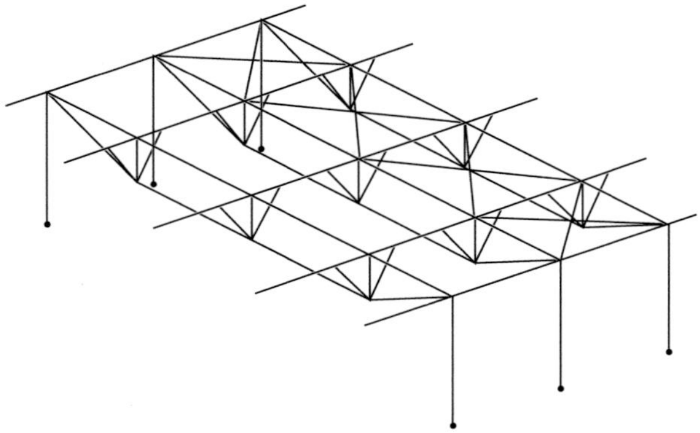

Figure D9-18 Lateral support of tension chords using knee-bracing against purlins. (STEP 1995 Article D9)

The key function of the main load-bearing elements within a structural system is to transmit vertical forces, e.g. self-weight and snow on roofs, while bracing structures are designed to absorb horizontal wind and stabilisation forces (buckling forces from the main load-bearing element). The structural engineer would normally handle these actions separately and design main elements and bracing in a range of steps. In reality, however, the system is actually a three-dimensional structure (Figure D9-19). In this case, the connection of two main members, the two simple triangular trusses shown in Figure D9-19, with one upper and two lower purlins, adding two diagonal bracing elements initially suffices. If the support reactions are now counted, a total of eight emerge, one of which is required to secure the existing hinged square at eaves level. This leaves seven support reactions with six possible equilibrium conditions in space. Thus already this simple system remains statically indeterminate.

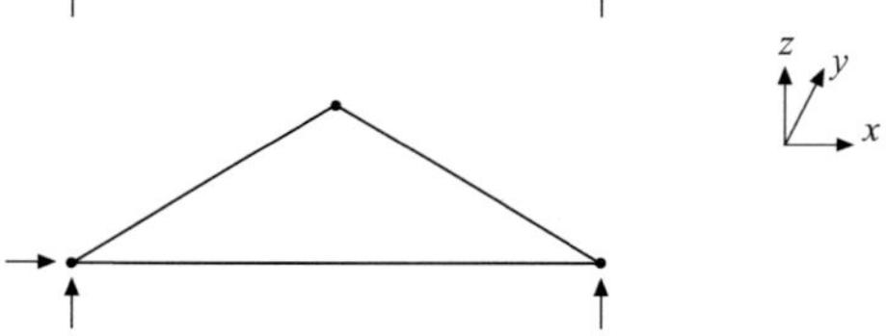

Figure D9-19 Support reactions of the three-dimensional braced system. (STEP 1995 Article D9)

The engineer simplifies the calculation, for example, by disregarding a support reaction in the *x*-direction, see Figure D9-19, which means the horizontal forces generated from the bracing are thus transferred to the two supports in the *y*-direction. This only holds, however, if the actual displacements of the supports deemed to be fixed in the *y*-direction are nearly the same; otherwise the "omitted" support will have to resist forces in the *x*-direction resulting from the rotation of the structure in plan. Therefore, this assumption is only possible, because the deformations of the chords are negligible and result only in the bracing being displaced in the *y*-direction. If the length of the posts and diagonals change and a connection slip in the truss joints occurs, this results in shear deformation as shown in Figure D9-20. Accordingly, the vertical element at the support will remain vertical without causing any horizontal reaction in the "omitted" support.

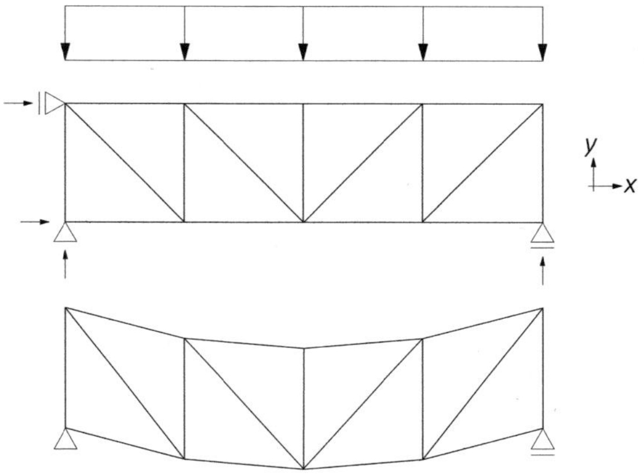

Figure D9-20 Deformation of a bracing structure. (STEP 1995 Article D9)

In sloping roofs, the span of the bracing is assumed to equal to the length of the roof area and is assumed to be a planar structure. The spatial load-bearing behaviour generates deviation forces in the apex zone, which should also be taken into consideration (Figure D9-21).

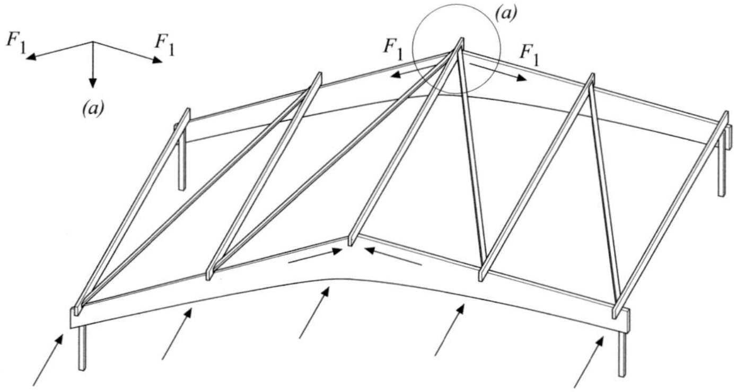

Figure D9-21 Vertical load of beams caused by horizontal actions if the beams are not straight. (STEP 1995 Article D9)

For the calculation model, it is both practical and advisable for roofs with a significant incline to disregard any transfer of shear forces in the apex zone, leaving two cantilevers (Figure D9-22). If the system and loading are symmetrical and the deformation of the chords is disregarded, the horizontal displacement of the apex is identical on both bracings. No deviation forces resulting from the forces in the chords of continuous bracings are exerted at the end of the cantilevers in this case. What do emerge, however, are two sets of horizontal support forces in the *x*-direction, which must be monitored and offset within the structure (internal forces).

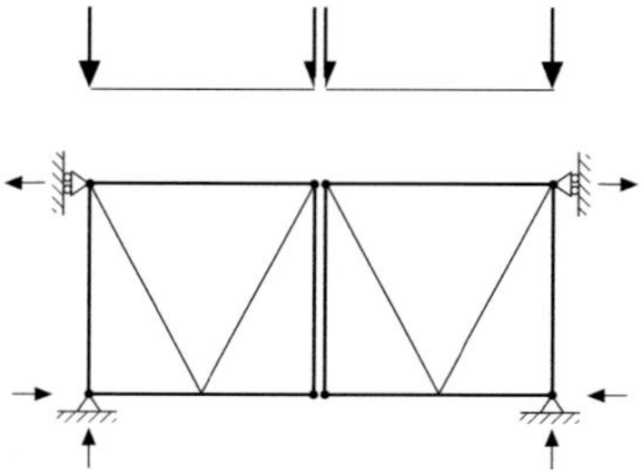

Figure D9-22 Support reactions of system, considered as cantilever. (STEP 1995 Article D9)

It may also be advisable to provide additional supports to the bracing system at the apex and/or at intermediate locations, which transfer the lateral forces towards the eaves. Given a sufficiently rigid support, this results in fewer deflections due to the shortened span of the bracing and thus reduced lateral forces.

This measure makes particular sense, for example, in constructions with trusses with punched metal plate fasteners as the main structural elements, which are usually narrowly spaced. This reduced spacing between the trusses does not allow for any bracing elements with sufficient beam depth (stiffness) between the compression chords of the trusses, if the span is assumed to be that of the main structural elements (Figure D9-23).

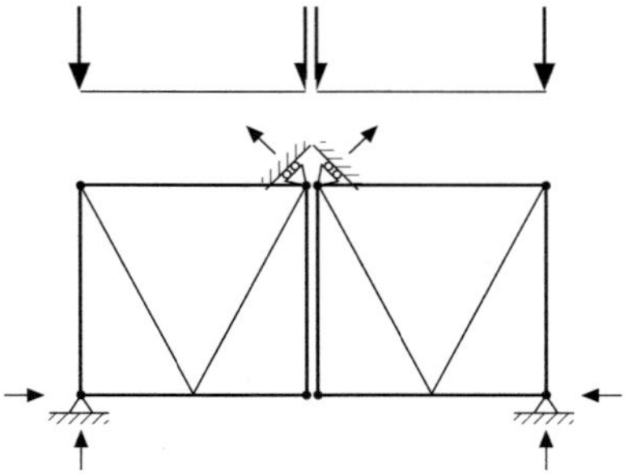

Figure D9-23 Support of the apex (additional support to shorten the span of the bracing). (STEP 1995 Article D9)

Details of bracing

Types of bracing

Bracing systems are mainly formed as trusses by adding diagonal members to the main structural elements, e.g. beams and purlins carrying vertical actions. Prefabricated trusses can also be used and installed between the members to be braced. Although this article only covers trusses, beams, diaphragms or individual supports can also be used for bracing purposes. In most cases, the chords of the bracing are formed by the main structural elements and if the latter are trusses, the compression chord to be supported should be part of the bracing. If the main structural elements are beams, the bracing system should be placed in their compression zone, which is then supported against lateral deviations. Existing purlins may be bracing elements and diagonals are additionally installed. At this point, there is also scope to select a range of geometries, which are linked to various characteristics.

Crossed diagonals resisting only tensile forces (Figure D9-24)

- Generally made of steel, e.g. as round steel bars with turnbuckles,
- Simple assembly,
- Purlins will be subject to additional loads, when, as often happens, they are used as posts in bracing trusses.

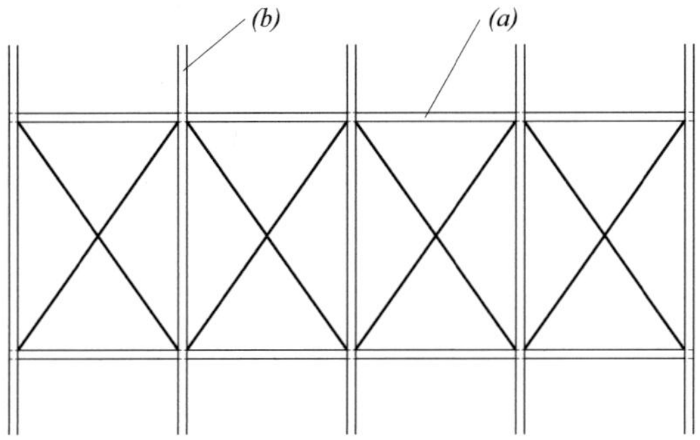

Figure D9-24 Crossed diagonals resisting only tensile forces. (*a*) Braced beam, (*b*) purlin. (STEP 1995 Article D9)

W-trusses (Figure D9-25)

- No additional forces in the purlins,
- Due to generally fluctuating stress directions, all diagonals must resist to both tension and compression, which is why timber members are recommended,
- Securing the diagonals subject to compression to the purlins can help reduce their buckling lengths.

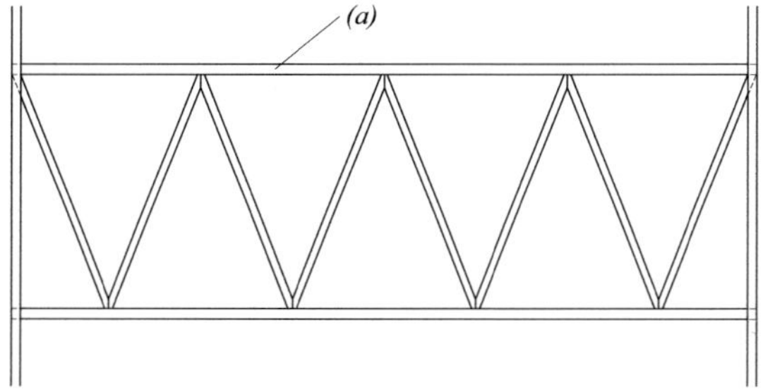

Figure D9-25 W-truss. (*a*) Braced beam. (STEP 1995 Article D9)

N-trusses (Figure D9-26)

- The application is recommended for places where one loading direction generates larger bracing forces than another,
- The vertical posts subject to compression here have lower buckling lengths than the diagonals (here, subject to tension).

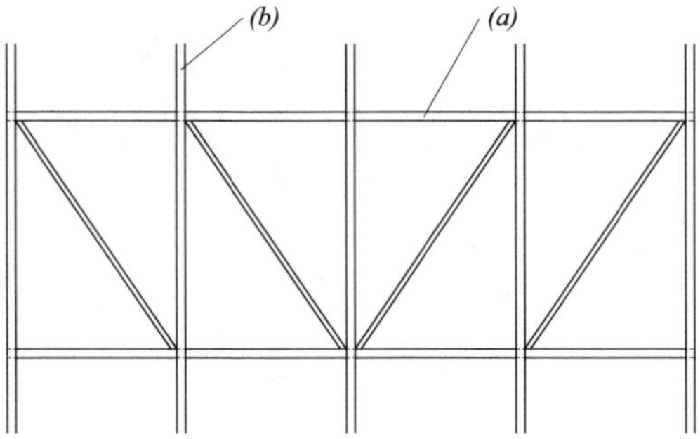

Figure D9-26 N-truss. (*a*) Braced beam, (*b*) purlin. (STEP 1995 Article D9)

K-trusses (Figure D9-27)

- The forces in the struts are halved, with unchanged strut inclination,
- Many connections are required,
- Buckling lengths are comparatively low,
- They offer larger openings if used as vertical bracing systems (Figure D9-28),
- The diagonals support the purlins at midpoint, reducing the buckling length in the plane of bracing.

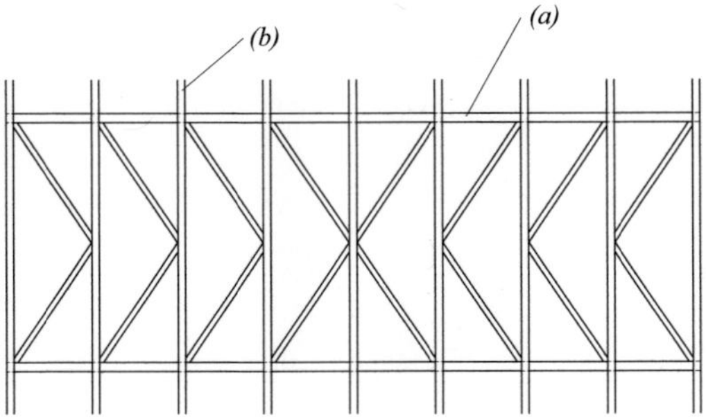

Figure D9-27 K-truss. (*a*) Braced beam, (*b*) purlin. (STEP 1995 Article D9)

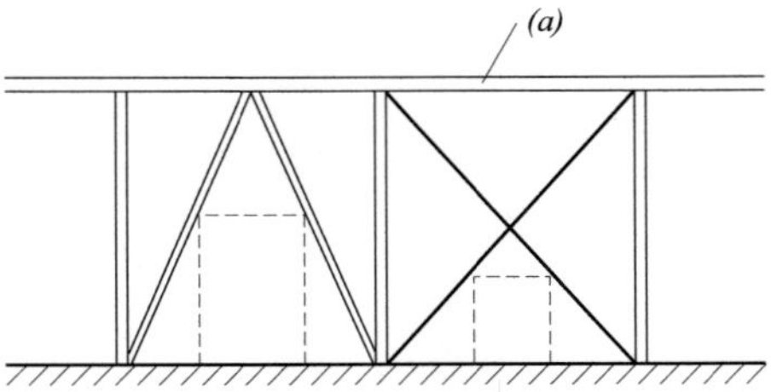

Figure D9-28 Wall bracing. (STEP 1995 Article D9)

Trussed beam (Figure D9-29)

- Simple assembly, e.g. flat steel sections can be screwed or nailed onto the purlins,
- The trussed beam can be optimally exploited by selecting a parabolic shape, since the tensile force remains constant with a uniformly distributed load exerted,
- Only effective in one loading direction,
- The ability to accommodate non-uniformly distributed loads must be ensured using specific measures.

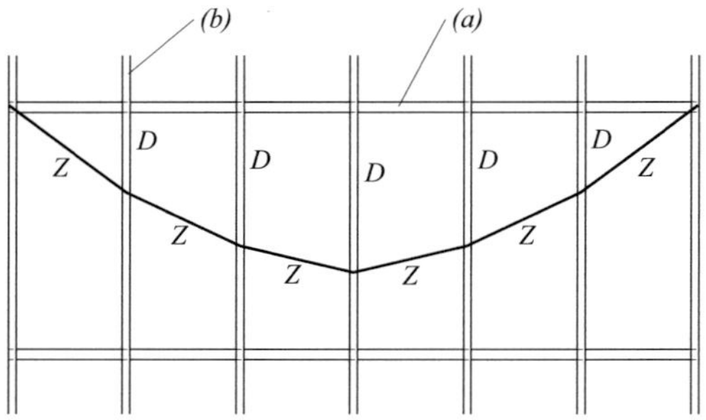

Figure D9-29 Trussed beam. (*a*) Braced beam, (*b*) purlin. (STEP 1995 Article D9)

Connections

Connections can be detailed in a number of ways and examples shown in Figure D9-30 to Figure D9-32 have worked well in practice. Timber diagonals can be connected with steel plates using nails or dowels as appropriate fastener types, as in Figure D9-30. For slotted-in steel plates, timber member and steel plate should be collectively predrilled beforehand and here, since more closely spaced nails are possible as compared to non-predrilled nail holes, this paves the way for compact, rigid and high load-bearing connections. When the forces in the diagonals are lower, unilaterally applied nailing plates can also be used. However, since predrilling is normally not used, the connection area is around three times as large as for slotted-in plates with predrilled nailing. Moreover, the diagonals are subject to bending moments due to the eccentric connection, which should be taken into consideration at the time of verification. Steel diagonals can be connected via cold-formed steel connectors like angle brackets or flat bars, as in Figure D9-31. The example in Figure D9-32 is particularly easy to assemble. The round steel diagonals can be guided through the chords and connected on the rear side. For this purpose, steel parts allowing a wide-ranging angle of inclination of the diagonal are available.

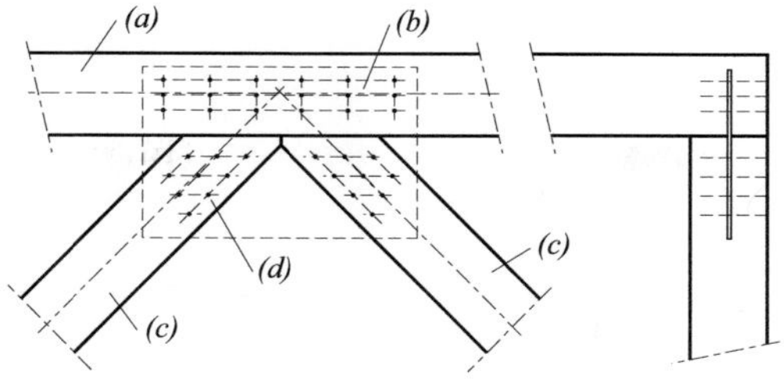

Figure D9-30 Timber diagonals connected via slotted-in steel plate. (*a*) Braced beam, (*b*) slotted-in steel plate, (*c*) timber diagonal, (*d*) predrilled nailed connection. (STEP 1995 Article D9)

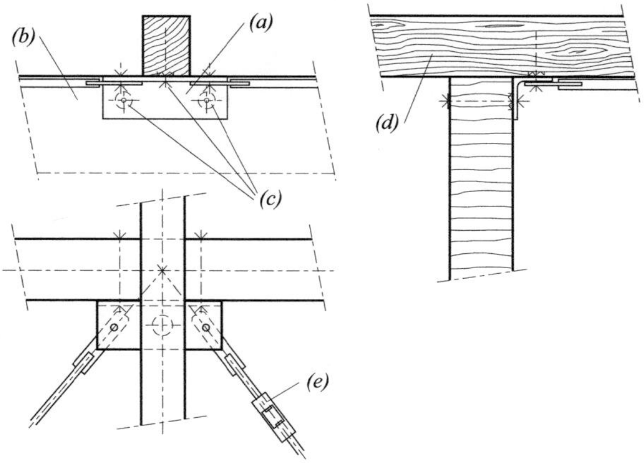

Figure D9-31 Connection of steel diagonals with angle brackets and flat bars. (*a*) Angle bracket, (*b*) braced beam, (*c*) fastener, (*d*) purlin, (*e*) turnbuckle. (STEP 1995 Article D9)

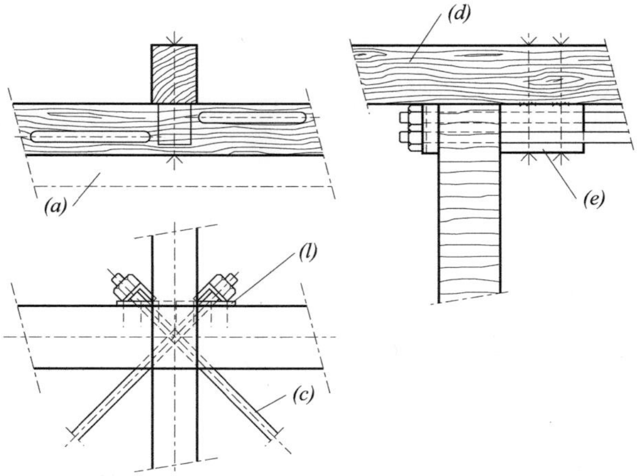

Figure D9-32 Connection of steel diagonals on the rear side of the chords. (*a*) Braced beam, (*l*) nailed steel fitting, (*c*) steel rod, (*d*) purlin, (*e*) connection with timber block and fasteners. (STEP 1995 Article D9)

D9.4 Literature

T. Alsmarker, H. Brüninghoff, S. Winter, original Articles B13, B15, D9, E14, STEP 1995.

Brüninghoff H. (1983). Determination of Bracing Structures for Compression Members and Beams. Paper 16-15-1, CIB-W18 Meeting 16, Lillehammer.

Brüninghoff H. et al. (1989). Holzbauwerke: Eine ausführliche Erläuterung zu DIN 1052, Teil 1 bis 3. Deutsches Institut für Normung e.V., Deutsche Gesellschaft für Holzforschung e.V., 1. Auflage. Beuth Verlag, Berlin.

Burgess H.J. (1989). Suggested Changes in Code Bracing Recommendations for Beams and Columns. Paper 22-15-1, CIB-W18 Meeting 22, Berlin.

Heimeshoff B. (1983). Einspannung von Stützen aus Brettschichtholz durch Verguß in Betonfundamenten. Holzbau Statik Aktuell, Ausgabe Juli 1983/7.

Möhler K. and Schelling W. (1968). Zur Bemessung von Knickverbänden und Knickaussteifungen im Holzbau. Der Bauingenieur 43(2).

Steinmetz (1992). Die Aussteifung von Holzhäusern am Beispiel des Holzrahmenbaues. Holzbau Statik Aktuell, Ausgabe Juli 1992/1.

D10 Timber-concrete composite structures

Original article: A. Ceccotti

Timber-concrete composite structures represent a form of mixed construction, in which the beams or columns are made of timber and concrete connected via semi-rigid joints using mechanical fasteners. The majority of timber-concrete composite structures feature concrete in the compression zone and timber in the tension zone of a beam allowing the most advantageous features of each of these construction materials to be utilized. The strength and stiffness of the concrete peak when exposed to compressive stress. Meanwhile, if timber is used in the tensile zone, this eliminates the need for concrete in the same, as is normally included in reinforced concrete cross-sections, which cannot bear stress when cracked and therefore represents only additional load. This enables cross-sections which retain high stiffness but relatively low self-weight. The load-bearing capacity is virtually double that of an all-timber floor and there is potential to obtain three to four times the bending stiffness. Figure D10-1 shows a comparison of the dead weights of three different floor constructions; depending on the span for an imposed load of q = 2.5 kN/m^2.

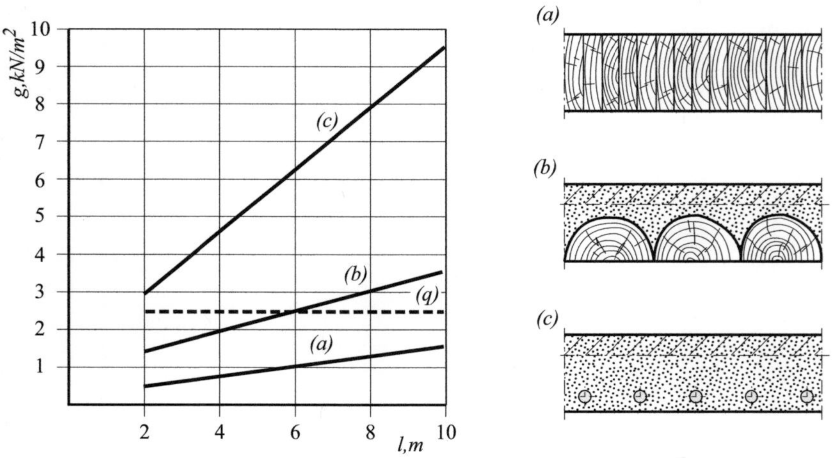

Figure D10-1 Self-weight depending on the span for variable action of q = 2.5 kN/m^2. (*a*) all-timber floor, (*b*) timber-concrete floor, (*c*) reinforced concrete floor. (according to Natterer, 1993, STEP 1995 Article E13)

The use of timber-concrete composite floors can effectively reduce floor deflection, a common problem affecting all-timber floors. The fact that it is also far easier to dampen vibrations in timber-concrete composite floors than in all-timber equivalents also facilitates compliance with serviceability limit state conditions during verifications.

The in-plane rigidity of the concrete compression zone of a timber-concrete composite floor is often considered indefinite compared to a timber floor, which is a key benefit when the building is exposed to earthquake loads in particular, since rigid floor diaphragms help the building retain its shape. For timber floors, of course, this cannot be generally guaranteed. The prerequisite for this favourable behaviour in response to earthquake load is effective and force-locking anchoring of the timber beams and the concrete slab in the surrounding masonry (see Figure D10-2).

Applying a concrete slab to existing timber floors is an approach which can also improve the sound insulation properties of floors. Firstly, the significant increase in mass reduces the airborne sound transmission, while the greater damping values of the composite floor also mean lower structural sound transmission. Moreover, the concrete slab represents an effective barrier to limit the spread of fire and means such constructions have greater fire resistance than floors made of timber beams alone. The timber beams themselves outperform corresponding steel or reinforced concrete precast beams in fire resistance terms.

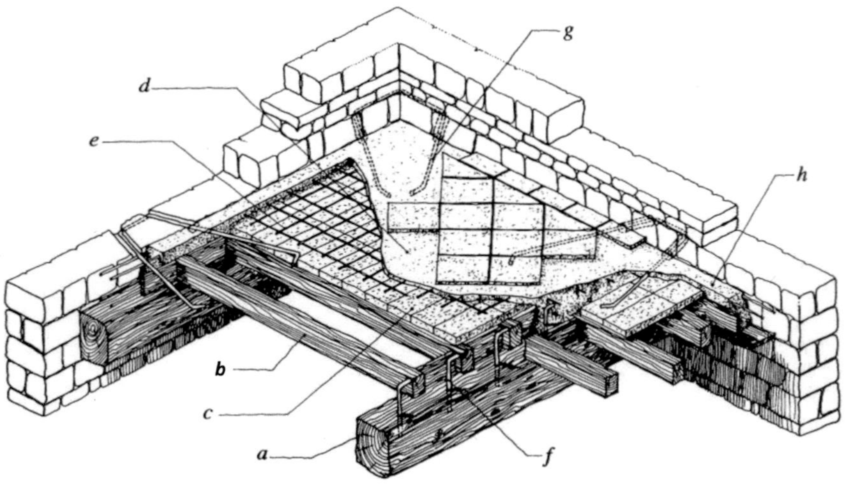

Figure D10-2 Example of a timber-concrete composite floor in a masonry construction subject to earthquake load. (a) Main beam, (b) secondary beam, (c) brick tiles, (d) concrete slab, (e) reinforcement, (f) glued-in dowels using epoxy resin, (g) steel stirrups connecting concrete slab with masonry, (h) ring beam made of reinforced concrete. (STEP 1995 Article E13)

For this reason, timber-concrete floors represent a cost-effective alternative to those made of reinforced concrete, particularly when renovating older buildings, since this approach eliminates the need to replace existing timber floor beams. Generally speaking, reinforcing timber beam floors which no longer meet present-day requirements by applying an additional concrete slab will suffice. The scope of any economic comparison in this case has to go beyond the square-metre costs of any alternative, since potential savings can emerge elsewhere, e.g. in foundations, lower earthquake loads exerted due to the lighter construction or less formwork and equipment required. Accordingly, timber-concrete composite structures are becoming a particularly popular option in some European countries, e.g. to reinforce existing timber beam floors in masonry structures, floors in new constructions or when making bridges. In Scandinavia, meanwhile, prefabricated timber-concrete elements are also used as wall panels, with the concrete slab giving the necessary racking strength and the timber studs installed for the required out-of-plane bending stiffness and necessary stiffening against buckling.

Experiences of the past 20 years show that despite variation in how both construction materials respond, timber-concrete composite structures exhibit outstanding load-bearing and deformation behaviour. During design and construction, there is a need to focus on how the timber and concrete respond, particularly in terms of creep and shrinking. Subsequent sections will set out the main areas to scrutinise when designing such constructions.

D10.1 Connection types

Figure D10-3 shows an overview of the most frequently used connections between concrete slab and timber beams and in this case, the joints have been classified in terms of stiffness. Nailed, screwed and dowelled connections, employing fasteners perpendicular to the joint line and as shown in Figure D10-3 (a) 1 to 3, show the lowest levels of stiffness. Screws arranged at an angle to the joint line (Figure D10-3 (a) 4) or surface connectors such as split ring, shear plate, toothed-plate connectors or punched metal plate fasteners, which do not penetrate the wood as deeply, are stiffer than connections with dowel-type fasteners loaded perpendicular to the fastener axis (Figure D10-3 (b)). Even greater stiffness is possible with shear keys where pre-bored notches in the wood are filled with concrete and which function as dowels after curing accordingly (Figure D10-3 (c)). Basically rigid connections are formed with glued joints in the manner shown in Figure D10-3 (d).

The effective bending stiffness $E \cdot I_{ef}$ of the composite cross-section for the various connections of Figure D10-3 amounts to around 50% (Figure D10-3 (a)) and up to 100% (Figure D10-3 (d)) respectively of the bending stiffness of a cross-section with a rigid connection. In the event of unlimited connection stiffness, linear strain distribution over the cross-sectional height can be assumed. The various moduli of elasticity of timber and concrete can thus be taken into consideration, as is done when calculating composite beams made of solid timber and wood-based materials (see Articles D6 and D7).

Where the semi-rigidity of the timber-concrete connection can no longer be ignored, the slip in the joint line must be taken into account as part of the stress analysis. In this case, the composite cross-section no longer remains plane when shear loads are exerted (see Figure D10-4), whereupon verification of the composite structure can be performed via the γ-method shown in Article D7. This is generally required for the types of connection shown in Figure D10-3 (a) to (c). For semi-rigid connections, it is also important to note that concrete in the tension zone is prone to cracking, which will render it no longer able to exhibit bending or axial stiffness or cross-sectional load-bearing behaviour. This can be taken into consideration using an iterative calculation, in which a very low modulus of elasticity is assigned to the concrete tensile area.

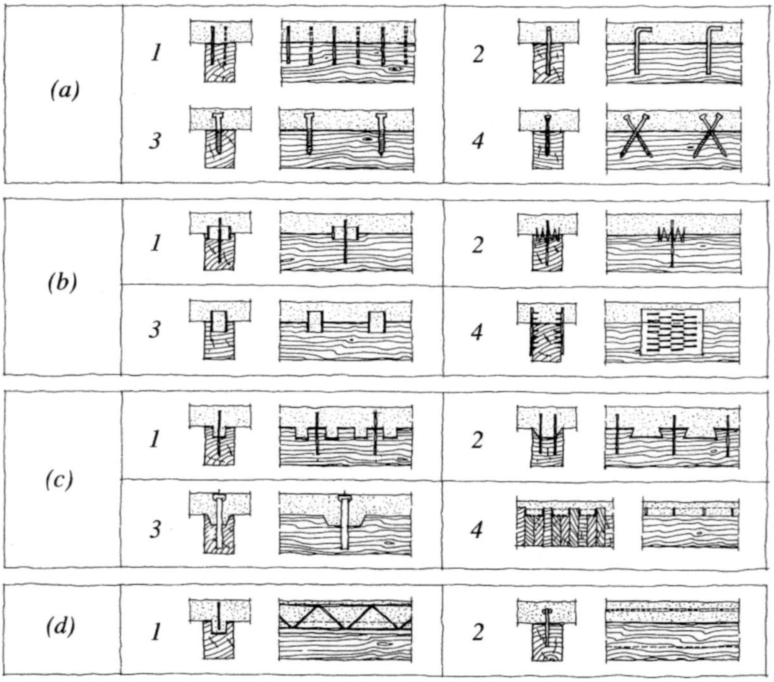

Figure D10-3 Examples of timber-concrete connections. (a1) Nails, (a2) dowels made of reinforcing steel bars, (a3, a4) screws, (b1) split ring connector, (b2) toothed-plate connector, (b3) steel tubes, (b4) punched metal plate fasteners, (c1) pre-bored notches with fasteners preventing uplift, (c2) dovetail-shaped notches with fasteners, (c3) notches with pre-tensioned steel rods, (c4) dowel-laminated timber (Brettstapel) with steel shear plates, (d1) truss made of steel rods glued to timber beam, (d2) steel plate glued to timber beam. (STEP 1995 Article E13)

D10.2 Mechanical performance

The material properties of the concrete required for the calculation are the characteristic strength value f_{ck}, the mean modulus of elasticity E_{cm} and the creep coefficient φ; numeric details of which are included in EC 2. For timber members, meanwhile, the characteristic bending strength $f_{m,k}$, modulus of elasticity $E_{0,mean}$ and the deformation factor k_{def} have to be known. However, the key property for the load-bearing behaviour is the stiffness of the timber-concrete connection, as identified by the slip modulus K_{ser} or K_u. The size of the slip modulus determines the stress distribution over the height of the composite section.

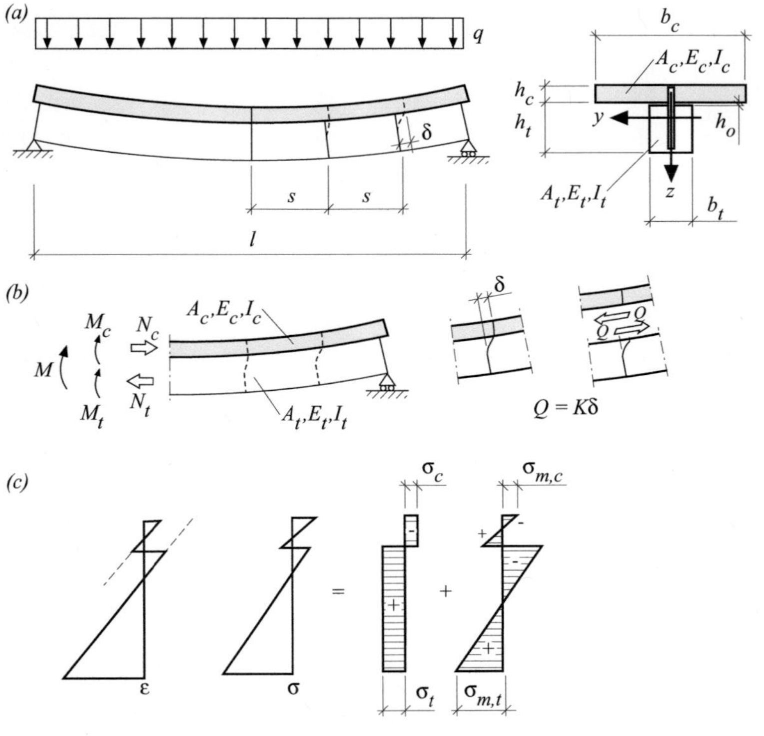

Figure D10-4 Load-bearing behaviour of a timber-concrete composite beam with semi-rigid connections. Above: composite cross-section does not remain plane; middle: concrete slab under compressive and bending stress, timber beams under tensile and bending stress and joint under shear load; below: constant increase in strain distribution, normal stress distribution comprising longitudinal and bending stresses. (STEP 1995 Article E13)

The load-bearing capacity and stiffness of the timber-concrete connection are to be determined by testing in accordance with EN 26891. In short-term tests, the load-deformation behaviour of connections can be determined, also taking into consideration any interim layers (see h_o in Figure D10-4) between the concrete and timber. An example test set-up is shown in Figure D10-5. The short-term tests conducted in accordance with EN 26891 can be used to determine the slip modulus K_{ser} (in EN 26891 namely k_s).

Even when plastic deformations may actually occur when reaching the ultimate limit state of the concrete and joints, linear-elastic behaviour can still be assumed when determining the internal forces and moments. To determine plastic deformations as accurately as possible, a nominal secant modulus is used for the modulus of elasticity of the concrete (see EC 2) and an equivalent secant modulus for the slip modulus of the joints. Meanwhile, the stiffness values of the concrete to be applied when determining the internal forces and moments are calculated while assuming an uncracked cross-section, although when verifying stresses of the concrete cross-section, the tensile strength of the concrete is not taken into account. On the compression side, the compressive strength is assumed to equate to the limit value for compressive stress which applies in the event of plastic deformations of the concrete (see EC 2). If not all the concrete cross-section is under compressive stresses, there is a need to use either a corresponding reinforcement or omit the cracked cross-sectional part. The value used for the slip modulus of the joint is K_u, whereby $K_u = 2/3 \cdot K_{ser}$. In this case, the value K_u represents an average and not a 5% quantile value.

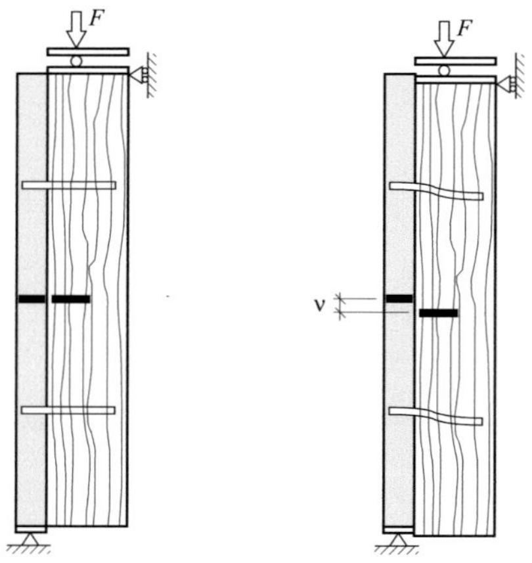

Figure D10-5 Possible test arrangement to determine the load-deformation behaviour of a timber-concrete connection. (STEP 1995 Article E13)

Accordingly and also when verifying the ultimate limit state, in the process of determining internal forces and moments, average values for the stiffness properties of the construction materials and connections should be used. EC 2 specifies only nominal values for the modulus of elasticity, which are considered to constitute average values. If average values were used for the slip moduli of the connections and the modulus of elasticity of the concrete collectively with 5% quantile values for the modulus of elasticity of timber, the stresses determined for the timber cross-sections would be underestimated – namely on the unsafe side. For verification of stresses, of course, the characteristic strength values have to be used.

Since both concrete and timber are prone to creep and concrete in particular is subject to longitudinal shrinkage, this must be taken into consideration when verifying composite sections, e.g. by reducing the stiffness properties of the concrete, timber and connection as applicable. For timber members with typical semi-rigid connections, the impact of creep on the distribution of internal forces and moments in the composite section can be disregarded, since similar creep behaviour is assumed for both timber and connections. For timber-concrete composite structures conversely, the individual cross-sections respond differently to changes in the prevailing climate. The reduction in stiffness properties represents an approximation, which equates to an overestimation of the actual deformation occurring. Accordingly, it is conservative from a structural engineer's perspective.

Shrinkage of the concrete in the longitudinal direction reduces the overall deformation and thus the load exerted on the fasteners, although this exacerbates the deflection of the composite beam at the same time. This additional deflection can be offset by installing a precamber on the construction before applying the concrete. Since most of the shrinkage deformation has already occurred during the process of curing the concrete and the composite construction is supported during this period, the impact of said shrinkage is considerably less. Still further reduction is possible via the cracks in the concrete, which arise from tensile stresses that developed when shrinking is prevented. The way in which concrete and timber behave varies respectively in the event of prevailing climatic changes – concrete is particularly sensitive to changes in temperature, while timber is more significantly affected by changes in moisture content – but this is only problematic if the connections between the timber and concrete are very rigid and the members concerned are particularly long.

D10.3 Recommendations for design and execution

The following list includes certain basic recommendations for design and execution, which should also help avoid errors:

- Dry wood should be used in principle. In addition, propping should be provided for longer after the concreting than would apply for all-concrete slabs.
- Fasteners must be protected against corrosion depending on the service class. This can be done by hot-dip galvanising, passivating, coating or otherwise using stainless steel fasteners.
- When thick concrete slabs are used, reinforcement is required to avoid wider cracks developing on the tension side of the concrete cross-section. In addition, the cracked zone must be taken into consideration when determining internal forces and moments.
- When applying the concrete, the wood must be protected against humidity. This can be done, for example, by using protective films or mixing additives into the concrete, which enable the water content in the concrete to be minimised. This also, moreover, reduces the shrinkage of the concrete. Some wood species, e.g. larch, contain extractives which slow down the curing process or may prevent it on a localised basis.
- When a large span is involved, softer semi-rigid joints are preferable, since this facilitates efforts to dissipate the constraint actions caused by the shrinkage of the concrete.

D10.4 Literature

A. Ceccotti, original Article E13, STEP 1995.

RILEM TC 111 CST (1992). Behaviour of timber-concrete load-bearing structures. Proc. of ACMAR-Ravenna International Symposium, Dept. of Civil Engineering, University of Florence.

Natterer J. (1993). Constructions en bois II. Notes du cours IBOIS, EPFL Lausanne.

D11 System strength

Original article: H. J. Blass

For structural systems comprising multiple similar members, positioned equidistantly and connected via a continuous load distribution system, EC 5 allows the design load-bearing capacity values to be increased by a load distribution factor. One example of such a structural system is a timber beam floor, in which the beams are linked by panel sheathing. The load distribution factor takes into consideration two different impacts, which boost the strength of members within a system:

- The low likelihood that a member with low strength will be installed at the specific point where stresses tend to peak,
- The positive correlation between the strength and stiffness of timber members.

This positive correlation explains why more rigid members, which, by nature, are exposed to above-average load, can also accommodate this load. In contrast, less rigid members, most of which are also weaker, have to carry less load. Accordingly, the load distribution acts to offset the impact of the comparatively large natural scatter of timber properties.

Both for concentrated as well as uniformly distributed loads, the load distribution leads to higher characteristic values for the load-bearing capacity of members within a structural system. For concentrated loads, the load distribution system dissipates part of the load onto neighbouring members and thus eases the forces exerted on the member under most strain. Figure D11-1 shows the cross-section of a timber beam floor under the impact of a concentrated load.

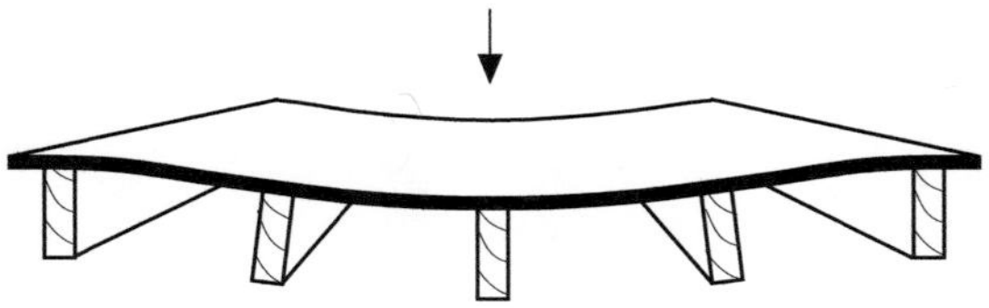

Figure D11-1 Floor construction under the impact of a concentrated load. (STEP 1995 Article B16)

When the load is uniformly distributed, the effect of the load distribution is less clear. When members have identical stiffness, they would tend to deform to the same extent, even in the absence of any load distribution system. However, since stiffness values tend to vary in reality, less stiff members are prone to more extensive deformations than stiffer equivalents. Line (b) in Figure D11-2 shows the various deflections of timber beams in a floor without a load distribution system. If a load distribution system is present, there will be less overall variation of deflection (see line (a) in Figure D11-2). In this case, the load distribution system dissipates part of the stress from the softer to the more rigid members.

This also applies when the behaviour of individual members is no longer linear. If, due to microscopic cracks or plastic deformations, member stiffness declines, forces are redistributed to the members remaining intact or those located in the elastic area. The damaged member can thus still – albeit to a more limited extent – contribute to the overall load-bearing capacity of the system.

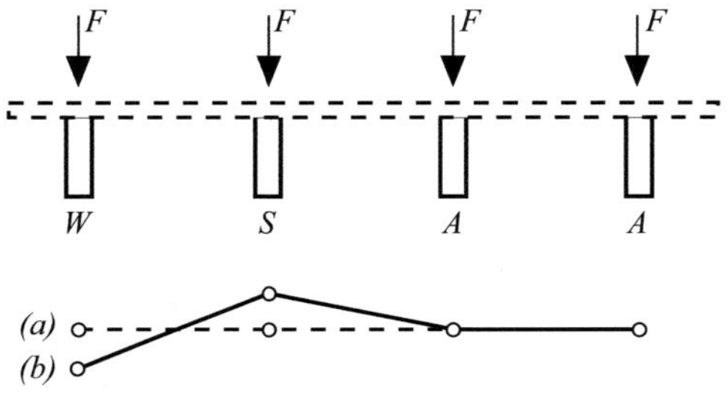

Figure D11-2 Effect of a load distribution system on deflection of timber beams. (a) with and (b) without load distribution system; W: member with lower stiffness; S: member with higher stiffness; A: member with medium stiffness. (STEP 1995 Article B16)

D11.1 Application in EC 5

The system strength of such similar members, which are connected by a continuous load distribution system and at equidistant intervals, is taken into consideration in EC 5 using a coefficient k_{sys}, with which the relevant strength parameters can be increased. The basis assumption is $k_{sys} = 1.1$, provided the continuous load distribution system is in a position to accommodate the forces exerted by one member on another. Verification of compliance with the ultimate limit state of the load distribution system should be ensured for short load duration. For laminated timber decks, meanwhile, with prestressed or glued laminations, a higher value may be used for k_{sys}, Figure D11-3.

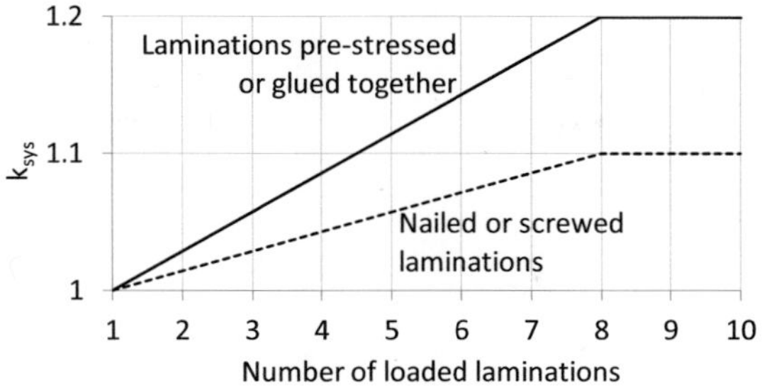

Figure D11-3 System strength factor k_{sys} for laminated deck plates of solid timber or glued laminated members.

D11.2 Load distribution in various structural systems

Floors and flat roofs

Foschi, Folz and Yao (1989) derived load distribution (system strength) factors for use in designing beams in floors and flat roofs. In the process, they assumed stress was uniformly distributed and the ratio between permanent and variable loads was varied. The behaviour of the members and joints was also assumed to be linear-elastic.

The system strength factor was derived using a reliability analysis applied to an individual beam within a construction, which involved examining the change in the probability of failure of an individual beam, when the beam functioned as part of a load-bearing system. Parameter studies could illustrate the influence of the various variables.

The support conditions, size, number and interval distance of the beams and the ratio between permanent and variable loads did not impact significantly on the system strength factor k_{sys}, while the following parameters increased the system strength factor and enhanced the load-bearing capacity of the system in the process:

- Larger stiffness of the load distribution system in relation to beam stiffness,
- Larger variance in the modulus of elasticity of the beams,
- A closer correlation between modulus of elasticity and bending stiffness.

The variance in bending strength of the beams also impacts significantly on the size of the system strength factor. Very small or large values for the coefficient of variation (COV) respectively result in only marginal values for the system strength factor, the value of which peaks for COV values between 0.20 and 0.30. For a typical floor structure, k_{sys} = 1.15 was determined, which correlates well to the value of 1.1 in EC 5.

Trusses with punched metal plate fasteners

Wolfe and McCarthy (1989) and Wolfe and LaBissoniere (1991) examined load distribution in roof trusses with punched metal plate fasteners. Tests were performed with three different truss configurations; initially with the individual trusses, followed by the entire roof structure, where a roof sheathing made of 12 mm-thick plywood acted as the load distribution system. The interaction of the individual trusses within the roof structure was estimated, which involved comparing the load-deformation behaviour of the truss as an individual beam subjected to linear load with the corresponding behaviour of the same truss as part of the roof structure subject to a uniformly distributed load.

As expected, the interaction of the trusses in the roof structure increased the load-bearing capacity, while variation in deflection declined. Once individual trusses had reached their load-bearing capacity and been partially damaged, they were still capable of accommodating a portion of their original load, since the latter was redistributed to the still undamaged trusses. Where only one of the individual trusses in the system was exposed to a load along its top chord, 40 to 70% of the load was transferred via the sheathing to the neighbouring trusses. The load distribution meant the ratio of the load-bearing capacity of the roof structure relative to that of the weakest truss went from 1.09 to 1.47. These values, which depend on the effectiveness of the load distribution system and the truss positioning within the roof structure, show that the design values of the load-bearing capacity of the entire truss, namely the members and connections, could be increased by 10% thanks to the load distribution effect. For most structures, an acceptably conservative estimate would be $k_{sys} = 1.1$.

Sheet piling

If the planks of a retaining wall are connected, e.g. by a tongue and groove joint, this will increase the load-bearing capacity of the retaining wall compared to that of the individual planks. In this case, the connection between the individual planks functions as a load distribution system. This connection also ensures virtually uniform deflection of the planks, however large the scatter in bending stiffness.

A numerical analysis (Van der Linden et al., 1994) was performed with properties of ekki (*Lophira alata*), including moisture content exceeding 30% ($f_{m,mean}$ = 103 N/mm^2 and $E_{0,mean}$ = 17600 N/mm^2). The coefficient of correlation between the bending strength and modulus of elasticity was assumed to be 0.73, while the coefficient of variation for bending strength and modulus of elasticity was assumed at 15%. The plastic deformation capacity of the timber was taken into consideration when exposed to high bending compressive stresses and a related reduction in stiffness before reaching the bending strength limit. Using a non-linear finite-element program, the load-bearing capacities of simulated walls were calculated assuming wide-ranging plank properties, while the system strength factor was determined as the ratio of the characteristic load-bearing capacity of a sheet piling system with ten planks to the characteristic load-bearing capacity of the single planks. This resulted in $k_{sys} = 1.15$, although this is only applicable for bending strength, since the load distribution system for normal forces in the planks is ineffective.

D11.3 Literature

H.J. Blass, original Article B16, STEP 1995.

Foschi R.O., Folz B.R. and Yao F.Z. (1989). Reliability-based design of wood structures. Structural Research Series, Report No. 34, University of British Columbia.

Van der Linden M.L.R., Van de Kuilen J.-W.G. and Blass H.J. (1994). Application of the Hoffman yield criterion for load sharing in timber sheet piling. Pacific Timber Engineering Conference, Gold Coast.

Wolfe R.W. and LaBissoniere T. (1991). Structural performance of light-frame roof assemblies. II. Conventional truss assemblies. Research Paper FPL-RP-499, Forest Products Laboratory, USA.

Wolfe R.W. and McCarthy M. (1989). Structural performance of light-frame roof assemblies. I. Truss assemblies with high truss stiffness variability. Research Paper FPL-RP-492, Forest Products Laboratory, USA.

E

Joints

E1 Joints in timber structures

Original article: P. Racher

Basic tasks of any structural design include the conceptual design of the structure, the choice of the bracing system, structural design of the individual members and the ease with which the structure itself can be formed. However, the joints are particularly important, since those used in timber construction tend to be weaker than the members being joined. The serviceability limit state is also very largely dictated by the mostly semi-rigid joints. Selecting a joint design within a timber construction requires consideration of more than just the load and the load-bearing capacity. Other aspects to be taken into account include the visual appeal, cost efficiency and production process, as well as the erection method and the preferences of the structural engineer or architects (Natterer et al., 1991). This precludes a one-size-fits-all approach to decide on the optimal joint for a timber construction. Instead, key is to use the simplest possible joints and minimise the fastener types used for a better final construction result.

The main joints differentiated are glued joints, carpentry joints and joints using various metal fasteners. Rigid glued joints include, for example, glued-in steel rods, which are focused on in more detail in Article E10 and also include the finger joints used in glulam production. Carpentry joints or contact connections meanwhile, such as step joints or mortise and tenon joints are introduced in Article E9. This article focuses on joints using metal fasteners. The key fasteners are introduced, while important design-relevant points are summarised. Since all featured joints are explained in more detail in subsequent articles, the current scope merely features an overview and generally applicable design rules, which should facilitate understanding of all following articles.

E1.1 Types of metal fasteners used in timber joints

Traditional metal fasteners can be divided into two main groups, based on the type of force transmission between the members. The first features dowel-type fasteners, whereby forces exerted generate bending and tensile stresses in the fasteners as well as embedment and shear stresses in the wood along the shank. This category includes nails, staples, bolts, screws, dowels and threaded rods and these are the most frequently used fasteners. The second group includes "surface-type" fasteners such as split ring and toothed-plate connectors and punched metal plate fasteners, in which most of the force transmitted is focused on the surface area of the member.

Dowel-type fasteners

Nails

Nails, a popular choice for construction parts such as horizontal and vertical diaphragms and trusses, are the most frequently employed fasteners in timber construction and come in wide-ranging sizes, shapes and materials (see Figure E1-1). The most common types used in timber construction are round smooth shank nails, although nails with a square cross-section or profiled shank are also available. The dimensions of nails correspond to a range of standardised measurements in European countries, but are normally 2 to 8 mm in diameter and 40 to 200 mm long. When producing nailed joints, predrilling of nail holes may be required to prevent the timber splitting or make it easier for the nails to penetrate timber with a greater density. In timber constructions, meanwhile, nails tend to be used in single-shear joints with timber, steel or wood-based materials. For more on nailed joints, please refer to Article E3.

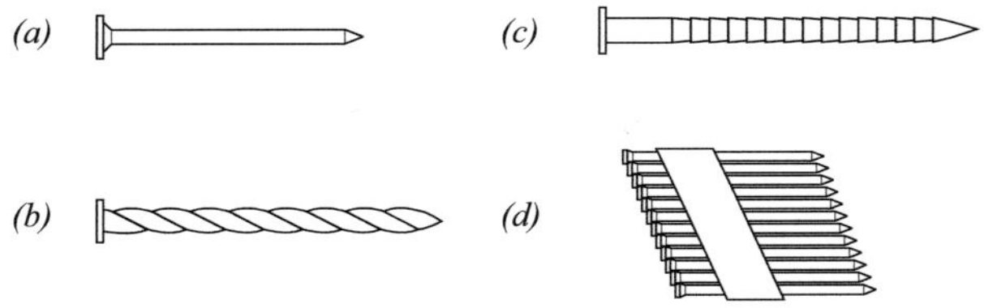

Figure E1-1 Nail shapes. (*a*) Round smooth shank nail, (*b*) spiral nail, (*c*) ringed shank nail, (*d*) machine driven nails. (STEP 1995 Article C1)

Staples

Staples as shown in Figure E1-2 are the classic fastener choice for timber frame buildings (at least in German-speaking countries), given the rapidity with which such constructions can be completed. Staples are made of high-tensile and ductile steel, since the production process imposes a reshaping angle of 90° on the staple material, which it must withstand undamaged. Staples are generally resinated and can only be processed using special fastener tools, while their very slender nature precludes any hammering in. Stapled joints are covered in more detail in Article E3.

Figure E1-2 Staples.

Bolts, dowels and threaded rods

Bolts (Figure E1-3 on the left) are generally made of steel, including hexagonal or square heads and nuts and are between 12 and 30 mm in diameter. To make it easier to drive home the bolts, as specified in EC 5, predrilled holes 1 mm larger than the bolt diameter can be used. This hole tolerance reduces the stiffness of bolted joints. Accordingly and with the visual appearance in mind, dowels or fitted bolts without clearance are often preferred to bolts. Dowels are steel rods with a round cross-section and driven into suitably predrilled holes, see Figure E1-3 on the right. Bolted and dowelled joints are covered in E4. Joints with threaded rods are designed as bolted joints (with the external thread diameter).

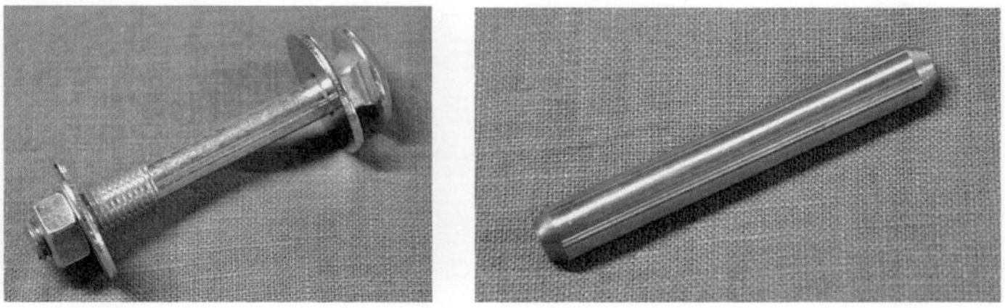

Figure E1-3 Left: bolt, right: dowel.

Screws

The fastest evolving fastener is the self-tapping wood screw. Some examples are illustrated in Figure E1-4, which also differentiates the two basic types of such screws, namely "partially threaded" (with a partially smooth shank) and "fully threaded", the latter of which come in sizes up to an external thread diameter of 14 mm and up to 1.5 m long. Depending on the technical approval or ETA and the timber type used, it may even be possible to use self-tapping screws of 14 mm in diameter without predrilling. However, such predrilling is possible, and can be done, for example, using the core (= inner thread) diameter. Other examples of screws include those with a thread shape in accordance with DIN 7998, which, from a diameter of 8 mm onwards, can only be screwed into predrilled holes. However, the advent of self-tapping screws has seen these traditional screws used less and less frequently, so they will not be covered in any more detail at this point (design of screwed joints with threads in accordance with DIN 7998 resembles to that of self-tapping screws). Screwed joints are covered in Article E5.

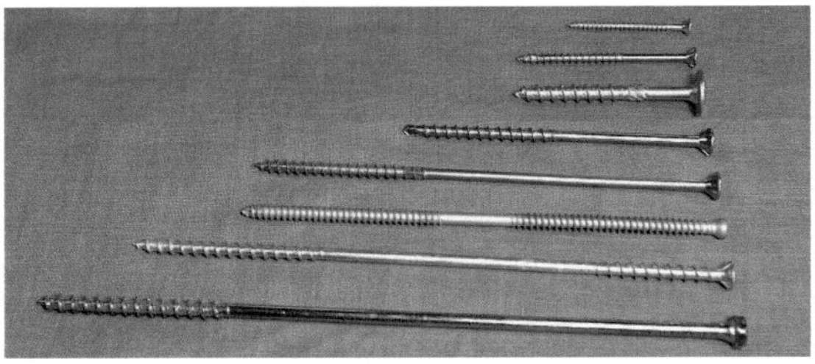

Figure E1-4 Examples of self-tapping screws.

Surface-type fasteners

Connectors

Increasing the load-bearing area allows connectors to accommodate considerable loads. In truss joints, for example, this paves the way for virtually perfectly hinged joints, if a single connector is employed for each joint, instead of multiple dowel-type fasteners. Figure E1-5 shows typical forms of split ring, shear plate and toothed-plate connectors, whereby the installation and load capacity depend largely on the accuracy of grooving and boring. Split ring and shear plate connectors are made of cast aluminium, cast steel or steel and are between 60 and 260 mm in diameter. The second connector type is the toothed-plate connector, made of cast steel or hot dip galvanised steel sheet and between 38 and 165 mm in diameter. Larger connectors are used when connecting glulam members and connector diameters of up to 75 mm for solid timber. To absorb the moments generated in the joints, which press the timber members apart, connector joints must be secured using fasteners such as bolts or screws, which are capable of accommodating such tensile forces. The diameter of the bolt washers should correspond to around half the diameter of the connector. Double-sided connectors are used for timber-to-timber joints. They absorb the force exerted directly between the surfaces of the timber members to be joined and such joints are generally established at the construction site. Single-sided connectors can be used for both steel-to-timber joints as well as timber-to-timber joints. They allow joints to be prefabricated, meaning only the bolts need be installed on-site. When these single-sided connectors are used, the bolts loaded perpendicular to their axis accommodate the force transmitted by the base plate of the connector. Connectors are covered in Article E6.

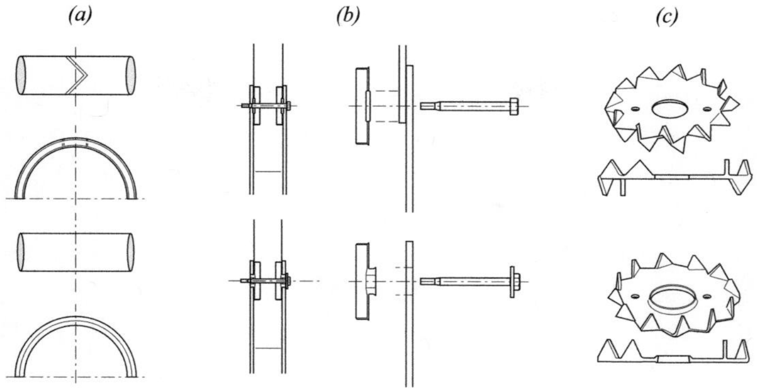

Figure E1-5 Common connectors: (*a*) double-sided split ring connector, (*b*) single-sided shear plate connector, (*c*) double- and single-sided toothed-plate connector. (STEP 1995 Article C1)

Punched metal plate fasteners

Punched metal plate fasteners (Figure E1-6) allow joints to be established between two pieces of wood in the same plane and are made from galvanised steel sheets or stainless steel, 0.9 to 3.0 mm thick. Producing joints with punched metal plate fasteners requires the use of special factory tools. They are mainly used for lightweight timber trusses, with individual timber members each at least 35 mm thick. Since these trusses tend to have a very low stiffness perpendicular to the truss axis, careful transport and assembly are required to avoid damaging the joints. A range of punched metal plate fasteners from different manufacturers is available and here, the structural engineer should refer to the corresponding manufacturer details, which are recognised by a certification body. The relevant standard in this case is EN 14250 "Timber structures – Product requirements for prefabricated structural members assembled with punched metal plate fasteners". Punched metal plate fasteners are covered in Article E7.

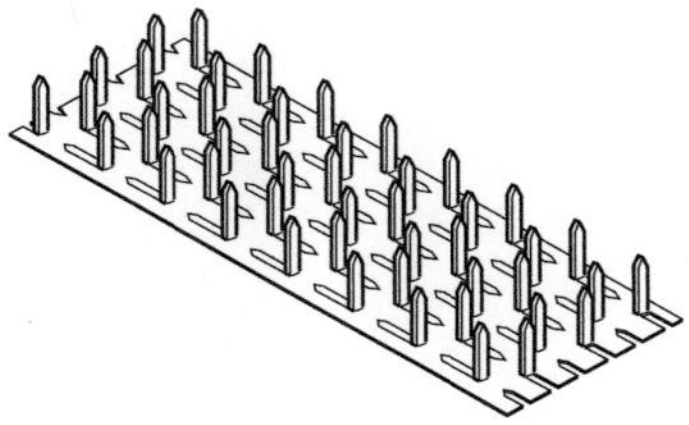

Figure E1-6 Typical punched metal plate fastener. (STEP 1995 Article C1)

E1.2 Load-deformation behaviour of joints

For design purposes, overall timber structural calculations have to be combined with localised joint analysis and the key consideration here is joint behaviour, which impacts on both the diagrams of internal forces and moments and the deformation of the structure. The behaviour of a joint can be determined by tests carried out in accordance with EN 26891 "Joints made with mechanical fasteners. General principles for the determination of strength and deformation characteristics." Figure E1-7 shows experimental load-deformation curves of joints with various fasteners loaded perpendicular to their axis, in which the load *F* is defined as the load per shear plane.

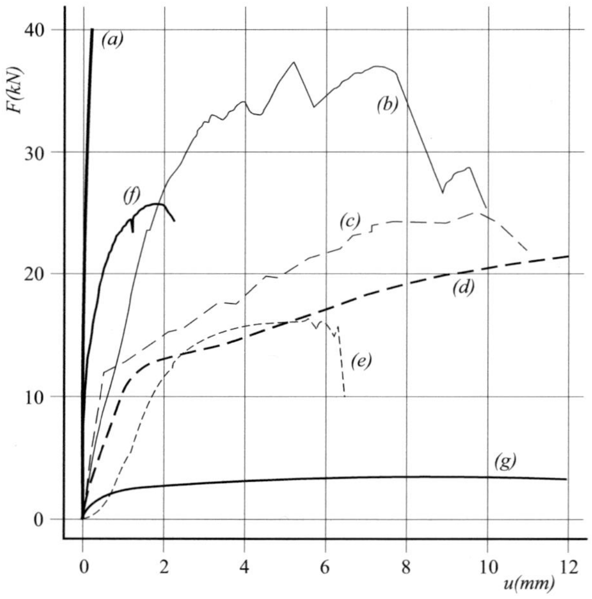

Figure E1-7 Experimental load-deformation curves for joints exposed to tensile load parallel to the grain. (*a*) glued joint ($12.5 \cdot 10^3$ mm^2), (*b*) split ring connector (100 mm), (*c*) double-sided toothed-plate connector (62 mm) (Hirashima, 1990), (*d*) dowel (14 mm), (*e*) bolt (14 mm), (*f*) punched metal plate fastener (10^4 mm^2), (*g*) nail (4,4 mm). (STEP 1995 Article C1)

Undoubtedly, the key feature, as is clearly shown in Figure E1-7, is that laterally loaded joints with mechanical fasteners, unlike rigid glued joints, undergo significant deformations, which structural engineers must take into consideration.

However, the increased calculation effort due to the semi-rigid joints can be deemed positive overall:

- Safety aspect: a semi-rigid joint can withstand a higher degree of deformations before failure → failure is signalled.
- Load distribution aspect: plastic deformation or creep may result in load redistribution, which can help relieve highly stressed areas.

The elastic-plastic, ductile behaviour of laterally loaded timber joints with mechanical fasteners, particularly dowel-type fasteners, is due to the interaction between the plastic deformation of the fasteners and the crushing (so-called embedment) of the wood underneath the fasteners, if specific edge and end distances and spacings are maintained. **The required minimum distances guarantee that almost all brittle failure modes like splitting can be prevented, which paves the way for ductile behaviour.**

However, the ductile behaviour of laterally loaded joints with dowel-type fasteners depends not only on the specific embedment strength of the timber, the yield moment of the fastener and the geometric boundary conditions (edge, end distances, spacing), but also on how slender the fastener used actually is. Figure E1-8 shows one example of a dowelled joint, where the slenderness of the timber-to-timber joint with one dowel and two shear planes can be defined as the ratio of the thickness t_2 of the middle timber member and the diameter d of the dowel.

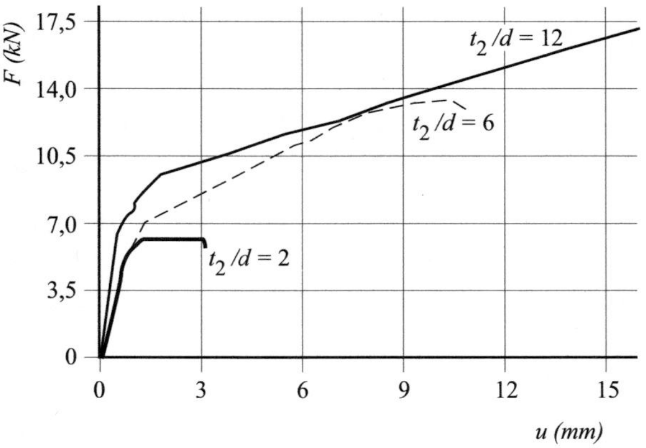

Figure E1-8 Influence of slenderness on the load-deformation behaviour of a timber-to-timber joint loaded in tension parallel to the grain. (STEP 1995 Article C1)

The direction of the forces transmitted also impacts on the joint behaviour. If a single fastener is used, this influence is dictated by the fastener diameter and the grain direction of the timber. Tests (Smith and Whale, 1986) have shown that the load-bearing capacity of joints with fasteners of diameter up to around 8 mm is independent of the angle between the force and grain directions. Depending on this angle, however, stresses perpendicular to the grain may reduce the ductility and load-bearing capacity of the joint by causing the timber to split prematurely. To prevent brittle fracture and splitting however, it is possible to enhance the ductility of a joint by reinforcing the timber member in the area of the joint. Effective reinforcements may include punched metal plate fasteners or wood-based materials, which can be pressed or glued onto both internal sides of the timber members, or the insertion of fully threaded screws (see Article E12).

E1.3 Design of joints with dowel-type fasteners

The key metal fasteners from an engineer's perspective are those in the dowel-type category, such as nails, staples, screws, bolts and dowels, to which the above statements concerning load-bearing capacity and deformation behaviour apply. In this section, we now reconcile these statements in terms of general design principles and equations, which help clarify the individual articles on fasteners.

Minimum distances

For all dowel-type fasteners, minimum values have been defined for edge and end distances and spacings, most of which are based on experience or comprehensive experimental investigations. The minimum distances defined in Figure E1-9 vary according to the fasteners involved and constitute a prerequisite for the design equations specified in EC 5. Certain fundamental and significant aspects of fastener distances are as follows:

- The spacing parallel to the grain a_1 influences the effective number of fasteners and hence the overall load-bearing capacity of the joint, since an insufficient spacing a_1 may trigger premature splitting of the joint along the set of fasteners in the grain direction.
- The required spacing perpendicular to the grain a_2 does not exceed a_1, where however no splitting will take place perpendicular to the grain.
- The required distance to the unloaded edge/end does not exceed that to the loaded edge/ end, where insufficient distances to the loaded edge/end lead to premature splitting.
- The spacings a_1 and a_2 (and the end distance a_3) influence the block shear behaviour of timber members in the area of the joint; the smaller these distances, the greater the likelihood of a block shear failure (Article E13).

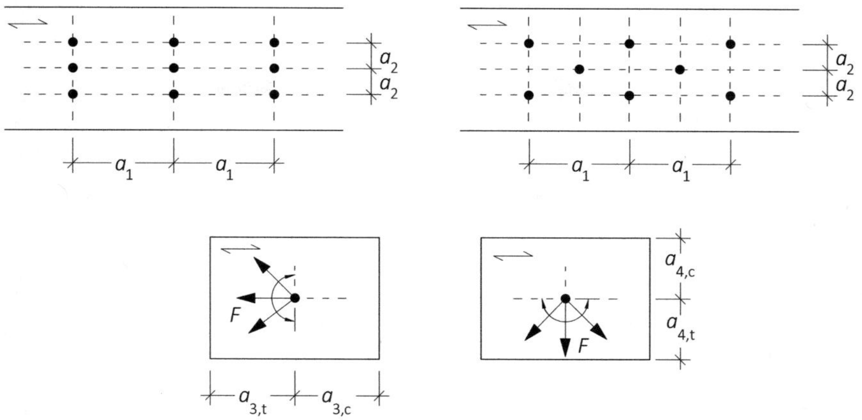

Figure E1-9 Spacing parallel to the grain a_1, spacing perpendicular to the grain a_2, loaded end distance $a_{3,t}$, unloaded end distance $a_{3,c}$, loaded edge distance $a_{4,t}$, unloaded edge distance $a_{4,c}$.

Types of loading and failure modes

Dowel-type fasteners are subject to two kinds of loading; firstly in shear, when a force perpendicular to the fastener axis is exerted and secondly withdrawal/pushing in, when force is exerted in the direction of the fastener axis. Combined stresses are also possible.

Joints with laterally loaded fasteners

The behaviour of joints with laterally loaded fasteners was discussed in Section E1.2 and, assuming compliance with minimum distances, is mainly dictated by the embedment strength, the yield moment of the fastener and the joint geometry. For example, possible failure modes of a timber-to-timber joint with two shear planes are shown in Figure E1-10. For these modes, the load-bearing capacity $F_{v,R}$ can be calculated by force and moment equilibrium conditions. Johansen (1949) was first to publish this method for timber-to-timber joints, which is why the equations used to determine the load-bearing capacity bear his name.

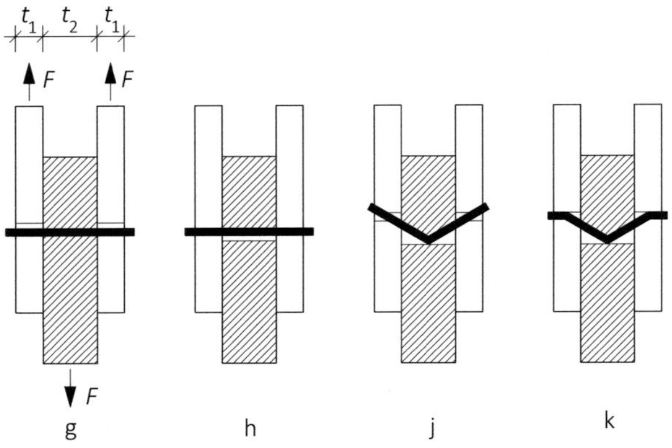

Figure E1-10 Possible failure modes of a timber-to-timber joint with two shear planes.

Failure modes j and k are characterised by an inclination of the fastener due to the emergence of one or more plastic hinges, from where the rope effect emerges. This generates (tensile) forces along the shank of the fastener, the intensity of which depend on how deeply the fastener is anchored in the timber and which press the joint members against each other. A smooth shank nail will generate less rope effect than a self-tapping screw. The tensile forces F_{ax} generated by the rope effect in the shank direction can be taken into account during design. The contribution of the rope effect (see expression $F_{ax}/4$ in equations (E1-3) and (E1-4)) to the load-bearing capacity is limited in EC 5, depending on the fastener type used, to the following percentages of the Johansen part:

- Round, smooth shank nails: 15%
- Square, smooth shank nails: 25%
- Ringed shank or spiral nails: 50%
- Screws: 100%
- Bolts: 25%
- Dowels: 0%

This means that for screws, ¼ of the withdrawal capacity F_{ax} can be taken into account, but a maximum of 100% of $F_{v,R}$. Accordingly, if the Johansen load-bearing capacity of a screwed joint is $F_{v,R} = 2$ kN and the withdrawal capacity F_{ax} of the screw is 6 kN, then 6 kN/4 = 1.5 kN < $F_{v,R} = 2$ kN and the total load-bearing capacity of the joint is 2 kN + 1.5 kN = 3.5 kN. Conversely, with withdrawal capacity of 10 kN, this results in $F_{ax}/4 = 2.5$ kN > $F_{v,R} = 2$ kN. In this case, only 2 kN can be considered as rope effect, resulting in a total load-bearing capacity of 4 kN. However, for round, smooth shank nails, if the Johansen part amounts to e.g. $F_{v,R} = 1$ kN, the rope effect part F_{ax} of a round, smooth shank nail can be considered only with 0.15 · 1 kN = 150 N, even if ¼ of the withdrawal capacity of the nail F_{ax} is higher. Dowels only exacerbate the rope effect in terms of friction generated between the timber and the dowel shank, but this cannot be taken into account.

For all the failure modes shown in accordance with Figure E1-10, using the equilibrium condition of forces and moments, allows the load-bearing capacity $F_{v,R}$ per fastener and shear plane to be determined (derivations of which are shown in Article E2). The final equations of EC 5 for double-shear timber-to-timber joints, which also encompass the rope effect (⟶ $F_{ax,Rk}$, origin of the ¼ factor, see Article E2), read as follows:

Failure mode g: Embedment strength of the side members is reached:

$$F_{v,Rk} = f_{h,1,k} \cdot t_1 \cdot d \tag{E1-1}$$

Failure mode h: Embedment strength of the middle member is reached:

$$F_{v,Rk} = 0.5 \cdot f_{h,2,k} \cdot t_2 \cdot d \tag{E1-2}$$

Failure mode j: Embedment strength in all members is reached and a plastic hinge is formed:

$$F_{v,Rk} = 1.05 \cdot \frac{f_{h,1,k} \cdot t_1 \cdot d}{2+\beta} \cdot \left[\sqrt{2 \cdot \beta \cdot (1+\beta) + \frac{4 \cdot \beta \cdot (2+\beta) \cdot M_{y,Rk}}{f_{h,1,k} \cdot t_1^2 \cdot d}} - \beta \right] + \frac{F_{ax,Rk}}{4} \tag{E1-3}$$

Failure mode k: Embedment strength in all members is reached and three plastic hinges are formed:

$$F_{v,Rk} = 1.15 \cdot \sqrt{\frac{2 \cdot \beta}{1+\beta}} \cdot \sqrt{2 \cdot M_{y,Rk} \cdot f_{h,1,k} \cdot d} + \frac{F_{ax,Rk}}{4} \tag{E1-4}$$

where

$F_{v,Rk}$	Characteristic capacity per fastener and shear plane
t_i	Member thickness, i = 1 or 2
$f_{h,i,k}$	Characteristic embedment strength in timber member i
d	Fastener diameter
$M_{y,Rk}$	Characteristic yield moment of fastener
$F_{ax,Rk}$	Characteristic withdrawal capacity of fastener
β	Ratio of embedment strength values, $\beta = f_{h,2,k}/f_{h,1,k}$

The pre-factors 1.05 or 1.15 in equations (E1-3) and (E1-4) counterbalance the different partial safety factors γ_M for steel and wood and the modification factor k_{mod}, which is applied to determine the design load-bearing capacity of the joint, although the moisture content and load duration only influence the embedment strength of the timber, not the yield moment of the fastener. Since $F_{v,Rd}$ is determined with k_{mod} and a partial safety factor of $\gamma_M = 1.3$ (in Germany), this adversely affects the steel components ($\gamma_M = 1.00$) in joints, offset by an increase of 5 or 15% respectively. Figure E1-11 shows the load-bearing capacities in accordance with Johansen (namely without the rope effect) for a double-shear timber-to-timber joint, where the respective failure mechanisms in accordance with Figure E1-10 are specified depending on the timber thicknesses. A clear interaction between timber thicknesses and failure mechanisms is seen, whereby the thicker the timber members, the more plastic hinges emerge. A ductile failure mode, involving up to two plastic hinges per fastener and shear plane, can only be guaranteed with sufficient timber thicknesses.

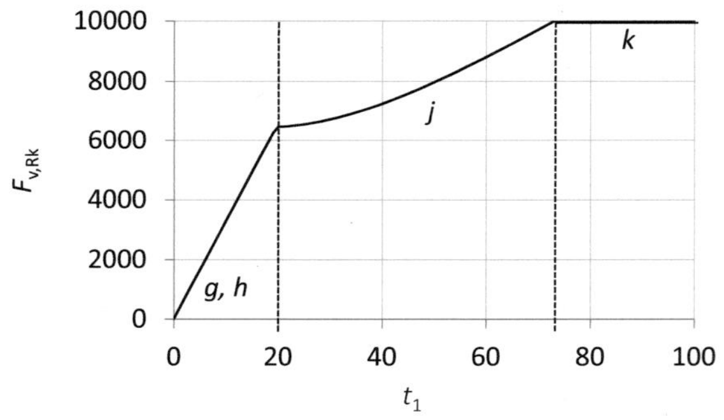

Figure E1-11 Load-bearing capacities according to Johansen in [N] depending on the timber thickness in [mm], equations (E1-1) to (E1-4), with dowel $d = 12$ mm and $F_{ax,Rk} = 0$, $t_2 = 2 \cdot t_1$, $f_u = 600$ MPa, $\rho_k = 380$ kg/m^3. Failure modes g to ka see Figure E1-10.

Joints with axially loaded fasteners

Dowel-type fasteners may also be exposed to systematic load in the direction of their axis; namely tensile load when pulled out and compressive load when pushed in, the latter of which can be exploited using self-tapping screws e.g. to increase compression perpendicular to the grain resistance (Article E5). In this respect, the load-bearing capacity in the shank direction depends on the surface geometry of the fasteners and the withdrawal resistance of a self-tapping screw far exceeds that of a ringed shank nail, while dowels have no appreciable withdrawal resistance. The shank direction in comparison to the grain direction of the timber is also a key variable, since the withdrawal resistance of a fastener arranged in parallel to the grain is far lower than when perpendicular to the same.

As a general rule, axial load may generate the following failure mechanisms, which should be verified:

- Withdrawal failure (fastener pulled out of the timber),
- Head pull-through failure (fastener head is pulled through the timber),
- Tensile failure of the fastener (rupture).

There is also a need to examine whether groups of fasteners exposed to withdrawal may be at risk of block shear failure (Article E13). When loaded in compression parallal to the shank axis, i.e. when the fastener is pushed in, meanwhile, the fasteners may fail due to buckling.

Influence of loading direction

The influence of the angle between the force and grain directions is very high, due to anisotropy of the wood. The greater the angle between the force and grain directions, the greater the tensile stress perpendicular to the grain and, in turn, the higher the risk of brittle failure of the timber member. This must be taken into consideration by the designer and is outlined in more detail in Article E11. Brittle failure triggered by a tensile force perpendicular to the grain is unrelated to other brittle failure mechanisms, which may, for example, occur in groups of fasteners or joints with low timber thicknesses loaded parallel to the grain, although tensile stresses perpendicular to the grain and shear stresses peaking are responsible for the brittle behaviour in the latter case.

Embedment strength declines alongside the rise in tensile stress perpendicular to the grain with increasing angle between force and grain directions. However, since the embedment strength is a system rather than material parameter and the fastener diameter impacts on the embedment strength, this effect can be practically ruled out for fasteners with smaller diameters ($d < 6$ mm). Accordingly, for dowelled and bolted joints, plus screwed joints with diameters $d > 6$ mm, the embedment strength is determined as a factor of the angle α between force and grain directions, while for nails, staples and screws where $d \leq 6$ mm, the embedment strength is determined independently of α. In all cases, however, a distinction is made between the embedment strengths with or without predrilling.

Moreover, depending on the loading situation of the fasteners, lateral and axial loads may be exerted simultaneously, whereupon both load components are taken into consideration via a linear or quadratic interaction.

Simplified design

The NA sets out simplified design rules, which can be used as an alternative to Johansen equations, subject to compliance with specific required timber thicknesses. The reasons behind these regulations are as follows:

- The targeted failure mechanism results in two plastic hinges per shear plane.
- Plastic behaviour is contingent on greater penetration depths of the fasteners.
- Slender fasteners result in more ductile joints than sturdy fasteners.
- Premature splitting determines the load-bearing capacity, particularly in joints with many fasteners.
- Simplified equations elicit more positive results for slender fasteners.

If we take the example of a double-shear timber-to-timber joint (Figure E1-10 and equations (E1-1) to (E1-4)), this helps clarify a simplified design according to NA. NA states that the load-bearing capacity per fastener and shear plane can be determined as follows:

$$F_{v,Rk} = \sqrt{\frac{2 \cdot \beta}{1+\beta}} \cdot \sqrt{2 \cdot M_{y,Rk} \cdot f_{h,1,k} \cdot d} \qquad \text{(E1-5)}$$

Contingent on compliance with the following required minimum timber thicknesses:

$$t_{1,req} = 1.15 \cdot \left(2 \cdot \sqrt{\frac{\beta}{1+\beta}} + 2 \right) \cdot \sqrt{\frac{M_{y,Rk}}{f_{h,1,k} \cdot d}} \qquad \text{Side member} \qquad \text{(E1-6)}$$

$$t_{2,req} = 1.15 \cdot \frac{4}{\sqrt{1+\beta}} \cdot \sqrt{\frac{M_{y,Rk}}{f_{h,2,k} \cdot d}} \qquad \text{Middle member} \qquad \text{(E1-7)}$$

It is very easy to derive equations (E1-5) to (E1-7) based on Figure E1-11. The required timber thickness corresponds to the specific thickness, at which failure mode j moves to failure mode k. This means that equations (E1-3) and (E1-4) are equated and solved in accordance with *t*. A simplified design means that only failure mode k is taken into consideration in accordance with Figure E1-10 with two plastic hinges per shear plane or, in other words, failure mode k is forced by the required timber thicknesses. Provided the required timber thicknesses in accordance with equations (E1-6) and (E1-7) are complied with, this eliminates any need to take failure modes g, h and j into consideration, allowing the load-bearing capacity to be directly determined from the equation (E1-5). However, the rope effect part $F_{ax,Rk}$ can also be taken into account. For nailed joints, these NA regulations have been even further simplified. The embedment strength is specified independently of the angle between the force and grain directions and the minimum timber thicknesses are only specified as a function of the nail diameter. Simplified design is no longer addressed in Article E2 and the individual articles on the various types of fasteners, but even so, according to NA, it can be applied for all dowel-type fasteners.

Joints with multiple shear planes

Joints with more than two shear planes can be calculated using Johansen equations, whereby each shear plane is deemed to constitute a part of a series of double-shear timber-to-timber or steel-to-timber joints. Here, it should be noted that not all failure modes are prone to occur within a joint with multiple shear planes or rather that the failure modes in the respective shear planes must be reciprocally compatible. Accordingly, in a timber-to-timber joint, a mode j failure in accordance with Figure E1-10 is only possible in both external shear planes. Article E15 takes a closer look at joints with multiple shear planes.

Stiffness

The structural engineer must also ensure consideration of the semi-rigidity of timber joints with mechanical fasteners as required and this semi-rigidity is determined via the slip modulus K_{ser}. The slip modulus of a joint in the ultimate limit state K_u is determined as follows:

$$K_u = \frac{2}{3} \cdot K_{ser} \qquad \text{(E1-8)}$$

K_u is required, for example, to determine the effective bending stiffnesses of mechanically jointed beams (e.g. nailed I-beams). The values for K_{ser} depend on the fastener type and diameter and on the density. The equations specified in Table E1-1 for K_{ser} apply per shear plane and fastener of timber-to-timber joints and are based on comprehensive test results. For steel-to-timber joints meanwhile, the values have to be multiplied by the factor 2.0. If the mean densities ρ_m of the connected timber members differ, K_{ser} should be calculated with the geometric mean of both densities.

Table E1-1 Values for K_{ser} in N/mm for some fastener types. See also EN 1995-1-1:2010 Table 7.1.

Fastener type	K_{ser} (d in mm, ρ_m in kg/m^3)
Dowels, screws, nails (predrilled), Bolts with or without clearance*	$K_{ser} = \frac{d \cdot \rho_m^{1.5}}{23}$
Nails (non-predrilled)	$K_{ser} = \frac{d^{0.8} \cdot \rho_m^{1.5}}{30}$
Staples	$K_{ser} = \frac{d^{0.8} \cdot \rho_m^{1.5}}{80}$

* The clearance should be added separately to the deformation.

E1.4 Additional general rules for dowel-type fasteners

For joints with dowel-type fasteners, certain additional and generally valid regulations and statements apply, covering various aspects of a joint. These aspects, the most important of which will be summarised in the following section, concern both influences on the load-bearing capacity through the shape of the joint, which can generate additional stresses in the members, as well as execution rules, although the allocation may not always be clear in the latter case.

Execution rules

Predrilling

Dowel-type fasteners can be driven into members made of timber or wood-based materials with or without predrilling of holes. While bolted and dowelled joints always require predrilling of timber members, nails and screws may be usable without such predrilling, depending on the wood species and fastener diameter. It follows that joints with hardwood and densities exceeding 500 kg/m^3 must always be predrilled[4]. Such predrilling is also advisable for nails of diameter exceeding 6 mm. In principle, staples are driven in without predrilling. For nails, the drilling diameter is around $0.8 \cdot d$ and for self-tapping screws, the core diameter (= inner thread diameter) is the basis for predrilling. Steel plates in steel-to-timber joints can be predrilled for bolted and dowelled joints with up to d + 1 mm. Meanwhile, self-tapping dowels are also available and can be driven in without predrilling of steel

[4] Recent research projects deal with a new generation of self-tapping screws for hardwoods that can be inserted without predrilling.

plates. For joints with bolts and threaded rods, there is also scope to drill holes in the timber members with d + 1 mm, although this clearance has to be taken into consideration when calculating the deformations. The timber members of joints with dowels and fitted bolts, conversely, are predrilled with the nominal diameter of the fasteners.

Minimum penetration depth

Minimum penetration depths help ensure ductile failure behaviour and must also be taken into account when using nails and screws in members with the tip of the nail or screw. This means that e.g. for smooth shank nails in timber-to-timber joints, a minimum penetration depth of $8 \cdot d$ applies, while for non-smooth nails, the figure is $6 \cdot d$. Since these minimum penetration depths represent robustness requirements, sufficient load-bearing capacity can still be guaranteed, even in the event of unwanted gaps present between the members e.g. due to climate-related deformations.

Minimum number of fasteners

For all dowel-type fasteners, robustness requirements dictate that at least two fasteners be used for load-bearing joints (with a few exceptions, such as fixing casings or battens). For dowelled joints meanwhile, the joint must include at least four shear planes as well as two dowels. Another advisable step, particularly for a dowelled joint, involves replacing a certain number of dowels with fitted bolts, to hold the joint together.

Fire resistance and corrosion

Another challenge imposed on the structural engineer is compliance with fire resistance requirements, one of the solutions to which involves “concealing” fasteners in the timber. This is also what underpins visually appealing solutions. Further information on the fire resistance of joints is included in Article G1.

When designing joints subject to harsh environmental conditions, the risk of corrosion must also be borne in mind. Initially, the design focus should be on ensuring that no water can ingress on the connection face. For unprotected constructions, applying a cover may help protect against sun and water effectively. Under sub-optimal conditions, corrosion can be prevented by coatings or using stainless metals, while the planner should also consider the compatibility of the metal with wood preservatives. For this reason, e.g. contact between aluminium or steel and wood that has been treated with wood preservatives containing copper can be problematic.

Influences on load-bearing capacity

Groups of fasteners

Timber joints with mechanical fasteners tend to incorporate more than just one fastener, which means an uneven load distribution between them, even when the load withstood by the joint is exerted at the centre of gravity of the joint. The load-bearing capacity of a joint corresponds to the total collective forces accommodated by the individual fasteners in the failure state. If the loads on the individual fasteners now differ significantly when the joint fails and the load exerted on some of the fasteners is far below their capacity, as is possible with the brittle behaviour of a joint, the load-bearing capacity of the joint is deemed lower than the sum of capacities of the individual fasteners. This explains the reduction in load-bearing capacity per fastener in joints with multiple fasteners for specific fastener types. In principle, the various influences on the load distribution in joints impact on both joints with multiple identical fasteners as well as those with multiple different fasteners. Accordingly, the effective characteristic capacity of a joint with a row of fasteners $F_{v,ef,Rk}$, featuring fasteners consecutively arranged in force and grain direction, is determined with an effective number of fasteners n_{ef}:

$$F_{v,ef,Rk} = n_{ef} \cdot F_{v,Rk} \tag{E1-9}$$

The effective number n_{ef} was determined using regression analyses on numerous experimental test results for each fastener type and is explained in more detail in the individual articles and in Article E13.

Another failure mode affecting groups of fasteners is block shear. If many fasteners are closely aligned or arranged linearly, the load-bearing capacity of the joint may be limited due to part of the member rupturing (Figure E1-12). The block shear failure of a fastener group is attributable to reaching the shear strength at member planes parallel to the force and grain directions and the tensile strength at a member plane perpendicular to the same. The failure is initiated in the form of a rupture in one of the resistance areas S, followed by failure in the area perpendicular to the first rupture area. Given that timber is a brittle material, the two strengths cannot be collectively pooled, which means the member strength has to be determined separately for the net cross-section S_t for tensile stress and S_c for shear respectively. The design block shear strength is the greater of the two values (see Article E13).

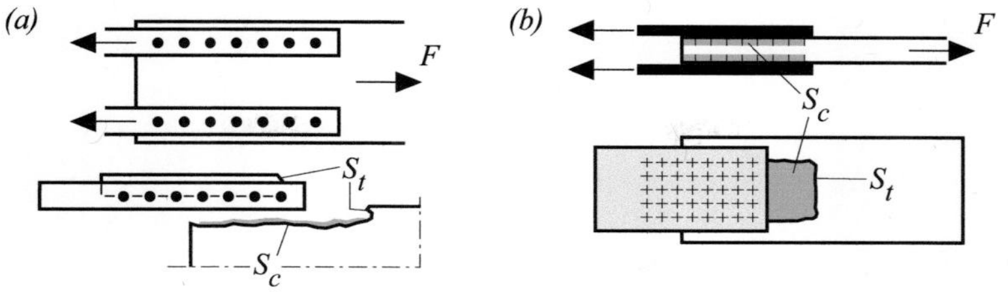

Figure E1-12 Blockshear failure in a joint: (*a*) Tension failure in the net area S_t, (*b*) shear failure in the net area S_c. (STEP 1995 Article C1)

Swelling and shrinkage

Particular attention should be paid to cross-sectional, dimensional changes of the timber caused by varying moisture levels. Areas in which fasteners prevent swelling and shrinkage may be subject to tensile stresses perpendicular to the grain, which lead to members rupturing (Figure E1-13 (a)). To prevent or limit such rupture, the area in which swelling and shrinkage and thus strains are restrained must be limited, with fasteners clustered on one side of the connected members (Figure E1-13 (b)). Fasteners solely intended to hold the joint together should be driven into elongated holes. In other cases, e.g. moment-resisting frame corners, the area where strains are restrained should be limited to around a metre and/or reinforcements should be applied.

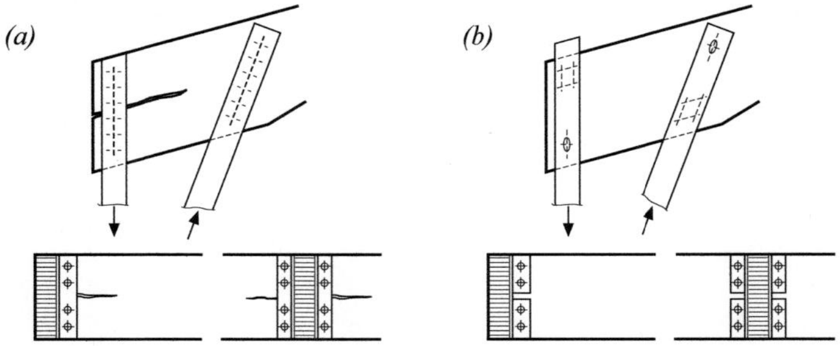

Figure E1-13 Joint details: (*a*) Splitting due to shrinkage, (*b*) correctly executed joint with oval holes. (STEP 1995 Article C1)

Eccentricities

Joints and members within structures should be arranged as symmetrically and concentrically as possible, particularly if subject to considerable stresses. Even so, eccentricities may occur for the following reasons (Figure E1-14):

- Type of fastener used,
- Layout of the joint,
- Layout of the structural system.

In eccentrically arranged fasteners, such as connectors, the impact of bending moment on unilaterally stressed members must be limited through fasteners under tensile load, such as bolts with washers. However, eccentricities can often be eliminated by arranging fasteners and members correctly, as shown in Figure E1-14 (b) and Figure E1-14 (d). In other cases, secondary forces arising from eccentricities (moment, shear, tension) which act on the fasteners and members must also be taken into account during the design.

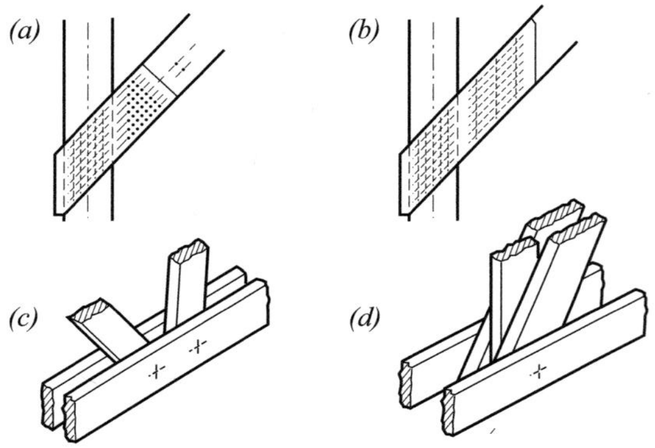

Figure E1-14 Eccentricities in structures generated by arrangements of fasteners (*a*) or members (*c*). Modified arrangements, to avoid eccentricities (*b*), (*d*). (STEP 1995 Article C1)

Combination of different types of fasteners

At times, particularly in additionally reinforced joints, different types of fasteners may be used in a joint to transfer forces. In such cases, the load-bearing capacity should be determined taking the stiffness of the various fasteners into consideration. To prevent overloading due to excessive differences in stiffness, however, gluing and mechanical fasteners must not be combined.

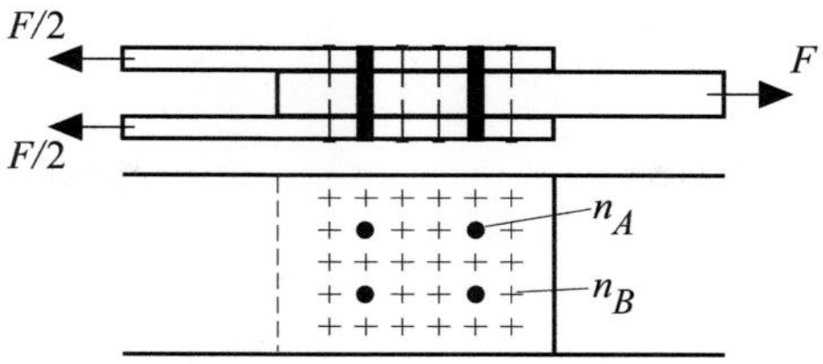

Figure E1-15 Joint with a combination of fasteners: n_A dowels and n_B nails. (STEP 1995 Article C1)

Conservatively, a joint is designed assuming elastic behaviour of fasteners and the design load F_d is allocated via the slip moduli of the fastener. If a force is exerted on two types of fasteners (see Figure E1-15), the equilibrium condition and compatibility of the deformation can be determined as follows:

$$F = \sum_{i=1}^{n} F_i = n_A \cdot F_A + n_B \cdot F_B \qquad \text{and} \qquad u = \frac{F_A}{K_{u,A}} = \frac{F_B}{K_{u,B}} \tag{E1-10}$$

where $K_{u,A}$ and $K_{u,B}$ are the slip moduli in the ultimate limit state.

Consistency of units

Finally, a statement valid for all design situations, but particularly relevant here, given the numerous empirical equations involved in joint design. One of the key control mechanisms during the design process is checking of units, which is imperative. This approach also allows errors to be detected very quickly when e.g. converting area loads into line- and individual loads.

However, the many empirical equations used in timber construction compound the difficulty of unit checking. Many equations have been derived from regression analyses based on numerous test results with a large scatter, which require parameters to be deployed in specific units and where the unit of the result is already defined. One example would be equation (E1-11), which is used to determine K_{ser} for a staple:

$$K_{ser} = \frac{d^{0.8} \cdot \rho_m^{1.5}}{80} \tag{E1-11}$$

The exponents clearly show that equation (E1-11) is empirical and based on regression analyses. Diameter d in [mm] and the mean density ρ_m in [kg/m^3] have to be implemented, to determine K_{ser} in [N/mm]; no derogation from these units is possible.

E1.5 Literature

P. Racher, original Article C1, STEP 1995.

Hirashima Y. (1990). Lateral resistance of timber connector joints parallel to grain direction. International Engineering Conference, Vol. 1:254-261, Tokyo.

Johansen K.W. (1949).Theory of timber connections. International Association of Bridge and Structural Engineering (IABSE), Publication 9, Basel.

Natterer J., Herzog T. and Volz M. (1991). Holzbau Atlas Zwei. Edition française, Presses polytechniques et universitaires romandes, Le Mont-sur-Lausanne.

Smith I. and Whale L.R.J. (1986). Mechanical timber joints. TRADA Research Report 18/86, Hughenden Valley, England.

E2 Johansen model

Original article: B. O. Hilson

Figure E2-1 shows joints with laterally loaded dowel-type fasteners. Examples of dowel-type fasteners include nails, staples, screws, dowels or bolts.

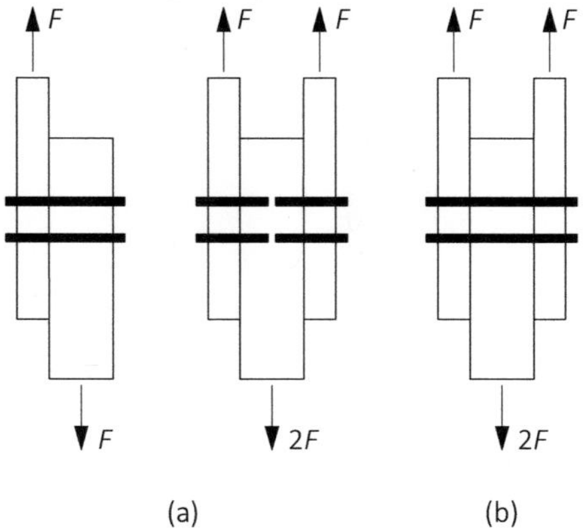

Figure E2-1 Laterally loaded joints with dowel-type fasteners: (a) single-shear (one shear plane per fastener); (b) double-shear (two shear planes per fastener).

Allowable stresses within such joint types were previously determined based on short-term tests with comparatively few tested joints. A common method used to determine allowable values consisted of e.g. specifying the mean value of the load-bearing capacity of multiple similar test specimens and then dividing this by an overall safety factor, to record the influences of variations in loads and resistances, on the degree of craftmanship as well as the load duration. The results obtainable from such tests, however, do not generally suffice to enable reliable estimates of the characteristic load-bearing capacities, as is required for EC 5. Similarly, tests with representatively selected joints to determine characteristic values are infeasible, due to the great variety of joints and correspondingly huge number of tests, which would be required.

Accordingly, mechanical models are used, so that characteristic load-bearing capacities can be calculated based on specific material properties and joint dimensions. The equations contained in EC 5 to calculate the capacity of joints are based on the work of Johansen (1949) and dictate that capacity is limited to the point where the embedment strength is reached in at least one of the connected timber members and in specific cases by the simultaneous emergence of plastic hinges in the fastener. The failure mechanism depends on joint geometry as well as material properties, particularly the plastic bending moment of the dowel or the embedment strength of the timber or wood-based material. Numerous tests were performed, to verify the validity of the Johansen equations. All these tests, e.g. conducted by Möller (1951), Aune and Patton-Mallory (1986) or Hilson et al. (1990) revealed effective conformity between test and theory, provided that frictional influences and tensile forces in the fasteners excluded from the theoretical scope could be limited to a negligible level.

E2.1 Material properties

As already shown in Article E1 (equations (E1-1) to (E1-4)), to determine the capacity of a laterally loaded joint, the properties required are "embedment strength f_h" and "yield moment M_y". All other required parameters are geometric details such as timber thicknesses $t_{1/2}$ or fastener diameter d. The property still required for equations (E1-3) and (E1-4), namely "withdrawal capacity $F_{ax,R}$" is a parameter added at a later stage, which is independent of the Johansen model and which takes the increased capacity due to the rope effect into consideration. The rope effect can only occur in fasteners, which end up inclined when failing, as is explained in more detail in Section E2.6.

The embedment strength f_h of the timber or wood-based material is defined as the peak stress calculated in an embedment test and a property related to a system rather than a material property, dictated by how the fastener and timber interact. A typical test arrangement is shown in Figure E2-2. To ensure uniform impression (embedment) in the timber or wood-based material, it is important to avoid any deformation of the fastener as far as possible. This is generally done by clamping the ends of the fastener into the lateral steel plates as well as limiting the specimen thickness t, generally to twice the fastener diameter.

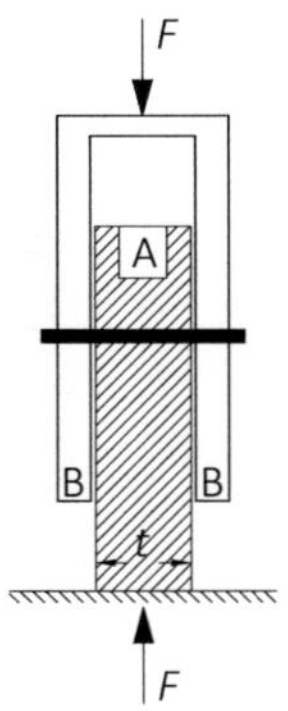

Figure E2-2 Typical embedment test arrangement. Clamping of the fastener into lateral steel plates (B), obtaining uniform embedment of the fastener in the test specimen (A). (STEP 1995 Article C3)

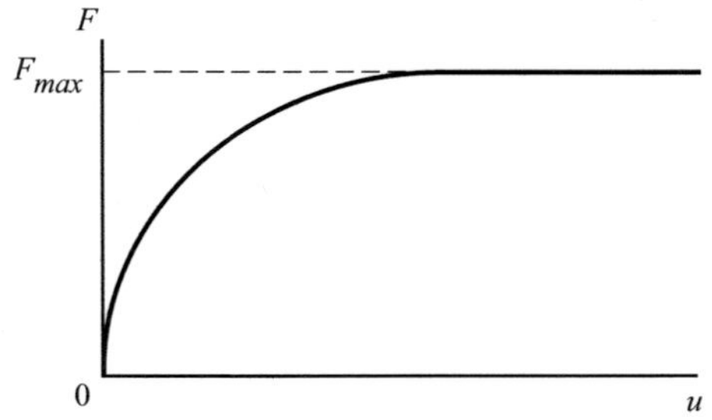

Figure E2-3 Typical load-embedment relationship. (STEP 1995 Beitrag C3)

A typical load-embedment relationship is as shown in Figure E2-3. The embedment strength is defined as the load F reached before a displacement of 5 mm in the test divided by the projected area of the fastener in the test specimen:

$$f_h = \frac{F}{d \cdot t} \quad \text{(E2-1)}$$

In the above, t is the specimen thickness and d the fastener diameter.

The embedment u is the relative displacement between the fastener and specimen, namely between BB and A in Figure E2-2. Additional indications to define the embedment strength are included in EN 383 "Timber structures – test methods – determination of embedment strength and foundation values for dowel-type fasteners", while experimental devices to define the embedment strength are described in Rodd et al. (1987). Furthermore, the yield moment M_y of the fastener is a key parameter for the Johansen equations. The procedure for determining the yield moment is described in EN 409 "Timber structures – test methods – determination of the yield moment of dowel-type fasteners".

E2.2 Johansen equations for single-shear timber-to-timber joints

One prerequisite for deriving the Johansen equations is the rigid-plastic material behaviour assumed for the fastener when exposed to bending stress and for the timber when exposed to embedment stress. The assumed load-embedment relationship of the fastener in the timber is shown in Figure E2-4. This approximation simplifies the calculation but has virtually no impact on the final result. To derive design equations, the first task is to identify the possible failure modes, so that a design model can then be derived. In timber-to-timber joints with fasteners in single-shear, analogously to Figure E1-10, the failure modes shown in Figure E2-5 can occur in accordance with Johansen.

Figure E2-4 Simplified load-embedment relationship. (STEP 1995 Article C3)

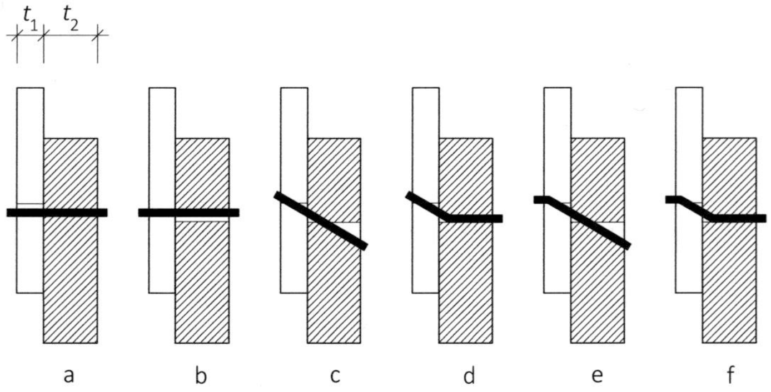

Figure E2-5 Possible failure modes of timber-to-timber joints with a laterally loaded fastener in single-shear.

The following section shows examples of how two failure mechanisms can be derived; cases a and d in accordance with Figure E2-5. The derivation of further failure mechanims is shown in Annex 6. The following designations are used:

- t_1 and t_2 are the timber thicknesses or penetration depths of the fastener in member 1 or 2
- $f_{h,1,k}$ is the characteristic embedment strength of member 1
- $f_{h,2,k}$ is the characteristic embedment strength of member 2
- d is the fastener diameter
- $M_{y,Rk}$ is the characteristic yield moment of the fastener
- $F_{v,Rk}$ is the characteristic load-bearing capacity per shear plane
- β is the ratio of the embedment strengths: $\beta = f_{h,2,k}/f_{h,1,k}$

Failure mechanism a – embedment failure in member 1:

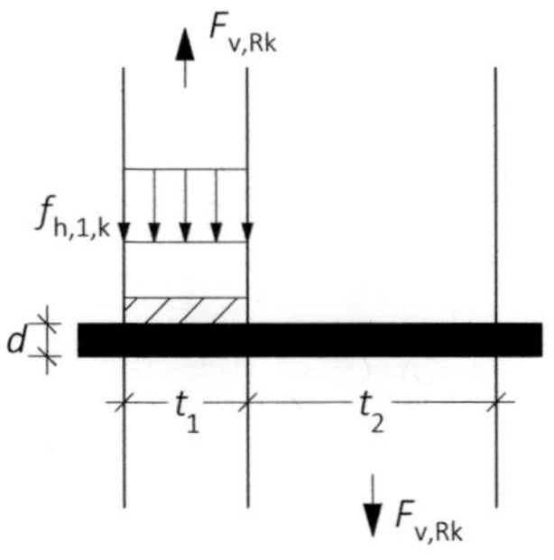

Figure E2-6 Failure mechanism a.

In the event of a mode a failure as shown in Figure E2-6, the embedment strength in member 1 is the key parameter. The equilibrium of forces $\Sigma V = 0$ in member 1 results in:

$$F_{v,Rk} = f_{h,1,k} \cdot d \cdot t_1 \tag{E2-2}$$

Equation (E2-2) corresponds precisely to the equation for failure mechanism a of a timber-to-timber joint in EC 5.

Failure mechanism d – embedment failure in both members and one plastic hinge in member 2:

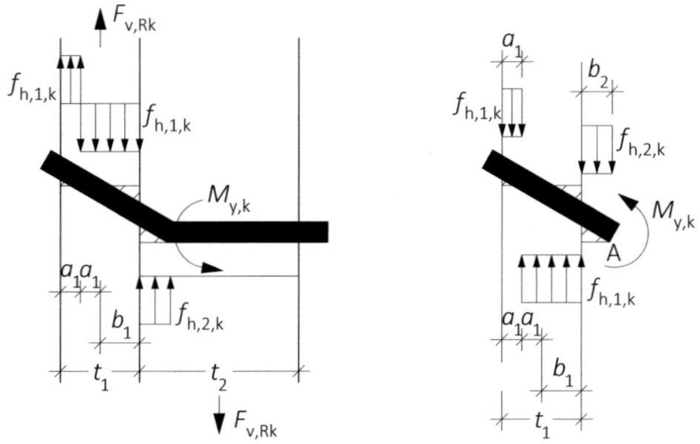

Figure E2-7 Failure mechanism d. Right: free-body diagram of fastener.

In the event of a mode d failure as shown in Figure E2-6, both members reach the embedment strength and a plastic hinge forms. The following applies:

$$\beta = \frac{f_{h,2,k}}{f_{h,1,k}}$$

The shear force in the fastener is equal to zero at the point where the bending moment peaks:

$$f_{h,2,k} \cdot d \cdot b_2 - f_{h,1,k} \cdot d \cdot b_1 = \beta \cdot f_{h,1,k} \cdot d \cdot b_2 - f_{h,1,k} \cdot d \cdot b_1 = 0$$

Consequently:

$$\beta \cdot b_2 = b_1$$

The equilibrium of moments at the fastener $\Sigma M^A = 0$ is:

$$M_{y,k} = -f_{h,2,k} \cdot d \cdot b_2 \cdot \frac{b_2}{2} + f_{h,1,k} \cdot d \cdot (b_1 + a_1) \cdot \left(b_2 + \frac{b_1 + a_1}{2} \right) -$$

$$- f_{h,1,k} \cdot d \cdot a_1 \cdot \left(b_2 + b_1 + a_1 + \frac{a_1}{2} \right)$$

Now $f_{h,2,k}$ is replaced by $\beta \cdot f_{h,1,k}$, b_2 by b_1/β and $a_1 = (t_1\text{-}b_1)/2$ and the equation is transformed:

$$b_1^2 + t_1 \cdot \frac{2 \cdot \beta}{2+\beta} \cdot b_1 - \frac{\beta \cdot t_1^2}{2+\beta} - \frac{4 \cdot \beta}{2+\beta} \cdot \frac{M_{y,k}}{f_{h,1,k} \cdot d} = 0$$

Solving the quadratic equation reveals:

$$b_1 = \frac{t_1}{2+\beta} \cdot \left[\sqrt{2 \cdot \beta \cdot (1+\beta) + \frac{4 \cdot \beta \cdot (2+\beta) \cdot M_{y,k}}{f_{h,1,k} \cdot d \cdot t_1^2}} - \beta \right] \tag{E2-3}$$

In addition, the equilibrium of forces $\Sigma V = 0$ of the fastener in member 1 equates to:

$$F_{v,Rk} = f_{h,1,k} \cdot d \cdot b_1$$

Inserting equation (E2-3) reveals:

$$F_{v,Rk} = \frac{f_{h,1,k} \cdot d \cdot t_1}{2+\beta} \cdot \left[\sqrt{2 \cdot \beta \cdot (1+\beta) + \frac{4 \cdot \beta \cdot (2+\beta) \cdot M_{y,k}}{f_{h,1,k} \cdot d \cdot t_1^2}} - \beta \right] \tag{E2-4}$$

Equation (E2-4) (E2-4) is reflected in the equation for failure mechanism d of a timber-to-timber joint in EC 5. Equation (E2-4) corresponds to the Johansen part in EC 5, the contribution of the rope effect as well as the pre-exponential factor 1.05 are missing (explanation of factor 1.05 see Article E1).

E2.3 Johansen equations for double-shear timber-to-timber joints

The same approach employed for single-shear joints can also be used to derive Johansen equations for double-shear joints and possible failure modes are shown in Figure E2-8 (see also Figure E1-10). When deriving the equations, it is important to note that the Johansen equations are determined per shear plane, whereby the joints are conceptually separated in the middle (namely at $t_2/2$) in accordance with Figure E2-8. Accordingly, the Johansen equations for double-shear joints are merely expansions of the single-shear joint equations.

In EC 5, the capacities according to equations (E2-7) and (E2-8) are further expanded, to take account of (i) the rope effect $F_{ax,Rk}/4$ (see Section E2.6) and (ii) the pre-factors 1.05 or 1.15 are applied, which take into account the varying partial safety factors γ_M and k_{mod} for steel ($M_{y,k}$) and timber ($f_{h,k}$). Equations (E2-5) and (E2-6) conversely do not reveal any changes in comparison to EC 5, since (i) both failure modes described do not lead to any inclination of the fastener and hence no rope effect takes place and (ii), in terms of material parameters, only reveal embedment failure (with γ_M and k_{mod} for timber) and no plastic hinge (with γ_M for steel).

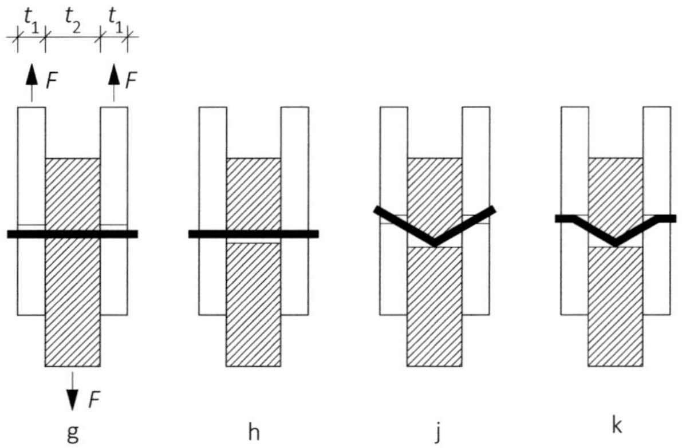

Figure E2-8 Possible failure modes of timber-to-timber joints with a laterally loaded fastener in double-shear.

The resulting equations are as follows:

Failure mechanism g (see Figure E2-6 and equation (E2-2)):

$$F_{v,Rk} = f_{h,1,k} \cdot d \cdot t_1 \tag{E2-5}$$

Failure mechanism h (see Annex 6, failure mechanism b, equation (20), here with half width $t_2/2$):

$$F_{v,Rk} = f_{h,2,k} \cdot d \cdot \frac{t_2}{2} \tag{E2-6}$$

Failure mechanism j (see Figure E2-7 and equation (E2-4) or Annex 6, failure mechanism e):

$$F_{v,Rk} = \frac{f_{h,1,k} \cdot d \cdot t_1}{2+\beta} \cdot \left[\sqrt{2 \cdot \beta \cdot (1+\beta) + \frac{4 \cdot \beta \cdot (2+\beta) \cdot M_{y,k}}{f_{h,1,k} \cdot d \cdot t_1^2}} - \beta \right] \tag{E2-7}$$

Failure mechanism k (see Annex 6, failure mechanism f, equation (25)):

$$F_{v,Rk} = \sqrt{2 \cdot M \cdot f_{h,1,k} \cdot d} \cdot \sqrt{\frac{2 \cdot \beta}{1+\beta}} \tag{E2-8}$$

E2.4 Johansen equations for steel-to-timber joints with thick steel plates

In steel-to-timber joints with sufficiently thick steel plates, the steel plate acts as a clamped restraint for the fastener. The embedment strength of the steel plate considerably exceeds the equivalent figure for timber or wood-based materials and is assumed to be indefinite when deriving Johansen equations. This explains why plastic hinges occur in the shear plane between steel plate and timber in the case of failure modes developing plastic hinges. For a steel plate thickness at least equivalent to the fastener diameter, the formation of plastic hinges in the shear plane between steel plate and timber can be assumed. The derivations for steel-to-timber joints resemble those for timber-to-timber joints, while possible failure mechanisms in accordance with Johansen for steel-to-timber joints are shown in Figure E2-9.

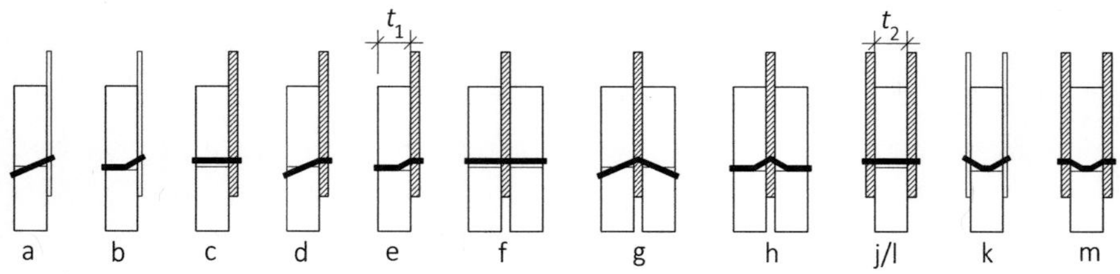

Figure E2-9 Possible failure modes of steel-to-timber joints.

The derivation is shown as an example for failure modes d and e of a single-shear steel-to-timber joint.

Failure mechanism d

In the event of failure in accordance with mode d, as in Figure E2-9 and Figure E2-10, embedment strength is reached in the timber member and a plastic hinge forms in the shear plane.

The equilibrium of moments at the fastener $\Sigma M^A = 0$ is:

$$M_{y,k} + f_{h,1,k} \cdot d \cdot a_1 \cdot \left(b_1 + a_1 + \frac{a_1}{2} \right) - f_{h,1,k} \cdot d \cdot (b_1 + a_1) \cdot \left(\frac{b_1 + a_1}{2} \right) = 0$$

Inserting $a_1 = (t_1 - b_1)/2$ and transforming reveals:

$$b_1^2 + 2 \cdot t_1 \cdot b_1 - t_1^2 - \frac{4 \cdot M_{y,k}}{f_{h,1,k} \cdot d} = 0$$

Consequently:

$$b_1 = t_1 \cdot \left[\sqrt{2 + \frac{4 \cdot M_{y,k}}{f_{h,1,k} \cdot d \cdot t_1^2}} - 1 \right]$$

In addition, the equilibrium of forces $\Sigma V = 0$ of the fastener in the timber member equates to:

$$F_{v,Rk} = f_{h,1,k} \cdot d \cdot b_1$$

Inserting of b_1 results in:

$$F_{v,Rk} = f_{h,1,k} \cdot d \cdot t_1 \cdot \left[\sqrt{2 + \frac{4 \cdot M_{y,k}}{f_{h,1,k} \cdot d \cdot t_1^2}} - 1 \right] \qquad \text{(E2-9)}$$

Equation (E2-9) is included in the equation for the failure mechanism d of a steel-to-timber joint in EC 5 and expanded there with the rope effect, but incorrectly without the pre-factor, which takes into consideration the differing partial safety factors and k_{mod} for steel and timber.

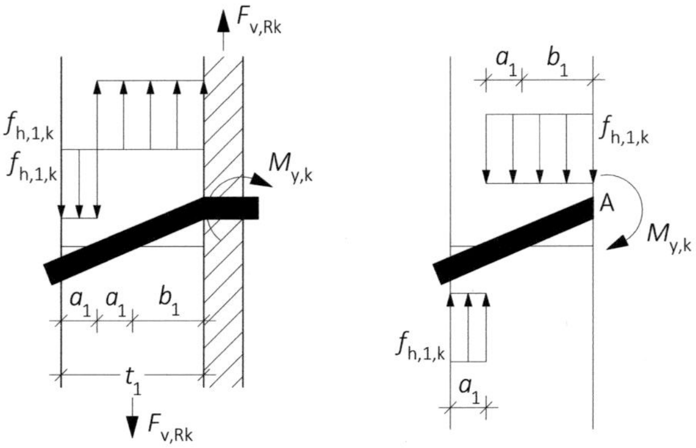

Figure E2-10 Failure mechanism d. Right: free-body diagram of fastener.

Failure mechanism e

In the event of failure in accordance with mode e, as in Figure E2-9 and Figure E2-11, the embedment strength has been reached in the timber member, and two plastic hinges form. The equilibrium of moments at the fastener $\Sigma M^A = 0$ is:

$$2 \cdot M_{y,k} - f_{h,1,k} \cdot d \cdot b_1 \cdot \frac{b_1}{2} = 0$$

Solving for b_1:

$$b_1 = 2 \cdot \sqrt{\frac{M_{y,k}}{f_{h,1,k} \cdot d}}$$

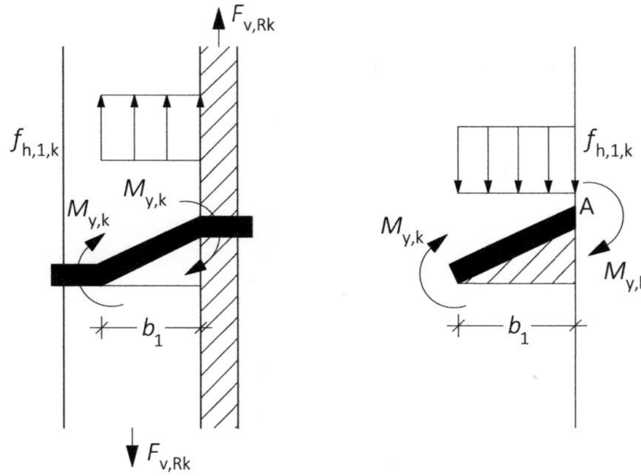

Figure E2-11 Failure mechanism e. Right: free-body diagram of fastener.

In addition, the equilibrium of forces $\Sigma V = 0$ of the fastener in the timber member equates to:

$$F_{v,Rk} = f_{h,1,k} \cdot d \cdot b_1$$

Inserting b_1:

$$F_{v,Rk} = f_{h,1,k} \cdot d \cdot 2 \cdot \sqrt{\frac{M_{y,k}}{f_{h,1,k} \cdot d}} = 2 \cdot \sqrt{M_{y,k} \cdot f_{h,1,k} \cdot d} \qquad \text{(E2-10)}$$

Equation (E2-10) is reflected in the equation for the failure mechanism e of a steel-to-timber joint in EC 5 and expanded there with the rope effect and the pre-factor 1.15 ($2 \cdot 1.15 = 2.3$).

E2.5 Johansen equations for steel-to-timber joints with thin steel plates

External thin steel plates cannot provide a clamped restraint of the fastener in the steel plate, hence the emergence of plastic hinges in the shear planes between external steel plates and the middle timber member. Accordingly, the corresponding equations were derived assuming a pinned restraint of the fasteners in the steel plates. A steel plate is deemed thin, if the thickness of the steel plate does not exceed half the fastener diameter *d*. For **double-shear joints**, if the middle section comprises a thin steel plate, the same equations apply as are used in the case of thick steel plates. The connection geometry allows the fastener to form a plastic hinge in the steel plate area, although this presupposes sufficient load-bearing capacity of the steel plate. If the thickness of the steel plate is between $0.5 \cdot d$ and d, the load-bearing capacity of the joint is determined from a linear interpolation between the load-bearing capacity values for thin and thick steel plates. All possible failure modes of a steel-to-timber joint with thin steel plates are shown in Figure E2-9. For failure mode a, as in Figure E2-9 and Figure E2-12, embedment failure of the timber without a plastic hinge forming in the fastener, the load-bearing capacity in accordance with Johansen is derived here as an example.

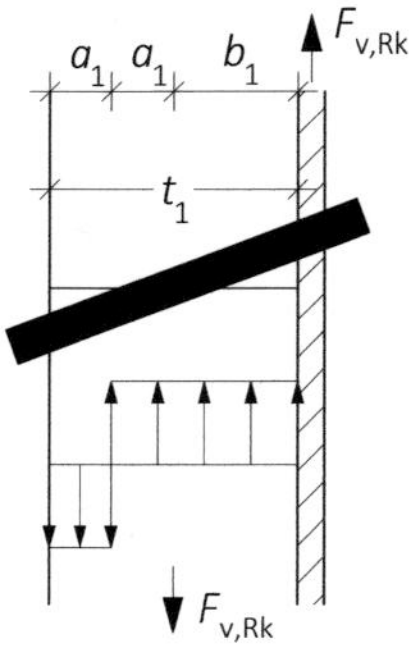

Figure E2-12 Failure mechanism a of a single-shear steel-to-timber joint with thin steel plate.

In the shear plane, the moment = 0:

$$f_{h,1,k} \cdot d \cdot a_1 \cdot \left(b_1 + a_1 + \frac{a_1}{2} \right) - f_{h,1,k} \cdot d \cdot (b_1 + a_1) \cdot \left(\frac{b_1 + a_1}{2} \right) = 0$$

Inserting $a_1 = (t_1 - b_1)/2$ and transforming:

$$b_1^2 + 2 \cdot t_1 \cdot b_1 - t_1^2 = 0$$

Consequently:

$$b_1 = t_1 \cdot \left(\sqrt{2} - 1\right) = 0{,}4 \cdot t_1$$

In addition, the equilibrium of forces $\Sigma V = 0$ of the fastener in the timber member equates to:

$$F_{v,Rk} = f_{h,1,k} \cdot d \cdot b_1$$

Inserting b_1:

$$F_{v,Rk} = 0{,}4 \cdot f_{h,1,k} \cdot d \cdot t_1 \qquad \text{(E2-11)}$$

Equation (E2-11) corresponds to the equation for the failure mode a of a steel-to-timber joint in EC 5.

E2.6 Rope effect

The equations in accordance with Johansen apply for joints with laterally loaded dowel-type fasteners, in which failure occurs due to embedment failure in the timber or wood-based material or due to a combination of embedment failure and plastic hinges in the fasteners. When deriving the equations, it is assumed that no normal forces in the direction of the axis occur in the fasteners. For fasteners capable of transmitting normal forces, however, this is incorrect. In all failure mechanisms, which lead to the fasteners inclining, normal forces are exerted and, at times, can lead to considerable load increases in comparison to pure Johansen cases. Unlike fasteners with threads or nuts and washers (bolts), smooth shank fasteners such as dowels or smooth shank nails can only withstand low normal forces. Figure E2-13 shows an example failure of a nailed joint, in which the tip of the nail being pulled out is obvious. Figure E2-14 shows the same failure mechanism in a fully threaded screw, in which no pulling out of the screw tip is visible.

Figure E2-13 The nail is pulled out of the timber.

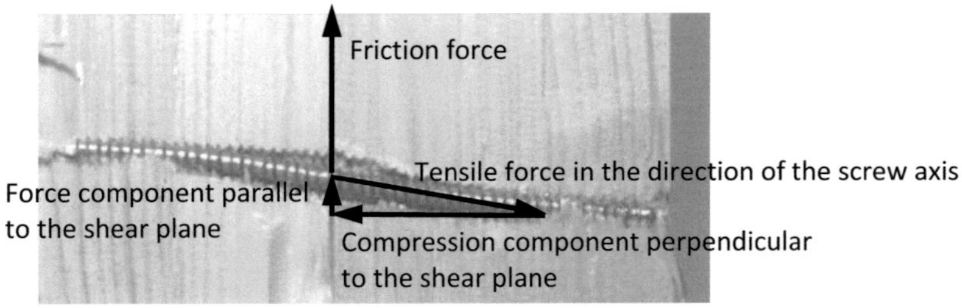

Figure E2-14 The screw tip is not pulled out of the timber.

The anchoring generates a tensile force in the fasteners in the direction of its axis. In Figure E2-14 meanwhile, the triangle of forces emerging at the current deformation is marked. The emerging tensile force can be divided into two components parallel and perpendicular to the shear plane respectively, while the compressive force component on the shear plane generates frictional force between both timber members, which has a positive effect on the load-bearing capacity of the joint. As explained in the following section, the force component parallel to the shear plane is disregarded.

Even without any plastic deformation of the fasteners, contact between the timber surfaces generates compressive forces, which lead to friction and thus a rope effect. Bejtka and Blass (2002) determined the load-bearing capacity taking into consideration the rope effect and originating from a screw inserted at an angle α. In the derivation of the rope effect shown here, an **undeformed state** is actually expected. In the deformed state, namely where the fastener is inclined due to significant joint deformations, the actual load-bearing capacities are higher, since the friction between the shank of the fastener and the surrounding timber may generate additional normal forces along the fastener axis, through which, as shown in Figure E2-14, a force component parallel to the shear plane can emerge. However, the additional force component parallel to the shear plane can only develop with the fastener considerably inclined, meaning significant joint deformation. Since such significant joint deformation cannot be practically guaranteed, these additional force components parallel to the shear plane are not taken into consideration.

An additional important note reveals that the frictional forces can only form when compressive stress is exerted perpendicular to the joint (see force components H_1 and H_2 in Figure E2-15). With this in mind, the rope effect does not apply to laterally loaded joints, which are also subject to tensile forces perpendicular to the shear plane. One example of a joint subject to tensile and shear loads is the connection of a tension diagonal to the chord of a truss.

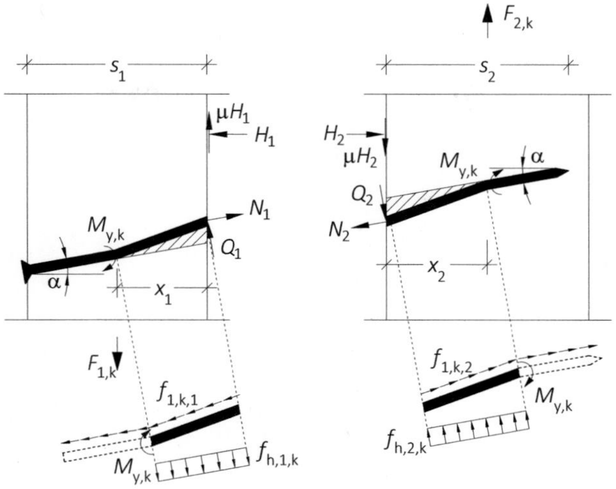

Figure E2-15 Equilibrium of forces for a timber-to-timber joint with fully threaded screws inserted at an angle α and for the failure mechanism with two plastic hinges per shear plane.

$f_{1,k,1}$ and $f_{1,k,2}$ are the characteristic values of the withdrawal parameter in members 1 and 2.

The moment in the shear plane is:

$$M_1 = M_{y,k} - \frac{f_{h,1,k} \cdot d \cdot x_1^2}{2 \cdot \cos^2 \alpha} \quad \text{and} \quad M_2 = -M_{y,k} + \frac{f_{h,2,k} \cdot d \cdot x_2^2}{2 \cdot \cos^2 \alpha}$$

With $M_1 = M_2$ and $f_{h,2,k} = \beta \cdot f_{h,1,k}$:

$$x_1^2 + \beta \cdot x_2^2 = \frac{4 \cdot M_{y,k} \cdot \cos^2 \alpha}{d \cdot f_{h,1,k}} \tag{E2-12}$$

Equilibrium of forces $\Sigma V = \Sigma H = 0$ in the shear plane:

$$F_{1,k} = Q_1 \cdot \cos\alpha + N_1 \cdot \sin\alpha + \mu \cdot H_1 \tag{E2-13}$$

$$F_{2,k} = Q_2 \cdot \cos\alpha + N_2 \cdot \sin\alpha + \mu \cdot H_2 \tag{E2-14}$$

$$H_1 = -Q_1 \cdot \sin\alpha + N_1 \cdot \cos\alpha \tag{E2-15}$$

$$H_2 = -Q_2 \cdot \sin\alpha + N_2 \cdot \cos\alpha \tag{E2-16}$$

where:

$$Q_1 = \frac{f_{\text{h,1,k}} \cdot d \cdot x_1}{\cos\alpha} \tag{E2-17}$$

$$Q_2 = \frac{f_{\text{h,2,k}} \cdot d \cdot x_2}{\cos\alpha} \tag{E2-18}$$

$$N_1 = \frac{f_{\text{1,k,1}} \cdot d \cdot s_1}{\cos\alpha} \tag{E2-19}$$

$$N_2 = \frac{f_{\text{1,k,2}} \cdot d \cdot s_2}{\cos\alpha} \tag{E2-20}$$

Inserting equations (E2-15), (E2-17) and (E2-19) in equation (E2-13), the following is obtained:

$$F_{\text{1,k}} = f_{\text{1,k,1}} \cdot d \cdot s_1 \cdot (\mu + \tan\alpha) + f_{\text{h,1,k}} \cdot d \cdot x_1 \cdot (1 - \mu \cdot \tan\alpha) \tag{E2-21}$$

Likewise, inserting equations (E2-16), (E2-18) and (E2-20) in equation (E2-14):

$$F_{\text{2,k}} = f_{\text{1,k,2}} \cdot d \cdot s_2 \cdot (\mu + \tan\alpha) + f_{\text{h,2,k}} \cdot d \cdot x_2 \cdot (1 - \mu \cdot \tan\alpha) \tag{E2-22}$$

The equilibria of forces $F_{1,k} = F_{2,k}$ and $f_{h,2,k} = b \cdot f_{h,1,k}$ elicit:

$$\frac{\left(f_{\text{1,k,1}} \cdot s_1 - f_{\text{1,k,2}} \cdot s_2\right) \cdot (\mu + \tan\alpha)}{f_{\text{h,1,k}} \cdot (1 - \mu \cdot \tan\alpha)} = \beta \cdot x_2 - x_1 \tag{E2-23}$$

Equations (E2-12) and (E2-23) elicit:

$$x_1 = \frac{\left(f_{\text{1,k,2}} \cdot s_2 - f_{\text{1,k,1}} \cdot s_1\right) \cdot (\mu + \tan\alpha)}{f_{\text{h,1,k}} \cdot (1 - \mu \cdot \tan\alpha) \cdot (1 + \beta)} + \\ + \sqrt{\frac{2 \cdot \beta}{1 + \beta}} \cdot \sqrt{\frac{2 \cdot M_{\text{y,k}} \cdot \cos^2\alpha}{f_{\text{h,1,k}} \cdot d} - \frac{(\tan\alpha + \mu)^2 \cdot \left(f_{\text{1,k,1}} \cdot s_1 - f_{\text{1,k,2}} \cdot s_2\right)^2}{2 \cdot f_{\text{h,1,k}}^2 \cdot (1 + \beta) \cdot (1 - \mu \cdot \tan\alpha)^2}} \tag{E2-24}$$

Based on the equilibrium of forces $F_{1,k} = F_{2,k}$, the horizontal forces H_1 and H_2 in the shear plane are equivalent: -(E2-15) =(E2-16) with equations (E2-17) to (E2-20):

$$\frac{f_{1,k,1} \cdot s_1 - f_{1,k,2} \cdot s_2}{f_{h,1,k} \cdot \tan\alpha} = x_1 - \beta \cdot x_2 \tag{E2-25}$$

Equating equations (E2-23) and (E2-25):

$$-\frac{\left(f_{1,k,1} \cdot s_1 - f_{1,k,2} \cdot s_2\right)}{f_{h,1,k} \cdot \tan\alpha} = \frac{\left(f_{1,k,1} \cdot s_1 - f_{1,k,2} \cdot s_2\right) \cdot \left(\mu + \tan\alpha\right)}{f_{h,1,k} \cdot \left(1 - \mu \cdot \tan\alpha\right)} \tag{E2-26}$$

Equation (E2-26) only has a solution, if $f_{1,k,1} \cdot s_1 = f_{1,k,2} \cdot s_2$, since otherwise, the solution would be $-1 = \tan^2\alpha$. This means that the screw tensile force in members 1 and 2 must be identical. Accordingly, the lower withdrawal capacity is the decisive parameter. For $f_{1,k,1} \cdot s_1 = f_{1,k,2} \cdot s_2$, equation (E2-24) can be adapted:

$$x_1 = \sqrt{\frac{2 \cdot \beta}{1 + \beta}} \cdot \sqrt{\frac{2 \cdot M_{y,k} \cdot \cos^2\alpha}{f_{h,1,k} \cdot d}} \tag{E2-27}$$

Now, x_1 from the simplified equation (E2-27) is inserted in equation (E2-21):

$$F_{1,k} = f_{1,k,1} \cdot d \cdot s_1 \cdot \left(\mu + \tan\alpha\right) + \left(1 - \mu \cdot \tan\alpha\right) \cdot \sqrt{\frac{2 \cdot \beta}{1 + \beta}} \cdot \sqrt{2 \cdot M_{y,k} \cdot f_{h,1,k} \cdot d \cdot \cos^2\alpha} \tag{E2-28}$$

The load-bearing capacity $F_{v,Rk}$ for timber-to-timber joints with fully threaded screws inserted at an angle α (or dowel-type fasteners capable of withstanding axial tensile forces and inserted at an angle α) for the failure mode with two plastic hinges per shear plane amounts to:

$$F_{v,Rk} = F_{ax,Rk} \cdot \left(\mu + \tan\alpha\right) + \left(1 - \mu \cdot \tan\alpha\right) \cdot \sqrt{\frac{2 \cdot \beta}{1 + \beta}} \cdot \sqrt{2 \cdot M_{y,k} \cdot f_{h,1,k} \cdot d \cdot \cos^2\alpha} \tag{E2-29}$$

With the withdrawal capacity $F_{ax,Rk}$ and the penetration lenghts s_1 and s_2 in members 1 and 2:

$$F_{ax,Rk} = \min\begin{cases} \dfrac{1}{\cos\alpha} \cdot f_{1,k,1} \cdot d \cdot s_1 \\ \dfrac{1}{\cos\alpha} \cdot f_{1,k,2} \cdot d \cdot s_2 \end{cases} \tag{E2-30}$$

Fasteners are often inserted perpendicular to the grain direction with $\alpha = 0°$ whereupon equation (E2-29) is simplified as follows:

$$F_{v,Rk} = F_{ax,Rk} \cdot \mu + \cdot \sqrt{\frac{2 \cdot \beta}{1+\beta}} \cdot \sqrt{2 \cdot M_{y,k} \cdot f_{h,1,k} \cdot d} \tag{E2-31}$$

Equation (E2-31) now comprises the Johansen part of the load-bearing capacity, see also equation (E2-8), and the part of the rope effect $\mu\ F_{ax,Rk}$. The load-bearing capacity in accordance with Johansen accordingly increases with the rope effect by $\mu \cdot F_{ax,Rk}$. In EC 5, the friction coefficient is assumed to be $\mu = 0.25 = ¼$.

E2.7 Literature

B.O. Hilson, original Article C3, STEP 1995.

Aune P. and Patton-Mallory M. (1986). Lateral load-bearing capacity of nailed joints based on the yield theory – theoretical development and experimental verification. Forest Products Laboratory, Research Papers FPL 469 & 470.

Bejtka I. and Blass H.J. (2002). Joints with inclined screws. Paper 35-7-4, CIB-W18 Meeting 35, Kyoto.

Hilson B.O., Whale L.R.J. and Smith I. (1990). Characteristic properties of nailed and bolted joints under short-term lateral load. Part 5 – Appraisal of current design data in BS5268:Part 2:1984 Structural Use of Timber. Journal of the Institute of Wood Science 11(6):208-212.

Johansen K.W. (1949). Theory of timber connections. International Association of Bridge and Structural Engineering. Publication No. 9:249-262. Bern.

Möller T. (1951). En ny metod för beräkning av spikförband. Report No 117, Chalmers University of Technology.

Rodd P.D., Anderson C., Whale L.R.J. and Smith I. (1987). Characteristic properties of nailed and bolted joints under short term lateral load. Part 2 – Embedment test apparatus for wood and wood-based sheet materials. Journal of the Institute of Wood Science 11(2):60-64.

E3 Joints with nails and staples

Original articles: B. O. Hilson

Nails are the most frequently used fasteners in timber constructions and come in a wide range of lengths, diameters and surface finishes. Nailed joints are easy to manufacture and generally exhibit ductile behaviour, which facilitates a balanced load distribution in nail groups. In addition, the ductile behaviour helps dissipate energy when cyclic loads such as earthquakes are imposed, which may improve the load-bearing behaviour of the construction. The most commonly used type is the smooth shank nail with a circular cross-section, which is produced from a wire with a minimum tensile strength of 600 N/mm^2 and with standard diameters ranging between 2 and 8 mm. The nails can be blank or protected against corrosion e.g. by galvanising. The nail head is generally flat and circular with a diameter of around double the nail shank diameter. However, certain types of nails have smaller heads, so they can be driven in flush to the timber surface. Certain countries also manufacture nails with a square cross-section. Nails can be driven in manually or with pneumatic nail guns.

The load-bearing behaviour of a nail can be improved, in terms of withstanding both forces perpendicular to the nail axis as well as in a shank direction (withdrawal) by modifying the nail surface. One approach would be to deform the surface of smooth shank nails by rolling on grooves (rings) or deforming to a helical thread (spirals). Another would involve helically twisting nails with a square cross-section. This process not only alters the nail area but also hardens the steel and thus boosts the yield strength. Hot-dip galvanizing, etching or coating with resin or plastics are further means of improving the load-bearing behaviour.

Since staples are very similar fasteners to nails, there is very little difference in their load-bearing behaviour. However, there are two main things to bear in mind. One is the fact that the steel grade used for staples tends to be far higher than that for nails, as reflected in the yield moment. The other is the need to monitor the angle between the staple crown and the grain direction of the wood. Nails and staples are shown in Figure E3-1, while the backgrounds, design equations and derivations thereof can be taken from Articles E1 and E2. A nailed joint must comprise at least two nails.

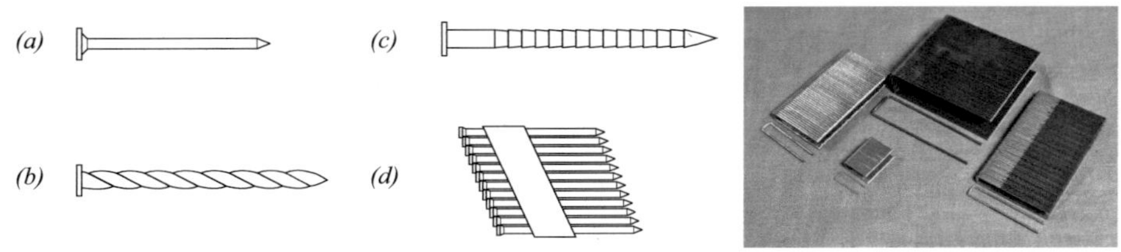

Figure E3-1 Left: Nail shapes. (*a*) Round smooth shank nail, (*b*) spiral nail, (*c*) ringed shank nail, (*d*) machine driven nails. (STEP 1995 Article C1) Right: Staples.

E3.1 Predrilling

If nails are driven into wood with a higher density, there is a risk of the wood splitting or the nail buckling. Accordingly, timber members must be predrilled with around 80% of the nail diameter, either when the timber has a characteristic density from 500 kg/m^3 or the nail diameter exceeds 6 mm. There are three key advantages to predrilling:

- The embedment strength and hence the capacity when the nails are laterally loaded are increased,
- The nail spacings and edge distances can be reduced, which allows more compact joints to be achieved,
- Joints are subject to less displacement movement when loaded.

One further important factor is the thickness of the timber members to be connected, given that the splitting risk rises with decreasing thickness. This is why both EC 5 and the NA (NDP to 8.3.1.2) stipulate minimum timber thicknesses depending on density.

E3.2 Nail distances and spacings

Nails must be arranged at sufficient distances from the end grain and edges as well as each other, to prevent splitting. Figure E1-9 shows how a range of distances are defined. The end is deemed loaded, when the force transmitted by the nail has a component pushing in the direction of the end. Otherwise, the end is deemed unloaded. The increased splitting risk means greater end distance requirements are imposed on the loaded end than on the unloaded end. Similarly, the edges may be loaded or unloaded. According to EC 5, the minimum nail distances and spacing are based on many years of experience and specified in Table E3-1 for timber-to-timber joints. The values for minimum nail distances differ from one wood species to the other and are mainly a factor of the splitting risk, shear strength, wood density and nail diameter. Predrilling considerably reduces the tendency of the wood to split and thus allows denser clustering of such nails.

Table E3-1 Minimum nail distances and spacings, d = nail diameter in mm, $0° \leq \alpha \leq 90°$ = angle between force and grain directions. See also EN 1995-1-1:2010 Table 8.2.

Distances and spacings	Minimum distances and spacings		
	Non-predrilled		Predrilled
	$\rho_k \leq 420$ kg/m³	420 kg/m³ $< \rho_k \leq 500$ kg/m³	
Spacing a_1 (parallel to the grain)	$d < 5$ mm: $(5 + 5 \cdot \lvert\cos\alpha\rvert) \cdot d$ $d \geq 5$ mm: $(5 + 7 \cdot \lvert\cos\alpha\rvert) \cdot d$	$(7 + 8 \cdot \cos\alpha) \cdot d$	$(4 + \cos\alpha) \cdot d$
Spacing a_2 (perpendicular to the grain)	$5 \cdot d$	$7 \cdot d$	$(3 + \sin\alpha) \cdot d$
Distance $a_{3,t}$ (loaded end)	$(10 + 5 \cdot \lvert\cos\alpha\rvert) \cdot d$	$(15 + 5 \cdot \cos\alpha) \cdot d$	$(7 + 5 \cdot \cos\alpha) \cdot d$
Distance $a_{3,c}$ (unloaded end)	$10 \cdot d$	$15 \cdot d$	$7 \cdot d$
Distance $a_{4,t}$ (loaded edge)	$d < 5$ mm: $(5 + 2 \cdot \lvert\sin\alpha\rvert) \cdot d$ $d \geq 5$ mm: $(5 + 5 \cdot \lvert\sin\alpha\rvert) \cdot d$	$d < 5$ mm: $(7 + 2 \cdot \sin\alpha) \cdot d$ $d \geq 5$ mm: $(7 + 5 \cdot \sin\alpha) \cdot d$	$d < 5$ mm: $(3 + 2 \cdot \sin\alpha) \cdot d$ $d \geq 5$ mm: $(3 + 4 \cdot \sin\alpha) \cdot d$
Distance $a_{4,c}$ (unloaded edge)	$5 \cdot d$	$7 \cdot d$	$3 \cdot d$

For joints between timber and wood-based materials and steel-to-timber joints, smaller nail distances can be assumed than for timber-to-timber joints. This is due to the generally lower risk of the wood splitting and this, in turn, is down to the reinforcing effect of the wood-based materials or steel plates. In addition, nails in such joints generally have one shear plane, meaning they do not completely penetrate the timber member and further reducing the risk of any split in the member in question. These favourable boundary conditions were confirmed on tests performed with nails. One particularly relevant problem concerning wood-based panels including the nail tip is the cracks on the underside (where the nails protrude), which can significantly reduce the load-bearing capacity of the joint. For joints between wood-based materials and timber, EC 5 allows the minimum nail distances specified in Table E3-1 to be reduced by 15%, but not the edge distances. For plywood, specific minimum edge distances apply; $3 \cdot d$ for the unloaded edge and $(3 + 4\,\lvert\sin\alpha\rvert) \cdot d$ for the loaded edge. Similar modifications are proposed for steel-to-timber joints, whereby the recommended reduction factor for the minimum spacings is 0.7. Nails driven in parallel to the grain (namely in the end grain) are deemed non load-bearing. While some exceptions are allowed according to EC 5, subject to compliance with specific conditions, this provision is deemed invalid in the NA.

E3.3 Embedment strength

EC 5 recommends the following characteristic embedment strength values for timber-to-timber nailed joints with nail diameter up to 8 mm, regardless of the angle between the force and grain directions.

Timber (solid timber, glued laminated timber) and laminated veneer lumber (LVL):

Non-predrilled:

$$f_{h,k} = 0.082 \cdot \rho_k \cdot d^{-0.3} \tag{E3-1}$$

Predrilled:

$$f_{h,k} = 0.082 \cdot \rho_k \cdot (1 - 0.01 \cdot d) \tag{E3-2}$$

where ρ_k is the characteristic density in kg/m^3 and d is the nail diameter in mm.

For joints between timber and wood-based materials, additional equations are provided in EC 5 and in the NA to determine the embedment strength of wood-based panels (plywood, hardboards, particleboards, OSB, gypsum plasterboards and cement-bonded particleboards, for equations, see also Section E12.1).

The equations for the characteristic values of embedment strength are based on regression analyses and were derived from numerous embedment tests, which encompassed wide-ranging timber densities, wood species and nail diameters. An illustration of the origin of equations (E3-1) and (E3-2) is shown in Whale et al. (1989).

E3.4 Yield moment

EC 5 includes equations for the characteristic yield moment $M_{y,k}$ (in Nmm) for smooth shank nails, which have been manufactured from wire with a minimum tensile strength of 600 N/mm^2.

For smooth shank nails with a round cross-section, the following applies:

$$M_{y,k} = 0.3 \cdot f_u \cdot d^{2.6} \tag{E3-3}$$

And for nails with a square cross-section:

$$M_{y,k} = 0.45 \cdot f_u \cdot d^{2.6} \tag{E3-4}$$

where f_u is the wire tensile strength in N/mm^2 and d is the nail diameter in mm (side length for square nails).

Like equations (E3-1) and (E3-2), which are used to define embedment strengths, equations (E3-3) and (E3-4) are also based on the regression analyses of numerous test results. Exponent 2.6 is based on the observations of Werner and Siebert (1991), who noted how the nail yield stress declines with increasing diameter.

Other nail types must be tested in accordance with EN 409 "Timber structures – Test methods – Determination of the yield moment of dowel-type fasteners", to determine applicable values for $M_{y,k}$. These values for $M_{y,k}$ can be obtained from the declarations of performance (DOPs, see Article A1) for the individual nail types.

E3.5 Deformations in nailed joints

Nailed joints deform under loads, just like joints established with other mechanical fasteners. This is shown in Figure E3-2 using a typical load-deformation diagram showing a test on a laterally loaded nailed joint with two shear planes.

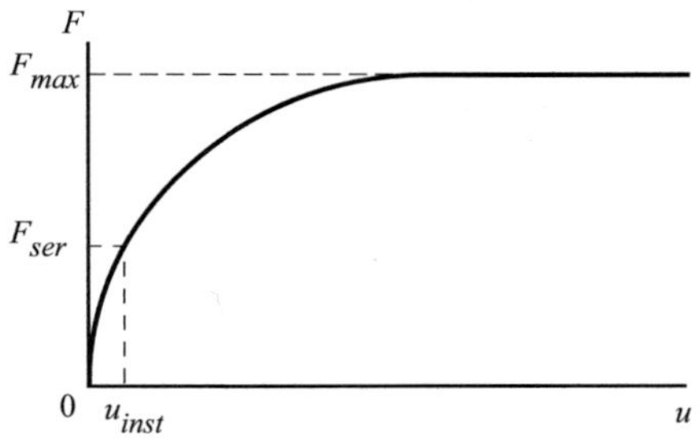

Figure E3-2 Typical load-deformation diagram of a nailed joint. F_{max} is the maximum load, F_{ser} is the load at serviceability limit state and u_{inst} is the elastic initial deformation. (STEP 1995 Article C4)

The elastic initial deformation, which occurs when exposed to serviceability load F_{ser}, can be estimated using the slip modulus K_{ser} in accordance with EC 5. The following applies:

Non-predrilled:

$$K_{ser} = \frac{d^{0.8} \cdot \rho_m^{1.5}}{30} \tag{E3-5}$$

Predrilled:

$$K_{ser} = \frac{d \cdot \rho_m^{1.5}}{23} \tag{E3-6}$$

In this respect, ρ_m is the mean density in kg/m^3, d the diameter in mm and K_{ser} in N/mm.

Consequently:

$$u_{inst} = \frac{F_{ser}}{K_{ser}} \tag{E3-7}$$

Due to creep, the final deformation in nailed joints may exceed the level of initial elastic deformation. The final deformation can be estimated from (see also Article F1):

$$u_{fin} = u_{inst} \cdot (1 + k_{def}) \tag{E3-8}$$

Equations (E3-5) and (E3-6) apply for timber-to-timber joints, while for steel-to-timber joints, the determined value should be doubled, leading to deformations which amount to only half those of timber-to-timber joints. When the mean densities among the members to be connected vary, the geometric mean for ρ_m is used to determine K_{ser}.

Deformations in joints often represent a significant proportion of the overall deformations in a nailed timber construction under a serviceability load and should thus be taken into account during calculations. Within a nailed truss, the portion of deflection due to semi-rigidity of the joints often exceeds that due to elastic deformation of the truss members. In nailed composite beams, meanwhile, simple beam theory cannot be applied. For stresses and deflections, displacements must also be taken into consideration, e.g. between the flange and web of a nailed I-joist, which can be done using the calculation methods described in Article D7. The semi-rigidity also impacts on the moment-rotation behaviour of joints with nailed-on plywood panels in portal frames.

E3.6 Axially loaded nails – withdrawal

The load-bearing capacity of axially loaded smooth shank nails (Figure E3-3) is low and the long-term behaviour unfavourable. Accordingly, instead of smooth shank nails only ringed shank nails may be used for permanent or long-term loading in an axial direction. The point at which the withdrawal resistance peaks is when nails are driven in perpendicular or at an angle to the grain. Nails arranged in parallel to the grain direction, meanwhile, generally offer negligible withdrawal resistance and are unsuitable for transferring forces in axial direction. In addition, varying the wood moisture content reduces the withdrawal resistance of smooth shank nails.

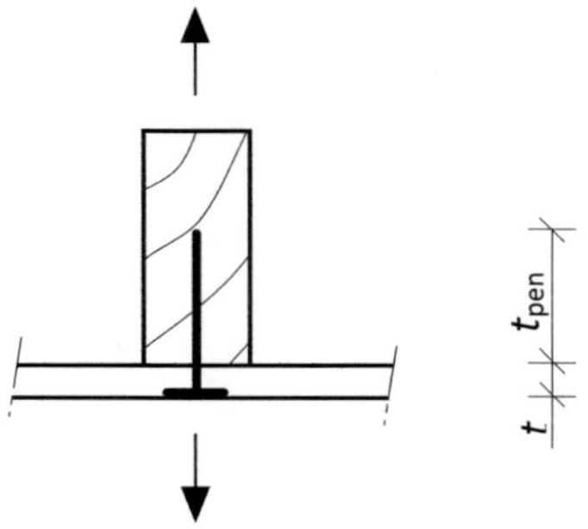

Figure E3-3 Axially loaded nail arranged perpendicular to the grain. (STEP 1995 Article C5)

In addition to a tensile failure of the nail, the following possible failure modes for the nailed joint shown in Figure E3-3 must be taken into consideration:

- Withdrawal of the nail from the timber member containing the nail tip (withdrawal parameter $f_{ax,k}$),
- Pull-through of the nail head through the plate material (head pull-through parameter $f_{head,k}$).

The NA regulates the use of smooth shank nails in more detail and states that smooth shank nails in predrilled holes should not be subject to axial loads. The NA also defines load-bearing capacity classes for nails that determine the parameters $f_{ax,k}$ (withdrawal parameter) and $f_{head,k}$ (head pull-through parameter) required to determine the withdrawal resistance. It can be assumed that smooth shank and ringed shank nails of load-bearing capacity class 1 can be exposed to 60% of the withdrawal resistance, if permanently exposed to axial loads in connections in coupled purlins in roofs with a maximum 30% incline. The experimental determination of the parameters and limit values for classifying the nail types in load-bearing capacity classes is defined in E DIN 20000-6:2012 (draft).

Other factors which impact on the withdrawal resistance of nails include the density of the wood into which the nail is driven and the nail surface quality. Cement-coated or hot-dip galvanised nails, ringed shank and spiral nails and nails with square cross-sections perform more favourably in response to withdrawal loads than uncoated or galvanised smooth shank nails. Another advantage of ringed shank and spiral nails lies in the fact that their withdrawal resistance is hardly affected by changes in wood moisture content.

Empirical equations to determine withdrawal resistance were determined for a number of combinations. For nails, the characteristic withdrawal resistance $F_{ax,Rk}$ should be assumed as the smallest of the following values:

For nails without smooth shank:

$$F_{ax,Rk} = \min\begin{cases} f_{ax,k} \cdot d \cdot t_{pen} \\ f_{head,k} \cdot d_h^2 \end{cases} \tag{E3-9}$$

For smooth shank nails:

$$F_{ax,Rk} = \min\begin{cases} f_{ax,k} \cdot d \cdot t_{pen} \\ f_{ax,k} \cdot d \cdot t + f_{head,k} \cdot d_h^2 \end{cases} \tag{E3-10}$$

Here, d = diameter, d_h = head diameter of the nail, t = thickness of the timber member on the nail head side and t_{pen} = penetration depth on the nail tip side. The characteristic values of the withdrawal parameter $f_{ax,k}$ and the head pull-through parameter $f_{head,k}$ are functions of the characteristic density of the wood used and are specified in the NA or the declarations of performance of the nail manufacturer.

The EC 5 also includes minimum values for the penetration depth t_{pen}. Also noteworthy is the fact that when considering the rope effect for nails loaded laterally, not the entire withdrawal resistance may be taken into account, but only specific portions of the load-bearing capacity calculated in accordance with Johansen theory: a maximum of 15% for smooth shank, round nails and 25% for smooth shank, square nails. For non-smooth shank nails, the limit is at 50%.

E3.7 Effective number of nails

The load-bearing capacity of a series of n nails arranged in the grain direction must be reduced by an effective number n_{ef}, if the nails are not arranged offset with respect to each other by at least $1 \cdot d$ perpendicular to the grain direction (see also Article E13 on n_{ef}). This kind of offset arrangement is often practically applied, which is why n_{ef} is usually irrelevant for nailed joints. In addition, the need for n_{ef} for nailed and above all stapled joints is disputed; in DIN 1052 nailed and stapled joints need not be subject to any reduction in the actual number n.

In EC 5, the following equation applies with k_{ef} from Table E3-2:

$$n_{ef} = n^{k_{ef}} \tag{E3-11}$$

Table E3-2 Values for k_{ef}, interim values may be linearly interpolated.

Nail spacing	k_{ef} Non-predrilled	Predrilled
$a_1 \geq 14 \cdot d$	1.0	1.0
$a_1 = 10 \cdot d$	0.85	0.85
$a_1 = 7 \cdot d$	0.7	0.7
$a_1 = 4 \cdot d$	-	0.5

E3.8 Combined loads

When a nail is subject to simultaneous lateral load $F_{v,Ed}$ and withdrawal load $F_{ax,Ed}$, both load components overlap. For smooth shank nails, a linear interaction applies. However, according to NA, for smooth shank nails in connections in coupled purlins, the following less conservative equation may be applied:

$$\left(\frac{F_{v,Ed}}{F_{v,Rd}}\right)^{1.5} + \left(\frac{F_{ax,Ed}}{F_{ax,Rd}}\right)^{1.5} \leq 1 \tag{E3-12}$$

For non-smooth shank nails, a quadratic interaction of both load components can be used.

E3.9 Staples

Staples are very similar fasteners to nails, which means correspondingly little variation in their load-bearing behaviour. However, two things must be taken into consideration above all. The first is the fact that the steel quality used for staples is usually far higher than that for nails (minimum tensile strength of steels for staples of 800 N/mm^2), which is reflected in the yield moment. Secondly, the angle between the staple crown and the grain direction of the wood must be taken into account. The basic rule is that the load-bearing capacity of stapled joints can be assumed as equivalent to that of two nails of equivalent diameter, provided the angle between the staple crown and grain direction is at least 30°. For an angle below 30°, the load-bearing capacity of laterally loaded stapled joints must be reduced by 70%. A stapled joint must comprise at least two staples and the penetration depth must be at least $14 \cdot d$.

The characteristic yield moment is (EC 5/A2:2014):

$$M_{\mathrm{y,Rk}} = 150 \cdot d^3 \qquad \text{(E3-13)}$$

The applicable minimum distances differ from those for nails, details are specified in EC 5. There is no effective number of fasteners: $n_{\mathrm{ef}} = n$.

E3.10 Literature

B.O. Hilson, original Articles C4, C5, STEP 1995.

Werner H. and Siebert W. (1991). Neue Untersuchungen mit Nägeln für den Holzbau. Holz als Roh- und Werkstoff 49:191-198.

Whale L.R.J., Smith I. and Hilson B.O. (1989). Characteristic properties of nailed and bolted joints under short term lateral load. Part 4 – The influence of testing mode and fastener diameter upon embedment test data. Journal of the Institute of Wood Science 11(5):156-161.

E4 Joints with bolts and dowels

Original articles: J. Ehlbeck, H. Werner

Dowels are slender, cylindrical fasteners made of steel with smooth or even slightly grooved surfaces. Based on this definition, each dowel has a diameter of between 6 and 30 mm, with diametric tolerance defined as -0.1/+0.5 mm. Holes for dowels must be predrilled with the nominal diameter. The holes in steel plates for steel-to-timber joints may be drilled 1 mm larger than the dowel diameter; resulting additional deformations have to be taken into consideration. Similarly, bolts are dowel-type fasteners made of steel, which include a head and nut. They are inserted into predrilled holes and then tightened such as to ensure that the wood or steel parts are secured in close proximity. When necessary, the bolts must also be retightened, if the wood has reached its equilibrium moisture content. The predrilled holes may be up to 1 mm larger than the diameter of the bolt. For joints with fitted bolts, the bolts are driven into a hole, which is of equivalent dimensions to the diameter of the bolt. Each of the bolt also has washers on both sides, the side lengths or diameter of which should be at least $3 \cdot d$ and they should be at least $0.3 \cdot d$ thick (d is the bolt diameter). The washers should have a full contact surface with the wood.

Dowelled joints have excelled when it comes to transferring large forces. Not only is this joint type economic, it is also easy to set up. However, dowelled joints require at least one dowel to be replaced with a fitted bolt, to guarantee the joint cohesion or help prevent a gap between the individual joint members. In comparison to normal bolted joints, dowelled joints are more rigid and accordingly, bolted joints should be avoided at any point where good dimensional stability and high construction rigidity is required. Joints with fitted bolts, conversely, like dowelled joints, do not feature any predrilled holes of excessive size and can thus be as rigid and dimensionally stable as dowelled joints. Bolted and dowelled joints can be executed as purely timber-to-timber joints, but also in the form of joints connecting timber and wood-based materials or steel-to-timber joints.

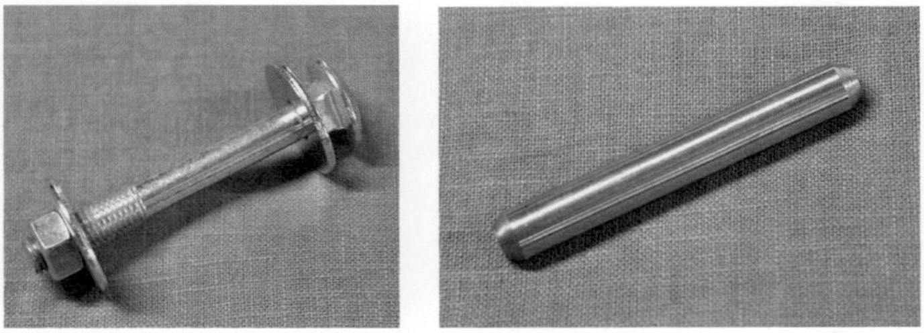

Figure E4-1 Left: bolt. Right: dowel.

A dowel and a bolt are shown in Figure E4-1. Background details and design equations for laterally loaded joints and derivations thereof can be taken from Articles E1 and E2. The NA stipulates that a load-bearing bolted or dowelled joint should have at least four shear planes and at least two fasteners. If only one fastener is used, only half the load-bearing capacity may be assumed. Another execution rule is that for external steel plates, fitted bolts must be used instead of dowels.

E4.1 Fastener distances and spacings

Similarly, as for nails, dowels and bolts must be arranged at sufficient distances from the end grain, edges and each other, to avoid the risk of the wood splitting. Limited distances between fasteners or excessive proximity to the edges will lead to premature failure of wood due to brittle failure, e.g. splitting. Figure E1-9 shows how the different distances are defined. The minimum spacings of bolts and dowels differ, due primarily to the size of the washers. However, regulations governing the minimum distances for bolts and dowels are complex, since the end distance $a_{3,c}$ to the unloaded end depends on the size of the angle α between the force and grain directions. The end distance can be reduced if α is smaller. The minimum distances for bolts are specified in Table E4-1 and those for dowels in Table E4-2.

Table E4-1 Minimum bolt distances and spacings, d = bolt diameter in mm, $0° \leq \alpha \leq 90°$ = angle between force and grain directions. See also EN 1995-1-1:2010 Table 8.4.

Distances and spacings	Angle	Minimum values
Spacing a_1 (parallel to the grain)	-	$(4 + \cos\alpha) \cdot d$
Spacing a_2 (perpendicular to the grain)	-	$4 \cdot d$
Distance $a_{3,t}$ (loaded end)	-	$\max(7 \cdot d; 80\ \text{mm})$
Distance $a_{3,c}$ (unloaded end)	$\alpha < 30°$	$4 \cdot d$
	$30° \leq \alpha \leq 90°$	$(1 + 6 \cdot \sin\alpha) \cdot d$
Distance $a_{4,t}$ (loaded edge)	-	$\max[(2 + 2 \cdot \sin\alpha) \cdot d; 3 \cdot d]$
Distance $a_{4,c}$ (unloaded edge)	-	$3 \cdot d$

Table E4-2 Minimum dowel distances and spacings, d = dowel diameter in mm, $0° \le \alpha \le 90°$ = angle between force and grain directions. See also EN 1995-1-1/A2:2014 Table 8.5.

Distances and spacings	Angle	Minimum values
Spacing a_1 (parallel to the grain)	-	$(3 + 2 \cdot \cos\alpha) \cdot d$
Spacing a_2 (perpendicular to the grain)	-	$3 \cdot d$
Distance $a_{3,t}$ (loaded end)	-	$\max(7 \cdot d; 80\ \text{mm})$
Distance $a_{3,c}$ (unloaded end)	$\alpha < 30°$ $30° \le \alpha \le 90°$	$\max(3.5 \cdot d; 40\ \text{mm})$ $a_{3,t} \cdot \sin\alpha$
Distance $a_{4,t}$ (loaded edge)	-	$\max[(2 + 2 \cdot \sin\alpha) \cdot d; 3 \cdot d]$
Distance $a_{4,c}$ (unloaded edge)	-	$3 \cdot d$

E4.2 Embedment strength

For bolts and dowels up to a diameter of 30 mm, the characteristic embedment strength $f_{h,0,k}$ when loaded parallel to the grain is (in N/mm²):

$$f_{h,k} = 0.082 \cdot (1 - 0.01 \cdot d) \cdot \rho_k \tag{E4-1}$$

where ρ_k the characteristic density in kg/m³ and d the fastener diameter in mm.

Equation (E4-1) corresponds to equation (E3-2) for predrilled nailed joints. Unlike nailed joints, however, for bolted and dowelled joints, the influence of the angle between the force and grain directions on embedment strength has to be taken into consideration, since for large fastener diameters, the embedment strength declines with increasing angle between the force and grain directions:

$$f_{h,\alpha,k} = \frac{f_{h,0,k}}{k_{90} \cdot \sin^2\alpha + \cos^2\alpha} \tag{E4-2}$$

where α = angle between the force and grain directions.

In equation (E4-2), the Hankinson equation (3) in Annex 2 re-emerges, see also Figure D1-5 and equation (D1-1), which defines the design format for compressive stress at an angle to the grain direction. The influence of the angle α between the force and grain directions on the embedment strength is shown in Figure E4-2. The reduction in embedment strength with increasing angle α is more prominent in softwoods than hardwoods.

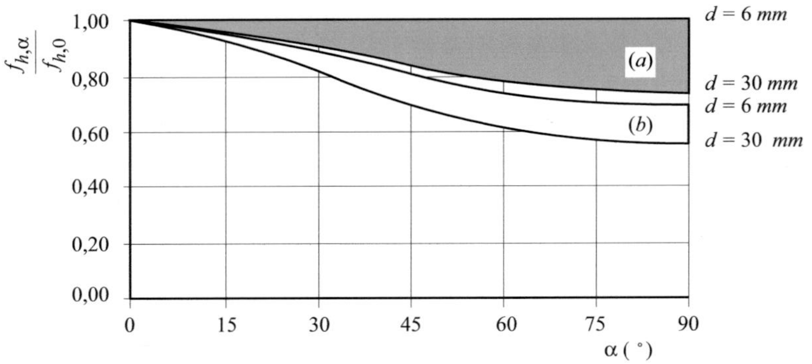

Figure E4-2 Ratio $f_{h,a}$ / $f_{h,0}$ depending on the angle α between the force and grain directions; (*a*): Hardwood; (*b*): Softwood. (STEP 1995 Article C6)

The coefficient $k_{c,90}$, which resembles to the factor k_{90} from equation (E4-2), is already known from Article D1 (cf. equation (D1-4)). There, the verifications for compressive stress at an angle to the grain direction were introduced as well as the factor $k_{c,90}$, to take the type of loading, splitting risk and degree of compressive deformation into consideration. Compressive forces also appear under the fasteners in joints subject to lateral loads. Factor k_{90} is defined in EC 5 as follows:

$$k_{90} = \begin{cases} 1.35 + 0.015 \cdot d & \text{for softwoods} \\ 1.30 + 0.015 \cdot d & \text{for LVL} \\ 0.09 + 0.015 \cdot d & \text{for hardwoods} \end{cases} \qquad \text{(E4-3)}$$

Additional equations to determine embedment strength when using wood-based panels (plywood, OSB, particleboards) are specified in EC 5 (equations see also Section E12.1).

E4.3 Yield moment of the fasteners

The following relationship is applied in EC 5 to determine the characteristic value of the yield moment of round steel bolts and steel dowels:

$$M_{y,Rk} = 0.3 \cdot f_u \cdot d^{2.6} \qquad \text{(E4-4)}$$

where f_u is the tensile strength of the fastener.

Equation (E4-4) corresponds to equation (E3-3) for smooth shank nails, although the background of both equations is different (Blass et al., 2000). Similarly, for bolts and dowels, the yield moment is generally determined in accordance with EN 409.

E4.4 Verification of serviceability

The load-deformation behaviour of joints with dowel-type fasteners can be shown using load-slip curves. Figure E4-3 specifies idealised load-slip curves for bolted and dowelled joints with around the same load-bearing capacities, where $F_{max,est}$ indicates the estimated maximum load.

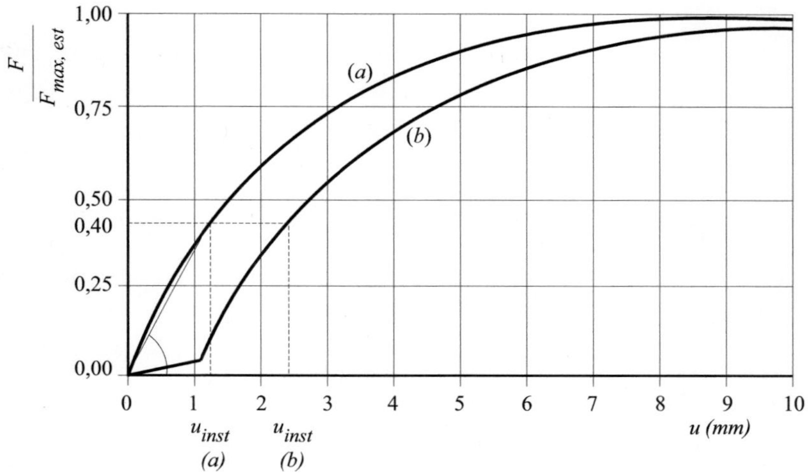

Figure E4-3 Idealised load-slip curves of a (*a*) dowelled and a (*b*) bolted joint. (STEP 1995 Article C7)

The slip modulus K_{ser} derived from the curve is a characteristic parameter of the overall joint. Based on numerous test results, EC 5 gives a slip modulus K_{ser} for bolted and dowelled joints per shear plane and fastener under a serviceability load as follows:

$$K_{ser} = \frac{d \cdot \rho_m^{1.5}}{23} \tag{E4-5}$$

where K_{ser} in Nmm, ρ_m in kg/m³ and d in mm.

For bolted joints with clearance, the clearance must be added to the displacements (see Figure E4-3 curve (b)). If the two timber members to be connected have differing mean densities, the geometric mean for ρ_m should be applied:

$$\rho_m = \sqrt{\rho_{m,1} \cdot \rho_{m,2}} \tag{E4-6}$$

Equation (E4-5) applies for timber-to-timber joints. For steel-to-timber joints, the value for K_{ser} determined using equation (E4-5) must be multiplied by factor 2.0.

Accordingly, the following applies, similar to nailed joints:

$$u_{inst} = \frac{F_{ser}}{K_{ser}} \tag{E4-7}$$

The final deformation can be estimated as follows:

$$u_{fin} = u_{inst} \cdot (1 + k_{def}) \tag{E4-8}$$

E4.5 Axially loaded fasteners

Dowels cannot be exposed to withdrawal loads, while for bolted joints, for the rope effect portion, up to 25% of the portion in accordance with Johansen theory may be applied. The withdrawal resistance is calculated from the minimum value of the tensile strength of the bolt and the load-bearing capacity of the washer, which corresponds to the compression strength perpendicular to the grain between the washer and timber surface.

E4.6 Interaction between multiple fasteners in a row

For joints exposed to load in the grain direction, the force exerted is distributed unevenly on the fasteners in the direction of force (especially if brittle failure occurs). This means that the load-bearing capacity of a joint with multiple fasteners in a row does not equate to the sum of the load-bearing capacities of the individual fasteners. The load-bearing capacity of such a joint with n dowels or bolts is calculated with an effective number n_{ef}:

$$n_{ef} = \min \begin{cases} n \\ n^{0.9} \cdot \sqrt[4]{\dfrac{a_1}{13 \cdot d}} \end{cases} \tag{E4-9}$$

Equation (E4-9) is based on comprehensive experimental results. Provided there is no premature splitting of the wood and plastic deformations are possible, the forces can be distributed between the individual fasteners, as applies, for example, for joints exposed to compression stress perpendicular to the grain, in wich $n_{ef} = n$ applies. For dowelled joints, according to NA, a further rule applies to determine the effective number of fasteners. For moment-resisting joints with multiple dowel circles, $n_{ef} = 0.85 \cdot n$ applies. Additional information and background details to the effective number of fasteners are included in Article E13.

E4.7 Literature

J. Ehlbeck, H. Werner, original Article C6, C7, STEP 1995.

Blass HJ, Bienhaus A and Krämer V (2000) Effective bending capacity of dowel-type fasteners. CIB-W18 Meeting 33, Paper 33-7-5. Delft, the Netherlands.

E5 Joints with self-tapping screws

Original article: J. Ehlbeck, W. Ehrhardt

Great progress has been made in recent decades in the development of wood screws. Meanwhile, timber construction mainly features the use of self-tapping screws, which, as circumstances allow and depending on the declaration of performance and density, can be deployed up to a diameter of 14 mm without predrilling. If predrilling takes place, the drilling diameter is roughly equivalent to the inner thread diameter. Self-tapping screws have differing diameters, but tend to have a shank diameter that is smaller than the external thread diameter, but exceeds that of the inner thread (= core diameter). The nominal diameter of a self-tapping screw is generally the external thread diameter. Such screws come in numerous varieties (Figure E5-1): partially threaded self-tapping screws with a smooth shank area, fully threaded screws, various screw heads and threads or steel varieties.

Screwed joints are particularly ideal for connecting steel plates, steel members and wood-based materials to timber, but can also be used for purely timber-to-timber joints. In this case, most of the joints feature one shear plane only. One additional and important area of use is that of reinforcements (Articles D8 and E12), in which self-tapping screws are used as reinforcements in joints loaded perpendicular to the grain, notched beams, openings and curved beams or as reinforcements in beam supports (see Section E5.3). Reinforcements almost exclusively feature the use of fully threaded screws, the use of which is explicitly regulated in the NA.

The European product standard EN 14592 "Timber structures – Dowel-type fasteners – Requirements" also applies to self-tapping screws. Due to the variety of screws and the associated variation in steel properties, diameters and threads, there is a need to have screws checked by a notified body, whereupon the results will be declared by the manufacturer as characteristic values of the yield moment, tensile capacity or withdrawal parameter (declaration of performance). Alternatively, the characteristic values used to calculate the load-bearing capacity of screwed joints can also be specified in a European Technical Assessment (ETA) (see also Article A1).

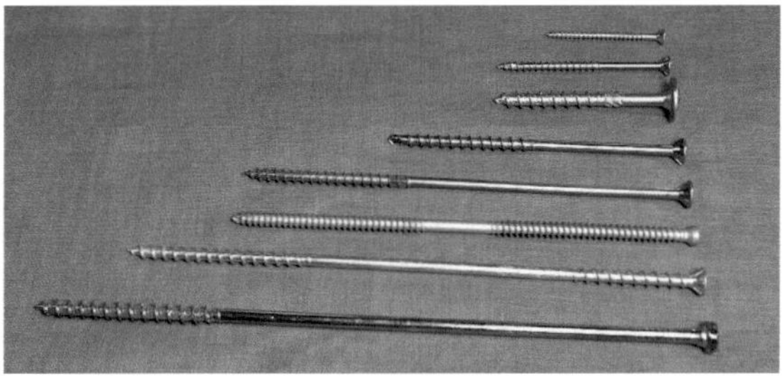

Figure E5-1 Typical self-tapping screws.

This means that the parameters required for joint design, such as yield moment $M_{y,k}$, withdrawal parameter $f_{ax,k}$ and diameter d required to determine the embedment strength for modern self-tapping screws are stipulated in European Technical Assessments (ETA) or declared by the manufacturer in declarations of performance (DOP). Accordingly, the engineer must consult the technical documents for the screw types used (ETA or DOP).

One additional key detail in the ETAs concerns the torque moment. Self-tapping screws come with special drive systems (usually Torx), to initiate the required torque levels in the screws to insert them. However, these torques must not be at a level which leads to the screws being damaged due to the torque level initiated. The ratio of torque failure moment to insertion moment must be at least 1.5 for the longest screws (see EN 14592:2009).

Background details, design equations for laterally loaded joints and derivations thereof can be taken from Articles E1 and E2. The NA sets out that a screwed joint must comprise at least two screws, except when used to secure casings, battens, wind braces, rafters, purlins and similar.

E5.1 Load-bearing behaviour

Self-tapping screws in timber joints can transmit high forces in the direction of the screw axis thanks to their thread, which ensures an effective bond between fasteners and the surrounding wood, as well as the high steel quality involved and this is a quality structural engineers can exploit. Self-tapping screws, which are subject to axial load when pulled out, often allow far higher forces to be transmitted than screws subject to lateral load. When subject to lateral load, screws may have lower load-bearing capacities than nails or dowels of the same nominal diameter (and the same tensile strength), since the bending resistance of the threaded part is far lower, due to the lower core area. For fully threaded screws however, when exposed to lateral loads, the rope effect can result in a load-bearing capacity for the screw, which exceeds that of a corresponding dowel of equivalent nominal diameter. For small fastener diameters (e.g. nails), the angle between the force and grain directions of the wood has hardly any influence on embedment strength and hence load-bearing capacity. For larger fastener diameters meanwhile (e.g. bolts or dowels), a noticeable impact on the embedment strength of the members to be connected is apparent. This means that screwed joints where $d < 6$ mm can be designed like nailed joints, while for $d \geq 6$ mm, the decrease in embedment strength with increasing angle between the force and grain directions has to be considered. Even so, most ETAs for self-tapping screws allow any reduction in embedment strength where $\alpha > 0°$ to be ignored for screws where $d \geq 6$ mm.

E5.2 Design of joints with self-tapping screws

The regulations of EC 5 on joints with screws are confusing and sometimes outdated. The basic regulations for screws subject to lateral load still originate from the period before the introduction of modern, self-tapping screws and only apply to partially threaded screws, namely those with a partially smooth shank and an external thread diameter equivalent to that of the smooth shank. As withdrawal resistance has become an increasingly important parameter with the emergence of modern screws, the regulations applied in EC 5 for this type of loading are unclear and often contradict the respective declarations of performance or ETAs of the manufacturer. The latter often also include more favourable values for the respective screws than EC 5, while it is preferable to reference regulations on minimum distances from the individual ETAs rather than EC 5. Since both the Johansen model and the rope effect are relevant for modern self-tapping screws with partial or full threads (Article E2), the simplest approach involves extracting the parameters required to calculate the load-bearing capacity, such as yield moment, embedment strength or withdrawal resistance, directly from the ETAs or the declarations of performance of the used screws rather than using the equations of the EC 5 concerning withdrawal or head pull-through resistances. The very high load-bearing capacity in the screw axis means the portion of the rope effect to be taken into account for screws may be up to 100% of the total according to Johansen theory. The wide variety of available screws (differing threads, head shapes, steel qualities, inner thread diameter, external diameter, etc.) also mean that in the foreseeable future, the resistances of the individual screw types, e.g. the withdrawal resistance, must be defined in tests and consequently taken from the technical documents.

Laterally loaded screws

Depending on the joint type (e.g. a single- or double-shear timber-to-timber joint), the load-bearing capacities are to be determined using the Johansen model with rope effect as shown in Article E2. The required values such as yield moment, withdrawal parameter or minimum distances are to be taken from the ETAs or the declarations of performance.

As already discussed in Section E5.1, the embedment strength of screws of nominal diameter up to 6 mm is determined with the equations for (non-predrilled) nails (equation (E3-1)) and that for screws where $d > 6$ mm is determined with the equations for bolts (equations (E4-1) and (E4-2)). When multiple screws are arranged in a row in the grain direction, the risk of splitting means an effective number $n_{ef} < n$ must be applied. This effective number is, in turn, calculated with the equation for nails (screws with a nominal diameter of up to 6 mm, equation (E3-11)) or those for bolts ($d > 6$ mm, equation (E4-9)).

Axially loaded screws

Wood screws are particularly ideal when it comes to absorbing greater withdrawal loads and are thus often deployed in joints where the fasteners are arranged in an inclined or cross-wise set-up. An inclined arrangement means the screws are subject to biaxial stresses; in the screw axis (when pulled out) and perpendicular to the screw axis (Johansen theory), whereby, given the high load-bearing capacity in the screw axis, high-load bearing joints can be formed. The vast majority of the load is accommodated in the longitudinal screw direction in this case, given the clearly higher stiffness than when laterally loaded. The lateral load-bearing capacity is not considered for screws arranged at an angle, since the equations used to define the embedment strength in EC 5 only apply to fasteners arranged in a direction perpendicular to the grain. However, the influence of friction on the shear plane can be taken into consideration for design purposes. For axially loaded screws, only the penetration depth of the threaded part can be taken into account. The following failure modes must be checked (the head pull-off resistance must exceed the tensile strength of the screw):

- Withdrawal failure of the threaded part in the member with the screw tip (parameter f_{ax}),
- Pull-through failure of the screw head or withdrawal failure of the threaded part in the member with the screw head (parameter f_{head} or f_{ax}),
- Tensile failure of the screw (tensile strength f_{tens}),
- Buckling failure of the screw when exposed to axial compression (see Section E5.3),
- Block shear failure (see Article E13).

The required parameters plus the minimum spacings and end and edge distances are specified in the technical documents. The minimum penetration depth of the thread on the screw tip side should usually be at least $6 \cdot d$ and the screwing angle between the screw axis (force) and grain direction must generally be at least 30°. There are also ETAs for which the angle between the screw axis and grain direction can be between 0 and 90°.

An effective number of screws n_{ef} must also be taken into consideration. This reflects the fact that due to the risk of brittle failure, the load-bearing capacity of a group of n screws is lower than the total of the individual load-bearing capacities of the screws (N. B.: n_{ef} is applicable to groups of screws, which are loaded by a systematic tensile force acting in the axial direction, but not for the rope effect):

$$n_{ef} = n^{0.9} \tag{E5-1}$$

E5.3 Buckling under compressive load

The verification of a screwed joint when exposed to compression is determined by the minimum resistance value against pushing-in and buckling. This verification is carried out e.g. when using fully threaded screws as reinforcement in beam supports, if the load-bearing capacity of a support perpendicular to the grain is supposed to be increased, see Figure E5-2 on the left (Bejtka and Blass, 2006). For cross-wise arranged screws, buckling analysis must be performed for the compressed screw. The resistance against pushing-in can be determined with the withdrawal parameter $f_{ax,k}$, while the buckling verification is performed in accordance with EC 3. The relative slenderness ratio is determined by taking the elastic foundation of the screw in the timber into consideration. Figure E5-2 on the right shows test specimens with buckled fully threaded screws.

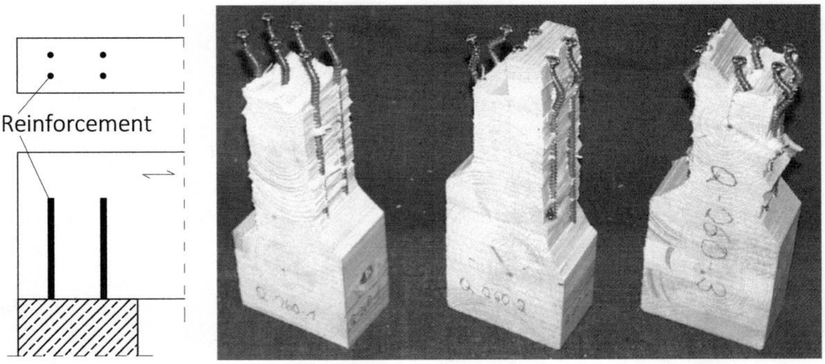

Figure E5-2 Left: Example of a reinforcement in a beam support. Right: Test specimens with buckled screws (Bejtka, 2005).

The design load-bearing capacity $F_{ax,Rd}$ in the direction of the screw axis is determined as follows:

$$F_{ax,Rd} = \min\left(f_{ax,d} \cdot d \cdot \ell_{ef}\, ;\, \kappa_c \cdot N_{pl,d}\right) \tag{E5-2}$$

The first term in equation (E5-2) shows that the resistance of a screw against being pushed in corresponds to the withdrawal resistance, since the withdrawal parameter $f_{ax,d}$ is also used to determine the push-in resistance.

The resistance against buckling in accordance with EC 3 is determined as follows (partial safety factor here is $\gamma_{M1} = 1.1$ (Germany) for steel in the buckling analysis):

$$\kappa_c = \begin{cases} 1 & \text{for } \bar{\lambda}_k \leq 0.2 \\ \dfrac{1}{k + \sqrt{k^2 - \bar{\lambda}_k^2}} & \text{for } \bar{\lambda}_k > 0.2 \end{cases} \tag{E5-3}$$

where

$$k = 0.5 \cdot \left[1 + 0.49 \cdot \left(\bar{\lambda}_k - 0.2\right) + \bar{\lambda}_k^2\right] \tag{E5-4}$$

and with the relative slenderness ratio

$$\bar{\lambda}_k = \sqrt{\frac{N_{pl,k}}{N_{ki,k}}} \tag{E5-5}$$

Here:

$N_{pl,k} = \dfrac{\pi \cdot d_1^2}{4} \cdot f_{y,k}$	Characteristic, plastic normal force, related to the inner thread diameter
$f_{y,k}$	Characteristic yield strength of the screw in the core area
$N_{ki,k} = \sqrt{c_h \cdot E_S \cdot I_S}$	Characteristic elastic buckling load, depending on the inner thread diameter, characteristic density and elastic foundation of the screw in the timber
$c_h = (0.19 + 0.012 \cdot d) \cdot \rho_k \cdot \left(\dfrac{90^\circ + \alpha}{180^\circ}\right)$	Elastic foundation modulus
ρ_k	Characteristic density of the timber
$E_S \cdot I_S = \dfrac{210\,000 \cdot \pi \cdot d_1^4}{64}$	Bending stiffness of the inner thread diameter
d_1	Inner thread diameter of the fully threaded screw
α	Angle between screw axis and grain directions

The design load-bearing capacity $F_{90,Rd}$ of a reinforced support amounts to:

$$F_{90,Rd} = \min \begin{cases} n \cdot F_{ax,Rd} + k_{c,90} \cdot \ell_{ef} \cdot b \cdot f_{c,90,d} \\ \ell_{ef,2} \cdot b \cdot f_{c,90,d} \end{cases} \qquad \text{(E5-6)}$$

$k_{c,90}$ is the coefficient for compression perpendicular to the grain, cf. equation (D1-4). The effective contact length ℓ_{ef}, which is applied during compressive stress exposure perpendicular to the grain to determine the effective contact surface, can also be taken from Article D1.

However, failure due to compression perpendicular to the grain may occur not only on the lower edge of the timber beam, but also on the level of the screw tips. Compression perpendicular to the grain at the level of the screw tip, meanwhile, is verified with the lower term of equation (E5-6), where $A_{ef} = \ell_{ef,2} \cdot b$ represents the effective area at the level of the screw tips, which is subject to compression perpendicular to the grain. A_{ef} depends on the type of load (direct or indirect), the type of load introduction (single- or double-sided) and the resulting load distribution and is explained in Figure E5-3 and Figure E5-4. Comprehensive simulations have shown that the stress distribution is approximately linear when a direct load is introduced, which means $\ell_{ef,2}$ can be calculated directly from the thread length ℓ_s (Figure E5-3). When an indirect load is applied, conversely, simulations show the emergence of exponential stress distribution (Figure E5-4). Accordingly, for indirect loading, the following equations derived from regressions apply (Bejtka and Blass, 2006):

$$\begin{aligned} &\text{Single-sided load introduction:} \quad && \ell_{ef} = \ell + 0.25 \cdot \ell_s \cdot e^{3.3 \cdot \frac{\ell_s}{h}} \\ &\text{Double-sided load introduction:} \quad && \ell_{ef} = \ell + 0.58 \cdot \ell_s \cdot e^{3.6 \cdot \frac{\ell_s}{h}} \end{aligned} \qquad \text{(E5-7)}$$

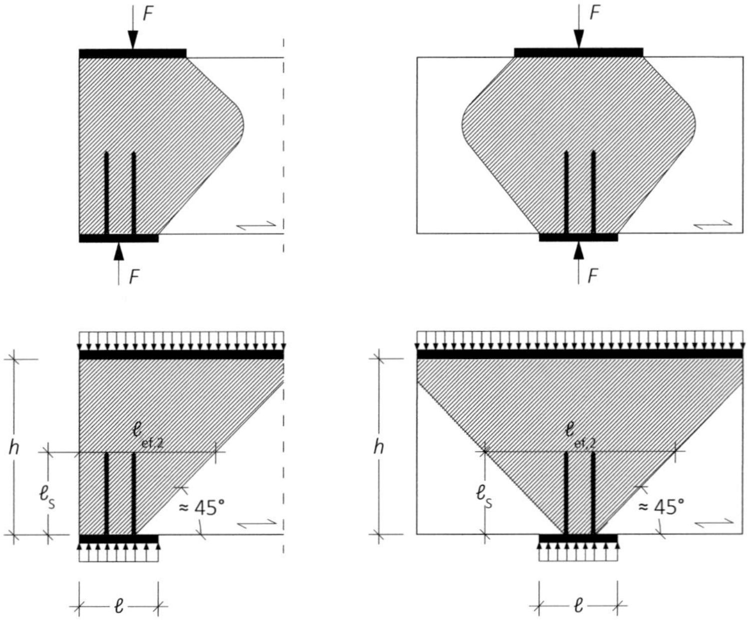

Figure E5-3 Above: Load distribution with direct load introduction (Bejtka, 2005). Below: Effective area at level of screw tips with direct load introduction (Bejtka and Blass, 2006).

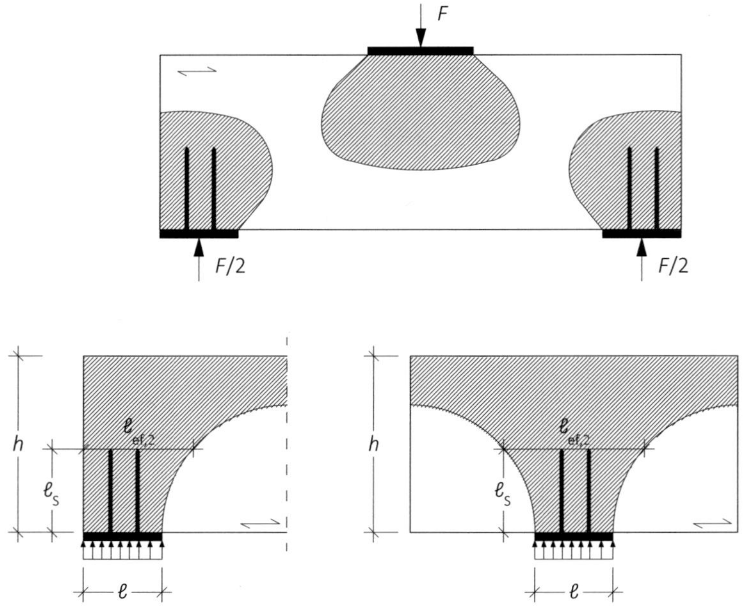

Figure E5-4 Above: Load distribution with indirect load introduction (Bejtka, 2005). Below: Effective area at level of screw tips with indirect load introduction (Bejtka and Blass, 2006).

E5.4 Joint stiffness

For laterally loaded screws, equations to determine K_{ser} can be taken from the provisions for nails (for screws of up to 6 mm (external thread diameter)) or for bolts (for screws from 6 mm in diameter). For axially loaded screws, empirical equations to determine $K_{ser} = K_{ax}$ are specified in the technical documents of the individual screws.

One example shown here is how the stiffness parallel to the shear plane can be calculated for a joint with cross-wise arranged screws, as shown in Figure E5-5. For joints with cross-wise arranged screws, one screw is pushed in and the other pulled out.

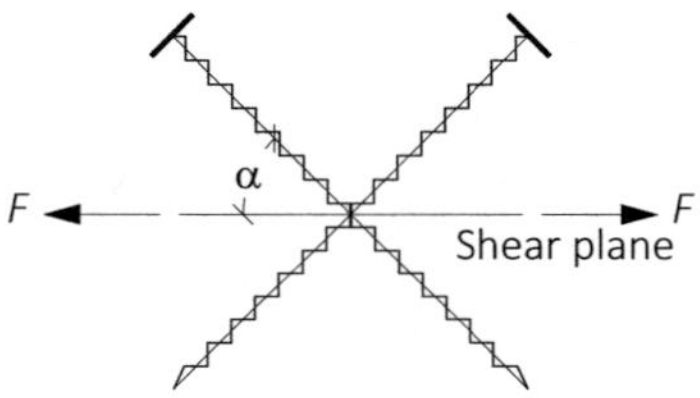

Figure E5-5 Cross-wise arranged screws, inclined by angle α.

The stiffness of the joint in the direction of the shear plane K_{ser} can be determininned from the rigidities K_{ax} in the direction of the screw axes using the principle of virtual work, while waiving the portions due to lateral load and friction.

Normal force N in screw axis due to force F ($F/2$ per screw):

$$N = \frac{F}{2 \cdot \cos\alpha} \tag{E5-8}$$

Normal force in screw axis due to virtual force "1":

$$\bar{N} = \frac{1}{2 \cdot \cos\alpha} \tag{E5-9}$$

Accordingly, it is now possible to calculate the deformation w in the shear plane direction for one side of the joint (for n = 2 screws and assuming the normal force is evenly distributed over the thread length → superimposition of two rectangles):

$$w=\sum_{n=1}^{2}\frac{N\cdot\overline{N}}{K_{ax}}=\frac{N_1\cdot\overline{N}_1}{K_{ax}}+\frac{N_2\cdot\overline{N}_2}{K_{ax}}=2\cdot\frac{1}{K_{ax}}\cdot\frac{F}{2\cdot\cos\alpha}\cdot\frac{1}{2\cdot\cos\alpha}=\frac{F}{2\cdot K_{ax}\cdot\cos^2\alpha} \quad \text{(E5-10)}$$

The stiffness K_{ser} of the joint is as follows for each side of the shear plane:

$$K_{ser,SF}=\frac{F}{w}=2\cdot K_{ax}\cdot\cos^2\alpha \quad \text{(E5-11)}$$

Consequently, the overall stiffness with equivalent thread lengths on both sides of the joint is as follows:

$$K_{ser}=K_{ax}\cdot\cos^2\alpha \quad \text{(E5-12)}$$

E5.5 Combined loading

When screws are simultaneously subject to lateral loads $F_{v,Ed}$ and axial loads $F_{ax,Ed}$, the following equation must be satisfied:

$$\left(\frac{F_{v,Ed}}{F_{v,Rd}}\right)^2+\left(\frac{F_{ax,Ed}}{F_{ax,Rd}}\right)^2\leq 1 \quad \text{(E5-13)}$$

E5.6 Literature

J. Ehlbeck, W. Ehrhardt, original Article C8, STEP 1995.

Bejtka I. (2005). Verstärkung von Bauteilen aus Holz mit Vollgewindeschrauben. Dissertation. Universität Karlsruhe.

Bejtka I. and Blass H.J. (2006). Self-tapping screws as reinforcement in beam supports. Paper 39-7-2, CIB-W18 Meeting 39, Florence.

E6 Joints with connectors

Original articles: H. J. Blass

The term 'connectors' is applicable to so-called split ring, shear plate and toothed-plate connectors. **Split ring and shear plate connectors**, as shown in Figure E6-1, are used in both timber-to-timber as well as steel-to-timber joints; usually together with bolts. While split ring connectors can only be used in timber-to-timber joints, shear plate connectors are also suitable for joining wood with steel. Shear plate connectors are used in timber-to-timber joints, if e.g. the connectors are to be pre-installed prior to assembly or if the construction has to be assembled and disassembled multiple times. Split ring and shear plate connectors come in wide-ranging shapes and sizes and with diameters ranging between 60 and 260 mm. They are always circular, since they are inserted into pre-milled depressions in the timber and can be made of cast aluminium alloys, steel, cast steel or solid oak. Most split ring and shear plate connectors used in Europe are described in EN 912 "Timber fasteners – Specifications for connectors for timbers". EN 912 sets out descriptions of split ring connectors made of metal as type A, shear plate connectors as type B and split ring connectors made of solid oak as type D.

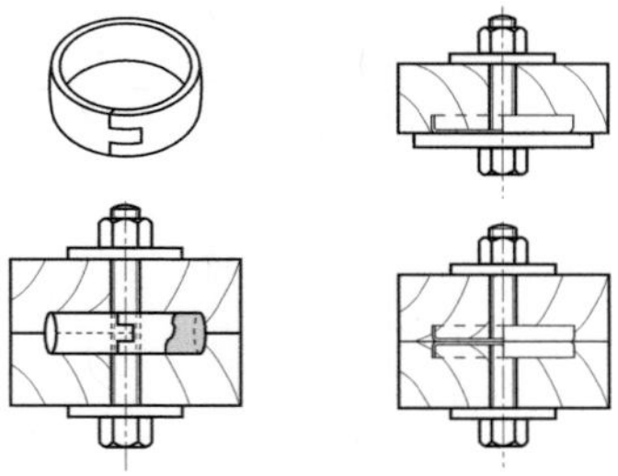

Figure E6-1 Joint with a split ring (left) and a shear plate connector (right). (STEP 1995 Article C9)

Multiple procedures are required when it comes to fabricating a joint using split ring or shear plate connectors. The first involves milling the depression for the connector, which is usually done at the same time as drilling the bolt hole (see Figure E6-2 on the left). Next, the connector is placed in the corresponding depressions and the timber members to be connected are assembled. Finally, the bolts are slid into the holes and the nuts tightened (see Figure E6-2 on the right). If no bolts are usable, e.g. for connectors on the upper side of deep glulam beams, screws can also be used instead of bolts.

Figure E6-2 Left: Drilling the bolt hole and milling the depression for the connector. Right: Assembly of a joint with split ring connectors. (STEP 1995 Article C9)

Toothed-plate connectors as shown in Figure E6-3 are used similarly to the split ring and shear plate connectors in timber-to-timber and steel-to-timber joints, normally together with bolts. While split ring and shear plate connectors are inserted into pre-milled depressions in the timber, toothed-plate connectors are pressed into the pieces of timber to be connected. To avoid any gap due to the base plate (up to 3 mm) of the toothed-plate connectors between the timber members, the base plates can also be placed in pre-milled depressions in the wood. While double-sided toothed-plate connectors are only usable in timber-to-timber joints, single-sided toothed-plate connectors can also be used when connecting timber and steel. Like shear plate connectors, single-sided toothed-plate connectors are also used in timber-to-timber joints, e.g. if the connector is to be installed already prior to assembly or if the construction has to be assembled and disassembled multiple times. Given the need to press the "teeth" into the pieces of wood, which becomes increasingly difficult with rising density, toothed-plate connectors can only be installed in timber members with a characteristic density of less than around 500 kg/m^3.

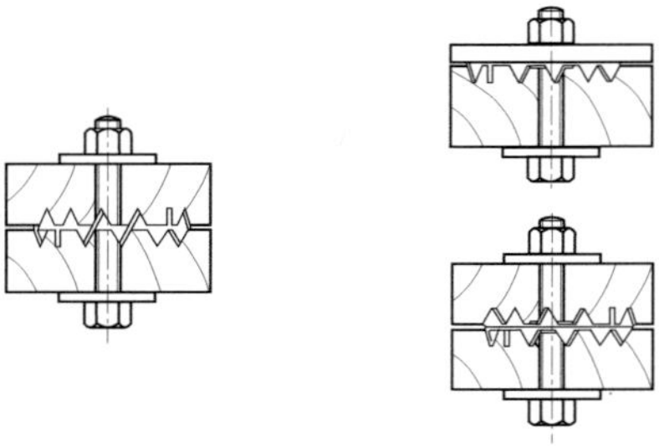

Figure E6-3 Joint with double-sided (left) and single-sided toothed-plated connector (right). (STEP 1995 Article C10)

Toothed-plate connectors come in many shapes and sizes, with diameters between 38 and 165 mm. Although they tend to be circular, square and oval shapes are also possible. Toothed-plate connectors comprise cold-formed steel strip, hot-dip galvanised steel or malleable cast iron. Most of the toothed-plate connectors used in Europe are also described in EN 912 "Timber fasteners – Specifications for connectors for timbers". EN 912 describes toothed-plate connectors as type C.

Joints with toothed-plate connectors are manufactured similarly to bolted joints. First of all, the holes for the bolts are drilled. The connectors are then placed between the members to be connected and pressed together, during which the teeth of the toothed-plate connectors engage with the wood. Since the pressing-in process requires considerable force, either a special high-strength bolt or a hydraulic press is used for this process. The bolts required for the joint can only be used for the task of pressing in for small connector diameters of up to around 65 mm. To avoid excessive indentations perpendicular to the grain, when pressing in with bolts, oversized washers should be used to distribute the load. After being pressed in, the final bolts are assembled and tightened. As an alternative to bolts, screws can also be used, as is done for joints with split ring and shear plate connectors.

For both connector types, cross-sectional weaknesses must be taken into consideration when verifying the timber members. Here, the NA specifies so-called shortfall areas for all common connector types used in Germany.

E6.1 Load-bearing behaviour

For double-sided connectors (split ring and double-sided toothed-plate connector), the force is initially transmitted from one member via embedment stresses into the connector and from there, via the shear resistance of the connector to the second member. The bolts accommodate the resulting eccentricity and keep the joint together. For single-sided connectors (shear plate and single-sided toothed-plate connector), meanwhile, forces are distributed slightly differently: once the force has been transmitted to the connector, the bolt is exposed to embedment stress between the connector and bolt. Subsequently, the shear resistance of the bolt means that for timber-to-timber joints, the force is transmitted from the bolt to the second connector, while for steel-to-timber joints, it is transmitted directly to the steel member. The hole diameter in single-sided connectors must accordingly correspond to the bolt diameter plus a small additional tolerance. This tolerance means initial slip for joints with single-sided connectors.

Split ring and shear plate connectors

When performing tensile tests on joints with split ring or shear plate connectors, the most common cause of failure observed is a shear block failure of the timber in front of the connector. The calculation model used to determine the load-bearing capacity of such connector joints thus assumes the occurrence of such failure. Although unevenly distributed in reality, embedment stresses are assumed to be uniformly distributed and to act in parallel to the direction of the force applied. These are then, in turn, transferred to the member via shear stresses (Figure E6-4 centre). The sudden shear failure means such joints may fail, even if only very minor deformations are involved, which also means interaction of the bolt and connector cannot be guaranteed. The load-bearing capacity of the bolt is disregarded, since the bolt only begins to carry load following an initial slip, due to clearance. Figure E6-5 shows a tensile connection after shear failure in both the middle and one side member.

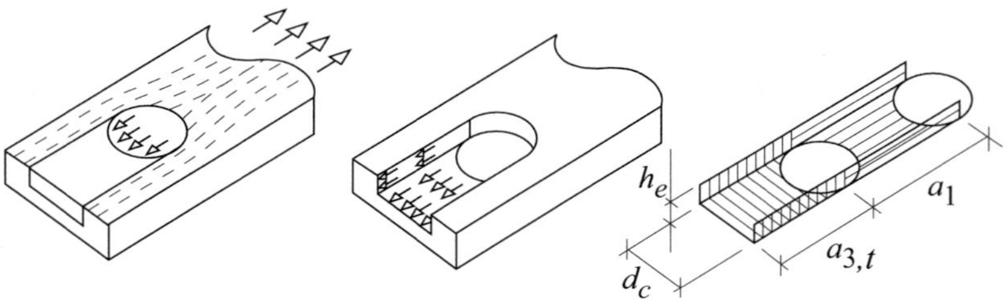

Figure E6-4 Stresses in a joint with a split ring or shear plate connector with associated shear areas. (STEP 1995 Article C9)

Figure E6-5 Shear failure in middle and side member of a joint with split rings subject to tension. (STEP 1995 Article C9)

Assuming that the shear failure of the wood is the key cause of failure for tensile joints, the load-bearing capacity depends on the shear area A_s in front of the connector and the shear strength of the wood itself. The shear area within the connector is not taken into consideration, since most tests showed that the wood core within the connector sheared off before the maximum load was reached, meaning it did not help boost the total load-bearing capacity in any way. The shear failure of the wood in front of the connector, however, only occurs when the wood has sufficient embedment strength. Accordingly, for larger distances to the loaded end $a_{3,t}$, embedment failure is the main factor dictating the load-bearing capacity of the joint (Figure E6-4 on the right).

If joints with connectors are loaded at an angle exceeding 30° or so to the grain direction or if loaded in compression, other failure mechanisms are triggered. Joints with angles between the force and grain directions of between around 30 and 150° tend to fail due to the wood splitting (Figure E6-6 on the left). Joints loaded in compression, meanwhile, exhibit combined embedment and splitting failure (Figure E6-6 on the right). In this case, however, the splitting only occurs following considerable embedment deformations under the connector and bolt. In a state of failure therefore, both connector and bolt transmit some of the force, although this interaction between connector and bolt only emerges in joints subject to compression. Compared to joints subject to tension or loads acting at an angle of less than 150°, which show brittle failure, greater plastic deformations occur in this case before the load peaks. Since the wood core within the connector of compressed joints is prone to shear failure before this point, the effective embedment area of the bolt is reduced by the area within the connector.

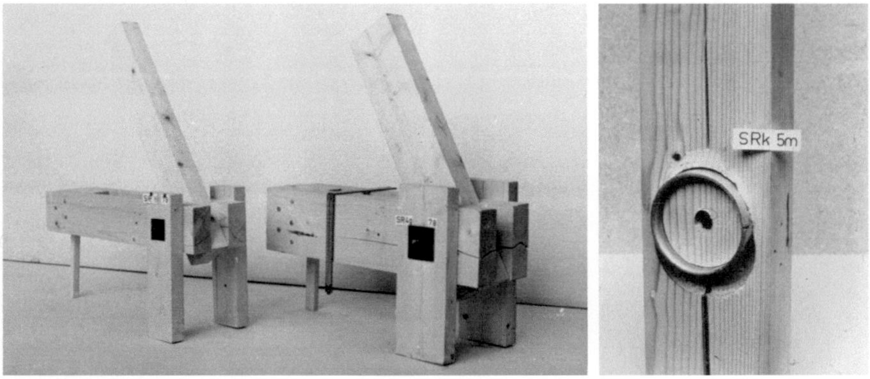

Figure E6-6 Left: Splitting failure in joints with a split ring loaded at an angle of 60° to the grain. (STEP 1995 Article C9). Right: Combined embedment-splitting failure in a joint with a split ring loaded in compression. (Photo: TU Delft, Vermeijden and Kuipers)

The load-bearing capacity $F_{v,0,R}$ of a joint with connectors subject to tensile load parallel to the grain is hence the smaller of the two values for resistance to shear and embedment respectively (definitions see also Figure E6-4 on the right):

$$F_{v,0,R} = \min \begin{cases} f_v \cdot A_s & \text{(shear)} \\ f_h \cdot d_c \cdot h_e & \text{(embedment)} \end{cases} \tag{E6-1}$$

where

$F_{v,0,R}$ Load-bearing capacity of a split ring or shear plate connector,

f_v Apparent or mean shear strength of the wood,

A_s Shear area per connector,

f_h Embedment strength of the wood,

d_c Connector diameter and

h_e Insertion depth of the connector in the wood.

As numerous tests conducted by Kuipers and Vermeyden (1964) have shown, the shear strength declines with increasing shear area, which explains the reference to 'apparent shear strength', as determined by many tests. However, as applies to all joints, certain minimum wood thicknesses are required, given the risk of splitting failure for smaller wood thicknesses, rather than shear or embedment failure, resulting in a lower load-bearing capacity (Scholten, 1944).

Although shear or embedment failure actually only affects joints loaded at an angle between the force and grain directions of up to around 30° and splitting failure, a different failure mechanism, occurs in joints loaded up to an angle of 150°, many tests showed that even for those joints loaded at an angle between the force and grain directions exceeding 30°, a calculation employing the approach of equation (E6-1), namely without taking splitting into consideration, would suffice. This can be explained by the fact that spacings and edge distances affect the load-bearing capacity in similar ways when various failure mechanisms occur. If splitting failure does occur when load is applied at an angle to the grain direction, increasing the distance to the loaded end results in the area of wood exposed to tension perpendicular to the grain expanding. The same applies when shear failure occurs in joints loaded parallel to the grain. Here, likewise, increasing the distance to the loaded end increases the size of the stressed areas in the wood, although shear stress in this case. Tests could show that the "higher load-bearing capacity due to greater end distance" was similar for both failure modes "tension perpendicular to the grain" and "shear". Accordingly, equation (E6-1) suffices as a means of verifying both failure modes. Only when the distance to the end is excessive and the maximum load is reached without the wood splitting or shearing off is no further increase in load-bearing capacity expected, even with increasing edge distance.

Toothed-plate connectors

In many tests with joints with toothed-plate connectors, the observed failure mode was embedment failure, under both connector teeth and bolts. The teeth were also bent in some types of toothed-plate connectors. Conversely, for tensile joints featuring short end distances, the failure modes involved splitting and shear failure of the wood. Since joints with toothed-plate connectors tend to undergo prominent plastic deformations before the maximum load is reached, interaction between connectors and bolts can be assumed. Figure E6-7 shows a joint after the maximum load has been reached, with both plastic embedment deformations of the wood under the connector teeth and bolt and plastic deformation of the bolt and connector teeth clearly apparent.

The load-bearing capacity of joints with toothed-plate connectors thus takes into consideration the interaction of connector and bolt:

$$F_{v,Rk} = F_{C,Rk} + F_{B,Rk} \tag{E6-2}$$

where

$F_{v,Rk}$ Characteristic load-bearing capacity of the joint with connector and bolt

$F_{C,Rk}$ Characteristic load-bearing capacity of the toothed-plate connector (type C)

$F_{B,Rk}$ Characteristic load-bearing capacity of the bolt in accordance with EC 5 with the characteristic values of embedment strength and yield moment of the bolt (Article E4)

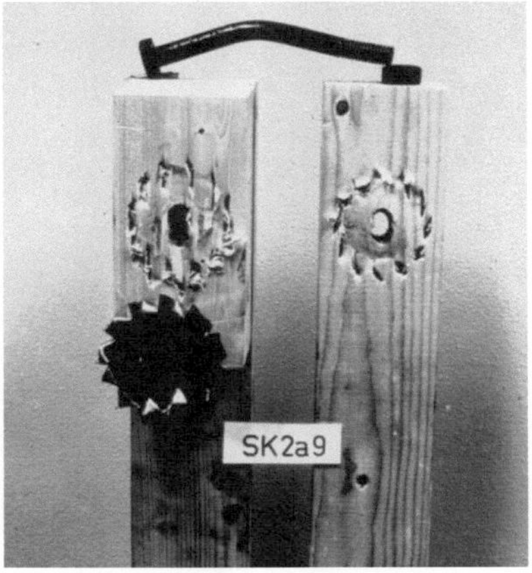

Figure E6-7 Embedment failure under the connector teeth and bolt. The teeth and bolt are deformed plastically. (STEP 1995 Article C10)

E6.2 Load-bearing capacity of split ring and shear plate connectors

The calculation model of equation (E6-1) was transposed via numerous tests to a design model, where, ensuring compliance with minimum distances and recommended minimum thicknesses, many different kinds of split ring and shear plate connectors, joint geometries, densities and angles between the force and grain directions can be calculated. To transpose equation (E6-1) to a design model, tests were used to determine regression functions for the apparent shear strength f_v and the embedment strength f_h:

$$\begin{aligned} f_{v,k} &= 20 \cdot A_s^{-0.25} \\ f_{h,k} &= 0.09 \cdot \rho \end{aligned} \quad \Rightarrow \quad F_{v,0,Rk} = \min \begin{cases} 20 \cdot A_s^{0.75} \\ 0.09 \cdot \rho \cdot d_c \cdot h_e \end{cases} \tag{E6-3}$$

As previously explained, shear strength decreases with increasing shear area (Kuipers and Vermeyden, 1964), and the embedment strength is determined as a function of density, since density is the key influencing parameter. The next step involves using Figure E6-4 to determine the shear area A_s, assuming $a_{3,t} = 2 \cdot d_c$ and $h_e = h_c/2 = d_c/8$ (h_c = connector height):

$$A_s = 2 \cdot a_{3,t} \cdot h_e + a_{3,t} \cdot d_c - \frac{1}{2} \cdot \frac{d_c^2}{4} \cdot \pi = 2 \cdot 2 \cdot d_c \cdot \frac{d_c}{8} + 2 \cdot d_c \cdot d_c - \frac{d_c^2}{8} \cdot \pi = 2.11 \cdot d_c^2 \tag{E6-4}$$

The reference density was $\rho_k = 350$ kg/m^3; which, together with equation (E6-4), results in:

$$F_{v,0,Rk} = \min \begin{cases} 35 \cdot d_c^{1.5} \\ 31.5 \cdot d_c \cdot h_e \end{cases} \tag{E6-5}$$

Equation (E6-5) shows the design model, although it only applies for the checked configuration. Correction coefficients are also required, which take the influence of differing densities into consideration (other than the reference density 350 kg/m^3), varying wood thicknesses and the influence of the distance to the loaded end $a_{3,t}$. The final design equation meanwhile, regulated in EC 5 for split ring and shear plate connectors, reads as follows:

$$F_{v,0,Rk} = \min \begin{cases} k_1 \cdot k_2 \cdot k_3 \cdot k_4 \cdot 35 \cdot d_c^{1.5} \\ k_1 \cdot k_3 \cdot 31.5 \cdot d_c \cdot h_e \end{cases} \tag{E6-6}$$

k_1 is the correction coefficient for varying wood thicknesses t_1 and t_2 and takes into consideration the unfavourable effect of thinner side or middle members on load-bearing capacity:

$$k_1 = \min\left(1; \frac{t_1}{3 \cdot h_e}; \frac{t_2}{5 \cdot h_e}\right) \tag{E6-7}$$

k_2 is the correction coefficient for the distance from the loaded end (where the above-mentioned assumption of $a_{3,t} = 2 \cdot d_c$ can be found back):

$$k_2 = \min\left(k_a; \frac{a_{3,t}}{2 \cdot d_c}\right) \tag{E6-8}$$

$k_a = 1.25$ for joints with one connector per shear plane and $k_a = 1.0$ for multiple connectors per shear plane. The angle α between the force and grain directions in this case peaks at 30°; for other values of α, $k_2 = 1.0$.

k_3 takes into consideration the increased load-bearing capacity for densities exceeding 350 kg/m^3, where characteristic densities over 610 kg/m^3 are no longer taken into consideration:

$$k_3 = \min\left(1.75; \frac{\rho_k}{350}\right) \tag{E6-9}$$

Finally, we have:

$$k_4 = \begin{cases} 1.0 & \text{for timber-to-timber joints} \\ 1.1 & \text{for steel-to-timber joints} \end{cases} \tag{E6-10}$$

When a load is exerted at an angle to the grain direction, the characteristic load-bearing capacity $F_{v,0,Rk}$ determined in accordance with equation (E6-6) must be reduced as follows (similar to determining the embedment strength of a bolted joint loaded at an angle to the grain, equation (E4-2)):

$$F_{v,\alpha,Rk} = \frac{F_{v,0,Rk}}{k_{90} \cdot \sin^2\alpha + \cos^2\alpha} \tag{E6-11}$$

where $k_{90} = 1.3 + 0.001 \cdot d_c$.

E6.3 Load-bearing capacity of toothed-plate connectors

Just as for those with split ring and shear plate connectors, the load-bearing capacity of joints with toothed-plate connectors was determined from numerous test results, which covered wide-ranging connector types, joint geometries, densities and angles between the force and grain directions, while complying with minimum distances and recommended minimum thicknesses. As equation (E6-2) shows, the load-bearing capacity of a joint with toothed-plate connectors is determined by adding the individual load-bearing capacities for connectors and bolts, since connectors and bolts work together in toothed-plate connectors, unlike split ring and shear plate connectors. The approach taken to determining the load-bearing capacity of the bolt was already explained in Article E4. Accordingly, reference will only be made at this point to determining $F_{C,Rk}$ of the toothed-plate connector as a portion of the total load-bearing capacity. When determining the load-bearing capacity of the connector, differences are established between two different types (note: connector types C6 to C9 are not normally used in Germany):

- Types C1 to C9 comprise plates, with edges, which are bent up into teeth.
- Types C10 and C11 comprise a plate with added conical teeth.

One difference emerging in comparison to split ring and shear plate connectors comes in how the connector diameter is defined. For toothed-plate connectors, the following applies (all dimensions in mm):

d_c: Diameter of connectors with teeth of types C1, C2, C6, C7, C10, C11

d_c: Side length of connectors with teeth of types C5, C8, C9

d_c: Root of the product of the side lengths of connectors with teeth of types C3, C4

Analogous to equation (E6-1), the calculation model of equation (E6-2) is transformed into a design model:

$$F_{C,Rk} = \begin{cases} k_1 \cdot k_2 \cdot k_3 \cdot 18 \cdot d_c^{1.5} & \text{for types C1 to C9} \\ k_1 \cdot k_2 \cdot k_3 \cdot 25 \cdot d_c^{1.5} & \text{for types C10 to C11} \end{cases} \tag{E6-12}$$

Coefficients k_1 to k_3 are very similar to those of the design equations for split ring and shear plate connectors and have the same significance. Coefficient k_1 is identical:

$$k_1 = \min\left(1; \frac{t_1}{3 \cdot h_e}; \frac{t_2}{5 \cdot h_e}\right) \tag{E6-13}$$

Coefficient k_2 differs for the two connector types, C1 to C9 or C10/C11. For connectors of types C1 to C9, meanwhile, the following applies:

$$k_2 = \min\left(1; \frac{a_{3,t}}{1.5 \cdot d_c}\right) \tag{E6-14}$$

where for $a_{3,t}$, the following value is taken with d = bolt diameter in mm:

$$a_{3,t} = \max\left(1.1 \cdot d_c; 7 \cdot d; 80\ \text{mm}\right) \tag{E6-15}$$

For toothed-plate connectors of types C10 and C11, the following applies for k_2 and $a_{3,t}$:

$$k_2 = \min\left(1; \frac{a_{3,t}}{2 \cdot d_c}\right) \quad \text{with} \quad a_{3,t} = \max\left(1.5 \cdot d_c; 7 \cdot d; 80\ \text{mm}\right) \tag{E6-16}$$

For coefficient k_3, only the upper limit of density changes in comparison to the split ring and shear plate connectors (610 kg/m^3 reduced to 525 kg/m^3):

$$k_3 = \min\left(1.5; \frac{\rho_k}{350}\right) \tag{E6-17}$$

E6.4 Minimum distances and spacings

As with all joint types, minimum distances must also be ensured for joints with connectors, which means the design equations are valid and no unwanted premature splitting takes place. The minimum spacings and distances to be maintained for split ring and shear plate connectors are specified in Table E6-1 (for definitions of the minimum distances, see Figure E1-9). Minimum distances for (toothed-plate) connectors of types C1 to C9 are specified in Table E6-2 and those for types C10 and C11 in Table E6-3. Notwithstanding the above, the corresponding minimum distances for the bolts must be observed in accordance with EC 5 (Table E4-1). EC 5 Section 10.4.3 also sets out requirements for bolt diameters in combination with connectors.

Table E6-1 Minimum distances of split ring and shear plate connectors, d_c = connector diameter in mm, $0° \leq \alpha \leq 90°$ = angle between force and grain directions. See also EC 1995-1-1/A2:2014 Table 8.7.

Distances and spacings	Angle	Minimum values
Spacing a_1 (parallel to the grain)	-	$(1.2 + 0.8 \cdot \cos\alpha) \cdot d_c$
Spacing a_2 (perpendicular to the grain)	-	$1.2 \cdot d_c$
Distance $a_{3,t}$ (loaded end)	-	$2.0 \cdot d_c$
Distance $a_{3,c}$ (unloaded end)	$\alpha < 30°$ $30° \leq \alpha \leq 90°$	$1.2 \cdot d_c$ $(0.4 + 1.6 \cdot \sin\alpha) \cdot d_c$
Distance $a_{4,t}$ (loaded edge)	-	$(0.6 + 0.2 \cdot \sin\alpha) \cdot d_c$
Distance $a_{4,c}$ (unloaded edge)	-	$0.6 \cdot d_c$

Table E6-2 Minimum distances of toothed-plate connectors of types C1 to C9, d_c = connector diameter in mm, $0° \leq \alpha \leq 90°$ = angle between force and grain directions. EC 1995-1-1/A2:2014 Table 8.8.

Distances and spacings	Angle	Minimum values
Spacing a_1 (parallel to the grain)	-	$(1.2 + 0.3 \cdot \cos\alpha) \cdot d_c$
Spacing a_2 (perpendicular to the grain)	-	$1.2 \cdot d_c$
Distance $a_{3,t}$ (loaded end)	-	$1.5 \cdot d_c$
Distance $a_{3,c}$ (unloaded end)	$\alpha < 30°$ $30° \leq \alpha \leq 90°$	$1.2 \cdot d_c$ $(0.9 + 0.6 \cdot \sin\alpha) \cdot d_c$
Distance $a_{4,t}$ (loaded edge)	-	$(0.6 + 0.2 \cdot \sin\alpha) \cdot d_c$
Distance $a_{4,c}$ (unloaded edge)	-	$0.6 \cdot d_c$

Table E6-3 Minimum distances of toothed-plate connectors of types C10 and C11, d_c = connector diameter in mm, $0° \leq \alpha \leq 90°$ = angle between force and grain directions. See EC 1995-1-1:2010 Table 8.9.

Distances and spacings	Angle	Minimum values
Spacing a_1 (parallel to the grain)	-	$(1.2 + 0.8 \cdot \cos\alpha) \cdot d_c$
Spacing a_2 (perpendicular to the grain)	-	$1.2 \cdot d_c$
Distance $a_{3,t}$ (loaded end)	-	$2.0 \cdot d_c$
Distance $a_{3,c}$ (unloaded end)	$\alpha < 30°$ $30° \leq \alpha \leq 90°$	$1.2 \cdot d_c$ $(0.4 + 1.6 \cdot \sin\alpha) \cdot d_c$
Distance $a_{4,t}$ (loaded edge)	-	$(0.6 + 0.2 \cdot \sin\alpha) \cdot d_c$
Distance $a_{4,c}$ (unloaded edge)	-	$0.6 \cdot d_c$

E6.5 Effective number of connectors

Similar to joints with dowel-type fasteners, for joints with connectors, an effective number n_{ef} also applies when more than two connectors are arranged in a row parallel to the grain and force:

$$n_{ef} = 2 + \left(1 - \frac{n}{20}\right) \cdot (n-2) \tag{E6-18}$$

in which n is the number of connectors arranged in a row parallel to the grain and force.

Joints with toothed-plate connectors are subject to ductile failure when complying with the minimum distances, which is why calculations for this connector type should take the actual number of fasteners ($n_{ef} = n$) into account.

E6.6 Joint stiffness

Stiffness values for mechanical joints are required to verify serviceability, as well as to calculate the load-bearing capacity of mechanically jointed beams. The ultimate slip modulus K_u used to verify load-bearing capacity includes two-thirds of the corresponding value of K_{ser}. Given that the stiffness values of tested joints cover wide-ranging limits, it is almost impossible to reliably estimate the influence of different variables on the deformation behaviour. Accordingly, a simple relationship was selected instead, showing the slip modulus as a function of the connector diameter and the mean density of the timber. The influence of the angle between the force and grain directions, the wood moisture content, wood thickness and the number of connectors per shear plane was disregarded. The following equations apply for determining K_{ser} in N/mm. Split ring and shear plate connectors and toothed-plate connectors of types C10 and C11:

$$K_{ser} = \frac{d_c \cdot \rho_m}{2} \tag{E6-19}$$

Toothed-plate connectors of types C1 to C9:

$$K_{ser} = 1.5 \cdot \frac{d_c \cdot \rho_m}{4} \tag{E6-20}$$

where d_c is the connector diameter in mm and ρ_m is the mean density of the wood in kg/m^3.

E6.7 Literature

H.J. Blass, original Articles C9, C10, STEP 1995.

Kuipers J. and Vermeyden P. (1964). Research on timber joints in the Netherlands. Rapport 4-64-15, Onderzoek v-7, Stevin-Laboratorium, Technische Hogeschool Delft.

Scholten J.A. (1944). Timber-connector joints – their strength and design. Technical Bulletin No. 865, USDA Forest Service, Washington, D.C..

E7 Punched metal plate fasteners

Original article: L. R. J. Whale

Punched metal plate fasteners are defined in EN 1075 "Timber structures – Test methods – Joints made with punched metal plate fasteners" as a fastener comprising a metal plate with teeth ("nails") punched in one direction, which stick out from the plate at right angles. With punched metal plate fasteners, two or more timber members of equivalent thickness can be connected, whereby at least one punched metal plate fastener must be attached to each member side (symmetrical joint). A typical punched metal plate fastener is shown in Figure E7-1. Punched metal plate fasteners are generally produced from galvanised or stainless steel plates, between 0.9 and 3.0 mm thick. Plates with punched-out integral "nails" were originally used in the USA in the late 50s, after refining conventional hand-nailed steel or plywood gusset plates. Both systems proved suitable for manufacturing joints in the same plane, but the punched metal plate fasteners were better suited when it came to prefabricating trusses in the factory. They could transfer forces via smaller joint areas, which also helped save on material.

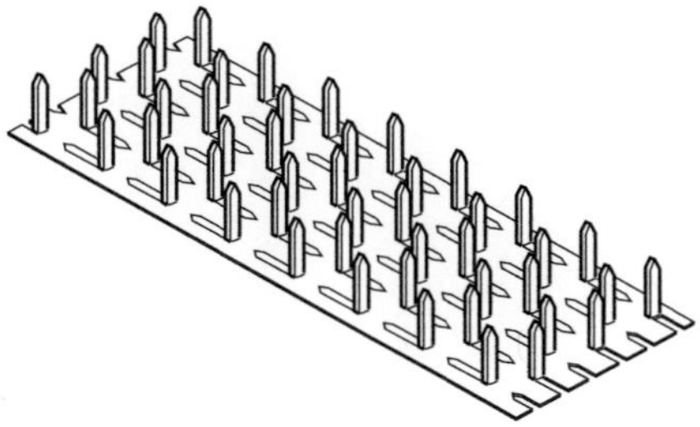

Figure E7-1 Typical punched metal plate fastener. (STEP 1995 Article C11)

Nowadays, punched metal plate fasteners are in widespread use for truss joints and joints in many other plane timber structures (Figure E7-2 (a)), which means that many different punched metal plate fasteners encompassing wide-ranging "nail" arrangements, lengths and shapes have been developed. However, the factors governing the load-bearing capacity of these different punched metal plate fasteners are all the same and this enables uniform design, as was accommodated in EC 5 for these fastener types.

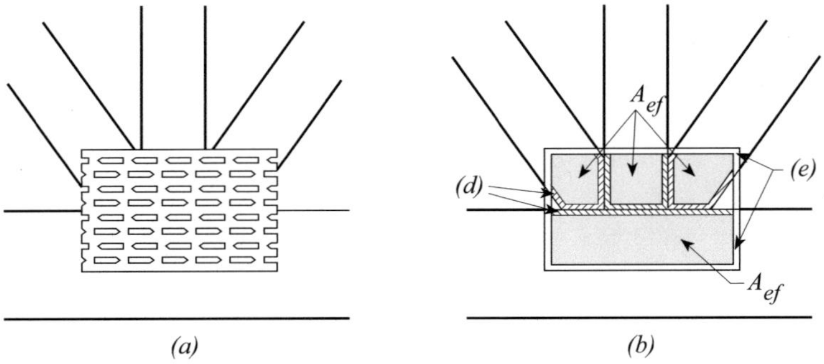

Figure E7-2 Typical joint: (*a*) view and (*b*) effective joint areas. (STEP 1995 Article C11)

E7.1 Parameters influencing the load-bearing capacity of connections with punched metal plate fasteners

In a connection with punched metal plate fasteners, the force is initially transferred from the timber member to the nails, then from the nails to the steel plate, via the joint line and finally, in turn, via the nails to the member on the other side of the joint. The load-bearing capacity of a connection with punched metal plate fasteners is thus reached when one of the following two criteria apply: either the load-bearing capacity of the nails (anchorage) in one of the connected members is attained, or the load-bearing capacity of the net cross-section of the steel plate in one of the joint lines between the members is crucial. The parameters influencing both these load-bearing capacity criteria can be summarised as follows (definitions see Figure E7-3).

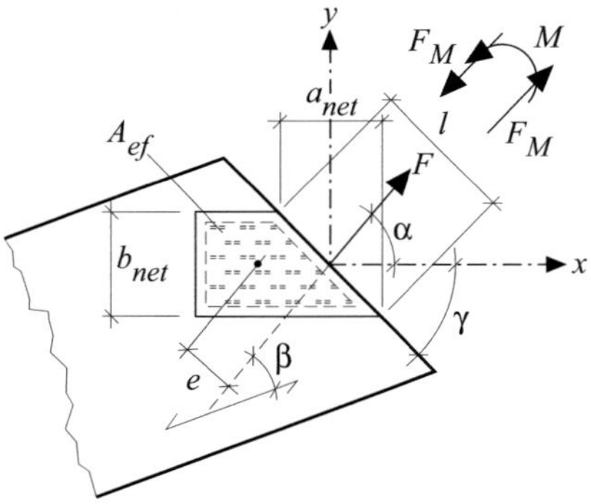

Figure E7-3 Connection with punched metal plate fastener – geometry and loading. (STEP 1995 Article C11)

Parameters influencing the load-bearing capacity of the nails (anchorage)

- α is the angle between the force direction and the main direction of the punched metal plate fastener (the main direction is that where the tensile capacity of the plate peaks, often parallel to the direction of the punched-out nails) and is generally the angle at which the individual nails are generally loaded. On the one hand, this influences the embedment area in the timber, which is stressed by a nail, secondly, the bending resistance of the nail itself, which often shows a rectangular cross-section.
- β is the angle between the force and grain directions; namely the angle to the wood grain direction, under which the nails exert stress on the wood.
- The wood species or strength class of the wood in the joint, namely its resistance against embedment or withdrawal stress as exerted via the punched-out nails.
- A_{ef} is the effective area of the punched metal plate fastener in each member, namely the contact area between the punched metal plate fastener and wood, taking tolerances for the positioning of the punched metal plate fastener into consideration. The effective joint area is defined as the smallest area that can be assumed, given that the plate is shifted arbitrarily from its correct position by a specific tolerance (± 5 mm to the timber edges and 6 x nominal plate thickness to the end grain) (see also Figure E7-2 (b)).
- r_{max} is the distance between the centre of gravity of the punched metal plate fastener and the point of A_{ef} situated furthest from the centre of gravity. I_p is the polar moment of inertia of the effective area A_{ef} about its centre of gravity. What must be noted is that r_{max} and I_p only have an influence, when moments are to be transferred over the punched metal plate fastener.

Parameters influencing the load-bearing capacity of the steel plate (steel capacity)

- γ is the angle between the main plate direction and the direction of the joint line between the timber members (this angle influences the net cross-section of the steel along the joint line of the connection, see Figure E7-3).
- ℓ is the length of the joint line covered by the plate. The corresponding projected lengths, parallel and perpendicular to the main plate direction, transfer the stresses in both orthogonal plate directions in the joint line between the timber members.
- The steel grade, from which the punched metal plate fastener was manufactured, namely the strength properties of the steel.

These parameters are included in design equations, with which the load-bearing capacity of joints can be determined depending on the specific strength properties of the punched metal plate fastener.

E7.2 Determining characteristic values for punched metal plate fasteners based on test results

According to EC 5, the following characteristic strength values are required for the design of connections with punched metal plate fasteners:

$f_{a,\alpha,\beta,k}$ is the characteristic nail load-bearing capacity per unit of area under angles α and β; requiring $f_{a,0,0,k}$ = nail load-bearing capacity for $\alpha = 0°$ and $\beta = 0°$ and $f_{a,90,90,k}$ = nail load-bearing capacity for $\alpha = 90°$ and $\beta = 90°$.

$f_{t,0,k}$ is the characteristic plate tensile capacity per length unit in the main plate direction ($\alpha = 0°$).

$f_{t,90,k}$ is the characteristic plate tensile capacity per length unit perpendicular to the main plate direction ($\alpha = 90°$).

$f_{c,0,k}$ is the characteristic plate compressive capacity per length unit in the main plate direction ($\alpha = 0°$).

$f_{c,90,k}$ is the characteristic plate compressive capacity per length unit perpendicular to the main plate direction ($\alpha = 90°$).

$f_{v,0,k}$ is the characteristic plate shear capacity per length unit in the main plate direction ($\alpha = 0°$).

$f_{v,90,k}$ is the characteristic plate shear capacity per length unit perpendicular to the main plate direction ($\alpha = 90°$).

Each of these properties is to be determined from tests described in EN 1075. The 5%-quantiles are converted into design values by multiplying by the modification factor k_{mod} and dividing by the partial safety factor of the material γ_M. The corresponding modification factors for timber are to be used to determine the nail load-bearing capacity, while for the plate load-bearing capacities, the coefficients $k_{mod} = 1.0$ and $\gamma_M = 1.25$ can be assumed. The nail load-bearing capacity $f_{a,\alpha,\beta,k}$ is to be determined over the entire range of α- and β-values. If a sufficient number of values in the $f_{a,\alpha,\beta,k}$-plane have been determined, interpolation between them is possible.

The **characteristic nail load-bearing capacity per plate in the grain direction** ($\beta = 0°$) $f_{a,\alpha,0,k}$ is obtained from tests on joints (Figure E7-4) with plate angles $\alpha = 0°$, 30°, 60° and 90°. A bilinear function can be adapted to these data (Figure E7-5), eliciting the coefficients k_1, k_2 and α_0, which are then used in the following equations:

$$f_{a,\alpha,0,k} = \begin{cases} f_{a,0,0,k} + k_1 \cdot \alpha & \text{for} \quad \alpha \leq \alpha_0 \\ f_{a,0,0,k} + k_1 \cdot \alpha + k_2 \cdot (\alpha - \alpha_0) & \text{for} \quad \alpha_0 \leq \alpha \leq 90° \end{cases} \qquad \text{(E7-1)}$$

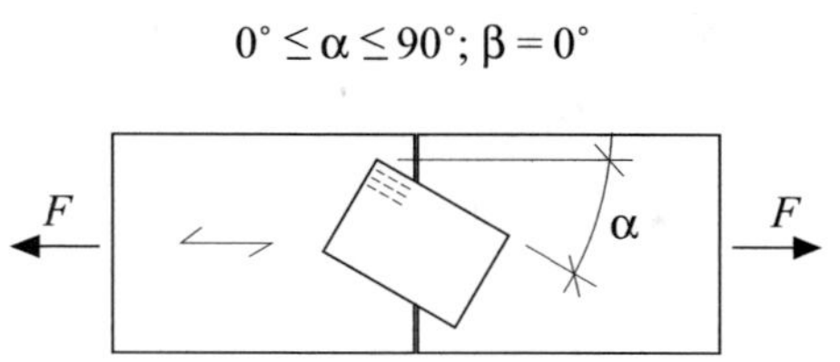

Figure E7-4 Standard test specimen $\beta = 0°$. (STEP 1995 Article C11)

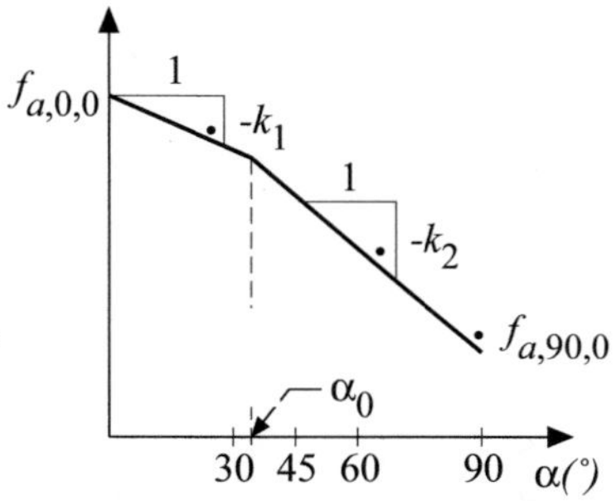

Figure E7-5 Derivation of coefficients k_1, k_2, α_0. (STEP 1995 Article C11)

The **characteristic nail load-bearing capacity per plate in the main plate direction** ($\alpha = 0°$) $f_{a,0,\beta,k}$ is obtained from tests with T joints (Figure E7-6 on the left) with grain angles $\beta = 45°$ and 90° ($\beta = 0°$ was already checked with the test setup of Figure E7-4). The test results reveal the interaction between the nail load-bearing capacity and β, Figure E7-6 on the right. For this diagram, test specimens with additional angles β were checked.

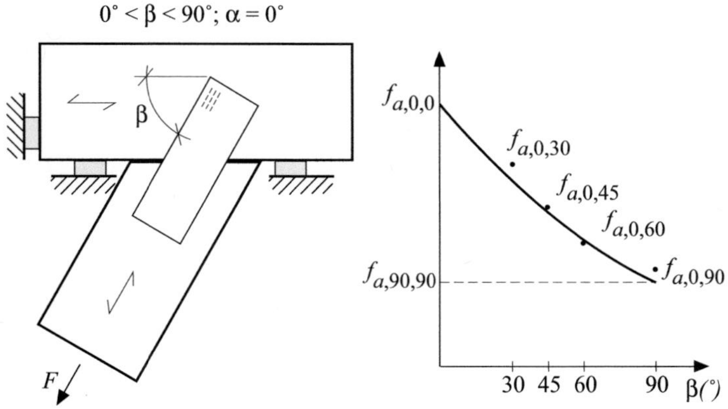

Figure E7-6 Standard test specimen $\alpha = 0°$. (STEP 1995 Article C11)

Once curves have been fitted, which represent the lower limits of the relationship between $f_{a,0,0}$ and $f_{a,90,0}$ (bilinear curve, Figure E7-5) for straight joints and between $f_{a,0,0}$ and $f_{a,0,90}$ or $f_{a,90,90}$ (sinusoidal curve, Figure E7-6 on the right) for T-joints, EC 5 specifies an interpolation method for arbitrary values $f_{a,\alpha,\beta,k}$ between these extreme values:

$$f_{a,\alpha,0,k} = \begin{cases} \max\begin{cases} f_{a,\alpha,0,k} + \left(f_{a,\alpha,0,k} - f_{a,90,90,k}\right)\cdot\dfrac{\beta}{45^\circ} \\ f_{a,0,0,k} + \left(f_{a,0,0,k} - f_{a,90,90,k}\right)\cdot\sin\left[\max\left(\alpha,\beta\right)\right] \end{cases} & \text{for} \quad \beta \leq 45^\circ \\ f_{a,0,0,k} + \left(f_{a,0,0,k} - f_{a,90,90,k}\right)\cdot\sin\left[\max\left(\alpha,\beta\right)\right] & \text{for} \quad 45^\circ \leq \beta \leq 90^\circ \end{cases} \tag{E7-2}$$

The characteristic course of the fitted area $f_{a,\alpha,\beta,d}$ in accordance with equation (E7-2) is compared with test data of a typical punched metal plate fastener in Figure E7-7.

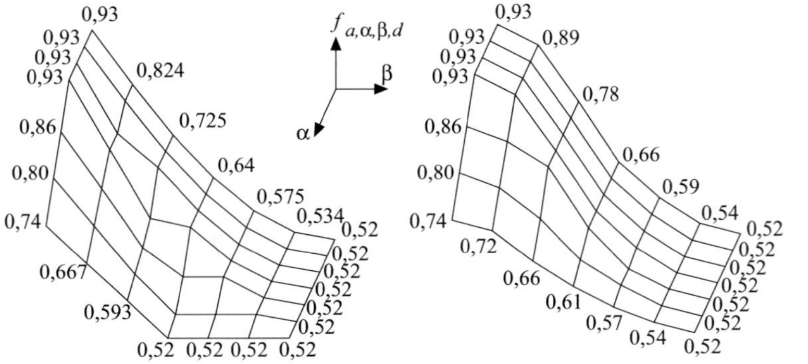

(a) (b)

Figure E7-7 Typical areas $f_{a,a,\beta,d}$: (a) theoretically and (b) experimentally determined. (STEP 1995 Article C11)

E7.3 Design of connections with punched metal plate fasteners

Nail load-bearing capacity (anchorage)

Induced stresses $\tau_{F,d}$ or $\tau_{M,d}$ from forces F_{Ed} and moments M_{Ed}, which act on the punched metal plate area, can be calculated as follows:

$$\tau_{F,d} = \frac{F_{A,Ed}}{A_{ef}} \tag{E7-3}$$

$$\tau_{M,d} = \frac{M_{A,Ed}}{W_p} \tag{E7-4}$$

where

$F_{A,Ed}$ Design value of the force acting on a punched metal plate fastener (corresponds to half the force in the member) in the centre of gravity of the effective joint area A_{ef} (A_{ef} defined in Section E7.1).

$M_{A,Ed}$ Design value of the moment acting on an effective joint area.

W_p Section modulus with r = distance from the centre of gravity of the punched metal plate fastener:

$$W_p = \int_{A_{ef}} r \cdot \mathrm{d}A$$

The section modulus W_p can also be calculated with a simplified, conservative assumption, without the integral of the effective joint area A_{ef} having to be assessed:

$$W_p = \frac{A_{ef} \cdot \sqrt{\left(\frac{A_{ef}}{h_{ef}}\right)^2 + h_{ef}^2}}{4}$$

where h_{ef} = the largest height of the effective joint area perpendicular to the longest side.

In joints subject to a net compressive force, only 50% of the force needs to be transferred through the plate, meaning that the design value $F_{a,Ed}$ may be halved. The remainder is transferred by direct timber bearing. Such direct timber bearing via contact may be taken into consideration, provided the average joint gap between the members does not exceed 1.5 mm.

The following ultimate limit state condition of the nail load-bearing capacity must be complied with for each joint area:

$$\left(\frac{\tau_{F,d}}{f_{a,\alpha,\beta,d}}\right)^2 + \left(\frac{\tau_{M,d}}{f_{a,0,0,d}}\right)^2 \leq 1 \tag{E7-5}$$

Plate load-bearing capacity (steel capacity)

To verify the plate load-bearing capacity in a joint line of a connection, in accordance with EC 5, all forces and plate load-bearing capacities are resolved into both orthogonal main directions *x* and *y* (Figure E7-3):

$$F_{x,Ed} = F_{Ed} \cdot \cos\alpha \pm 2 \cdot F_{M,Ed} \cdot \sin\gamma \tag{E7-6}$$

$$F_{y,Ed} = F_{Ed} \cdot \sin\alpha \pm 2 \cdot F_{M,Ed} \cdot \cos\gamma \tag{E7-7}$$

where

F_{Ed} Resulting force of an individual punched metal plate fastener in the shear plane (compression = negative).

$F_{M,Ed}$ Force caused by the moment in the shear plane where $F_{M,Ed} = 2 \cdot M_{A,Ed}/\ell$ (Figure E7-3).

The following ultimate limit state condition of the plate load-bearing capacity must be complied with in each joint area / shear plane:

$$\left(\frac{F_{x,Ed}}{F_{x,Rd}}\right)^2 + \left(\frac{F_{y,Ed}}{F_{y,Rd}}\right)^2 \leq 1 \tag{E7-8}$$

where

$F_{x,Ed}$, $F_{y,Ed}$ Design values of the forces in *x*- and *y*-directions, Equations (E7-6) and (E7-7)

$F_{x,Rd}$, $F_{y,Rd}$ Design values of the plate load-bearing capacities

The characteristic values of the plate load-bearing capacities in both these directions are determined as the maximum tensile and shear strengths of the plate as follows (here, only the equations for tensile forces $F_{x,Ed} > 0$ and $F_{y,Ed} > 0$ are given, whereas for compressive forces, the design values are determined in a similar manner):

$$F_{x,Rk} = \max \begin{cases} \left| f_{t,0,k} \cdot \ell \cdot \sin(\gamma - \gamma_0 \cdot \sin(2 \cdot \gamma)) \right| \\ \left| f_{v,0,k} \cdot \ell \cdot \cos\gamma \right| \end{cases} \tag{E7-9}$$

$$F_{y,Rk} = \max \begin{cases} \left| f_{t,90,k} \cdot \ell \cdot \cos\gamma \right| \\ k \cdot f_{v,90,k} \cdot \ell \cdot \sin\gamma \end{cases} \tag{E7-10}$$

Coefficient k in equation (E7-10) takes into consideration the influence of tensile or compressive force in the x-direction of the plate on the shear strength of the plate in the y-direction:

$$k = \begin{cases} 1 + k_{\mathrm{v}} \cdot \sin(2 \cdot \gamma) & \text{for} \quad F_{\mathrm{x,Ed}} > 0 \\ 1 & \text{for} \quad F_{\mathrm{x,Ed}} \leq 0 \end{cases} \tag{E7-11}$$

The constants γ_0 (equation (E1-9)) and k_{v} are to be determined from tests (similar to the coefficients k_1, k_2 and α_0) and can be taken from the technical documents for the plate types used.

E7.4 Additional rules

In addition to the abovementioned design rules, Section 10 of EC 5 also includes rules for trusses made with punched metal plate fasteners, concerning limit values for lateral deflections, distortions, curvatures or oblique positions during the assembly. The NA also requires verification of the transport and assembly conditions.

In addition, there is also a need to meet the requirements of EN 14250 "Timber structures – Product requirements for prefabricated structural members assembled with punched metal plate fasteners". As well as formulating requirements for materials and members (for example fire behaviour), information on the CE mark is also provided. In addition to this European regulation, there is also a national draft, E DIN 20000-6 "Application of building products in construction works – Part 6: Dowel-type and non dowel-type fasteners", describing further requirements, which are not regulated in the European standards. Among other things, for punched metal plate fasteners deployed in members with an overall length of > 12 m, the partial safety factors for the plate characteristics must be increased by a factor of 1.5. The reason for this regulation is that the new European standard includes less stringent requirements, which do not suffice to comply with the national standard to date.

Values for the slip modulus K_{ser} per mm^2 of the joint area can be taken from the technical documents for the punched metal plate fasteners.

E7.5 Literature

L.R.J. Whale, original Article C11, STEP 1995.

E8 Cold-formed steel connectors

Original article: E. Gehri

Cold-formed steel connectors such as joist hangers or joist anchors have largely replaced carpentry joints. This is primarily due to the ease of assembly of the steel connectors and the scope to eliminate complex machining of the timber elements to be connected. Figure E8-1 shows examples of timber joints with cold-formed steel connectors. The steel is generally between 1 and 4 mm thick and either galvanised or made of stainless steel. The joint between connector and timber is usually via ringed shank nails or self-tapping screws without any predrilling of the wood. The fact that the fastener holes have been pre-punched into the connectors guarantees swift and easy assembly at the construction site. For concealed beam hangers however, namely in which joists are connected via dowels, the joists must be split and drilled beforehand.

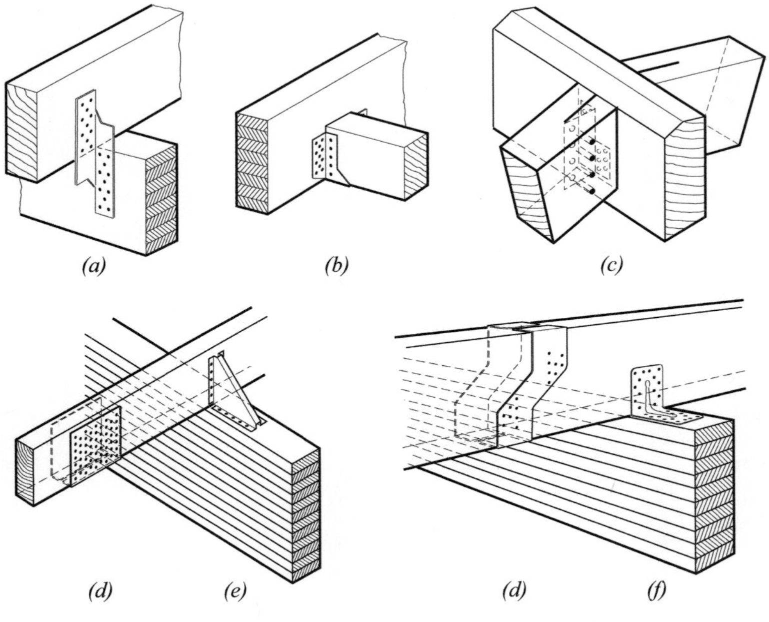

Figure E8-1 Cold-formed steel connectors. (*a*) Joist anchor, (*b*) joist hanger, (*c*) concealed beam hanger, (*d*) hinge connector, (*e*) cleat, (*f*) angle bracket. (STEP 1995 Article C13)

E8.1 Load-bearing behaviour

The load-bearing capacity of the joints using cold-formed steel connectors is not only influenced by the capacity of the steel-to-timber joint, but also by the capacity of the timber members and steel. In particular, tensile stresses perpendicular to the grain emerging in the timber beams may trigger failures, even before the load-bearing capacity of the nailed or screwed joint has been reached. Examples of joints in which tension failures perpendicular to the grain may occur include joist anchors and joist hangers as shown in Figure E8-1. Joints subject to tension perpendicular to the grain can be designed using one of the approaches outlined in Article E11.

For most cold-formed steel connectors, steel failure in the net cross-section is not the decisive criterion. This is due to limiting the number of pre-punched nail holes in the steel and hence limiting the force introduced by the fasteners. However, most joints including cold-formed steel connectors are subject to plastic deformations of the steel before the load-bearing capacity is actually attained.

Since many cold-formed steel connectors include at least two steel-to-timber joints in different planes, these joints tend to be eccentrically loaded, which means the fasteners concerned are often subject to combined lateral and axial loads.

Load-bearing capacity of the nailed steel-to-timber joint

In accordance with EC 5, the design value of the load-bearing capacity $F_{v,Rk}$ per nail for single-shear joints with thin steel plates (namely for $t \leq 0.5 \cdot d$ where t is the steel plate thickness) is the lower of the values obtained from the following equation (see also Article E2):

$$F_{v,Rk} = \min \begin{cases} 0.4 \cdot f_{h,k} \cdot t_1 \cdot d \\ 1.15 \cdot \sqrt{2 \cdot M_{y,Rk} \cdot f_{h,k} \cdot d} + \frac{F_{ax,Rk}}{4} \end{cases} \qquad \text{(E8-1)}$$

For joints with thick steel plates (namely for $t \geq d$), meanwhile, the design value of the load-bearing capacity can be obtained from the following equation:

$$F_{v,Rk} = \min \begin{cases} f_{h,k} \cdot t_1 \cdot d \\ f_{h,k} \cdot t_1 \cdot d \cdot \left[\sqrt{2 + \frac{4 \cdot M_{y,Rk}}{f_{h,k} \cdot d \cdot t_1^2}} - 1 \right] + \frac{F_{ax,Rk}}{4} \\ 2.3 \cdot \sqrt{M_{y,Rk} \cdot f_{h,k} \cdot d} + \frac{F_{ax,Rk}}{4} \end{cases} \qquad \text{(E8-2)}$$

For steel plate thicknesses of between $0.5 \cdot d$ and d, linear interpolation is permissible and any variation in load-bearing capacity according to equations (E8-1) or (E8-2) is based on the clamping effect of the nail in the steel plate. Tests with nailed steel-to-timber joints (Ehlbeck and Görlacher, 1982) showed, however, that also for steel plates of thickness $t = 2.0$ mm and ringed shank nails of diameter $d = 4.0$ mm, the steel plate acted as a clamped restraint for the nail, if the nails were conically shaped under the head (Figure E8-2) and driven into tight fitting holes in the steel plates. In these cases, the load-bearing capacities in accordance with EC 5 for thick steel plates were consistently attained.

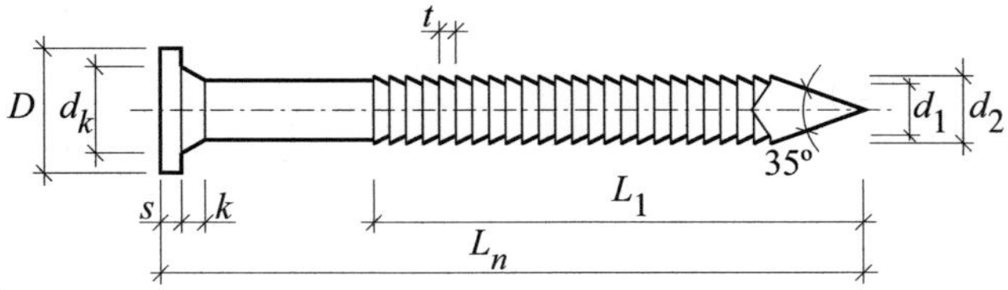

Figure E8-2 Ringed shank nails for steel-to-timber joints with a conical shank under the head. (STEP 1995 Article C13)

The characteristic value of the embedment strength $f_{h,k}$ depends on the nail diameter d in mm and the characteristic timber density ρ_k in kg/m^3 and is for non-predrilled timber (equation (E8-3) corresponds to equation (E3-1)):

$$f_{h,k} = 0.082 \cdot \rho_k \cdot d^{-0.3} \tag{E8-3}$$

Given the varying cross-section in the non-smooth section of the nail shank and the work hardening during manufacture, the characteristic value of the nail yield moment must be determined by tests. Calculating the yield moment from the plastic section modulus and the tensile strength of the nail wire is actually not applicable for ringed shank nails. European technical assessments (ETA) or declarations of performance (DOP) for ringed shank nails or screws reveal information about the yield moments to be applied or the characteristic values of load-bearing capacities $F_{v,Rk}$ (laterally loaded fasteners) and $F_{ax,Rk}$ (axially loaded fasteners) are already specified per fastener and shear plane. The values specified there are normally those derived for typical characteristic densities and must be reduced when using wood with a density lower than the reference density. The characteristic values of the load-bearing capacity of nailed joints in accordance with EC 5 are defined as for all fasteners under the prerequisite specific minimum nail distances. Since the holes in the steel are pre-punched, fastener distances are given. When designing joints with cold-formed steel connectors, however, the minimum distances must be verified.

E8.2 Joist hangers

Joist hangers are frequently used as supports for solid timber and glulam timber beams and are manufactured in wide-ranging shapes and dimensions. Figure E8-3 shows an example of a joist hanger for a timber-to-timber connection.

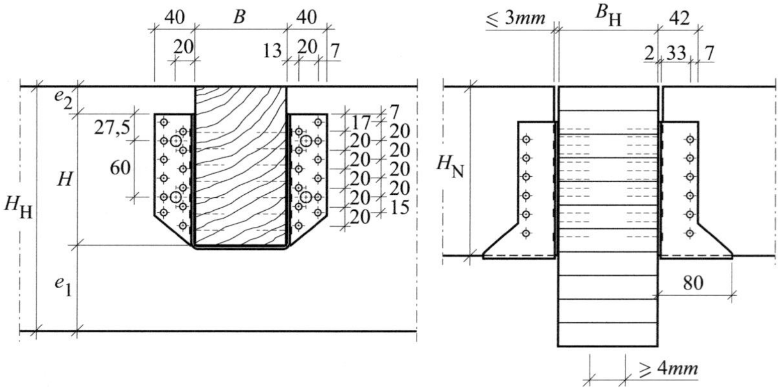

Figure E8-3 Joist hanger for a header-joist connection. (STEP 1995 Article C13)

When the joist hanger is exposed to a uniaxial load, the load acts on its plane of symmetry. The joist transmits the force via contact pressure into the base plate of the joist hanger and simultaneously via the steel-to-timber connection into the joist hanger and then, in turn, via the second steel-to-timber connection into the header. For joist hangers with relatively few fasteners in the joist connection, most of the load is transmitted via contact with the base plate. When designing a joist hanger, however, it is safe to assume that the force will act at the centre of gravity of the joist connection. However, if the transmission of force is mainly by contact with the base plate, the resulting force shifts away from the header. The joint between the joist hanger and header is accordingly stressed by an eccentric force, which exerts both lateral and axial loads on the fasteners.

Joist hangers with straps fixed onto the upper side of the header, tend to have fewer fasteners at the header and joist level. In this type of joist hanger, the load is normally transmitted via contact into the header. The load-bearing capacity of such joist hangers is almost impossible to calculate, hence the need for tests to determine the same. Unlike the joist hanger shown in Figure E8-3, tensile failure of the steel may also be a key criterion dictating load-bearing capacity, particularly for steel less than 2 mm thick.

Under normal circumstances, the nail distances to the end of the joist do not reach the minimum values defined in EC 5. For this reason, a lower load-bearing capacity would have to be assumed for these nails. As soon as the nails deform under load, however, an increasing portion of the load is transmitted to the base plate of the joist hanger via contact pressure. Riberholt (1975) introduced a mechanical model, with which the portion of the load in the base plate can be estimated. Since the load-bearing behaviour of the nailed joint and of the support on the base plate can be seen as plastic, the steel-to-timber joint capacity and of the contact pressure can be added.

E8.3 Joist anchors

Joist anchors are used to connect timber members arranged at right angles to each other, e.g. to transfer wind suction forces or suspend beams. Most connections feature two diagonally arranged joist anchors, although joist anchors with one and two rows of holes to accommodate nails both exist. Joist anchors are generally only exposed to tensile forces.

When designing joist anchors, three different components must be considered:

- The load-bearing capacity of the nailed joint. Assuming that the force is exerted on the corner of the joist anchor, the nailed joint is subject to eccentric load (Figure E8-4).
- The load-bearing capacity of the joist anchors in the net cross-section, which is also eccentrically loaded.
- The load-bearing capacity of the timber elements. Any stresses perpendicular to the grain can be verified using the design methods shown in Article E11.

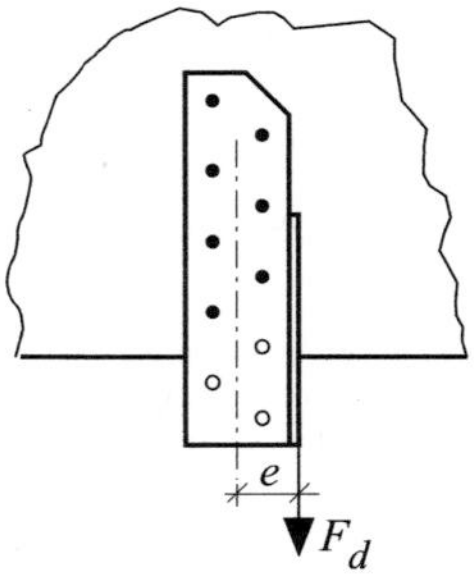

Figure E8-4 Tension load on a joist anchor. (STEP 1995 Article C13)

E8.4 Literature

E. Gehri, original Article C13, STEP 1995.

Ehlbeck J. and Görlacher R. (1982). Mindestnagelabstände bei Stahlblech-Holznagelung. Forschungsbericht, Versuchsanstalt für Stahl, Holz und Steine, Universität Karlsruhe.

Riberholt H. (1975). Berechnung von Stahlblech-Holz Verbindungsteilen in Dänemark. Bauen mit Holz 77:534-536.

E9 Contact joints (Carpentry joints)

Original article: J. Ehlbeck, M. Kromer

A timber construction that usually comprises bar-shaped members obtained from the tree stem is only considered effective, if the individual elements can be assembled together into a whole appropriately. Joints transmit internal forces generated by the external loads from one element to the other and two or more individual elements are united at the junctions of the construction. These joints are frequently contact joints (carpentry joints) working in compression, within which the forces to be absorbed are only transmitted by contact and, where applicable, friction. Certain joints also feature additional wooden or metal fasteners, intended to secure items into position, but may also contribute to the process of force transmission. Despite the numerous variations of traditional contact joints, they can all be traced back to specific basic forms. These forms, namely half-lap joint, step joint, mortise and tenon joint and cogged joint, are shown in Figure E9-1.

The specified basic forms for carpentry joints and variations thereof are used as longitudinal joints to extend timber elements in a direction parallel to the grain or as transverse joints; featuring bars at right angles or in a criss-cross pattern. Although carpentry joints are not regulated in EC 5, the NA to EC 5 (NCI NA.12) includes design equations and minimum dimensions for step, tenon and peg joints.

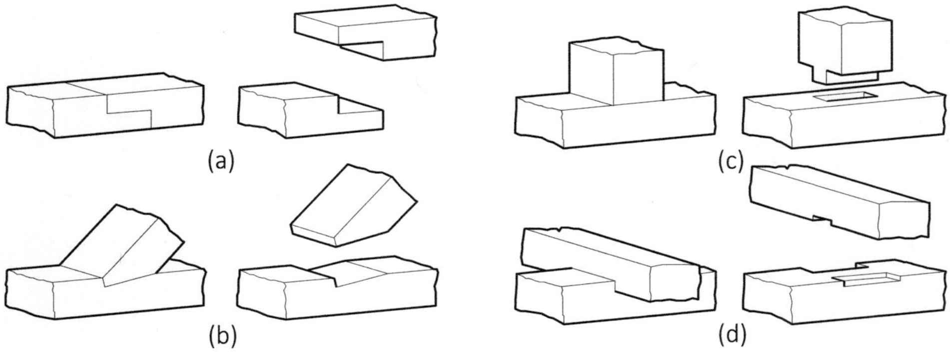

Figure E9-1 Basic forms: (a) half-lap joint, (b) step joint, (c) mortise and tenon joint, (d) cogged joint. (STEP 1995 Article C12)

E9.1 Step joints

A step joint is a connection designed to transmit compressive forces of inclined struts in timber chords. The compressive force of the strut is transmitted to the front surface of the joint by contact and then via shear stress in the loaded end to the chord. Previously, an additional tenon was often used to secure things in place, but today this is normally done using a rafter nail, wood screws or laterally nailed-on straps. The main timber engineering applications are the simple step joints shown in Figure E9-2, particularly single step and heel joints as well as both forms combined in the form of a double step joint.

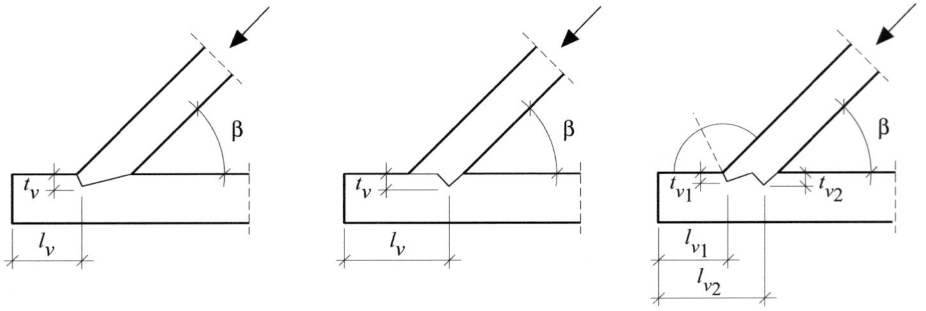

(a) (b) (c)

Figure E9-2 Implementation options for typical step joints: (a) single step - notch in front, (b) notch in heel, (c) double step - combination of both. (STEP 1995 Article C12)

When dimensioning a step joint, evidence is initially required to show that the load-bearing capacity of the existing contact surfaces suffices. For normal strut inclinations of 30 to 60°, only the front face is taken into consideration for the transmission of compressive forces and not the rear. The size of the front surface can be determined from the member width *b* and the notch depth t_v. The design stresses $\sigma_{c,\alpha,d}$, generated by the compressive force $F_{c,\alpha,Ed}$, act in the front surface *A* of the joint at an angle α to the grain direction and must meet the following condition:

$$\frac{\sigma_{c,\alpha,d}}{f_{c,\alpha,d}} \leq 1 \tag{E9-1}$$

where

$$\sigma_{c,\alpha,d} = \frac{F_{c,\alpha,Ed}}{A} \tag{E9-2}$$

and

$$f_{c,\alpha,d} = \frac{f_{c,0,d}}{\sqrt{\left(\frac{f_{c,0,c}}{2 \cdot f_{c,90,d}}\right)^2 \cdot \sin^4 \alpha + \left(\frac{f_{c,0,d}}{2 \cdot f_{v,d}}\right)^2 \cdot \cos^2 \alpha \cdot \sin^2 \alpha + \cos^4 \alpha}} \qquad \text{(E9-3)}$$

Equation (E9-3) shows that the resistance depends on compressive strengths parallel and perpendicular to the grain as well as the shear strength. The joint, which normally features the front surface at an angle to the chord (and strut), generates a combination of all three stress components (origin of equation (E9-3) see Annex 2, equation (5)). To maximise the compressive force transmitted via a single step joint, the inclination of the front surface should be selected such as to ensure that it corresponds to half the angle β between the strut and timber chord. If this is done, it is possible to ensure that the angle α between the force and grain directions of the wood is minimised; in both strut and chord alike ($\alpha = \beta/2$), see Figure E9-3.

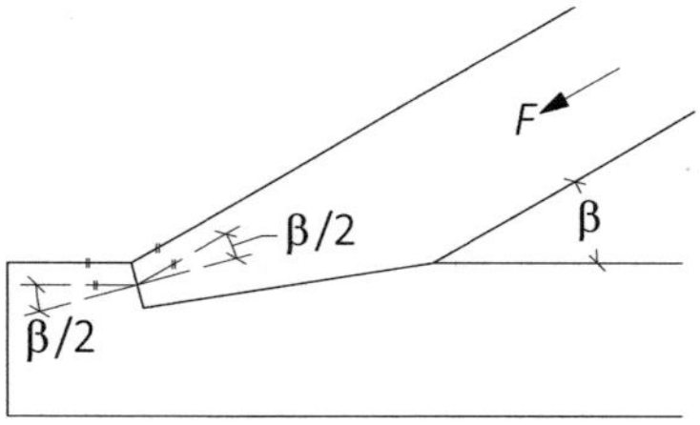

Figure E9-3 Definition of angles in a single step joint.

It is also advisable to ensure compliance with the following condition when selecting the notch depths t_v:

$$t_v \leq \begin{cases} h/4 & \text{for} \quad \beta \leq 50^\circ \\ h/6 & \text{for} \quad \beta > 60^\circ \end{cases} \qquad \text{(E9-4)}$$

The shear stresses in the chord can be assumed to be uniformly distributed, but loaded end lengths $> 8 \cdot t_v$ must not be taken into consideration when determining the shear area. For larger loaded end lengths, the distribution of shear stresses becomes increasingly uneven (peaking at the notch end), and expanding the shear area to a greater extent would lead to unrealistically high load-bearing capacities. In addition, the need for an effective width b_{ef} in accordance with equation (D1-17) must be taken into account, given the potential impact of cracks on shear strength. Step joints must also be secured against unintentional movement.

It should also be noted that the transmission of eccentric forces in the strut may generate **additional bending moments**, which must be taken into account during design. Likewise, the load-bearing capacity of the chord, which is subject to tensile and bending moment stresses, must be verified in the net cross-section.

Multiple step joint

It seems advisable to modify the existing joint geometry (single step and heel joint), to ensure the timber cross-sections can be used as efficiently as possible. Using a multiple step joint means the complete depth of the strut can be used to accommodate all the forces exerted. The interlocking between the strut and chord is achieved using multiple heel joints. A multiple step joint is illustrated in Figure E9-4 and is characterised by a low notch depth and as many heels as possible. The maximum heel number depends on the strut depth, notch depth and the angle between strut and chord, since the multiple step joint also requires an appropriate incline of the front surface, which should correspond to half the angle β between the strut and timber chord.

Figure E9-4 Multiple step joint. (Blass and Enders-Comberg, 2012)

Comprehensive experimental tests showed that while the multiple step joint has similar load-bearing capacities to the double step joint combining a single step and heel joint, only a third of the notch depth is required, which also helps mitigate any cross-sectional weakening of the chord (Blass and Enders-Comberg, 2012). Using hardwood (e.g. ash or beech) for the edge lamellae of the chord can significantly boost the load-bearing capacity and stiffness of the joint.

Design proposal

Using the multiple step joint focuses the resulting compressive force into the centre of the strut, meaning no additional moments are generated. When multiple step joints with the maximum number of heels were tested, this usually resulted in compression failure perpendicular to the grain in the chord (Blass and Enders-Comberg, 2012). Accordingly, for angles $\beta \geq 45°$, in the ultimate limit state design, compression failure perpendicular to the grain is assumed. With this in mind, the contact area calculated for the chord is expanded by 30 mm to the left and right. Since hybrid-glulam chords are prone to compression failure perpendicular to the grain in the softwood lamella underneath that made of beech, the compression strength perpendicular to the grain is determined in the area of transition from beech to softwood. Here, a load distribution angle of 45° is assumed, allowing for additional expansion of the cross-sectional area. The bending resistance of the beech lamella is not taken into consideration, but contingent on a sufficient loaded end length, so that compression failure perpendicular to the grain in the softwood is always decisive. Otherwise, the loaded end should be checked for possible shear failure. The following design proposal is included in Enders-Comberg and Blass (2014) and the corresponding geometric details can be taken from Figure E9-5.

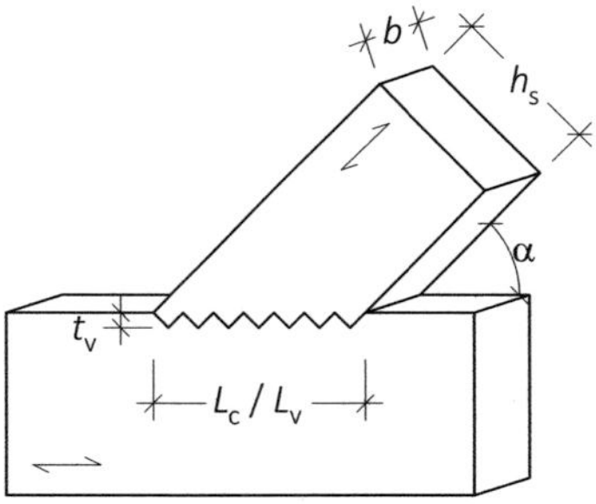

Figure E9-5 Multiple step joint, geometric details (Blass and Enders-Comberg, 2014).

Verification of compression capacity perpendicular to the grain $F_{c,90,R}$:

$$F_{c,90,R} = \frac{A_c \cdot f_{c,90} \cdot k_{c,90}}{\sin\alpha} \qquad \text{(E9-5)}$$

where the length of the compression area is L_c

$$L_c = \frac{h_s}{\sin\alpha} \qquad \text{(E9-6)}$$

and the compression area is A_c

$$A_c = b \cdot (L_c + 2 \cdot 30 \text{ mm}) \qquad \text{(E9-7)}$$

Verification of shear capacity $F_{v,R}$:

$$F_{v,R} = \frac{A_v \cdot f_v}{\cos\alpha} \tag{E9-8}$$

where the length of the shear area is L_v

$$L_v = \frac{h_s}{\sin\alpha} \tag{E9-9}$$

and the shear area is A_v

$$A_v = b \cdot L_v \tag{E9-10}$$

E9.2 Mortise and tenon joints

Mortise and tenon joints are used to connect individual members in floor, wall and roof constructions where the joint has to transfer shear forces. Nowadays, economic reasons mean it is only feasible to keep producing tenon joints using CNC-controlled machining equipment, which normally means straight tenons with central tenon position, low positioned tenons where the underside of the member runs through or an intermediate form with the tenon position varied. In this respect, central tenons are used to connect timber members of the same sectional height, while variable and low positioned tenons are used for mortises in trimmer joists or joists with higher cross-sections. Figure E9-6 shows a typical tenon with its geometric details. For manually processed joints, the tenon height h_z amounts to at least a third of the member depth h, while for newer constructions, although the tenon height tends to be in line with the latest machining equipment, it must be at least $h/6$. The minimum and maximum dimensions for tenon lengths ℓ_z are 15 and 60 mm according to NA and the entire length ℓ_z of the tenon must bear on the mortise. Mortises should only be arranged centrally or in the compressive bending stress area of the member. The cross-sectional weaknesses caused by mortises must be taken into consideration at the time of design, while the rule for the members with the tenon means that h_o must be at least as large as h_u. Further requirements include the lower notch h_u not exceeding $h/3$ and the members with the tenon requiring a height-width ratio of $1.5 \leq h/b \leq 2.5$. Figure E9-7 sets out the complex geometric boundary conditions in graphical form. The background to these geometric boundary conditions is the fact that the calculation model was validated using test specimens, with values within these limits. Any extrapolation may lead to non-conservative results.

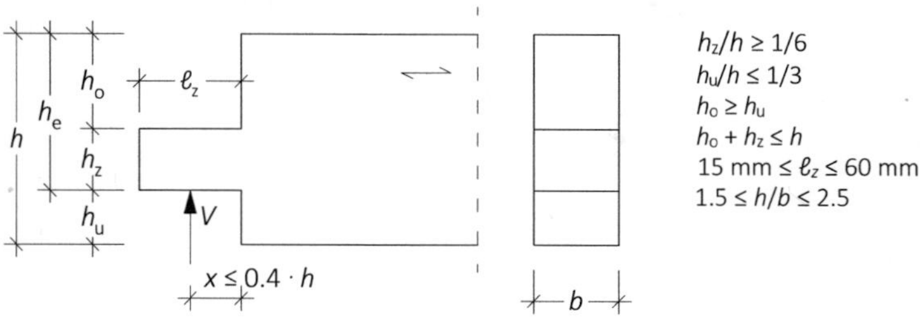

Figure E9-6 Tenon joint.

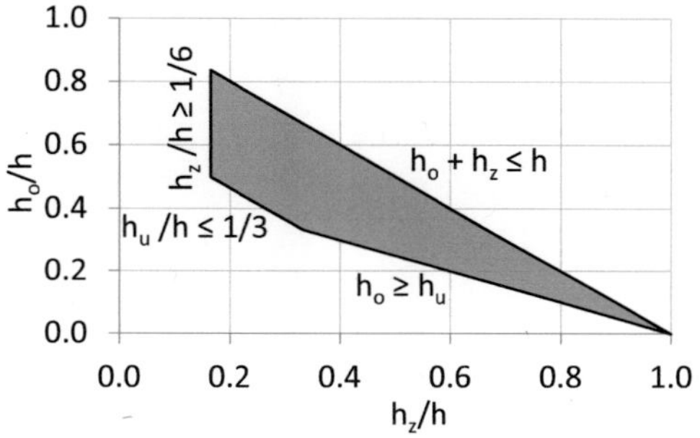

Figure E9-7 Geometric boundary conditions for tenons.

The design of tenon joints can be performed similarly to that for notched beams (Equation (D5-9) in Article D5), although the compression strength perpendicular to the grain $f_{c,90}$ must also be verified. The characteristic load-bearing capacity F_{Rk} of the tenon with a tenon width b is therefore determined as follows (dimensions in Figure E9-6), where $h \leq 300$ mm:

$$F_{Rk} = \min \begin{cases} \dfrac{2}{3} \cdot b_{ef} \cdot h_e \cdot k_z \cdot k_v \cdot f_{v,k} \\ 1.7 \cdot b \cdot \ell_{z,ef} \cdot f_{c,90,k} \end{cases} \qquad \text{(E9-11)}$$

The upper term in equation (E9-11) verifies tension perpendicular to the grain and has been reduced to shear stress verification. This verification equates to the failure mode "crack propagation in the grain direction" (for more background details, see Article D5). The effective width b_{ef} in accordance with equation (D1-17) takes into consideration the influence of cracks on shear strength, while the factor k_z is based on test results and takes influences from the tenon joint geometry into consideration:

$$k_z = \frac{h_z}{h_e} \cdot \left[1 + 2 \cdot \left(1 - \frac{h_z}{h_e} \right)^2 \right] \cdot \left(2 - \frac{h_e}{h} \right) \qquad \text{(E9-12)}$$

The coefficient k_v takes into consideration the risk of a crack developing and reduces the shear strength. k_v must be less than 1 and corresponds to the coefficient k_v for notched beams from equation (D5-10).

The lower term, conversely, represents a verification of compression perpendicular to the grain and is derived from test results, the pre-exponential factor 1.7 results from the adaptation to the test results. The pre-exponential factor corresponds to the coefficient $k_{c,90}$, which is applied when verifying compression perpendicular to the grain (see Article D1, equation (D1-4)) and takes into consideration, among other things, the fact that the compression perpendicular to the grain is distributed over areas outside the directly loaded partial area, whereby the compression strength perpendicular to the grain rises (see also Figure D1-3). The load distribution beyond the tenon length ℓ_z is ℓ_z also determined via the effective tenon length $\ell_{z,ef}$ (here too, similar to the verification of compression perpendicular to the grain, effective area A_{ef} in equation (D1-5)):

$$\ell_{z,ef} = \min\begin{cases} \ell_z + 30\,\text{mm} \\ 2 \cdot \ell_z \end{cases} \tag{E9-13}$$

In mortise and tenon joints, the members with the tenon may fail due to tension perpendicular to the grain, similarly to a notched beam, and due to compression perpendicular to the grain in the supported surface of the tenon and must accordingly be verified (equation (E9-11)). In mortise and tenon joints in floor constructions, however, the member with the mortise is also subject to tension perpendicular to the grain, since the tenon joint works like a joint loaded perpendicular to the grain in this case, meaning the member with the mortise is loaded perpendicular to the grain direction. The verifications required in this case are explained in Article E11.

E9.3 Dovetail joint

The emergence of computer-controlled systems (CNC machines) for use in fabricating engineered timber components means that even what were originally traditional carpentry joints are coming back into vogue. The joint referred to here, namely a dovetail joint, is regulated in a general national technical approval for a joint between headers and joists in service classes 1 and 2 (Z-9.1-649), Figure E9-8. The joist failure in the same, caused by a combination of tension failure perpendicular to the grain and shear failure in the area underneath the dovetail tenon, resembles the failure of a notched beam; whereby the load-bearing capacity of the header is mainly limited by the tension strength perpendicular to the grain, like the behaviour of a joint loaded perpendicular to the grain.

Test observations showed that the joist actually failed in a similar way to a notched support, the header like a member with a joint loaded perpendicular to the grain. The position of the cracks due to tension perpendicular to the grain in the joist differs from that for notched beams, the cracks are not at the tip of the notch, but rather at the transition between the straight tenon and lower fillet. This applies similarly to the header, for which cracks due to tension perpendicular to the grain also occur at the height of the transition between the straight tenon and lower fillet. Accordingly, the design method for notched beams or joints loaded perpendicular to the grain as included in the EC 5 and NA can be used to underpin the design of dovetail joints connecting headers and joists, while for equations, see also Z-9.1-649.

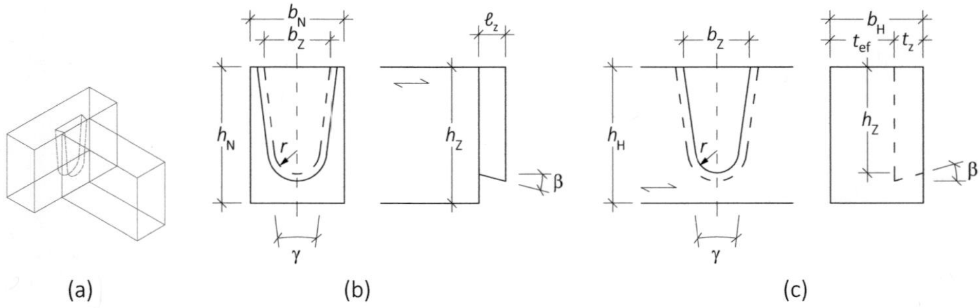

Figure E9-8 Single-sided dovetail joint, (a) 3D overview, (b) joist, (c) header.

The design load-bearing capacity $F_{90,Rd}$ of a dovetail joint subject to shear force acting in insertion direction amounts to the following per joint:

$$F_{90,Rd} = \min\begin{cases} k_{ab} \cdot \dfrac{h_Z}{h_Z - r} \cdot \left(6.5 + \dfrac{18 \cdot \left(h_H - h_Z + r\right)^2}{h_H^2} \right) \cdot \left(t_{ef} \cdot h_H\right)^{0.8} \cdot f_{t,90,d} \\ \dfrac{k_v \cdot b_N \cdot \left(h_Z - r\right)}{1.5} \cdot f_{v,d} \end{cases} \qquad \text{(E9-14)}$$

where (see also Figure E9-8):

h_H Header depth in mm, $h_H \geq 140$ mm

b_H Header width in mm, $b_H \geq 60$ mm

h_Z Tenon height parallel to the lateral surface of the header in mm

r Mortise radius parallel to the lateral surface of the header in mm, 15 mm $\leq r \leq$ 60 mm

t_{ef} Effective joint depth in mm, t_{ef} = min (b_H , 100 mm)

$f_{t,90,d}$ Design tension strength perpendicular to the grain (in Z-9.1-649: $f_{t,90,k}$ = 0.5 MPa)

k_{ab} Coefficient to consider single- or double-sided joints:
k_{ab} = 1 for single-sided joints

$$k_{ab} = \min\begin{cases} 1 \\ b_H/200 \end{cases} \text{ for double-sided joints}$$

h_N Joist depth in mm, 140 mm $\leq h_N \leq$ 280 mm

b_N Joist width in mm, $b_N \geq 60$ mm

ℓ_Z Tenon length in mm, 25 mm $\leq \ell_Z \leq$ 30 mm

$$k_v = \min\begin{cases} 1 \\ \dfrac{k_n}{\sqrt{h_N} \cdot \left(\sqrt{\alpha \cdot (1-\alpha)} + 0.4 \cdot \frac{\ell_z}{h_N} \cdot \sqrt{\frac{1}{\alpha} - \alpha^2} \right)} \end{cases}$$

α Notch ratio, $\alpha \geq 0.4$, $\alpha = \cos\delta \cdot (h_Z - r)/h_N$,

δ Joist inclination in °

$$k_n = \begin{cases} 5 & \text{for solid softwood} \\ 6.5 & \text{for glulam} \end{cases}$$

$f_{v,d}$ Design shear strength (in Z-9.1-649: $f_{v,k}$ = 2.5 MPa)

b_Z Tenon width where $b_Z \geq 0.8 \cdot b_N$

γ Tenon cone angle where 4° $\leq \gamma \leq$ 12°

β Dovetail milling angle where 10° $\leq \beta \leq$ 18°

The upper expression in equation (E9-14) outlines the verification of the header and its source resembles that shown in Article E11 for joints loaded perpendicular to the grain (equation (E11-35). The lower expression in equation (E9-14) includes verification of the joist and is along the lines of the verification outlined in Article D5 for non-reinforced notched beams (equation D5-9).

E9.4 Peg joints

EC 5 does not include any details of peg joints, although the NA does contain design rules for joints with solid oak pegs, which were derived from test results. The characteristic load-bearing capacity F_{Rk} in N of a peg loaded laterally and with a constant cross-section (normally round or octagonal with diameter d in mm) is determined as follows, where 20 mm ≤ d ≤ 30 mm (the only items checked were solid oak pegs between 20 and 30 mm in diameter, which means equation (E9-15) is only valid within these limits.):

$$F_{Rk} = 9.5 \cdot d^2 \qquad \text{(E9-15)}$$

The density of the timber members to be connected must be at least 350 kg/m^3, and the angle between the force and grain directions exerts no impact on F_{Rk}. The minimum wood thickness is $2 \cdot d$, while the minimum distances (edge and end distance and spacing) are also $2 \cdot d$.

E9.5 Literature

J. Ehlbeck, M. Kromer, original Article C12, STEP 1995.

Blass H.J. and Enders-Comberg M. (2012). Fachwerkträger für den industriellen Holzbau. Karlsruher Berichte zum Ingenieurholzbau, Band 22, KIT Scientific Publishing Karlsruhe, 161 S.

Enders-Comberg M. and Blass H.J (2014). Treppenversatz – Leistungsfähiger Kontaktanschluss für Druckstäbe. Der Bauingenieur 89(4):162-171.

E10 Glued-in steel rods

Original article: C. J. Johansson

Joints with glued-in rods are mainly used in structures using glued-laminated timber or laminated veneer lumber, some examples of which are shown in Figure E10-1 and Figure E10-2. The rods are used to prevent cracks in curved and notched beams due to tensile stresses perpendicular to the grain or transmit forces in a construction or portion thereof, e.g. when connecting a column to a foundation or a corner of a portal frame. Glued-in steel rods are also used for rehabilitation and repair, during which they may be subject to lateral or axial load or both combined. The manufacturer must also possess a proof of suitability for gluing load-bearing timber members (DIN 1052-10). Joints with glued-in steel rods are not included in EC 5, although the design and execution are regulated in the NA (NCI NA.11). The advantages of using glued-in rods are:

- Scope to transmit significant concentrated loads,
- Scope to arrange rods in parallel with the grain,
- Very rigid joint when exposed to axial load,
- Effective fire resistance, since the surrounding wood protects the steel.

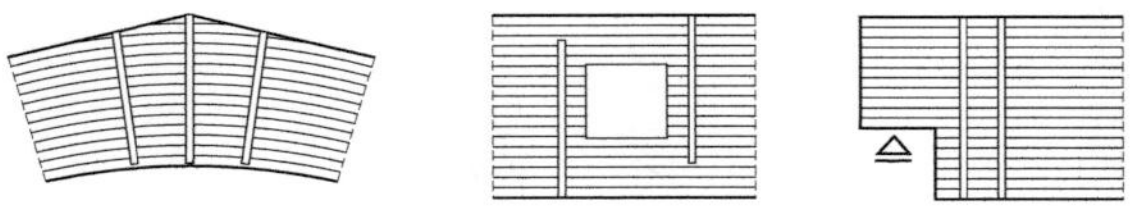

Figure E10-1 Glued-in rods to avoid cracks (pitched cambered beam, opening, notch). (STEP 1995 Article C14)

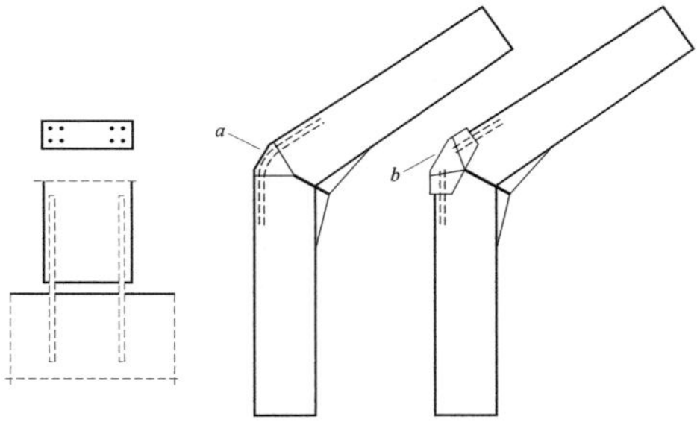

Figure E10-2 Glued-in rods in a clamped column end joint and in a moment-resisting joint; *a* filled with mortar, *b* steel fitting. (STEP 1995 Article C14)

Instead of glued-in rods, steel rods with a thread in accordance with DIN 7998 are an increasingly popular choice for reinforcements perpendicular to the grain direction; driven into predrilled holes like screws and hence mechanically connected with the wood via the thread. Glued-in steel rods, however, are best-suited for arrangements parallel to the grain. Reinforced joints are introduced in Article E12.

E10.1 Materials and manufacture

Ensuring reliable adhesion between a smooth steel surface and the adhesive is far from easy. Accordingly, to ensure mechanical interlocking between the rod and adhesive, threaded rods or RC rebars are preferred. All statements of the NA thus only apply for joints with threaded rods or rebars. Generally speaking, rods of diameter between 12 and 24 mm are used. The rods are glued-in, either by injecting adhesive (Figure E10-3) or pouring the adhesive into the hole and inserting the rod. The drilled holes are around 1 to 4 mm larger than the external diameter of the thread, to ensure sufficient room in which to inject the adhesive.

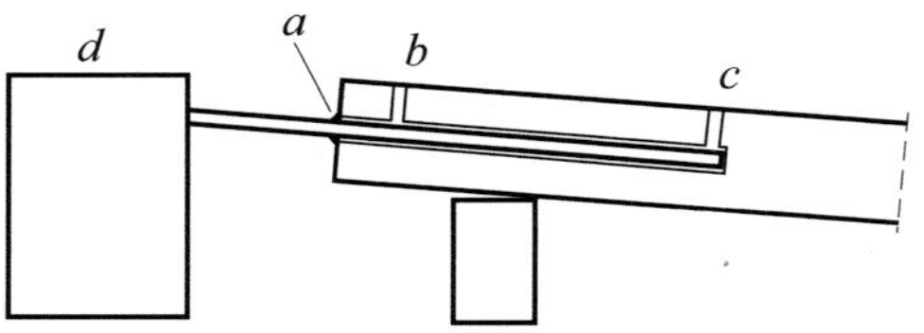

Figure E10-3 Injection of a joint with a glued-in rod; *a* sealing, *b* adhesive bleed, *c* adhesive ingress, *d* a block prevents the rod from being pushed out by the adhesive. (STEP 1995 Article C14)

Different adhesives are used, such as phenol resorcinol formaldehyde resin, dual-component polyurethanes and dual-component epoxy resins. The adhesive selection depends on the manufacturing method and the type of load to which the rods are subject. In Germany the adhesive requires a general national technical approval for use with glued-in steel rods. For hole diameters that exceed the rod diameter, the key factors determining the withdrawal strength of the rod are the strength and durability of the bond line.

Phenol resorcinol formaldehyde resins are a tried and tested option when it comes to structural timber, particularly e.g. glulam. Riberholt (1988), however, advocates that phenol resorcinol resins should not be used for joints manufactured using the injection method and holes of excess size, since the shrinking of the adhesive when it hardens reduces its strength. Dual-component polyurethane adhesives, conversely, frequently react sensitively to higher temperatures (Aicher, 1992) and should therefore only be used in this respect if a general technical approval has been issued. This is particularly true for joints which were established using the injection method and which are permanently subject to high loads.

For laterally loaded glued-in rods, the choice of adhesive is less important, since the loads are mainly transmitted via compressive – and at times tensile – stresses in the bond line and the mechanical behaviour resembles that of laterally loaded dowel-type fasteners.

E10.2 Laterally loaded glued-in rods

The lateral load-bearing capacity of rods can be determined using the Johansen equations for dowel-type fasteners (expanded to take the rope effect into consideration), cf. Articles E1 and E2. The provisions for bolts apply, including the steel rods with threads in accordance with DIN 7998 (not glued-in), Article E4.

Gluing-in the rods enables virtually infinite friction coefficients between steel and timber surfaces. Meanwhile, for rods glued-in perpendicular to the grain, Rodd et al. (1989) successfully demonstrated a considerable increase in load-bearing capacity and stiffness in comparison to non-glued-in rods. This also means that the embedment strength $f_{h,\alpha,k}$ determined with Equation (E4-2) can be increased by a factor of 1.25, if the steel rods are glued-in perpendicular to the grain. For rods glued-in in parallel to the grain, the embedment strength is very low and only 10% of the value for rods glued-in perpendicular to the grain. For angles between the force and grain directions of between 0 and 90°, the embedment strength can be determined by linear interpolation. The minimum distances for glued-in steel rods are specified in Table E10-1.

Table E10-1 Minimum distances for laterally loaded glued-in steel rods in accordance with NCI NA.11 (2013).

	Minimum distances
Steel rods glued-in in parallel to the grain: a_2 = spacing, $a_{2,c/t}$ = loaded/unloaded end distance	 $a_2 = 5 \cdot d$ $a_{2,c} = 2.5 \cdot d$ $a_{2,t} = 4 \cdot d$
Steel rods glued-in perpendicular to the grain	See dowels, Table E4-2

E10.3 Axially loaded glued-in rods

Axially loaded glued-in rods may fail in three different modes:

- Tensile failure of the steel rod,
- Failure of the bond line or the wood along the wall of the hole,
- Failure of the timber member.

If multiple glued-in steel rods of a joint are not uniformly stressed, the failure mode "tensile failure of the steel rod" has to become the key criterion. The plastic deformation of the most heavily loaded steel rod then makes it possible to redistribute load among the individual steel rods in the joint. Non-uniform tensile stresses in a group of steel rods can occur, e.g. due to inconsistent tightening of nuts or deformations in connecting steel parts. The latter failure type, of the timber member, can be checked by verifying the tensile stress of the member in the net cross-section or block shear. Both the other failure types are involved when determining the design withdrawal resistance $F_{ax,Rd}$ for a glued-in rod:

$$F_{ax,Rd} = \min \begin{cases} A_{ef} \cdot f_{y,d} \\ \pi \cdot d \cdot \ell_{ad} \cdot f_{k1,d} \end{cases} \quad \text{(E10-1)}$$

The load-bearing capacity for the tensile failure of the steel rod is determined via its effective cross-section A_{ef} and its yield strength $f_{y,d}$. The design of the bond line, meanwhile, resembles that of reinforcements with strength values for bond lines, see Article E12. ℓ_{ad} is the glued-in length of the steel rod, d the diameter and $f_{k1,d}$ the design strength for the bond line, the characteristic value of which is given in Table D8-1 (Article D8).

The required minimum distances for axially loaded rods are given in Table E10-2. Under specific boundary conditions, there is also scope to use lower minimum distances for steel rods glued-in in parallel to the grain in glulam (see general national technical approval No. Z-9.1-791).

The minimum glued-in length $\ell_{ad,min}$ for axially loaded rods is:

$$\ell_{ad,min} = \max \begin{cases} 0.5 \cdot d^2 \\ 10 \cdot d \end{cases} \quad \text{(E10-2)}$$

The minimum glued-in lengths should always be complied with, since tests have shown that shorter glued-in lengths are prone to brittle failure. Moreover, for multiple axially loaded glued-in steel rods, group effects may emerge, leading to a so-called "group tear-out" resembling a block shear failure and which lead to the group of fasteners being completely pulled out. There are no relevant indications contained in the NA. For groups of glued-in steel rods, therefore, there is a need to identify an **effective number** $n_{ef} = n^{0.9}$ as is done for axially loaded screw groups, if plastic steel failure is not the key criterion (for more details, see Article E13). Glued-in steel rods can also be used for joints loaded perpendicular to the grain, in which case the tensile stresses perpendicular to the grain must also be verified, as is explained in Article E11. Also important for axially loaded rods is preventing corrosion of the rods. Riberholt (1988) highlights the risk of seeing the bond line between the steel and wood destroyed due to the spread of rust, which is why glued-in rods must be protected by applying e.g. a zinc coating against corrosion. Riberholt (1986) stated that certain adhesives, such as epoxy resins, already provide effective corrosion protection.

Table E10-2 Minimum distances of axially loaded glued-in steel rods in accordance with NCI NA.11 (2013).

	Minimum distances
Steel rods glued-in parallel to the grain:	
Spacing	$a_2 = 5 \cdot d$
Edge distance	$a_{2,c} = 2.5 \cdot d$
Steel rods glued-in perpendicular to the grain:	
Spacing in beam width	$a_1 = 4 \cdot d$
Edge distance in beam width	$a_{1,c} = 2.5 \cdot d$
Spacing in beam length	$a_2 = 4 \cdot d$
End distance in beam length	$a_{2,c} = 2.5 \cdot d$

E10.4 Literature

C.J. Johansson, original Article C14, STEP 1995.

Aicher S. (1992). Testing of adhesives for bonded wood-steel joints. Meeting IUFRO S 5.02, Bordeaux.

Riberholt H. (1986). Glued bolts in glulam. Report No. R 210, Technical University of Denmark.

Riberholt H. (1988). Glued bolts in glulam: Proposal for CIB Code. Paper 21-7-2, CIB-W18 Meeting 21, Parksville.

Rodd P.D., Hilson B.O. and Spriggs R.A. (1989). Resin injected mechanically fastened timber joints. Second Pacific Timber Engineering Conference, Auckland.

E11 Joints loaded perpendicular to the grain

Original article: J. Ehlbeck, R. Görlacher

The term "joints loaded perpendicular to the grain" is used to refer to load introductions into a member with a force component at right angles to the member's grain direction. The load-bearing capacity of a joint loaded perpendicular to the grain is firstly limited by the load-bearing capacity of the joint itself (with various fasteners such as nails, screws, bolts, glued-in rods, punched metal plate fasteners or connectors), secondly when the tensile strength perpendicular to the grain of the timber member has been attained. Regardless of the fastener types used, however, joints loaded perpendicular to the grain generate tensile stresses perpendicular to the grain when introducing shear forces into the member. In timber construction, tensile stresses perpendicular to the grain are generally considered particularly serious, given the comparatively low strength and stiffness of the wood perpendicular to the grain direction. This means, for example, while the characteristic tensile strength parallel to the grain is 14 N/mm^2 for normal coniferous timber, the characteristic tensile strength perpendicular to the grain for the same is just 0.4 N/mm^2, namely less than 3% of the tensile strength parallel to the grain. In addition, wood is excessively brittle when tensile stress is applied at right angles to its grain direction. If the tensile strength perpendicular to the grain has been attained for a joint loaded perpendicular to the grain, this will lead to the wood splitting in the grain direction and ultimately to the failure of the joint. Despite all these issues, which can occur when load introductions are exerted perpendicular to the grain, joints loaded perpendicular to the grain are still frequently used in timber construction, whereby the tensile stresses generated perpendicular to the grain direction may be the key factor dictating the load-bearing capacity of the joint, regardless of the fastener. Some examples of typical joints loaded perpendicular to the grain with possible crack paths are shown in Figure E11-1.

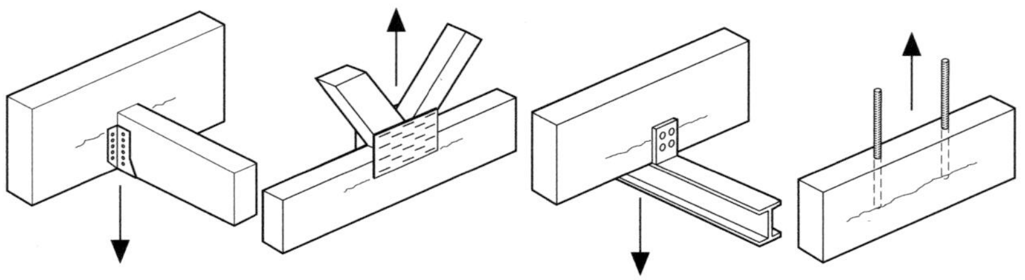

Figure E11-1 Examples of typical joints loaded perpendicular to the grain with possible crack paths. (*a*) joist hanger (steel-to-timber joint), (*b*) punched metal plate fasteners as gusset plate, (*c*) joints with dowels or connectors, (*d*) glued-in rods. (STEP 1995 Article C2)

The key tensile stresses perpendicular to the grain, which generally overlap with shear and longitudinal stresses, can be determined using the finite element method. Similar to the approach used to calculate notched beams, this results in stresses not directly comparable with the characteristic tension strength perpendicular to the grain, which is determined using prismatic test bodies. Accordingly, this is also an area in which fracture mechanics may prove useful, as was already described in relation to notched beams in EC 5 (Article D5).

Additional possibilities for handling issues with tensile stress perpendicular to the grain in joints include conventional stress criteria, based on the concept of an effective width. This approach is contained in the NA for joints loaded perpendicular to the grain with multiple fasteners and as with notches and openings (Article D5), the design of joints loaded perpendicular to the grain has seen fracture mechanics approaches become standard practice in addition to the more traditional stress criteria.

E11.1 Notes on reducing the risk of failure due to tension perpendicular to the grain

The following section outlines the key parameters influencing the load-bearing capacity of joints loaded perpendicular to the grain, which also includes structural detailing hints on how to reduce the risk of cracks due to tension perpendicular to the grain. Information on reinforced joints, meanwhile, is included in Article E12, while a joint loaded perpendicular to the grain with mechanical fasteners is schematically illustrated in Figure E11-2. The force F_{90} generating tension perpendicular to the grain is introduced via fasteners into a beam.

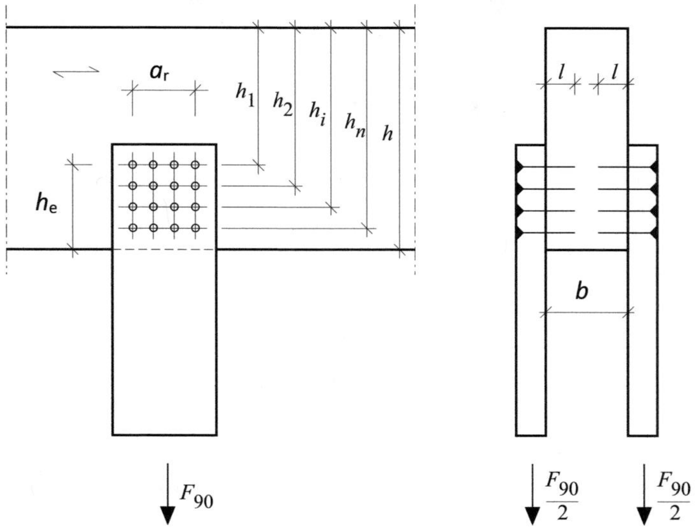

Figure E11-2 Joint perpendicular to the grain (designations). (STEP 1995 Article C2)

The following parameters influence the load-bearing capacity:

- The ratio of the distance h_e of the row of fasteners, which is furthest away from the loaded edge and the beam depth h. To reduce the risk of cracks due to tension perpendicular to the grain, the fasteners should be as close as possible to the unloaded edge, namely h_e/h should be maximised.
- Multiple fasteners in a row distribute the load over a larger joint area and reduce the maximum value of the tensile stresses perpendicular to the grain generated as a result. This influence is particularly helpful for larger joint widths or fasteners spaced relatively far apart.
- Expanding the beam width b or beam depth h (if h_e is increased) boosts the load-bearing capacity. Here, it is important to note that depending on the fastener type used, it is frequently only possible to employ part of the beam width to accommodate the tensile stresses perpendicular to the grain generated.
- Additional rows of fasteners reduce tensile stresses perpendicular to the grain in the uppermost row of fasteners, if a_r or h_e becomes larger.
- The incidence of brittle failure means the tensile strength perpendicular to the grain of the wood depends on the stressed volume, which means the load-bearing capacity of joints loaded perpendicular to the grain is also affected by beam size.

E11.2 Verification procedure based on fracture mechanics

The verification for joints loaded perpendicular to the grain is conducted in a manner resembling that for notched beams (Article D5), which is based on fracture mechanics. Based on linear-elastic fracture mechanics and the assessment of test results, the following verification was recorded for softwood subject to loads as shown in Figure E11-3 in EC 5 (Van der Put and Leijten, 2000; Leijten and Jorissen, 2001):

$$F_{90,Rk} = 14 \cdot b \cdot w \cdot \sqrt{\frac{h_e}{\left(1 - \frac{h_e}{h}\right)}} \tag{E11-1}$$

The following applies:

$F_{90,Rk}$ Characteristic tensile capacity perpendicular to the grain in N

b Beam width in mm

h_e Distance of the fasteners arranged furthest away from the loaded edge in mm

h Beam depth in mm

$$w = \begin{cases} \max\begin{cases} \left(\frac{w_{pl}}{100}\right)^{0.35} \\ 1 \end{cases} & \text{for punched metal plate fasteners} \\ 1 & \text{for all other joints} \end{cases}$$

where w_{pl} = width of the punched metal plate fastener parallel to the grain in mm

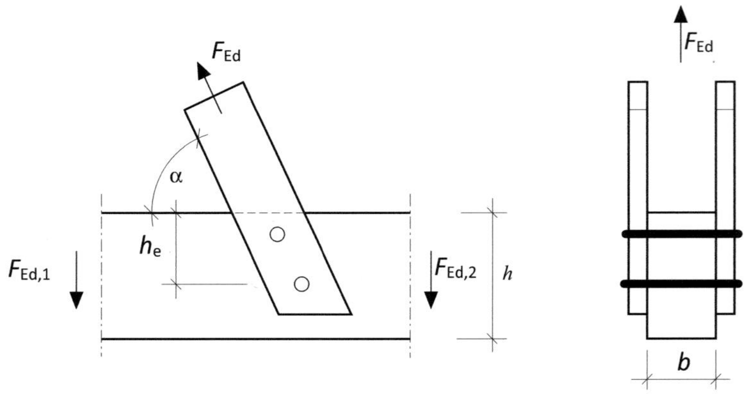

Figure E11-3 Load exerted at an angle to the grain direction. (STEP 1995 Article C2)

The basic approach used to derive equation (E11-1) can be taken from Article D5; only certain general notes will be made at this point. The (already simplified) original form of equation (E11-1) reads as follows (Van der Put and Leijten, 2000):

$$F_{90} = b \cdot \sqrt{\frac{G \cdot G_{c}}{0.6}} \cdot \sqrt{\frac{h_{e}}{\left(1 - \frac{h_{e}}{h}\right)}} \tag{E11-2}$$

The root terms in equations (E11-1) and (E11-2) are typical of linear-elastic fracture mechanics approaches and equation (E11-1) already includes key linear-elastic fracture mechanics parameters, such as the value of critical fracture energy G_c. This is why equation (E11-1) only applies for softwoods, since the only results obtained came from tests with softwood. Material properties, such as tensile strength perpendicular to the grain $f_{t,90}$ for example, are thus excluded from equation (E11-1) and only geometric information is present.

One issue with the abovementioned verification is the fact that the applicable scope is limited to the joint configuration shown in Figure E11-3. For joints in practice, however, most configurations include multiple rows of fasteners. Configurations other than that shown in Figure E11-3 are used, since one advantage of distributing fasteners over a longer length in the grain direction is reducing the peak tensile stresses perpendicular to the grain and hence increasing the load-bearing capacity. In such cases, the verification shown in equation (E11-1) does not apply. To remedy this loophole, the NA uses the verification already specified in DIN 1052:2008 for the configuration of choice and various wood species and types of wood-based materials, which is derived in the following section. This verification, as has already been noted, is based on the known concept of a stress criterion.

E11.3 Verification procedure in accordance with NCI NA.8.1.4

The stress verification in accordance with NCI NA.8.1.4 is based on theoretical and experimental examinations conducted by Ehlbeck and Görlacher (1983). A joint loaded perpendicular to the grain is dimensioned by comparing the calculated value of the tensile stress perpendicular to the grain with the characteristic tensile strength perpendicular to the grain; taking into consideration all impacts on load-bearing capacity. The failure behaviour of the wood when tensile load is applied perpendicular to the grain is particularly brittle, which means expanding the volume subject to tensile stress adversely affects strength. The tensile strength perpendicular to the grain and its characteristic value are determined by tests on comparatively small samples, which means recording the volume effect is not possible. This is a factor that must be taken into consideration, alongside the joint type and geometry when designing joints loaded perpendicular to the grain.

The load generating tension perpendicular to the grain as a portion of the total load

The design of a joint loaded perpendicular to the grain with respect to the tensile strength perpendicular to the grain of the member depends on the extent of the load generating tension perpendicular to the grain in the member. Since the load from such joint is introduced in the member by both compression as well as tension perpendicular to the grain, the first task involves defining the load component generating tension perpendicular to the grain. Accordingly, an equation to define this load as a portion of the total load from a joint is derived in the following section using the example of a single-span girder. The girder includes the span ℓ, depth h and width b and is subject to stress from the concentrated load F at mid-span. Now, the position of the load application point should be observed. If the load F is exerted on the top side of the beam, this will generate exclusively compression perpendicular to the grain in the beam. Conversely, if the load is applied to the bottom of the beam, only tension perpendicular to the grain is generated, as in Figure E11-4.

We cite the general case, in which the load F is applied at a point with an arbitrary distance a to the underside of the beam, as in Figure E11-5. The load F now generates both compression and tension perpendicular to the grain. The load portion generating tension perpendicular to the grain F_t is designated with η. The factor η is 1 if the load is exerted on the underside of the beam.

If the load is applied to the upper edge of the beam, $\eta = 0$ applies:

$$F_t = \eta \cdot F \tag{E11-3}$$

$$F_c = (1-\eta) \cdot F \tag{E11-4}$$

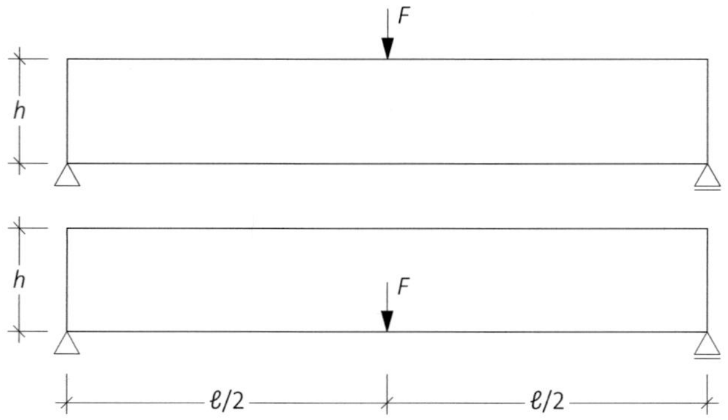

Figure E11-4 Single-span girder with load applied to the upper or lower edge.

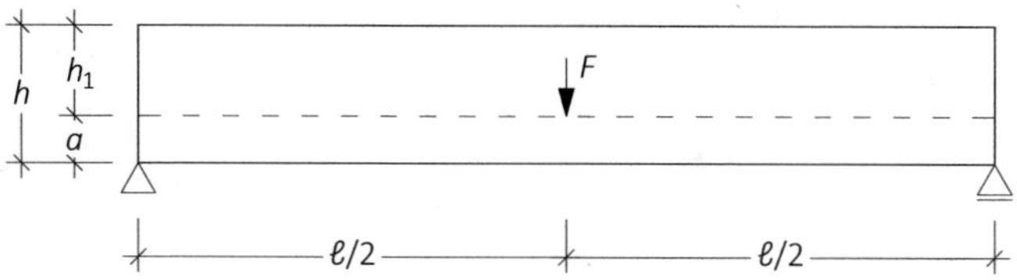

Figure E11-5 Load application at an arbitrary point at mid-span.

The factor η for load introduction between the upper and lower edges of the beam is not initially known and has to be determined. For this purpose, the beam is cut at the height of the load application point, namely at the distance h_1 from the upper edge of the beam, see free-body diagram in Figure E11-6. Subsequently, the deflection of the upper beam section shown in Figure E11-6 is determined, taking into consideration the internal forces, tensile force F_t and shear stress τ, which impact on the cut surface. Given the condition that the deflection of the cut beam portion must be equivalent to the deflection of the entire beam, an equation to define the load portion η generating tension perpendicular to the grain can be derived.

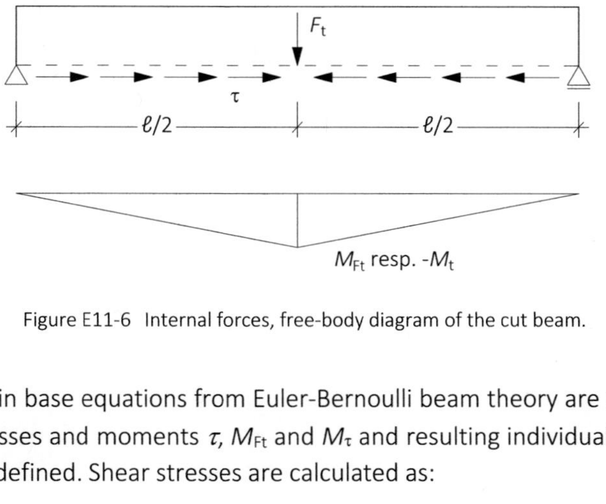

Figure E11-6 Internal forces, free-body diagram of the cut beam.

First of all, certain base equations from Euler-Bernoulli beam theory are required, so that the internal stresses and moments τ, M_{Ft} and M_τ and resulting individual deflection components can be defined. Shear stresses are calculated as:

$$\tau = \frac{V \cdot S}{I \cdot b}$$

The shear force amounts to:

$$V = \frac{F}{2}$$

The first moment of area of the cut cross-sectional part with cross-sectional area $a \cdot b$ amounts to:

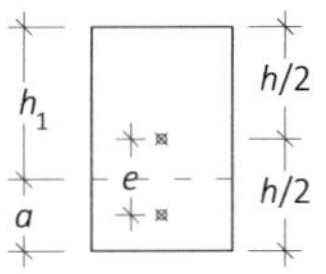

$$S = a \cdot b \cdot e = a \cdot b \cdot \left(\frac{h}{2} - \frac{a}{2}\right) = a \cdot b \cdot \left(\frac{h_1 + a}{2} - \frac{a}{2}\right) = a \cdot b \cdot \frac{h_1}{2}$$

The second moment of area for a rectangular cross-section b x h is:

$$I = \frac{b \cdot h^3}{12}$$

The above equations can be used to define the shear stress of the entire beam illustrated in Figure E11-6 at the height of the cut:

$$\tau = \frac{V \cdot S}{I \cdot b} = \frac{F}{2} \cdot a \cdot b \cdot \frac{h_1}{2} \cdot \frac{12}{b \cdot h^3 \cdot b} = \frac{3 \cdot F \cdot a \cdot h_1}{b \cdot h^3} \qquad \text{(E11-5)}$$

The cut beam part of height h_1 is subject to moments caused by the load component F_t and the shear stress τ, Figure E11-6. The moment M_{Ft} caused by the force F_t amounts to:

$$M_{Ft} = \frac{F_t \cdot \ell}{4} \qquad \text{(E11-6)}$$

The moment M_τ caused by shear stress τ results in:

$$M_\tau = -\frac{h_1}{2} \cdot \int_A \tau \mathrm{d}A = -\frac{h_1}{2} \cdot \tau \cdot b \cdot \frac{\ell}{2} \qquad \text{(E11-7)}$$

Inserting equation (E11-5) into equation (E11-7) allows the moment M_τ to be expressed in dependence of the force F:

$$M_\tau = -\frac{h_1}{2} \cdot b \cdot \frac{\ell}{2} \cdot \frac{3 \cdot F \cdot a \cdot h_1}{b \cdot h^3} = -\frac{3 \cdot F \cdot a \cdot h_1^2 \cdot \ell}{4 \cdot h^3} \qquad \text{(E11-8)}$$

With that, all the values required to determine the deflection of the beam are in place. Both the load component F_t and the shear stress τ result in a deflection of the beam part shown in Figure E11-6. The deflection caused by the load component F_t results from inserting F_t in the equation known from the Euler-Bernoulli beam theory to determine the deflection of a beam exposed to a concentrated load at mid-span:

$$f_{Ft} = \frac{F_t \cdot \ell^3}{48 \cdot E \cdot I_1} = \frac{\eta \cdot F \cdot \ell^3}{48 \cdot E \cdot I_1} \tag{E11-9}$$

Inserting equation (E11-6) into equation (E11-9) means that the deflection may also be be expressed in dependence of the moment M_{Ft}:

$$f_{Ft} = \frac{F_t \cdot \ell^3}{48 \cdot E \cdot I_1} = \frac{M_{Ft} \cdot 4}{\ell} \cdot \frac{\ell^3}{48 \cdot E \cdot I_1} = \frac{M_{Ft} \cdot \ell^2}{12 \cdot E \cdot I_1} \tag{E11-10}$$

The shape of the moment diagram caused by a concentrated load applied at mid-span corresponds to that of the shear stresses on the cut beam shown in Figure E11-6. The deflection f_τ of the cut beam of the height h_1 caused by these shear stresses can accordingly be determined by inserting (E11-8):

$$f_\tau = \frac{M_\tau \cdot \ell^2}{12 \cdot E \cdot I_1} = -\frac{3 \cdot F \cdot a \cdot h_1^2 \cdot \ell}{4 \cdot h^3} \cdot \frac{\ell^2}{12 \cdot E \cdot I_1} = -\frac{3 \cdot F \cdot a \cdot h_1^2 \cdot \ell^3}{48 \cdot h^3 \cdot E \cdot I_1} \tag{E11-11}$$

Adding deflection components $f_{Ft} + f_\tau$ allows the following equation:

$$f_1 = f_{Ft} + f_\tau = \frac{\eta \cdot F \cdot \ell^3}{48 \cdot E \cdot I_1} - \frac{3 \cdot F \cdot a \cdot h_1^2 \cdot \ell^3}{48 \cdot h^3 \cdot E \cdot I_1} = \frac{F \cdot \ell^3}{48 \cdot E \cdot I_1} \cdot \left(\eta - \frac{3 \cdot a \cdot h_1^2}{h^3} \right) \tag{E11-12}$$

The deflection of the beam at mid-span caused by the concentrated load F is independent of the distance a from the point at which the force is applied to the underside of the beam. The deflection f_1 for the cut beam under observation, in accordance with equation (E11-12), must therefore be equivalent to the total deflection f of the entire beam, whereby the total deflection is determined without taking shear deformations into consideration:

$$f = f_1 = \frac{F \cdot \ell^3}{48 \cdot E \cdot I} = \frac{F \cdot \ell^3}{48 \cdot E \cdot I_1} \cdot \left(\eta - \frac{3 \cdot a \cdot h_1^2}{h^3} \right) \tag{E11-13}$$

After transforming:

$$1=\frac{l}{l_1}\cdot\left(\eta-\frac{3\cdot a\cdot h_1^2}{h^3}\right)=\frac{h^3}{h_1^3}\cdot\left(\eta-\frac{3\cdot a\cdot h_1^2}{h^3}\right) \tag{E11-14}$$

By solving for η and replacing $h-a$ with h_1, equation (E11-14) can be transformed as follows:

$$\eta=\frac{h_1^3}{h^3}+\frac{3\cdot a\cdot h_1^2}{h^3}=\frac{(h-a)^3}{h^3}+\frac{3\cdot a\cdot (h-a)^2}{h^3}=\frac{h^3-3\cdot a^2\cdot h+2\cdot a^3}{h^3}$$

$$\eta=1-3\cdot\left(\frac{a}{h}\right)^2+2\cdot\left(\frac{a}{h}\right)^3 \tag{E11-15}$$

The portion η of load F generating tension perpendicular to the grain was determined based on the distance a of the load application point from the underside of the beam and the beam depth h. Equation (E11-15) also applies for loads, which are not applied at mid-span as during derivation, only the coefficients change, but are cancelled again. Superimposing concentrated loads allows any load distribution of choice to be attained. Equation (E11-15) is thus generally applicable, e.g. also for distributed loads.

Equation (E11-15) is also used when there is a need to determine the governing tensile stress components perpendicular to the grain of reinforced notches, Equation (D8-4) in Article D8.

Design based on a stress criterion

The stress state in a member is explained using a joint loaded perpendicular to the grain and with multiple fasteners as an example. The volume subject to tension perpendicular to the grain influences the stress distribution, which means the joint geometry must be taken into consideration when determining the stress. Figure E11-2 includes a definition of the geometric impact variables for a double-sided joint loaded perpendicular to the grain. For simplicity, it is assumed that the load F_{90} to be transmitted is uniformly distributed to all fasteners, although the actual load distribution is uneven. It varies depending on the local strength and stiffness properties of the wood as well as deformation of the members in the area of the joint. We now examine the stress distribution perpendicular to the grain direction in one row of fasteners and in the member subject to tension perpendicular to the grain (Figure E11-7).

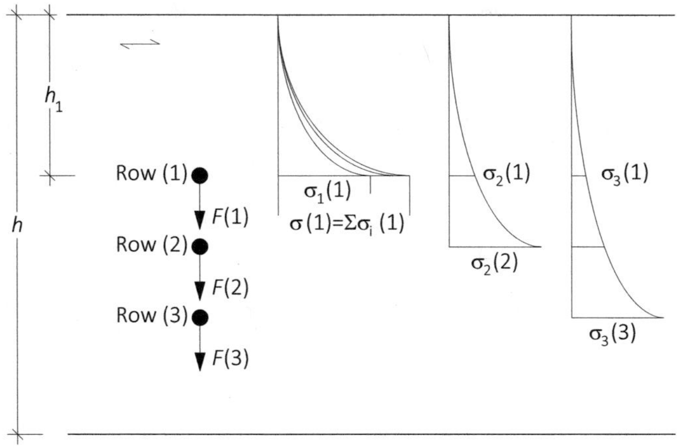

Figure E11-7 Distribution of tensile stresses perpendicular to the grain in multiple rows of fasteners.

The uppermost fastener in row (1) generates tension perpendicular to the grain $\sigma_1^{(1)}$ after force $F^{(1)}$ is introduced into the member. The tensile stress perpendicular to the grain peaks at the point at which the load is introduced and declines with increasing distance from the fastener in a parabolic shape. The tensile stress perpendicular to the grain is zero at the upper edge of the member, while the stresses perpendicular to the grain introduced by the fasteners in rows (2) and (3) lead to the stress distributions shown with the maxima $\sigma_2^{(2)}$ and $\sigma_3^{(3)}$. The respective stress distributions due to the different rows of fasteners must now be superimposed. To the stress exerted at the fastener position in row (1) caused by load $F^{(1)}$, the respective stresses at this position caused by $F^{(2)}$ and $F^{(3)}$ from fasteners in rows (2) and (3) have to be added. For the rows of fasteners underneath row (1), the stress components are also superimposed. Subsequently, however, comes the need to take compression perpendicular to the grain into consideration; that develop similar to tension perpendicular to the grain. The tensile and compressive stresses perpendicular to the grain superpose, the resulting tensile stresses perpendicular to the grain are therefore lower than the same in row (1). **The key criterion for tensile stresses perpendicular to the grain of a joint loaded perpendicular to the grain is thus always the fastener row furthest from the loaded member edge.** The sum of tensile stresses perpendicular to the grain of a joint loaded perpendicular to the grain with *n* rows of fasteners in the area of a fastener in row (1) amounts to:

$$\sigma^{(1)} = \sum_{i=1}^{n} \sigma_i^{(1)} = \sigma_1^{(1)} + \sigma_2^{(1)} + \sigma_3^{(1)} \qquad \text{(E11-16)}$$

The additional stress components from series (2) and (3) which are added to the tensile stress perpendicular to the grain $\sigma_1^{(1)}$ are initially unknown and depend on the respective stress distributions. The effective tensile stress components $\sigma_i^{(1)}$ in the uppermost fastener row from row (i) are also smaller by factor $(h_1/h_i)^2$ than the component $\sigma_1^{(1)}$ from the first row of fasteners:

$$\sigma_i^{(1)} = \sigma_1^{(1)} \cdot \left(\frac{h_1}{h_i} \right)^2 \tag{E11-17}$$

Using this relationship, which is determined by numerical investigations, the sum of tensile stresses perpendicular to the grain in row (1) can also be determined:

$$\sigma_{ges}^{(1)} = \sum_{i=1}^{n} \sigma_1^{(1)} \cdot \left(\frac{h_1}{h_i} \right)^2 \tag{E11-18}$$

To calculate the overall stress in row (1) in accordance with equation (E11-18), the size of the load $F_t^{(1)}$, that is generating tension perpendicular to the grain and that is a portion of the load $F^{(1)}$, is required, as well as the area subject to tension perpendicular to the grain. The load portion $F_t^{(1)}$ introduced by the fasteners in row (1) can be calculated using equations (E11-3) and (E11-15):

$$F_t^{(1)} = \eta \cdot F^{(1)} \tag{E11-19}$$

For joints loaded perpendicular to the grain with n rows of fasteners and assuming the force F is uniformly distributed over the rows of fasteners ($F^{(1)} = F/n$), the force $F_t^{(1)}$ of the fasteners in row (1) can be calculated as:

$$F_t^{(1)} = \frac{\eta \cdot F}{n} \cdot \sum_{i=1}^{n} \left(\frac{h_1}{h_i} \right)^2 \tag{E11-20}$$

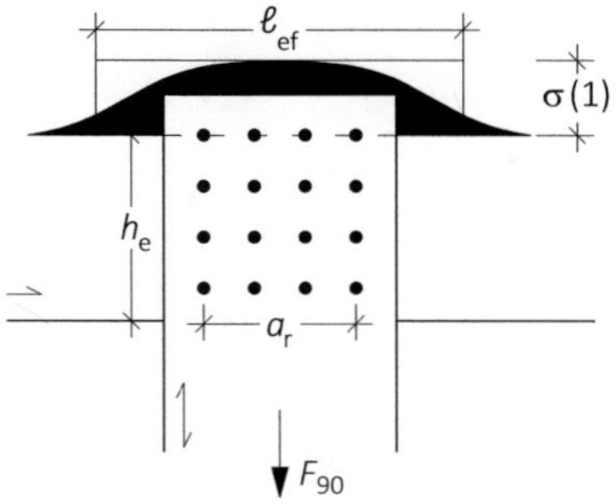

Figure E11-8 Distribution of tensile stresses perpendicular to the grain in longitudinal beam direction and with effective width ℓ_{ef}.

Now, the area exposed to tensile stress perpendicular to the grain is still missing. In the longitudinal beam direction, i.e. parallel to the grain direction of the member subject to tension perpendicular to the grain, however, the stresses are non-linearly distributed and decline with increasing distance from the joint (Figure E11-8). Accordingly, the effective area A_{ef} subject to tension perpendicular to the grain must be known to determine the tensile stresses perpendicular to the grain:

$$A_{ef} = \ell_{ef} \cdot t_{ef} \tag{E11-21}$$

The integral of the stress distribution corresponds to the area of a fictional stress distribution with constant tensile stresses perpendicular to the grain $\sigma^{(1)}$ and effective length ℓ_{ef}, rectangular area in Figure E11-8. The effective joint width ℓ_{ef} depends on geometric boundary conditions:

$$\ell_{ef} = \sqrt{a_r^2 + (c \cdot h)^2} \tag{E11-22}$$

The factor c was empirically determined using test results:

$$c = \frac{4}{3} \cdot \sqrt{\frac{h_e}{h} \cdot \left(1 - \frac{h_e}{h}\right)^3} \tag{E11-23}$$

The effective joint (penetration) depth t_{ef} depends on the respective fastener type.

Similarly to when the fasteners are arranged in multiple rows, distributing the fasteners in the grain direction over an extended length a_r serves to improve the load-bearing behaviour. However, this approach also increases the volume subject to tensile stresses, which, in turn, adversely affects the load-bearing capacity. Overall however, distributing the fasteners over a wide area is considered beneficial.

According to equation (E11-20), the key tensile stress perpendicular to the grain $\sigma_{t,90}$ in the fastener row placed furthest away from the loaded member edge is therefore (with the abbreviation k_1):

$$\sigma_{t,90} = \frac{F^{(1)}}{A_{ef}} = \frac{\eta \cdot F_{90}}{n \cdot A_{ef}} \cdot \sum_{i=1}^{n} \left(\frac{h_1}{h_i} \right)^2 = \frac{\eta \cdot F_{90}}{A_{ef}} \cdot k_1 \qquad \text{(E11-24)}$$

$$k_1 = \frac{1}{n} \cdot \sum_{i=1}^{n} \left(\frac{h_1}{h_i} \right)^2 \qquad \text{(E11-25)}$$

The factor k_1 considers the fact that, when using multiple rows of fasteners, the tensile stresses perpendicular to the grain are lower in the row of fasteners furthest away from the loaded member edge.

To verify the load-bearing capacity of a joint loaded perpendicular to the grain, the tensile stress must be compared with the tension strength perpendicular to the grain. Since tensile failure is classed as a brittle form of failure, the joint capacity cannot be readily verified using a stress criterion. The influence of volume on the tension strength perpendicular to the grain is taken into consideration, by multiplying the tension strength, benchmarked to a reference volume of 0.1 m^3 and with a coefficient k_{vol}:

$$\sigma_{t,90} = \frac{\eta \cdot F_{90}}{A_{ef}} \cdot k_1 \leq f_{t,90} \cdot k_{vol} \qquad \text{(E11-26)}$$

The coefficient k_{vol} depends on the size of the effective area A_{ef} subject to tension perpendicular to the grain in accordance with equation (E11-21):

$$k_{vol} = 13 \cdot A_{ef}^{-0.2} \quad \text{with} \quad A_{ef}^{-0.2} \text{ in mm}^2 \qquad \text{(E11-27)}$$

Since joints loaded perpendicular to the grain always involve concentrated loads being introduced in members, it is advisable to convert equation (E11-26) to a load F_{90}, which then allows the load which the joint can withstand to be calculated directly.

$$F_{90} \leq \frac{f_{t,90} \cdot k_{vol} \cdot A_{ef}}{k_1 \cdot \eta} = 13 \cdot \frac{f_{t,90} \cdot A_{ef}^{0.8}}{k_1 \cdot \eta} \qquad \text{(E11-28)}$$

The verification can also be expressed by inserting equation (E11-22):

$$F_{90} \le 13 \cdot \left(t_{ef} \cdot h\right)^{0.8} \cdot \frac{c^{0.8}}{\eta} \cdot f_{t,90} \cdot \frac{k_2}{k_1} \tag{E11-29}$$

The factor k_2 allows the impact of the number of adjacent fasteners on stress distribution to be taken into consideration:

$$k_2 = \left(1 + \left(\frac{a_r}{c \cdot h}\right)^2\right)^{0.4} \tag{E11-30}$$

An approximation function allows the factor k_2 to be expressed in simplified form:

$$k_2 = 0.7 + 1.4 \cdot \frac{a_r}{h} > 1 \tag{E11-31}$$

The expression $c^{0.8}/\eta$ in equation (E11-29) depends solely on the quotient h_e/h and can also be simplified by an approximation function:

$$\frac{c^{0.8}}{\eta} \approx 6.5 + 18 \cdot \left(\frac{h_e}{h}\right)^2 \tag{E11-32}$$

Equation (E11-29) is transformed accordingly:

$$F_{90,max} = 13 \cdot \frac{k_2}{k_1} \cdot \left(6.5 + 18 \cdot \left(\frac{h_e}{h}\right)^2\right) \cdot \left(t_{ef} \cdot h\right)^{0.8} \cdot f_{t,90} \tag{E11-33}$$

Equation (E11-33) for the maximum shear load $F_{90,max}$ allows the load-bearing capacity of a joint loaded perpendicular to the grain to be calculated assuming the failure of the member subject to tension perpendicular to the grain.

The design equation specified in the NA (NCI to 8.1.4) for joints loaded perpendicular to the grain is based on equation (E11-33) and reads as follows (for geometric details see Figure E11-2):

$$F_{\mathrm{v,Ed}} \leq F_{\mathrm{90,Rd}} \tag{E11-34}$$

where

$$F_{\mathrm{90,Rd}} = k_s \cdot k_r \cdot \left(6.5 + 18 \cdot \left(\frac{h_\mathrm{e}}{h}\right)^2\right) \cdot \left(t_\mathrm{ef} \cdot h\right)^{0.8} \cdot f_{\mathrm{t,90}} \tag{E11-35}$$

$$k_\mathrm{s} = \max\left(1; 0.7 + 1.4 \cdot \frac{a_\mathrm{r}}{h}\right) \tag{E11-36}$$

$$k_\mathrm{r} = \frac{n}{\sum_{\mathrm{i}=1}^{\mathrm{n}} \left(\frac{h_1}{h_\mathrm{i}}\right)^2} \tag{E11-37}$$

The following applies:

k_s Coefficient, which takes into account an increase in the tension strength perpendicular to the grain due to multiple adjacently arranged fasteners within a fastener group (corresponds to factor k_2 in equation (E11-33), see also equation (E11-31)).

k_r Coefficient k_r considers the fact that for joints with multiple (n) rows of fasteners, arranged on top of each other, lower tensile stresses perpendicular to the grain are exerted in the fastener row furthest away from the loaded member edge (corresponds to factor k_1 in equation (E11-33), see also equation (E11-25)).

t_ef Effective joint (penetration) depth in mm; different values apply for single- or double-sided cross joints as well as for different joint types, see NA.

The specified design of joints loaded perpendicular to the grain in equations (E11-35) to (E11-37) must only be implemented for ratios $h_\mathrm{e}/h \leq 0.7$. In the NA, no verifications are required for ratios $h_\mathrm{e}/h > 0.7$, while the key criterion for determining the load-bearing capacity of joints with dowel-type fasteners are the Johansen cases (with rope effect). Joints with $h_\mathrm{e}/h < 0.2$ may only be stressed by a short-term load. Additional regulations apply for joints with $a_\mathrm{r} / h > 1$ and $F_{\mathrm{v,Ed}} > 0.5 \cdot F_{\mathrm{90,Rd}}$; which always require reinforcement. In addition, rules apply for multiple groups of fasteners.

E11.4 Literature

J. Ehlbeck, R. Görlacher, original Article C2, STEP 1995.

Ehlbeck J. and Görlacher R. (1983). Tragverhalten von Queranschlüssen mittels Stahlformteilen, insbesondere Balkenschuhen, im Holzbau. Forschungsbericht der Versuchsanstalt für Stahl, Holz und Steine, Universität Karlsruhe.

Van der Put T.A.C.M. and Leijten A.J.M. (2000). Evaluation of perpendicular to grain failure of beams caused by concentrated loads of joints. Paper 33-7-7, CIB-W18 Meeting 33, Delft.

Leijten A.J.M. and Jorissen A.J.M. (2001). Splitting strength of beams loaded by connections perpendicular to grain, model validation. Paper 34-7-1, CIB-W18 Meeting 34, Venice.

E12 Reinforced joints

Original article: H. Werner

The load-bearing capacity of the joint area of timber members is the primary weak point when considering the load-bearing behaviour of the overall timber construction. Dowel-type fasteners such as nails, dowels or screws are often used to connect members, in configurations which could include timber-to-timber, timber-to-wood-based products or steel-to-timber joints. Taking the load-bearing capacity of the cross-section of the connected timber members as the benchmark, the effectiveness of such joints generally ranges from around 40 to 60%, which means the structural joint detailing has a crucial impact on the performance and thus efficiency of a structure. If the key influence parameters on the load-bearing and deformation behaviour of the joint as well as the different failure types are known, targeted changes made to the structural detailing may improve the load-bearing capacity of the joint, reducing the extent of the member cross-section required.

As was already shown in previous articles (primarily E1), joints with dowel-type fasteners may have a range of failure mechanisms. Embedment failure alone occurs if the timber or wood-based material under the fastener is completely plasticised and the fastener remains straight. The key parameters for this failure mode are the embedment strength of the timber or wood-based material plus the joint geometry (thickness of the members to be connected, spacings and end and edge distances of the fasteners). Another failure mode affecting joints with dowel-type fasteners is a bending failure of the fastener and simultaneous embedment failure of the timber or wood-based material. When such combined failure occurs, one or two plastic hinges emerge per shear plane and the fastener becomes inclined in the area of the shear plane leading to plastic embedment deformations. Here too, the embedment strength, joint geometry and yield moment of the fastener are the key parameters impacting on the load-bearing capacity. The third – brittle – failure mode occurs when the timber splits or in the event of block shear failure and mainly affects joints with multiple fasteners. Complying with the minimum fastener distances helps avoid premature splitting; the abovementioned ductile failure modes described (embedment or combined failures involving bending deformation of the fastener and embedment) are only seen when complying with minimum distances.

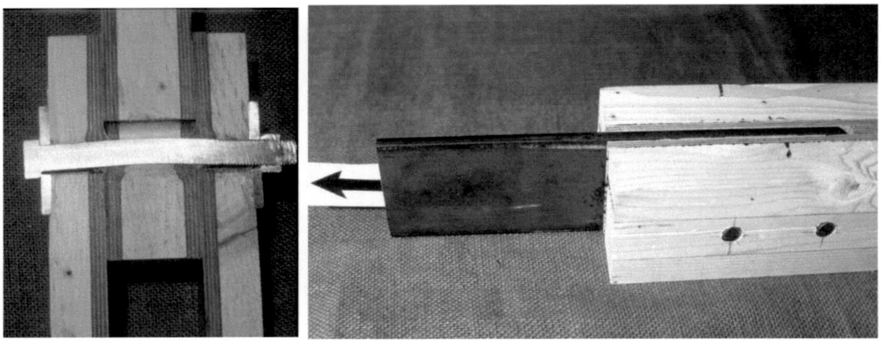

Figure E12-1 Double-shear timber-to-timber joint with glued-on plywood (left) and double-shear steel-to-timber joint with slotted-in steel plate and glued-on wood-based panels (right).

While using fasteners made of high-strength, ductile steel and timber of higher density can boost the load-bearing capacity of even non-reinforced joints, while reinforcement applied in the form of glued-on wood-based panels or fully threaded screws is intended to avoid premature splitting or local shear failure in the area of the joint. This approach can meet two aims: firstly, avoiding premature failure and secondly, in comparison to unreinforced joints, allowing lower fastener spacing. The group effect, which has a particularly adverse effect on joints with multiple fasteners arranged in a row in the force and grain direction (see also Article E13), can also be minimised. In addition, the embedment strength and hence load-bearing capacity of the individual fastener can be increased.

The first choice for glued-on reinforcements are wood-based panels with an embedment strength far exceeding that of the timber to be reinforced. Reinforcing joints with this approach not only allows embedment strengths close to shear planes of connected members to be increased, but also offers additional reinforcement against tension perpendicular to the grain in the area of the fasteners which, in turn, reduces susceptibility to splitting. The results obtained by Blass and Werner (1988) with glued-on beech plywood in the shear plane area of softwood joints showed that gluing-on panels significantly reduces the susceptibility of softwood to splitting and exploits the higher embedment strength of the plywood in the area of the joint where the loading stresses peak, thus increasing the overall load-bearing capacity of the joint. Two examples of such reinforcement are shown in Figure E12-1.

Other means of increasing the load-bearing capacity can be achieved with nail-plate connectors (Kevarinmäki et al., 1995; Blass and Schmid, 2001) or glued-on glass-grain mats (Haller et al., 1998). A nail plate is pressed in, in the area of the joint, where the central nail plate area is without punch-outs, whereupon the individual reinforced members are connected with a dowel-type fastener. If the joint is loaded laterally, both the nail plate and timber are exposed to embedment stresses. This method considerably increases the load-bearing capacity of the joint and helps prevent any splitting of the timber. The nail-plate connector can be considered a refined version of the single-sided toothed-plate connector (Article E6). In this case, however, the load-bearing behaviour differs due to the fact that for toothed-plate connectors, the critical parameter is the embedment strength of the timber in the area of the teeth, whereas for nail-plate connectors, the embedment strength of the nail plate itself is the key variable. A joint reinforced with nail-plate connectors as well as a close-up view of the cross-section through the joint, including bending deformation of the fastener and embedment deformation of the timber and nail plate, is shown in Figure E12-2. Since both the reinforcement methods mentioned, nail-plate connectors or glued-on glass grain mats, are not customary in Germany, no further detail will be provided in this article.

The wood in the joint area of connections with dowel-type fasteners can also be reinforced with self-tapping screws, arranged perpendicular to the grain and at right angles to the actual dowel-type fasteners in two possible configurations, as shown in Figure E12-3. Fully threaded screws can be arranged at a distance to the dowel-type fasteners in the grain direction (Figure E12-3 left), which ensures that any crack propagation will not extend beyond the fully threaded screws, although any increase in load-bearing capacity (beyond the Johansen case) will not be possible. Installing fully threaded screws directly underneath the fasteners (Figure E12-3 right), conversely, not only prevents splitting but also boosts load-bearing capacity due to the contribution of the screw acting as a "beam" supporting the fasteners. This kind of reinforcement therefore additionally increases the load-bearing capacity of a joint not at risk of splitting.

The following section explains reinforcement measures with glued-on wood-based panels and fully threaded screws in more detail.

Figure E12-2 A joint with tube fasteners that has been reinforced with nail-plate connectors (left), Close-up view of cross-section with reinforced joint and embedment deformations of the reinforcement (right).

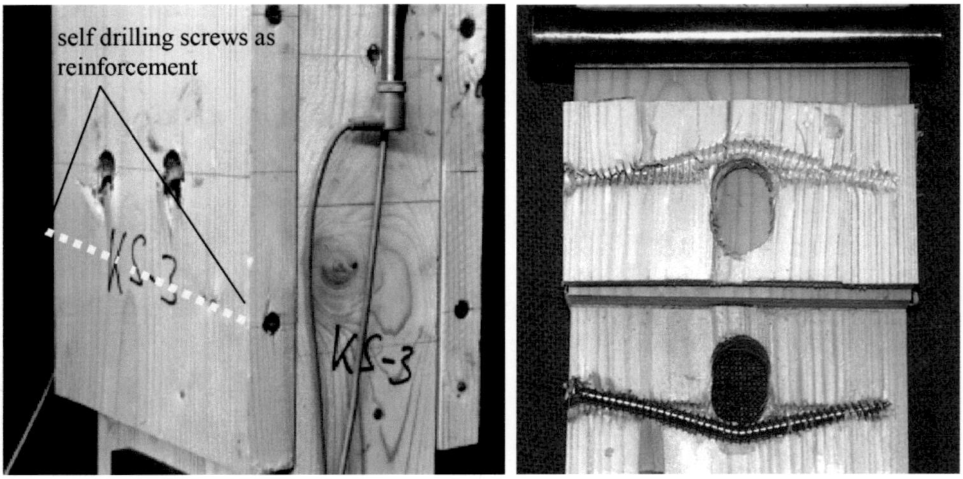

Figure E12-3 Joints reinforced with fully threaded screws. Left: reinforcement with a distance to the dowels. (Schmid, 2002) Right: opened specimen after test with reinforcement directly underneath the dowel. (Bejtka, 2005a)

E12.1 Joints reinforced with glued-on wood-based panels

Article E2 derived the Johansen equations for single- and double-shear timber joints and the load-bearing capacity of reinforced joints can be derived similarly, as will subsequently be shown in detail citing the example of a steel-to-timber joint.

Derivation of the load-bearing capacity $F_{v,s,R}$ of a reinforced joint

Taking the example of a laterally loaded steel-to-timber joint, featuring thick external steel plates and two shear planes (Figure E12-4), equations for the failure mode with two plastic hinges are derived for calculating reinforced joints. The input parameters for the calculation are the joint geometry (thickness of the timber member t_2, thickness of the reinforcement s, fastener diameter d), the embedment strengths of the timber f_h and the reinforcement $f_{h,s}$ as well as the yield moment M_y of the fastener. In all the specified equations below, if the thickness s of the reinforcement is set to zero, the Johansen equations for unreinforced joints from Article E2 emerge.

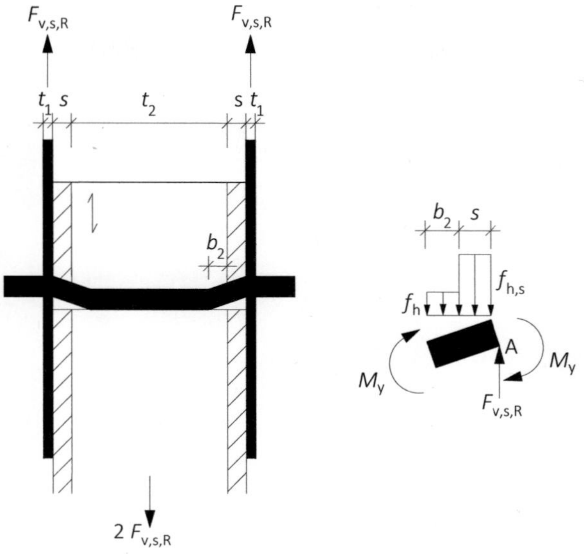

Figure E12-4 Left: double-shear steel-to-timber joint with thick, external steel plates, embedment strength in timber and reinforcement is exceeded and two plastic hinges have formed. Right: free-body diagram of the fastener.

The equilibrium of forces $\Sigma V = 0$ results in:

$$F_{v,s,R} - f_h \cdot d \cdot b_2 - f_{h,s} \cdot d \cdot s = 0 \iff F_{v,s,R} = f_h \cdot d \cdot b_2 + f_{h,s} \cdot d \cdot s \tag{E12-1}$$

The equilibrium of moments at the fastener $\Sigma M^A = 0$ is:

$$2 \cdot M_y - \left(f_{h,s} - f_h\right) \cdot d \cdot s \cdot \frac{s}{2} - f_h \cdot d \cdot \left(b_2 + s\right) \cdot \frac{b_2 + s}{2} = 0$$

Solving for b_2:

$$b_2 = \sqrt{\frac{4 \cdot M_y}{f_h \cdot d} - s^2 \cdot \left(\frac{f_{h,s}}{f_h} - 1\right)} - s \tag{E12-2}$$

Inserting equation (E12-2) in equation (E12-1) reveals the load-bearing capacity $F_{v,s,R}$:

$$F_{v,s,R} = f_{h,s} \cdot d \cdot s + f_h \cdot d \cdot \left[\sqrt{\frac{4 \cdot M_y}{f_h \cdot d} - s^2 \cdot \left(\frac{f_{h,s}}{f_h} - 1\right)} - s\right] \tag{E12-3}$$

The load-bearing capacity $F_{v,s,R}$ of a reinforced joint can be derived similarly for all other failure modes and joint typologies (see also Article E2 and Annex 6). In the following section, equations to calculate the load-bearing capacity for all failure mechanisms are specified for single- and double-shear timber-to-timber and steel-to-timber joints with reinforced joint areas. To simplify matters, equal thickness and embedment strength of the reinforcement panels arranged at the shear planes of a joint are assumed. The following designations apply:

t_1 , t_2	Thickness of the timber members or penetration depth
s	Thickness of the reinforcement
$f_{h,1}$, $f_{h,2}$	Embedment strength of the members
$f_{h,s}$	Embedment strength of the reinforcement
M_y	Yield moment of the fastener
d	Diameter of the fastener
β	Ratio of the embedment strengths of the members: $\beta = f_{h,2}/f_{h,1}$
η	Ratio of the embedment strength of the reinforcement and member 1: $\eta = f_{h,s}/f_{h,1}$

Reinforced timber-to-timber joints

Reinforced, single-shear joints made of timber or wood-based materials

The load-bearing capacity $F_{v,s,R}$ per fastener is determined as the lowest value from the following equations (E12-4) to (E12-9):

$$F_{v,s,R} = f_{h,1} \cdot d \cdot t_1 + f_{h,s} \cdot d \cdot s \tag{E12-4}$$

(embedment failure in member 1 (case a in Figure E2-5))

$$F_{v,s,R} = f_{h,2} \cdot d \cdot t_2 + f_{h,s} \cdot d \cdot s \tag{E12-5}$$

(embedment failure in member 2 (case b in Figure E2-5))

$$F_{v,s,R} = \frac{f_{h,1} \cdot d \cdot t_1}{1+\beta} \cdot \sqrt{\beta \cdot \left(1 - 4 \cdot \eta \cdot \frac{s^2}{t_1^2}\right) + 2 \cdot \beta^2 \cdot \left(1 + \frac{t_2}{t_1} + \frac{t_2^2}{t_1^2} + 4 \cdot \frac{s}{t_1} + 8 \cdot \frac{s^2}{t_1^2} + 4 \cdot \frac{s \cdot t_2}{t_1^2} - 2 \cdot \eta \cdot \frac{s^2}{t_1^2}\right) + \beta^3 \cdot \frac{t_2^2}{t_1^2}} - \frac{\beta \cdot f_{h,1} \cdot d \cdot t_1}{1+\beta} \cdot \left(1 + \frac{t_2}{t_1} + 4 \cdot \frac{s}{t_1}\right) + f_{h,s} \cdot d \cdot s \tag{E12-6}$$

(embedment failure in both members (case c in Figure E2-5))

$$F_{v,s,R} = \frac{\beta \cdot f_{h,1} \cdot d}{2+\beta} \cdot \left[\sqrt{(t_1 + 4 \cdot s)^2 + \frac{2+\beta}{\beta} \cdot \left(t_1^2 - 4 \cdot \eta \cdot s^2 + \frac{4 \cdot M_y}{f_{h,1} \cdot d} \right)} - (t_1 + 4 \cdot s) \right] + \\ + f_{h,s} \cdot d \cdot s \tag{E12-7}$$

(embedment failure in both members, plastic hinge in member 2
(case d in Figure E2-5))

$$F_{v,s,R} = f_{h,s} \cdot d \cdot s + \frac{\beta \cdot f_{h,1} \cdot d}{1+2 \cdot \beta} \cdot \\ \cdot \left[\sqrt{(t_2 + 4 \cdot s)^2 + (1+2 \cdot \beta) \cdot \left(t_2^2 - 4 \cdot \eta \cdot s^2 + \frac{4 \cdot M_y}{\beta \cdot f_{h,1} \cdot d} \right)} - (t_2 + 4 \cdot s) \right] \tag{E12-8}$$

(embedment failure in both members, plastic hinge in member 1
(case e in Figure E2-5))

$$F_{v,s,R} = \frac{2 \cdot \beta \cdot f_{h,1} \cdot d}{1+\beta} \cdot \left[\sqrt{s^2 - \frac{1+\beta}{2 \cdot \beta} \cdot \left(\eta \cdot s^2 - \frac{2 \cdot M_y}{f_{h,1} \cdot d} \right)} - s \right] + f_{h,s} \cdot d \cdot s \tag{E12-9}$$

(embedment failure and plastic hinges in both members (case f in Figure E2-5))

Reinforced, double-shear joints made of timber or wood-based materials

The load-bearing capacity $F_{v,s,R}$ per fastener and shear plane is determined as the lowest value from the following equations (E12-10) to (E12-13) (member 2 is at the centre):

$$F_{v,s,R} = f_{h,1} \cdot d \cdot t_1 + f_{h,s} \cdot d \cdot s \tag{E12-10}$$

(embedment failure in member 1 (case g in Figure E2-8))

$$F_{v,s,R} = 0{,}5 \cdot f_{h,2} \cdot d \cdot t_2 + f_{h,s} \cdot d \cdot s \tag{E12-11}$$

(embedment failure in member 2 (case h in Figure E2-8))

$$F_{v,s,R} = \frac{\beta \cdot f_{h,1} \cdot d}{2+\beta} \cdot \left[\sqrt{(t_1 + 4 \cdot s)^2 + \frac{2+\beta}{\beta} \cdot \left(t_1^2 - 4 \cdot \eta \cdot s^2 + \frac{4 \cdot M_y}{f_{h,1} \cdot d} \right)} - (t_1 + 4 \cdot s) \right] + \\ + f_{h,s} \cdot d \cdot s \tag{E12-12}$$

(embedment failure in both members, plastic hinge in member 2
(case j in Figure E2-8))

$$F_{v,s,R} = \frac{2 \cdot \beta \cdot f_{h,1} \cdot d}{1+\beta} \cdot \left[\sqrt{s^2 - \frac{1+\beta}{2 \cdot \beta} \cdot \left(\eta \cdot s^2 - \frac{2 \cdot M_y}{f_{h,1} \cdot d} \right)} - s \right] + f_{h,s} \cdot d \cdot s \qquad \text{(E12-13)}$$

(embedment failure and plastic hinges in both members
(case k in Figure E2-8))

Reinforced steel-to-timber joints

Reinforced, single-shear joints with thin steel plates

The load-bearing capacity $F_{v,s,R}$ per fastener for single-shear joints with thin steel plates (namely for a steel plate of thickness $t \leq 0.5 \cdot d$) is determined as the lowest value from the following equations (E12-14) and (E12-15):

$$F_{v,s,R} = f_{h,1} \cdot d \cdot \left[\sqrt{2 \cdot t_1^2 + 2 \cdot (2-\eta) \cdot s^2 + 4 \cdot s \cdot t_1} - (t_1 + 2 \cdot s) \right] + f_{h,s} \cdot d \cdot s \qquad \text{(E12-14)}$$

(embedment failure (case a in Figure E2-9))

$$F_{v,s,R} = f_{h,1} \cdot d \cdot \left[\sqrt{(1-\eta) \cdot s^2 + \frac{2 \cdot M_y}{f_{h,1} \cdot d}} - s \right] + f_{h,s} \cdot d \cdot s \qquad \text{(E12-15)}$$

(embedment failure and plastic hinge (case b in Figure E2-9))

Reinforced, single-shear joints with thick steel plates

The load-bearing capacity $F_{v,s,R}$ per fastener for single-shear joints with thick steel plates (namely for a steel plate thickness $t \geq d$) is determined as the lowest value from the following equations (E12-16) to (E12-18):

$$F_{v,s,R} = f_{h,1} \cdot d \cdot t_1 + f_{h,s} \cdot d \cdot s \qquad \text{(E12-16)}$$

(embedment failure (case c in Figure E2-9))

$$F_{v,s,R} = f_{h,1} \cdot d \cdot \left[\sqrt{2 \cdot t_1^2 + 2 \cdot (2-\eta) \cdot s^2 + 4 \cdot s \cdot t_1 + \frac{4 \cdot M_y}{f_{h,1} \cdot d}} - (t_1 + 2 \cdot s) \right] + \qquad \text{(E12-17)}$$

$$+ f_{h,s} \cdot d \cdot s$$

(embedment failure and plastic hinge (case d in Figure E2-9))

$$F_{\text{v,s,R}} = f_{\text{h,1}} \cdot d \cdot \left[\sqrt{(1-\eta) \cdot s^2 + \frac{4 \cdot M_{\text{y}}}{f_{\text{h,1}} \cdot d}} - s \right] + f_{\text{h,s}} \cdot d \cdot s \qquad \text{(E12-18)}$$

(embedment failure and 2 plastic hinges (case e in Figure E2-9))

Reinforced, double-shear joints with steel plate as central member

The load-bearing capacity $F_{v,s,R}$ per fastener and per shear plane is determined as the lowest value from the following equations (E12-19) to (E12-21):

$$F_{\text{v,s,R}} = f_{\text{h,1}} \cdot d \cdot t_1 + f_{\text{h,s}} \cdot d \cdot s \qquad \text{(E12-19)}$$

(embedment failure (case f in Figure E2-9))

$$F_{\text{v,s,R}} = f_{\text{h,1}} \cdot d \cdot \left[\sqrt{2 \cdot t_1^2 + 2 \cdot (2-\eta) \cdot s^2 + 4 \cdot s \cdot t_1 + \frac{4 \cdot M_{\text{y}}}{f_{\text{h,1}} \cdot d}} - (t_1 + 2 \cdot s) \right] + \\ + f_{\text{h,s}} \cdot d \cdot s \qquad \text{(E12-20)}$$

(embedment failure and plastic hinge (case g in Figure E2-9))

$$F_{\text{v,s,R}} = f_{\text{h,1}} \cdot d \cdot \left[\sqrt{(1-\eta) \cdot s^2 + \frac{4 \cdot M_{\text{y}}}{f_{\text{h,1}} \cdot d}} - s \right] + f_{\text{h,s}} \cdot d \cdot s \qquad \text{(E12-21)}$$

(embedment failure and 2 plastic hinges (case h in Figure E2-9))

Reinforced, double-shear joints with external, thin steel plates

The load-bearing capacity $F_{v,s,R}$ per fastener and per shear plane is determined as the lowest value from the following equations (E12-22) and (E12-23):

$$F_{\text{v,s,R}} = 0.5 \cdot f_{\text{h,2}} \cdot d \cdot t_2 + f_{\text{h,s}} \cdot d \cdot s \qquad \text{(E12-22)}$$

(embedment failure (case j/l in Figure E2-9))

$$F_{\text{v,s,R}} = f_{\text{h,2}} \cdot d \cdot \left[\sqrt{\left(1 - \frac{\eta}{\beta}\right) \cdot s^2 + \frac{2 \cdot M_{\text{y}}}{f_{\text{h,2}} \cdot d}} - s \right] + f_{\text{h,s}} \cdot d \cdot s \qquad \text{(E12-23)}$$

(embedment failure and plastic hinge (case k in Figure E2-9))

Reinforced, double-shear joints with external, thick steel plates

The load-bearing capacity $F_{v,s,R}$ per fastener and per shear plane is determined as the lowest value from the following equations (E12-24) and (E12-25) (whereby equation (E12-25) corresponds to the above-derived equation (E12-3)):

$$F_{v,s,R} = 0.5 \cdot f_{h,2} \cdot d \cdot t_2 + f_{h,s} \cdot d \cdot s \qquad \text{(E12-24)}$$

(embedment failure (case j/l in Figure E2-9))

$$F_{v,s,R} = f_{h,s} \cdot d \cdot s + f_h \cdot d \cdot \left[\sqrt{\frac{4 \cdot M_y}{f_h \cdot d} - s^2 \cdot \left(\frac{f_{h,s}}{f_h} - 1 \right)} - s \right] \qquad \text{(E12-25)}$$

(embedment failure and 2 plastic hinges (case m in Figure E2-9))

Verification of the bond line for glued-on wood-based panels

When reinforcing a joint with wood-based panels, the bonding surface A_L must also be verified and meet the following condition:

$$A_L \geq \frac{f_{h,s}}{f_v} \cdot n \cdot s \cdot d \qquad \text{(E12-26)}$$

where

$f_{h,s}$ Embedment strength of the wood-based panel
f_v Effective shear strength of the wood-based panel (out-of-plane, rolling shear) or timber, whichever is smaller
n Number of fasteners
s Panel thickness
d Diameter of the fastener

The gluing-on and the rigid joining it provides between timber and the reinforcement mean glued-on wood-based panels are the only type of reinforcement that can be relied on to also reduce the risk of block shear in joints with multiple fasteners (Article E13). This is not possible for mechanical reinforcements (fully threaded screws), which provide far greater semi-rigidity in the joint.

Embedment strength of reinforcing wood-based panels

Embedment strength of various wood-based panels

The embedment strength of wood-based panels is dictated by the type of panel and its build-up. For plywood, particleboards, OSB and hardboards, EC 5 specifies the following equations to calculate the characteristic embedment strength in N/mm^2 (where ρ_k is expressed in kg/m^3 and d and t in mm):

- Nails and screws in non-predrilled plywood: $f_{h,k} = 0.11 \cdot \rho_k \cdot d^{-0.3}$
- Dowels, bolts, nails and screws in predrilled plywood: $f_{h,k} = 0.11 \cdot (1 - 0.01 \cdot d) \cdot \rho_k$
- Nails and screws in non-predrilled particleboards or OSB: $f_{h,k} = 65 \cdot d^{-0.7} \cdot t^{0.1}$
- Dowels, bolts, nails and screws in predrilled particleboards or OSB: $f_{h,k} = 50 \cdot d^{-0.6} \cdot t^{0.2}$
- Nails in hardboards: $f_{h,k} = 30 \cdot d^{-0.3} \cdot t^{0.6}$

The best choice when selecting a reinforcement measure for joints is wood-based panels with an embedment strength that considerably exceeds that of the wood to be reinforced.

E12.2 Joints reinforced with fully threaded screws

Joints reinforced with fully threaded screws tend to be dowelled joints, in which timber members are subject to loading in the grain direction. The reinforcing elements (= fully threaded screws) are always arranged perpendicularly to the grain direction and the dowel axis, to prevent premature splitting (for joints with multiple fasteners), see Figure E12-3. If the reinforcement is arranged directly underneath the dowels rather than at a distance from them, however, the load-bearing capacity of the reinforced joint can be increased even further (increased embedment capacity).

Fully threaded screws, arranged at a distance from the dowels

When inserting fully threaded screws at a distance from the dowels (Figure E12-3 on the left), the sole function of the screws is to reinforce against tension perpendicular to the grain and prevent premature splitting and therefore premature failure at a lower load-bearing capacity than the respective Johansen case. The reinforcement prevents cracks extending any further than the fully threaded screws, although it is not possible to increase load-bearing capacity further (beyond the Johansen case). This means the effective number of fasteners can be reconciled with the actual number:

$$n_{ef} = n \qquad \text{(E12-27)}$$

This strengthening approach is comparable to installing fully threaded screws to reinforce notched beams or joints loaded perpendicular to the grain (Article D8). The screws are subject to axial load and have to absorb the tensile force perpendicular to the grain, which would otherwise trigger splitting in a non-reinforced joint. The calculation model derived by Schmid (2002) to determine the axial force component (tensile force) in the reinforcing screw depending on the external load component parallel to the grain applies only to joints with one fastener row in the grain direction and reinforcing screws installed perpendicular to the grain and fastener axis. Although this kind of reinforcement can reliably prevent splitting, it cannot eliminate block shear failure. It is advisable to dimension fully threaded screws exposed to withdrawal load and which act as a reinforcement against splitting for 30% of the load per dowel and per shear plane:

$$F_{ax,90,Rk} \leq 0.3 \cdot F_{v,Rk} \tag{E12-28}$$

whereby $F_{ax,90,Rk}$ corresponds to the characteristic withdrawal resistance of a fully threaded screw of penetration depth $\ell_{ef} = a_{4,c}$ (see Article E5) and $F_{v,Rk}$ is the characteristic load-bearing capacity of a dowel in accordance with Johansen (see Article E2) of a unreinforced joint.

Fully threaded screws, arranged directly underneath the dowels

As well as preventing the risk of splitting, installing fully threaded screws directly underneath the fasteners (Figure E12-3 on the right) also helps increase the load-bearing capacity, since the screw functions as a "beam", which supports the fasteners. This kind of reinforcement measure also means the load-bearing capacity of a joint not at risk of splitting can be increased beyond the Johansen load-bearing capacity. Similar to the process of extending Johansen equations for joints with glued-on wood-based panels or nail connectors, the Johansen model can also be extended for these reinforcement types (see also Figure E12-6). The variable required in addition to the Johansen parameters (embedment strength $f_{h,i}$, yield moment M_y, member thickness t_i, fastener diameter d) is the load-bearing capacity R_{VE} of the reinforcement element, whereby R_{VE} may be identical or greater than the force component F_{VE} from the dowel, Figure E12-5. Accordingly, a distinction is established between "soft" or "hard" reinforcement. In a "soft" reinforcement, the screw functions like a semi-rigid support for the dowel, whereas its role in a "hard" reinforcement is that of a rigid support. This is an important distinction, since depending on whether the support in question is semi-rigid or rigid (whether or not the reinforcement element has scope to bend), differing failure mechanisms may occur within the classic Johansen cases (involving no, one or two plastic hinges per shear plane). Figure E12-6 shows the range of different failure mechanisms, citing the example of a single-shear timber-to-timber joint for the Johansen case with two plastic hinges per shear plane. In the "hard" failure mechanism, a possible plastic hinge of the dowel will develop at the same position as the screw.

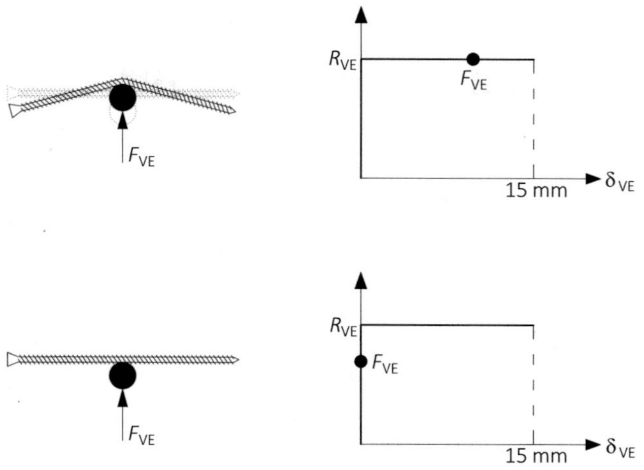

Figure E12-5 Above: „soft“ reinforcement, below: „hard“ reinforcement. (Blass et al., 2006)

Figure E12-6 also shows the internal forces and moments at the fastener belonging to the failure mechanisms and with which the Johansen equations can now be extended. The load-bearing capacity R_3 resulting from the equilibrium of forces and moments is derived as an example in the following case, featuring the "hard reinforcement" of a single-shear timber-to-timber joint with two plastic hinges per shear plane. Additional cases are contained in Bejtka (2005a) and Blass et al. (2006). The load-bearing capacity R_3 is determined from the equilibrium of forces and moments in the joint (for internal forces and moments, see the second case from above in Figure E12-6). The equilibrium of forces $\Sigma V = 0$ reveals:

$$R_3 - F_{1,\mathrm{VE},3} - f_{\mathrm{h},1} \cdot d \cdot p = 0 \quad \Leftrightarrow \quad R_3 = F_{1,\mathrm{VE},3} + f_{\mathrm{h},1} \cdot d \cdot p \qquad \text{(E12-29)}$$

The equilibrium of moments at the fastener $\Sigma M^A = 0$ reveals:

$$2 \cdot M_\mathrm{y} + f_{\mathrm{h},2} \cdot d \cdot p \cdot \frac{p}{2} - f_{\mathrm{h},1} \cdot d \cdot p \cdot \left(p + \frac{p}{2} \right) - F_{1,\mathrm{VE},3} \cdot 2 \cdot p = 0$$

Solving for $F_{1,\mathrm{VE},3}$ reveals the following, with $\beta = f_{\mathrm{h},2}/f_{\mathrm{h},1}$:

$$F_{1,\mathrm{VE},3} = \frac{M_\mathrm{y}}{p} - \frac{f_{\mathrm{h},1} \cdot d \cdot p}{4} \cdot (3 - \beta) \qquad \text{(E12-30)}$$

Inserting equation (E12-30) in equation (E12-29) results in the load-bearing capacity R_3:

$$R_3 = \frac{M_\mathrm{y}}{p} + \frac{f_{\mathrm{h},1} \cdot d \cdot p}{4} \cdot (1 + \beta) \qquad \text{(E12-31)}$$

However, equation (E12-31) only applies if the failure case is deemed "hard", namely, when the lateral load-bearing capacities of the reinforcement element $R_{1,VE}$ and $R_{2,VE}$ exceed the corresponding force components $F_{1,VE,3}$ and $F_{2,VE,3}$. The force components $F_{1,VE,3}$ and $F_{2,VE,3}$ are also derived from the equilibrium of forces and moments in the shear plane. For $F_{1,VE,3}$, equation (E12-30) applies, meaning:

$$R_{1,VE} > F_{1,VE,3} = \frac{M_y}{p} - \frac{f_{h,1} \cdot d \cdot p}{4} \cdot (3 - \beta) \tag{E12-32}$$

$F_{2,VE,3}$ is determined similarly. The equilibrium of moments at the fastener $\Sigma M^B = 0$ reveals:

$$2 \cdot M_y + f_{h,1} \cdot d \cdot p \cdot \frac{p}{2} - f_{h,2} \cdot d \cdot p \cdot \left(p + \frac{p}{2} \right) - F_{2,VE,3} \cdot 2 \cdot p = 0$$

Solving for $F_{2,VE,3}$ reveals the following, with $\beta = f_{h,2}/f_{h,1}$:

$$F_{2,VE,3} = \frac{M_y}{p} - \frac{f_{h,1} \cdot d \cdot p}{4} \cdot (3 \cdot \beta - 1) \tag{E12-33}$$

Consequently:

$$R_{2,VE} > F_{2,VE,3} = \frac{M_y}{p} - \frac{f_{h,1} \cdot d \cdot p}{4} \cdot (3 \cdot \beta - 1) \tag{E12-34}$$

Or with $R_{1,VE}$ and $\psi = R_{2,VE}/R_{1,VE}$:

$$R_{1,VE} > \frac{F_{2,VE,3}}{\psi} = \frac{M_y}{p \cdot \psi} - \frac{f_{h,1} \cdot d \cdot p}{4 \cdot \psi} \cdot (3 \cdot \beta - 1) \tag{E12-35}$$

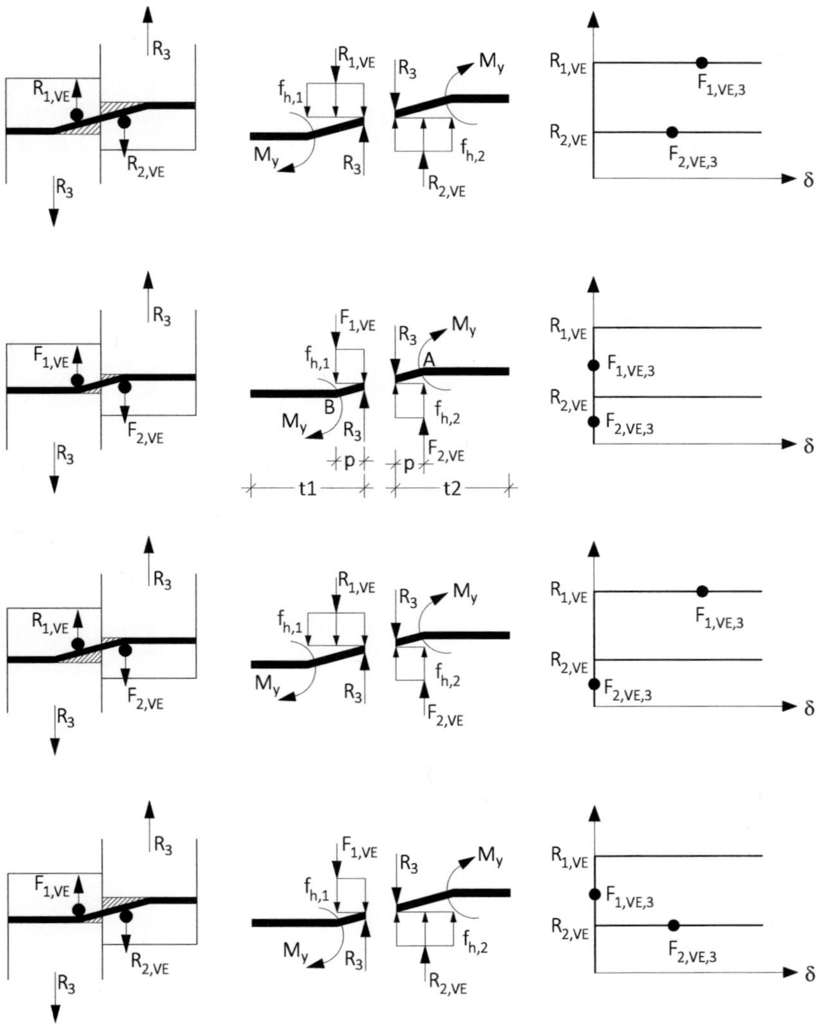

Figure E12-6 Failure mechanisms for the Johansen case with two plastic hinges per shear plane and internal forces and moments at the fastener, top down: „soft", „hard", „soft-hard", „hard-soft". (Blass et al., 2006)

The load-bearing capacity of a single-shear timber-to-timber joint is therefore determined for one of the possible failure cases (those involving a "hard" reinforcement) and for the Johansen case with two plastic hinges per shear plane with equations (E12-31), (E12-32) and (E12-35). Accordingly, the number of possible failure cases in a reinforcement featuring fully threaded screws arranged directly underneath the dowels clearly increases, since per Johansen case, different failure mechanisms of the reinforcement elements must be taken into consideration (see Figure E12-6, cases „soft", „hard", „soft-hard" and „hard-soft").

Load-bearing capacity of the reinforcement elements

To design a reinforced dowelled joint, with fully threaded screws arranged directly underneath the dowels, the only thing now missing is the value for the load-bearing capacity R_{VE} of the reinforcement elements and here, two cases can be distinguished, see Figure E12-7.

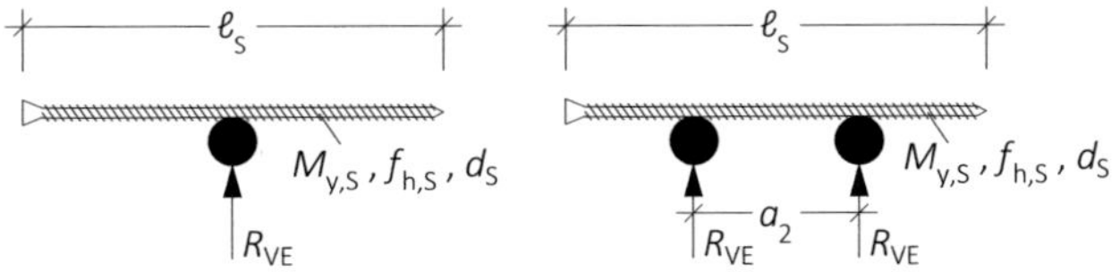

Figure E12-7 Left: one dowel, reinforced with a screw, right: two dowels, reinforced with a screw. (Blass et al., 2006)

The load-bearing capacity R_{VE} is calculated in the same way as the load-bearing capacity of a Johansen joint with slotted-in steel plate (= screw). The corresponding equations for the case of one dowel reinforced with a screw (Figure E12-7 on the left) are given in equation (E12-36); the equations for the second case can be referenced from literature (Bejtka, 2005b):

$$R_{VE} = \min \begin{cases} f_{h,S} \cdot d_S \cdot \ell_S \\ f_{h,S} \cdot d_S \cdot \ell_S \cdot \left[\sqrt{\dfrac{16 \cdot M_{y,S}}{f_{h,S} \cdot d_S \cdot \ell_S^2} + 2} - 1 \right] \\ 4 \cdot \sqrt{M_{y,S} \cdot f_{h,S} \cdot d_S} \end{cases} \tag{E12-36}$$

E12.3 Literature

H. Werner, original Article 9 (Band 3), STEP 1995.

Bejtka I. (2005a). Verstärkung von Bauteilen aus Holz mit Vollgewindeschrauben. Dissertation Universität Karlsruhe.

Bejtka I. (2005b). Self-tapping screws as reinforcements in connections with dowel-type fasteners. Paper 38-7-4, CIB-W18 Meeting 38, Karlsruhe.

Blass H.J., Bejtka I. and Uibel T. (2006). Tragfähigkeit von Verbindungen mit selbstbohrenden Holzschrauben mit Vollgewinde. Karlsruher Berichte zum Ingenieurholzbau Band 4. Universität Karlsruhe.

Blass H.J. and Schmid M. (2001). Verstärkung von Verbindungen. Bauen mit Holz 103:40-48.

Blass H.J. and Werner H. (1988). Stabdübelverbindungen mit verstärkten Anschlussbereichen. Bauen mit Holz 90:601-607.

Haller P., Wehsener J. and Chen C.J. (1998). Development of joints by compressed wood and glassgrain reinforcement. COST C1 – Control of the semi-rigid behaviour of civil engineering structural connections, International Conference, Liège.

Kevarinmäki A., Kangas J., Nokelainen T. and Kanerva P. (1995). Nail-plate reinforced bolt joints of Kerto-FSH structures. Publication 51, Helsinki University of Technology.

Schmid M. (2002). Anwendung der Bruchmechanik auf Verbindungen mit Holz. Dissertation Universität Karlsruhe.

E13 Joints with multiple fasteners

Original article: H. J. Blass

Timber joints with mechanical fasteners usually contain more than just a single fastener. However, this means the load distribution between multiple fasteners is always uneven, even if the load transmitted by the joint is applied to the centre of gravity of the joint. The load-bearing capacity of a joint corresponds to the sum of the loads withstood by the individual fasteners in the ultimate limit state. If now, in the ultimate limit state of the joint, the loads exerted on the fasteners vary and some fasteners are loaded with a load lower than their ultimate capacity, the load-bearing capacity of the joint is lower than the collective load-bearing capacities of the individual fasteners. This explains the reduction of the load-bearing capacity per fastener in joints with multiple fasteners for specific fastener types. As a general rule, the various influences on load distribution in joints act both on joints with multiple identical fasteners as well as on those with different fasteners. For joints with multiple fasteners, brittle failure mechanisms may also emerge, to which the Johansen model (Article E2) is inapplicable and which fail at lower loads than those calculated with the Johansen equations. In response, either joints can be reinforced (Article E12), an effective number $n_{ef} < n$ can be used or the brittle failure mechanisms can be separately verified. This article starts by examining the load distribution between individual fasteners in timber joints and the concept of the effective number of fasteners n_{ef} as well as illustrating the failure mode "wood failure in joints with multiple glued-in rods subject to axial stresses". Finally, the various failure modes occurring in joints with laterally loaded fasteners are shown, before explaining the verification approach for block shear failure.

E13.1 Load distribution in accordance with the theory of elasticity

Lantos (1969) developed a model to calculate the load distribution in timber joints, based on a linear-elastic load-deformation relationship of the individual fasteners without initial slip and assuming uniformly distributed normal stresses in the members to be connected. The validity of his model for laterally loaded fasteners is limited to the lower load range, in which the joint behaviour can be considered elastic, and to joints loaded parallel to the grain. Cramer (1968) adopts a similar approach when taking the non-uniform normal stress distribution in the cross-sections into consideration and its influence on the axial stiffness of the connected members. However, the elastic solution according to Lantos can be applied to calculate the load-bearing capacity of joints with linear-elastic load-deformation behaviour and multiple fasteners arranged in a row and in the direction of the load. Examples include timber-to-timber or steel-to-timber joints with screws arranged at an angle. The solution for the general case of non-linear load-deformation relationships is specified in Wilkinson (1986).

Figure E13-1 shows a single-shear timber-to-timber joint with nails in which the load is transferred between members 1 (M_1) and 2 (M_2) in discrete steps at the fastener locations. Here, each step represents the load transferred by the respective fasteners. In the deformed state, the load in member 1 between fasteners i and $i + 1$ corresponds to the total load in the joint reduced by the loads transmitted by fasteners 1 to i. The original length s between both fasteners is extended by $u_{1,i}$ in this case. Similarly, the corresponding section of member 2 is extended by s to $s + u_{2,i}$, while the forces transferred by fasteners i and $i + 1$ result in displacements of $u_{f,i}$ or $u_{f,i+1}$.

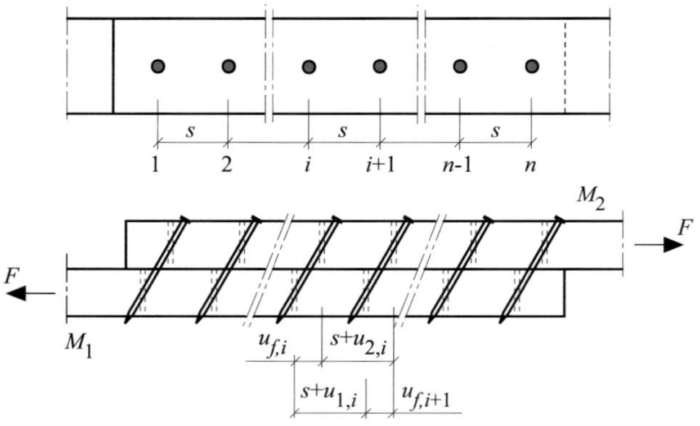

Figure E13-1 View of the undeformed (upper) and section of the deformed (lower) joint area. M_1: member 1; M_2: member 2. (STEP 1995 Article C15)

A comparison of the lengths and deformations in both members (see Figure E13-1) reveals that:

$$u_{f,i} + s + u_{2,i} = s + u_{1,i} + u_{f,i+1} \quad \text{(E13-1)}$$

The deformation in the individual fastener can be replaced by:

$$u_f = \frac{F_f}{K} \quad \text{(E13-2)}$$

where F_f is the load transferred by the fastener and K is the slip modulus.

The elongation of the members can be expressed by:

$$u = \frac{F_{\mathrm{m}} \cdot s}{E \cdot A} \tag{E13-3}$$

where F_{m} is the load present in the member between two fasteners.

If equations (E13-2) and (E13-3) are inserted into equation (E13-1), after some transformations, the resulting load on the most stressed fastener at the start or end of the fastener row is as follows (where the connected members have the same axial stiffness $E \cdot A$):

$$F_1 = F \cdot \left(1 - m_1 \cdot (1+\mu) + \mu + (m_1 - m_2) \cdot \frac{m_1^{\mathrm{n}} \cdot (1+\mu) - \mu}{m_1^{\mathrm{n}} - m_2^{\mathrm{n}}} \right) \tag{E13-4}$$

$$F_{\mathrm{n}} = F \cdot \left(-\mu + m_1^{\mathrm{n-1}} \cdot (1+\mu) - \left(m_1^{\mathrm{n-1}} - m_2^{\mathrm{n-1}} \right) \cdot \frac{m_1^{\mathrm{n}} \cdot (1+\mu) - \mu}{m_1^{\mathrm{n}} - m_2^{\mathrm{n}}} \right) \tag{E13-5}$$

where

$$\mu = -\frac{1}{1 + \frac{E_1 \cdot A_1}{E_2 \cdot A_2}} \tag{E13-6}$$

$$m_1 = \frac{\omega + \sqrt{\omega^2 - 4}}{2} \tag{E13-7}$$

$$m_2 = \frac{\omega - \sqrt{\omega^2 - 4}}{2} \tag{E13-8}$$

$$\omega = 2 + K \cdot s \cdot \left[\frac{1}{E_1 \cdot A_1} + \frac{1}{E_2 \cdot A_2} \right] \tag{E13-9}$$

Lantos demanded to design the most stressed fastener. Since the fasteners at the start or end of the fastener row are most stressed, they determine the load-bearing capacity of the joint. Based on this solution, the differences in individual fastener loads are dictated by the axial stiffnesses of the connected members, the number of fasteners in a row, the spacing among fasteners and the slip modulus. The linear-elastic approach by Lantos presented here remains relevant for joints, which respond in a linear-elastic manner up to failure. For joints with multiple fully threaded screws, arranged at an angle and in a row, the Lantos method can be used to identify and design the most stressed screw.

E13.2 Parameters influencing the load distribution in joints

In addition to the influence of varying elongation in the connected members, other parameters can also significantly influence the load distribution between the individual fasteners of a timber joint.

Plastic deformations and creep

Isyumov (1967) selected a more general approach to calculate the load distribution between individual fasteners arranged in a row and parallel to the load. His solution takes into consideration the non-linear load-deformation behaviour of the fasteners, which triggers load redistribution in the joint with increasing total load. As soon as the fastener subject to the greatest stress exhibits plastic deformations, its stiffness declines in comparison to other fasteners. Since more rigid elements in a parallel system are subject to comparatively higher stresses, this redistributes the load to the less heavily loaded fasteners in the middle of the row. These load redistributions offset the load peaks described by Lantos (1969) and increase load-bearing capacities in comparison to completely elastic joints. The same applies to the influence of time-dependent deformations in the joint. Creep also results in a relative loss of stiffness, which, in turn, leads to load redistributions. Given that the scale of the creep deformations rises with increasing loading, greater creep deformations are likely to occur at the start and end of a fastener row, which leads to a more uniform load distribution in the joint.

Plastic deformations of joints and the load redistributions they cause are actually the basis for design in accordance with EC 5. The calculation of the load-bearing capacity of joints with dowel-type fasteners in accordance with EC 5 is attributable to Johansen (1949). Johansen presumed ideal rigid-plastic behaviour for both fasteners and wood under embedment stress (see Article E2). Subject to sufficient edge distances and spacings and if the joint fails in accordance with one of the mechanisms described by Johansen, the joint shows plastic load-deformation behaviour. However, if splitting of the timber along the row of fasteners or block shear occurs at load levels well below the potential plastic capacity, a full redistribution of the load within the joint is prevented, which is why splitting or block shear significantly reduce the load-bearing capacity of joints with multiple fasteners. Sufficient edge distances and spacings lead to reduced splitting of the timber in the joint area, since the shear and tensile stresses perpendicular to the grain generated by the wedge effect of the fasteners decline with increasing distance. Accordingly, the greater the distances in the grain direction, the more favourable the plastic deformation behaviour when reaching the ultimate load, which, contrary to elastic theory, ends up increasing the load-bearing capacity of joints with multiple fasteners. Another means of preventing splitting and guaranteeing plastic failure instead is to reinforce the joint area e.g. by glued-on plywood or fully threaded screws (Article E12). Joints reinforced accordingly reach a load-bearing capacity which corresponds to the collective load-bearing capacities of the individual fasteners. No reduction need be taken into account due to the group effect in this case.

Considerable plastic deformations occur in the following timber joints with mechanical fasteners: joints with toothed-plate connectors, nailed joints and joints with other slender, dowel-type fasteners. Plastic deformations, which generally precede any failure in joints with toothed-plate connectors, explain why an interaction of the toothed-plate connector and bolt can be assumed for these joints. Conversely, for joints with split ring or shear plate connectors, which are often prone to brittle failure, the bolt is not taken into consideration when determining the load-bearing capacity (Article E6). Creep deformations, however, affect all types of timber joints with mechanical fasteners.

Fabrication tolerances

In joints with predrilled holes or pre-milled depressions such as dowelled joints or joints with connectors, differences in load-bearing behaviour between individual fasteners of a joint are exacerbated by manufacturing inaccuracies. Such manufacturing inaccuracies may include, for example, misalignment of the bolt holes, deviating boreholes in the wood, variations in the hole diameter or in the initial position of the bolts in the holes. Dannenberg and Sexsmith (1976) as well as Isyumov (1967) stress the key impact of such manufacturing inaccuracies on load distribution within joints with connectors (split ring, shear plate or toothed-plate connectors), which, in turn, affects their load-bearing capacity. According to Wilkinson (1986), manufacturing inaccuracies and differences in the load-deformation behaviour of individual fasteners are the main cause of non-uniform load distribution in timber joints with mechanical fasteners, while the influence of varying degrees of elongation of the connected members can be disregarded. Manufacturing inaccuracies such as misalignments of holes result in an initial slip for some fasteners within the joint. If the joint is then loaded, these fasteners only start to accommodate loads when the joint deformation exceeds the initial slip (see bolt No. 4 in Figure E13-2). If the joint fails before the plastic deformation stage due to splitting, the fasteners affected by initial slip do not carry any load at all. Tests conducted by Massé et al. (1989) show e.g. for bolted joints with Douglas fir glulam that the load-bearing capacity per bolt declined by more than 50%, when the number of bolts in the joint was increased from one to four. These results underline the need to ensure sufficient edge distances and spacings for joints with fabrication tolerances to guarantee plastic behaviour and hence allow the loads to be redistributed within the joint. Manufacturing inaccuracies can largely be avoided by ensuring precise manufacturing with CNC machinery.

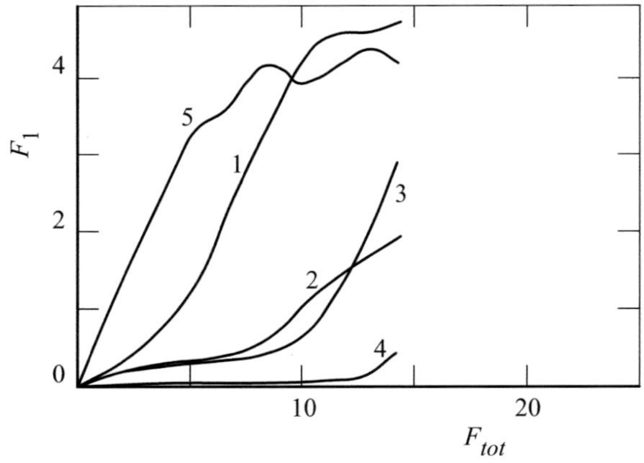

Figure E13-2 Example load distribution in a bolted joint according to Wilkinson (1986). F_1 is the load of the individual bolt and F_{tot} is the total load of the joint. (STEP 1995 Article C15)

Differences in the load-deformation behaviour of the individual fasteners

As well as fabrication tolerances, variation in the characteristics of the wood itself in the joint area impact on the load distribution. Knots, cracks, resin pockets, slope of grain or density variation all alter the load-deformation behaviour of the individual fasteners.

E13.3 Influence of the number of fasteners

In principle, when determining the load-bearing capacity, ideal-plastic behaviour is initially assumed, whereupon the influence of the number of fasteners is taken into consideration by reducing the load-bearing capacity per fastener. The so-called effective number n_{ef} of fasteners is then specified in EC 5 per fastener type. For nailed joints, meanwhile, the number n of fasteners arranged in a row is reduced by an exponent k_{ef}, which depends on nail spacing. For bolted and dowelled joints, however, the effective number n_{ef} depends on the diameter of the fastener as well as the spacing within a row of fasteners.

E13.4 Glued-in rods – group tear-out

One exceptional case for joints with multiple fasteners is that of joints with multiple glued-in rods subject to axial tensile stress. In such joints, a possible failure mode is the so-called group tear-out, which involves the entire group of fasteners, including the wooden block surrounding the rods, being torn out, as shown in Figure E13-3. This failure mode resembles the brittle failure modes mentioned above for joints with laterally loaded multiple fasteners arranged in a row along the grain direction. Group tear-out is also due to premature failure of the wood in response to tensile or shear stresses; exacerbated by varying loads in individual rods at the moment of failure. The total load-bearing capacity thus tends to be lower than the collective load-bearing capacities of the individual rods. As was already explained in Article E10, minimum bond lengths are required to ensure ductile failure, so that the brittle failure mode in question can be avoided.

Since this failure mode is not explicitly taken into account in EC 5, there is a need to consider an effective number of fasteners $n_{ef} = n^{0.9}$ as is done for joints with groups of screws subject to tensile stresses (EC 5, Equation 8.41), if plastic steel failure is not to prevail. The NA to EC 5, for example, stipulates that "the key criterion dictating the load-bearing capacity of the joint should be the load-bearing capacity of the rod and not the strength of the wood or bond line, if non-uniform loading cannot be ruled out." (NCI NA.11.2.3 (NA.2)); which means steel failure of the rods should be decisive. If this is not the case, the group tear-out should be verified, following the procedure used for block shear failure in Section E13.5.

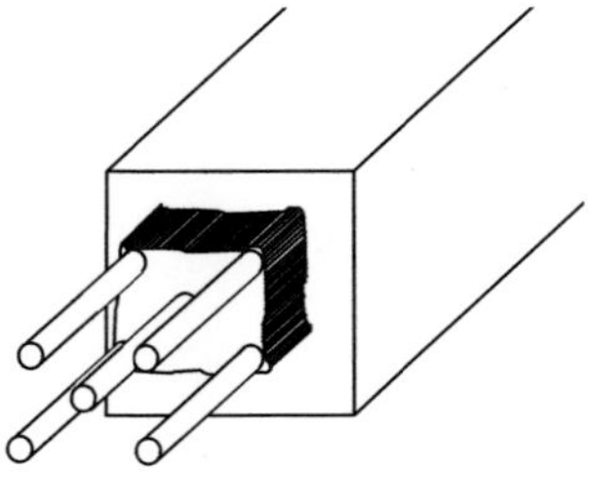

Figure E13-3 Failure mode "group tear-out" in a group of glued-in rods (Tlustochowicz et al., 2011).

E13.5 Block shear failure

Before explaining how block shear failure of timber joints is verified, as is set out in Annex A of EC 5, all possible failure modes of timber joints with laterally loaded fasteners will be covered. The method given in Annex A is limited to steel-to-timber joints, but block shear failure may also affect timber-to-timber joints.

Failure modes of joints with multiple fasteners and with timber members loaded parallel to the grain

The possible failure modes of joints with multiple fasteners and with timber members loaded parallel to the grain vary very widely. In addition to the ductile failure modes covered in the Johansen model and depending on the dimensions of the timber members, brittle modes such as splitting or shear failures may also occur. Joints with multiple dowels or bolts in a row are particularly prone to such brittle failure modes, even if the required minimum distances are complied with, reference to cases (i), (ii), and (iv) in Figure E13-4.

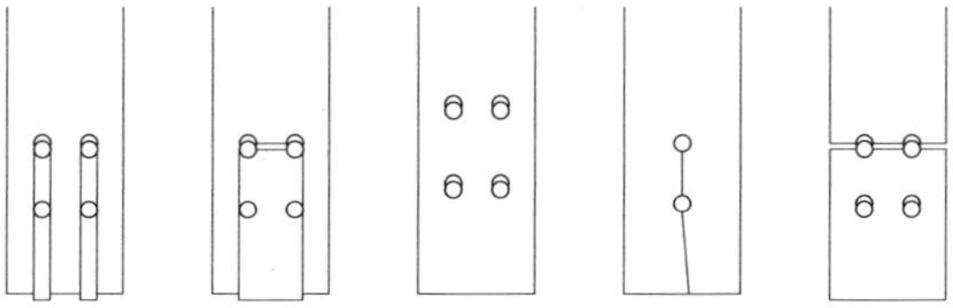

Figure E13-4 Possible failure modes of timber joints with dowels/bolts. (i) row shear-out, (ii) block shear, (iii) embedment/embedment and plastic hinge, (iv) splitting, (v) net tension failure. (Quenneville and Mohammad, 2000)

Splitting, as shown in case (iv), is considered by applying an effective number of fasteners n_{ef}, allowing the Johansen equations to be reused and the joint with a load-bearing capacity reduced by n_{ef}/n to be designed. One way to prevent such splitting and achieve the full potential load-bearing capacity in accordance with Johansen is to reinforce the joint (depending on the type of reinforcement, the load-bearing capacity may be increased even further, see Article E12). Failure modes (i) and (ii) in accordance with Figure E13-4, however, are only partially considered with the effective number n_{ef} and must still be separately verified. Row shear-out, case (i), is not explicitly regulated in EC 5, although block shear failure of a group of fasteners, namely multiple rows featuring multiple fasteners in a row, can be verified using Annex A of the EC 5. Shear failures (i) and (ii) can only, based on the current state of the art, be effectively prevented by gluing on wood-based panels, since only this ensures a sufficiently rigid joint between the timber member and reinforcing element.

Verification block shear failure

The "block shear" failure mode, an example of which is shown in Figure E13-5, primarily affects steel-to-timber joints with compact fastener arrangements. One of the first models used to verify this phenomenon was developed by Foschi and Longworth (1975) for nailed steel-to-glulam joints. The basic concept is that either the shear strength of the wood along the lateral and lower shear areas or the tensile strength of the wood in the external fastener row, furthest away from the end grain, is decisive, see failure areas in Figure E13-5 on the right. A third possible failure mechanism is ductile failure in accordance with Johansen. Since both shear and tensile failure parallel to the grain are brittle failure mechanisms, it is impossible to completely superpose both stress components. According to Foschi and Longworth (1975), the load-bearing capacity of a fastener group loaded parallel to the grain is the minimum value of the Johansen load-bearing capacity and the combined load-bearing capacity comprising shear capacity along the shear areas and tensile capacity parallel to the grain (with a failure area perpendicular to the grain) between both external rows of fasteners.

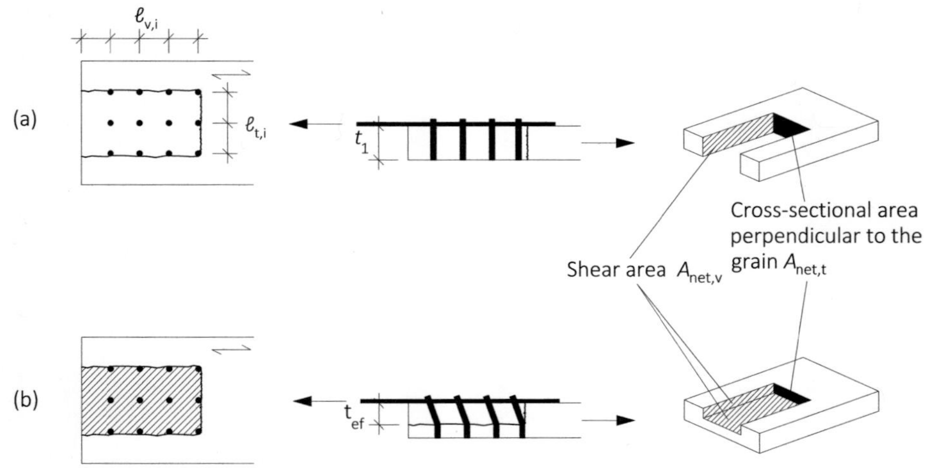

Figure E13-5 Block shear failure, (a) entire block sheared out, (b) block partially sheared out over the member thickness. The areas shown are either subject to shear out ($A_{net,v}$) or rupture ($A_{net,t}$).

While the EC 5 verification currently used for steel-to-timber joints is similar, in this case, the larger value of the shear or tensile capacity is decisive for block shear failure:

$$F_{bs,Rk} = \max \begin{cases} 1.5 \cdot A_{net,t} \cdot f_{t,0,k} \\ 0.7 \cdot A_{net,v} \cdot f_{v,k} \end{cases} \qquad \text{(E13-10)}$$

where

$F_{bs,Rk}$ Characteristic block shear load-bearing capacity

$A_{net,t}$ Net cross-section perpendicular to the grain

$f_{t,0,k}$ Characteristic tension strength parallel to the grain

$A_{net,v}$ Net shear area parallel to the grain

$f_{v,k}$ Characteristic shear strength

The required net cross-sections $A_{net,t}$ to determine tensile capacity and $A_{net,v}$ to determine shear capacity are calculated in simplified form in EC 5.

To calculate the **net cross-section $A_{net,t}$** perpendicular to the grain, as shown in Figure E13-5, the sum of the fastener spacings $\ell_{t,i}$ perpendicular to the grain are multiplied by the thickness of the timber block having sheared out. Depending on the failure mode of the joint, a timber block may shear out over the entire height of the timber member t_1 or only over a partial block of height t_{ef}:

$$A_{net,t} = \begin{cases} \sum_i \ell_{t,i} \cdot t_1 & \text{failure modes c, f, j/l} \\ \sum_i \ell_{t,i} \cdot t_{ef} & \text{all other failure modes} \end{cases} \qquad \text{(E13-11)}$$

where

t_1 Thickness of the timber member or penetration depth of fasteners

t_{ef} Effective thickness, depending on the failure mode

$\Sigma\ell_{t,i}$ Sum of fastener spacings perpendicular to the grain, see Figure E13-5 (a)

The effective thickness t_{ef} corresponds to the thickness over which the embedment strength is reached. Depending on the failure mode of the joint, a partial block as shown in Figure E13-5 (b) with an effective thickness smaller than the member thickness, $t_{ef} < t_1$, or a block over the entire timber member thickness t_1 may rupture or shear out, Figure E13-5 (a). If no plastic hinges occur and the fasteners do not rotate, the embedment strength is reached over the entire member thickness, which means that in the event of block shear failure, a block ruptures or shears out over the entire member thickness or penetration depth. The effective thickness corresponds, in turn, to the thickness t_1 of the timber member or penetration depth of the fastener and is specified in the upper term of equation (E13-11). For clarity, reference can be made to Figure E13-6, which shows all possible failure modes of steel-to-timber joints. A block shear failure over the entire thickness of the timber member t_1 can only occur in failure modes c, f and j/l, since this is the only configuration where the embedment strength is reached over the entire member thickness (in Figure E13-6, thicknesses t_1 or t_2). For all other failure modes shown in Figure E13-6, an effective thickness t_{ef}, smaller than that of the timber member thickness, must be used, as in the lower term of equation (E13-11).

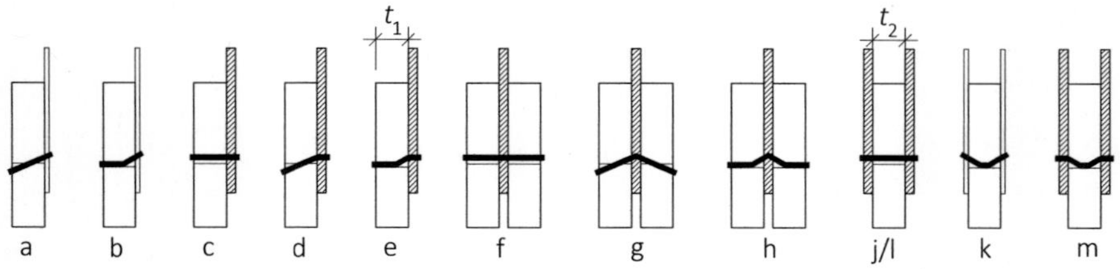

Figure E13-6 Failure modes according to Johansen for steel-to-timber joints.

The effective thickness t_{ef} is determined via the Johansen equations (Article E2). Via equilibrium considerations, the position of the plastic hinge can be determined, which, in turn, reveals the effective thickness t_{ef} (see schematic illustration in Figure E13-5 (b)).

As an example, t_{ef} shall be determined for a single-shear steel-to-timber joint with a thick steel plate if this joint fails with two plastic hinges per shear plane according to Johansen; reference is made to equation (E2-11) and failure mode e in accordance with Figure E13-6 or Figure E2-11:

$$F_{v,Rk} = f_{h,1,k} \cdot d \cdot b_1 = f_{h,1,k} \cdot d \cdot 2 \cdot \sqrt{\frac{M_{y,k}}{f_{h,1,k} \cdot d}} = 2 \cdot \sqrt{M_{y,k} \cdot f_{h,1,k} \cdot d} \qquad \text{(E13-12)}$$

where b_1 is the distance between both plastic hinges or the width of the area where embedment strength has been reached (Figure E2-11).

To now determine the block shear load-bearing capacity of such a joint with two plastic hinges per shear plane, $b_1 = t_{ef}$ applies:

$$t_{ef} = b_1 = 2 \cdot \sqrt{\frac{M_{y,k}}{f_{h,1,k} \cdot d}} \qquad \text{(E13-13)}$$

All other cases and the corresponding effective thicknesses t_{ef} can be taken from EC 5 Annex A.

The **net shear area $A_{net,v}$**, Figure E13-5, also depends on the failure mode of the fastener group, which impacts on load distribution within the group and is thus determined similarly to the net cross-section $A_{net,t}$ with t_1 or the effective thickness t_{ef}:

$$A_{net,v} = \begin{cases} \sum_i \ell_{v,i} \cdot t_1 & \text{failure modes c, f, j/l} \\ \frac{1}{2} \cdot \sum_i \ell_{v,i} \cdot \left(\sum_i \ell_{t,i} + 2 \cdot t_{ef} \right) & \text{all other failure modes} \end{cases} \qquad \text{(E13-14)}$$

where

t_1 Thickness of timber member

t_{ef} Effective thickness or penetration depth of fasteners, depending on the failure mode

$\Sigma \ell_{v,i}$ Sum of fastener spacings and end distances parallel to the grain, see Figure E13-5 (a), both external rows have to be summed up: in Figure E13-5 (a), i = 8

$\Sigma \ell_{t,i}$ Sum of fastener spacings perpendicular to the grain, see Figure E13-5 (a)

The lower term of equation (E13-14) means that in the event of one or more plastic hinge emerging or for a short penetration depth, the net shear area is assumed to be the sum of the lateral areas along the external rows of fasteners and the shear area underneath the fastener group, since such a case involves only a partial block shearing out over the entire thickness (Figure E13-5 (b)), through which the lower shear area is activated as well as the lateral shear areas. If no plastic hinges form and the fasteners remain straight, only the lateral shearing areas along the external rows of fasteners and over the member thickness or penetration depth of the fasteners are taken into consideration, see the upper term of equation (E13-14), since this involves a block to shear out over the entire member thickness or penetration depth t_1, meaning the lower shear area is not activated.

The fact that EC 5 determines the **block shear load-bearing capacity as the maximum value of the shear and tensile capacity of the net area** means that if the weaker component fails, the total load is absorbed by the other component. If the upper term of equation (E13-10), the net tensile capacity, is less than the lower term of equation (E13-10), the net shear capacity, this means that the joint may fail in the net cross-section perpendicular to the grain, in which case the entire load will be absorbed by the shear area.

Coefficients 1.5 and 0.7 in equation (E13-10) result from considerations of characteristic strengths. The characteristic tensile strength parallel to the grain $f_{t,0,k}$ of the timber member is increased by 50%, given the very low likelihood of large knots being present in the very restricted, localised area prone to failure, which would mean low tensile strength values. The increase thus corresponds to a volume effect with respect to tensile stress (Article D3). The shear strength, in turn, is reduced by 30%, because non-uniform shear stress distributions with peak stresses need to be taken into consideration.

E13.6 Literature

H.J. Blass, original Article C15, STEP 1995.

Bejtka I. (2005). Verstärkung von Bauteilen aus Holz mit Vollgewindeschrauben. Dissertation, Universität Karlsruhe. Band 2 der Karlsruher Berichte zum Ingenieurholzbau.

Cramer C.O. (1968). Load distribution in multiple-bolt tension joints. Journal of the Structural Division, ASCE 94(ST5):1101-1117.

Dannenberg L.J. and Sexsmith R.G. (1976). Shear-plate load distribution in laminated timber joints. Report No. 361, Department of Structural Engineering, Cornell University, Ithaca, New York.

Foschi R.O. and Longworth J. (1975). Analysis and design of Griplam nailed connections. Journal of the Structural Division 101(12):2537-2555.

Isyumov N. (1967). Load distribution in multiple shear-plate joints in timber. Forestry Branch Departmental Publication No. 1203, Department of Forestry and Rural Development, Ottawa.

Johansen K.W. (1949). Theory of timber connections. International Association of Bridge and Structural Engineering, Publication 9:249-262.

Lantos G. (1969). Load distribution in a row of fasteners subjected to lateral load. Wood Science 1(3):129-136.

Massé D.I., Salinas J.J. and Turnbull J.E. (1989). Lateral strength and stiffness of single and multiple bolts in glued-laminated timber loaded parallel to grain. Contribution No. C 029, Engineering and Statistical Research Centre, Agriculture Canada, Ottawa.

Quenneville J.H.P. and Mohammad M. (2000). On the failure modes and strength of steel-wood-steel bolted connections loaded parallel-to-grain. Canadian Journal of Civil Engineering 27:761-773.

Tlustochowicz G., Serrano E. and Steiger R. (2011). State-of-the-art review on timber connections with glued-in steel rods. Materials and Structures 44(5):997-1020.

Wilkinson T.L. (1986). Load distribution among bolts parallel to load. Journal of Structural Engineering 112(4):835-852.

E14 Moment-resisting joints

Original article: P. Racher

To create a timber structure, timber members have to be connected together and to other members. In many traditional structures (e.g. trusses, see also Article E15), joints are mainly subject to normal forces or also shear forces. In design, these joints are assumed to be a hinge, given that the fasteners are arranged within a limited space and the transmission of moments is limited. Depending on which static system is selected, the joints may also be capable of transferring bending moments in addition to normal and shear forces. The arrangement of joints in a configuration designed to transfer such bending moments paves the way for economic structures. For example, moment-resisting corners can be arranged in portal frame structures, allowing the use of three-hinged frames among others, for larger spaces with larger clearances and smaller foundations. Two configurations are possible when manufacturing moment-resisting joints. Glued joints can be established, e.g. in the form of large finger joints (Figure E14-1), whereby the members have finger joints over the entire member cross-section and are glued together under compressive force. Glued joints are considered rigid. The other option would be using mechanical fasteners, although joints with mechanical fasteners will always be semi-rigid.

Figure E14-1 Examples for glued joints, frame corners with large finger joints and an interim piece. (photo on the left: H. Damm, photo on the right: H. Brüninghoff)

The manufacture and load-bearing capacity of **large finger joints** is regulated in EN 14080. As well as setting out the requirements for adhesives, finger joints, devices and the manufacture of finger joints, the standard also encompasses requirements for the climatic boundary conditions to be complied with during the processing and curing of the adhesive. The expense also increases, when the size of the members or transport requirements dictate that the joints have to be manufactured on the construction site. In particular, when additional measures are needed to ensure compliance with the required climate constraints, such as a minimum temperature of 20°C, greater costs may be incurred. In Germany, glued joints can only be established by manufacturers which have a special "proof of suitability for gluing load-bearing wood members" (DIN 1052-10). The need to comply with such requirements explains why such glued joints usually cost far more to manufacture than joints with mechanical fasteners. Figure E14-2 shows examples of moment-resisting joints with **mechanical fasteners** and the corresponding load distribution, the latter of which depends on the joint geometry as well as the fastener arrangement. The axial and/or shear forces transferred by the fasteners are dictated by the position of the centre of rotation C, so that the resulting internal moments balance out the external moments. At the same time, the equilibrium of forces also has to be respected.

Figure E14-2 (a) shows how a guardrail post is connected to the cross beam of a pedestrian bridge. The individual fasteners of the group are axially loaded by the force F and the resulting moment M. When nails and screws are used, the axial load-bearing capacity depends on the withdrawal capacity of the fasteners, their tensile capacity and, where applicable, the head pull-through capacity. Figure E14-2 (b) shows a possible configuration for a column with a moment-resisting support at the base, which is subject to normal and shear loads as well as the moment itself. In the layout shown, the internal forces and moments are transferred to the steel parts in the foundation through normal and shear forces, while the fasteners are subject to lateral loads. The timber, meanwhile, is subject to forces acting both in parallel and perpendicular to the grain direction and limiting stresses on the timber exerted perpendicular to the grain in such joint configurations is a priority. One common solution used for frame corner joints with mechanical fasteners is shown in Figure E14-2 (c). Here, dowels are the first choice of fasteners and usually arranged in a circular form. In a moment-resisting frame corner, normal and shear forces must also be accommodated alongside the moment. The following sections focus on this type of moment-resisting joint in particular.

The prerequisite for the configuration shown in Figure E14-2 (*c*) is that the members to be connected must not be in the same plane, but either beams or columns are implemented comprising two components. In this case, filling elements are also often arranged between both load-bearing components. If joints are designed where the members to be connected have different angles to the grain, differences in the degree of shrinking and swelling of wood perpendicular and parallel to the grain must be considered. This applies in the configuration shown when connecting beams and columns. Varying moisture content can generate greater stresses perpendicular to the grain in the joint area which may result in splitting. Therefore, the magnitude of stresses perpendicular to the grain must be limited. This can be done, e.g. by limiting the height of members or gluing on wood-based panel materials. There is also scope to deploy self-tapping fully threaded screws to accommodate tensile stresses perpendicular to the grain and secure the corners or to use cross-laminated timber or crossbanded LVL members with reinforcing cross layers.

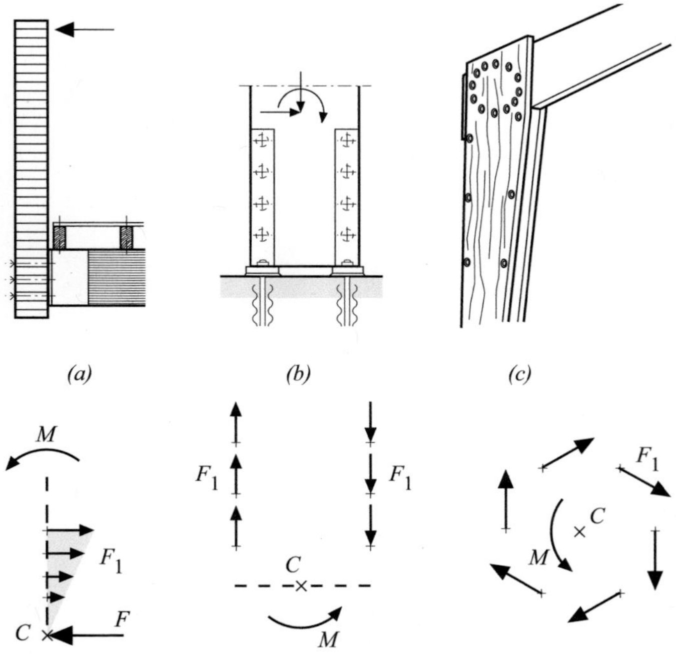

Figure E14-2 Examples of moment-resisting joints with mechanical fasteners (top) and load distribution on fasteners resulting from the moment only (bottom). (STEP 1995 Article C16)

Figure E14-3 and Figure E14-4 show additional options for moment-resisting frame corners with mechanical fasteners. In the layout shown in Figure E14-3, the corner moment is transferred via the lever arm *a* as a compressive and tensile force into the column. This joint typology features beams and columns lying in a single plane, meaning single-piece timber members can be used. This also favours the economical solution of connecting timber members with steel profiles, eliminating the need to arrange slotted-in plates. Likewise in this joint typology, the impact of tensile stresses perpendicular to the grain caused by climatic conditions must be taken into consideration by allowing the beam to shrink. Construction detailing must be of a nature to avoid restraining shrinkage in the joint of the beam to the column. The example in Figure E14-3 shows how a joint subject to tension transfers the force via two bolts and external steel plates in a manner that still allows for beam shrinkage due to moisture changes, see Figure E14-3 on the right. For this purpose, the tensile force is not parallel to the grain but results in compression perpendicular to grain stresses on the timber at the end of the nailed steel-to-timber connection. In addition, the upper edge of the beam in the joint area is notched to such an extent that restraints can be ruled out.

In the frame corner configuration shown in Figure E14-4, the moment is transferred in the form of compressive and tensile force into the column and tension strut respectively. Here, it is worth noting that the rafter is stressed by relatively high shear forces in the relatively confined space between column and strut.

Figure E14-3 Left: Division of the corner moment in tensile and compressive force. Right: The angle α between the grain direction of the beam and the system line of the hinged joint allows rotation of the steel plate when the beam depth changes (= shrinkage/swelling).

Figure E14-4 Frame corner in a portal frame structure. Division of the corner moment in a tensile and a compressive force. (photo: Finnforest)

Compared to glued or curved frame corners, in corner joints with mechanical fasteners, the transport dimensions may be far more compact, since any assembly can take place on the construction site itself. Here, the requirements in terms of assembly conditions and the assembly effort required are both significantly lower than for glued joints. It is also relatively easy to adapt beams and columns to the course of the internal moments and forces using tapered cross-sections. In comparison to curved frame corners, for example, this approach allows for larger clearances.

Moment-resisting joints, particularly with dowels arranged in a circular form which is important for practice, are not regulated in EC 5, although the NA provides some additional details, primarily concerning the effective number of fasteners. The design approach presented in this article and the design of reinforcements is based on Heimeshoff (1977) and not part of the NA.

E14.1 Influence of moment-resisting joints on the load-bearing behaviour

When calculating the internal forces and moments of a structure, joints with mechanical fasteners are often considered pinned or fixed. Figure E14-5 (a) and Figure E14-5 (b) show the course of the bending moment for the horizontal member for both assumptions.

In reality, joints with mechanical fasteners are semi-rigid with a stiffness K_r, see Figure E14-5 (c), which depends, among other things, on the arrangement, number and fastener types as well as the properties of the connected members. The stiffness of the joints influences the deformation behaviour of the structure and, for statically indeterminate systems, the distribution of internal forces and moments (Leijten, 1988; Komatsu, 1992).

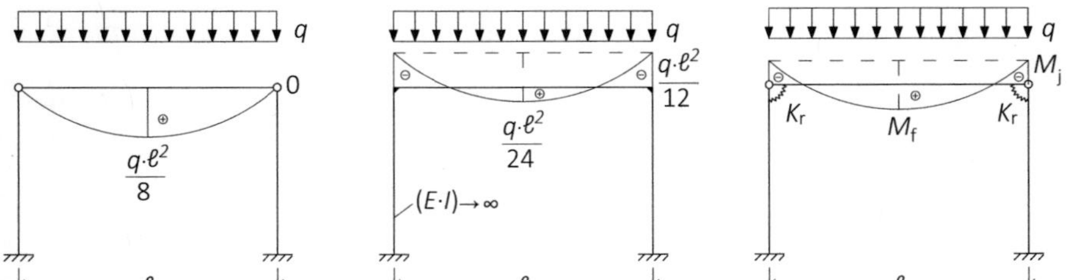

Figure E14-5 Bending moment diagram in a beam assuming pinned frame corners (a) compared with the bending moment diagram with fixed (moment-resisting) frame corners and infinitely rigid columns (index St, $(E \cdot I)_{St} \rightarrow \infty$) (b) and with semi-rigid frame corners having a rotational spring stiffness K_r,a corner moment M_j and a field moment M_f of the beam (c).

To take the influence of the rotation stiffness K_r of the joint on the distribution of internal forces and moments into consideration, a coefficient β_r is defined and functions as a benchmark for the ratio between rotation stiffness K_r of the joint and the bending stiffness $(E \cdot I)_{Ri}$ of the connected member:

$$K_r = \beta_r \cdot \frac{(E \cdot I)_{Ri}}{\ell} \quad \Leftrightarrow \quad \beta_r = K_r \cdot \frac{\ell}{(E \cdot I)_{Ri}} \tag{E14-1}$$

where

K_r Rotation stiffness of joint

$(E \cdot I)_{Ri}$ Bending stiffness of connected member (here: beam, index Ri)

ℓ Span of the connected member (here: beam)

The corner moment M_j shown in Figure E14-5 (c) of a frame with a semi-rigid corner joint will assume a value which, depending on the coefficient β_r, varies between the limit cases shown in Figure E14-5 (a) and (b). This corner moment can be determined using equation (E14-2) and disregarding longitudinal and shear deformations:

$$M_j = \frac{q \cdot \ell^2}{8} \cdot \frac{1}{1.5 + \alpha \cdot \frac{h}{\ell} + \frac{3}{\beta_r}} = \frac{q \cdot \ell^2}{8} \cdot \frac{1}{1.5 + \alpha \cdot \frac{h}{\ell} + \frac{3 \cdot (E \cdot I)_{Ri}}{K_r \cdot \ell}} \tag{E14-2}$$

where

h Frame height

ℓ Span or length of the beam

α Ratio of the bending stiffnesses of beam and columns, see also Figure E14-6 top

q Uniformly distributed load on beam

For the limit case “pinned“ ($K_r = 0 \rightarrow \beta_r = 0$), equation (E14-2) assumes the value $M_j = 0$ (Figure E14-5 (a)), for the limit case „fixed“ ($K_r \rightarrow \infty \rightarrow \beta_r \rightarrow \infty$) and with $\alpha = 0$ ($(E \cdot I)_{St} \rightarrow \infty$), the corner moment is $M_j = (q \cdot \ell^2)/12$ (Figure E14-5 (b)). Using the coefficient η_M, it is now possible to specify the influence of the joint stiffness on the corner moment M_j in relation to the corner moment with moment-resisting joint ($K_r \rightarrow \infty \rightarrow \beta_r \rightarrow \infty$):

$$\eta_M = \frac{M_j(\beta_r)}{M_j(\beta_r \rightarrow \infty)} \tag{E14-3}$$

While Figure E14-5 shows the values for the beam moments M_j and M_f for the lower and upper limits of K_r and β_r, Figure E14-6 shows an evaluation taking the stiffness K_r of the frame corner into consideration. The diagrams reveal the influence of rotational spring stiffness on the ratio η_M for various geometry and stiffness ratios.

When the values of the coefficient β_r are below 6, the corner moment is decreasing significantly. A joint is considered pinned if the ratio $\eta_M \leq 0.20$, which corresponds to a value for the coefficient β_r of 0.5. A fixed joint can be approximated at $\eta_M \geq 0.85$, whereby the coefficient β_r must reach values of at least 8 to 12. In all other cases ($0.5 < \beta_r \leq 8 - 12$), the semi-rigidity of the joint must be taken into consideration when determining the internal forces and moments.

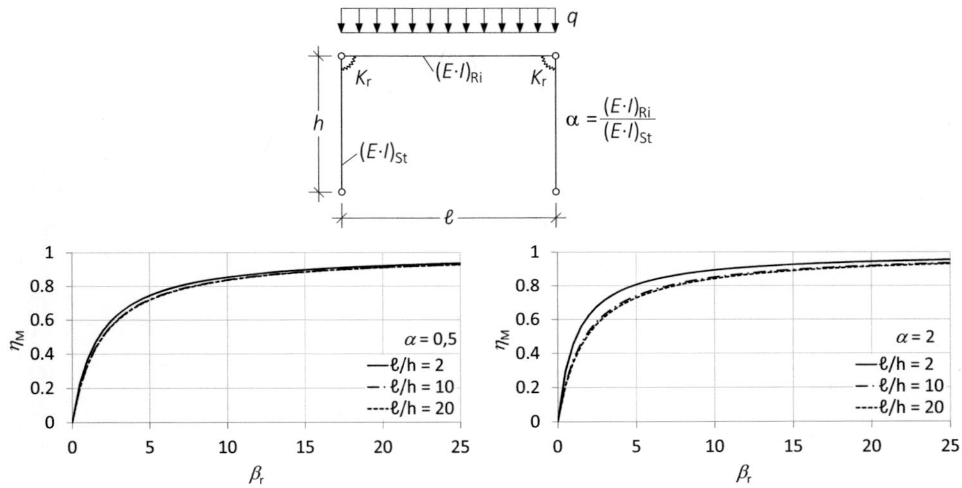

Figure E14-6 Above: frame geometry. Below: Influence of the joint stiffness β_r on the course of η_M for different ℓ/h-ratios, below left: $\alpha = 0.5$; below right: $\alpha = 2$.

In Figure E14-7, the ratio between the corner moment M_j and the field moment M_f of the beam, depending on ℓ/h and β_r, is specified (where $\alpha = 1.0$). An economical design of frame members is possible if the moments in the field and corners are approximately equivalent. The structural engineer can dictate the moment distribution accordingly when detailing the frame corner by selecting the joint stiffness appropriately. The rotational spring stiffness of the corner joint also dictates the deformation of the structure. The lower the joint stiffness, the larger the deflection in the centre of the beam span.

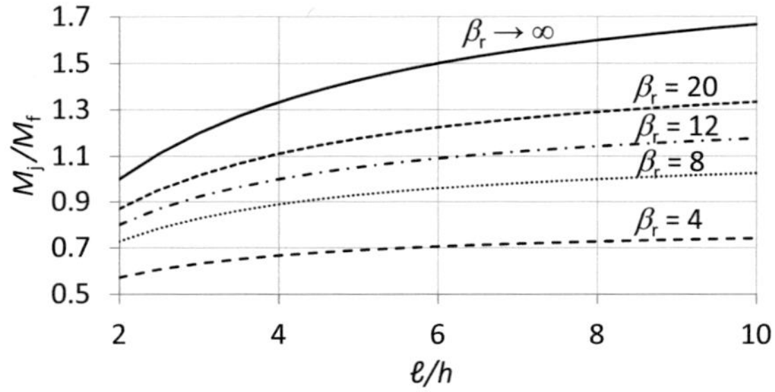

Figure E14-7 Influence of joint stiffness on the ratio between corner moment (M_j) and field moment (M_f) for $\alpha = 1$ and different values of β_r.

E14.2 Design of moment-resisting joints

When designing moment-resisting joints, the force distribution in the joint and/or fasteners must be taken into account and depends on the fastener configuration. The following section focuses on the mechanical behaviour of a widespread configuration of moment-resisting joints, in which dowel-type fasteners are arranged in circular form. The moment-resisting joint shown in Figure E14-8 is taken as an example. In the joint layout shown, the fasteners are subject to loads perpendicular to the fastener axis and in this case, the angle α_i between the force and grain directions varies for each of the individual fasteners. The behaviour of the wood in response to embedment stress, however, depends on the angle α_i to the grain direction, which results in the slip moduli for individual fasteners varying within the joint (Ohashi and Sakamoto, 1989). However, this dependency of slip moduli on the angle α_i between the force and grain directions is not taken into account in the design.

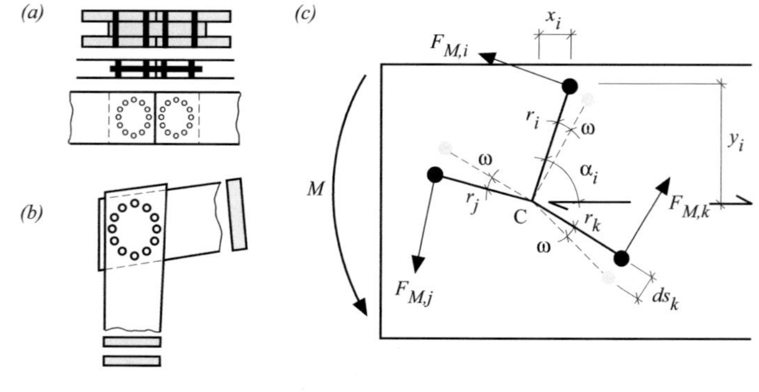

Figure E14-8 Moment-resisting joint: (*a*) joint in parallel members, (*b*) frame corner with dowels arranged in a circular form, (*c*) geometry and forces on fasteners. (STEP 1995 Article C16)

To determine the rotational spring stiffness K_r, the mean values for the initial slip moduli K_{ser} of timber-to-timber joints with mechanical fasteners specified in EC 5 are used and are independent of the angle between the force and grain directions (see also Table E1-1). For steel-to-timber joints, meanwhile, these values are doubled according to EC 5. The mean value of the slip modulus in the ultimate limit state $K_{u,mean}$ is $2/3 \cdot K_{ser}$ (equation (E1-8)). This value is then to be divided by the partial safety factor $\gamma_M = 1.3$. To determine the load distribution, the members in the joint area are assumed to be infinitely rigid, since they have far greater stiffness compared to fasteners. Considering the fastener displacement allows the rotation of the joint to be calculated, which, in turn, depends on the distance between the fasteners and the centre of rotation. This centre is generally the centre of gravity of the fasteners. The equilibrium of moments based on the centre of rotation C in accordance with Figure E14-8 (c) is:

$$M = \sum_{j=1}^{n} F_{M,j} \cdot r_j \tag{E14-4}$$

where

$F_{M,j}$ Load on the jth fastener

r_j Distance of the jth fastener to the centre of rotation C

n Number of fasteners

Assuming linear behaviour of the fasteners, the following relationships can be derived (small angle: $\tan\omega \approx \omega$), Figure E14-8 (c):

$$ds_k = \frac{F_{M,k}}{K_k} \quad \text{and} \quad \tan\omega \approx \omega = \frac{ds_k}{r_k} = \frac{F_{M,k}}{K_k \cdot r_k} = \frac{F_{M,j}}{K_j \cdot r_j} = \frac{F_{M,i}}{K_i \cdot r_i} \tag{E14-5}$$

where $K_{k/j/i}$ = slip modulus of kth, j^{th}, i^{th} fastener in the force direction.

The load $F_{M,i}$ of the ith fastener exerted by the moment can now be calculated using equations (E14-4) and (E14-5):

$$F_{M,i} = \frac{K_i \cdot r_i}{K_r} \cdot M \tag{E14-6}$$

where K_r is the rotational spring stiffness of the joint:

$$K_r = \sum_{j=1}^{n} K_j \cdot r_j^2 \tag{E14-7}$$

Considering the stiffness of the fasteners, the rotational spring stiffness of the joint can be calculated for different fastener arrangements and depending on the joint geometry. The rotational spring stiffnesses of the joints shown in Figure E14-9, each with equal fasteners, are specified in equations (E14-8) and (E14-9) while assuming linear-elastic load-deformation behaviour.

For a circular fastener arrangement, Figure E14-9 (a) applies:

$$K_r = K \cdot \left(n_1 \cdot r_1^2 + n_2 \cdot r_2^2\right) \tag{E14-8}$$

The rotational spring stiffness of the joint configuration in accordance with Figure E14-9 (b) is calculated as:

$$K_r = K \cdot \left(\sum_{i=1}^{n} x_i^2 + \sum_{i=1}^{n} y_i^2\right) \tag{E14-9}$$

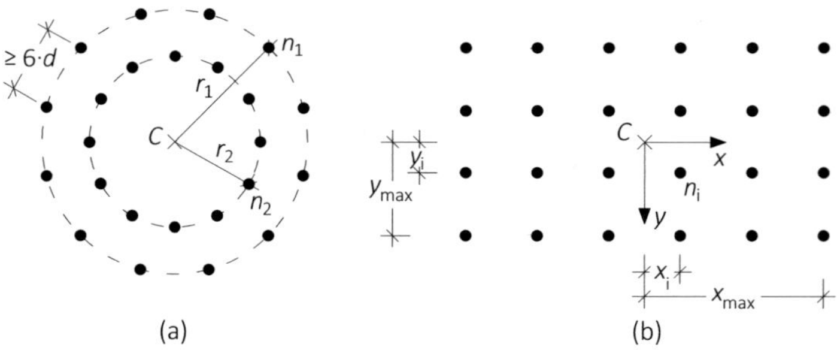

Figure E14-9 Geometry of different fastener arrangements in moment-resisting joints: (a) circular arrangement with two circles, (b) rectangular arrangement.

The moment M to be transferred generates a load F_M on the respective fastener, which acts perpendicular to the distance between the centre of rotation and the fastener. This load peaks for the circular arrangement shown in Figure E14-9 (a):

$$F_M = \frac{M \cdot r_1}{n_1 \cdot r_1^2 + n_2 \cdot r_2^2} \quad \left(= \frac{K \cdot r}{K_r} \cdot M = \frac{K \cdot r}{K \cdot \left(n_1 \cdot r_1^2 + n_2 \cdot r_2^2\right)} \cdot M \right) \tag{E14-10}$$

and for the rectangular arrangement shown in Figure E14-9 (*b*):

$$F_M = \frac{M \cdot \sqrt{x_{max}^2 + y_{max}^2}}{\sum_{i=1}^{n} x_i^2 + \sum_{i=1}^{n} y_i^2} \tag{E14-11}$$

The shear and normal force is uniformly distributed between the fasteners, Figure E14-10, meaning the following applies:

$$F_V = \frac{V}{n} \quad \text{and} \quad F_N = \frac{N}{n} \tag{E14-12}$$

where n = number of fasteners, $n = n_1 + n_2$ for two circles of fasteners, see Figure E14-9 (*a*).

The overall loading of the respective fastener can be calculated by the vector addition of F_M, F_V and F_N, Figure E14-10. The overall loading differs for each of the fasteners due to the changing directions of F_M. The greater the distance to the centre of rotation, the higher the load exerted by F_M on the respective fastener. Within a circle of fasteners meanwhile, the load on a fastener peaks if the direction of the resulting force of F_V and F_N corresponds to that of F_M, as in Figure E14-10 (*b*).

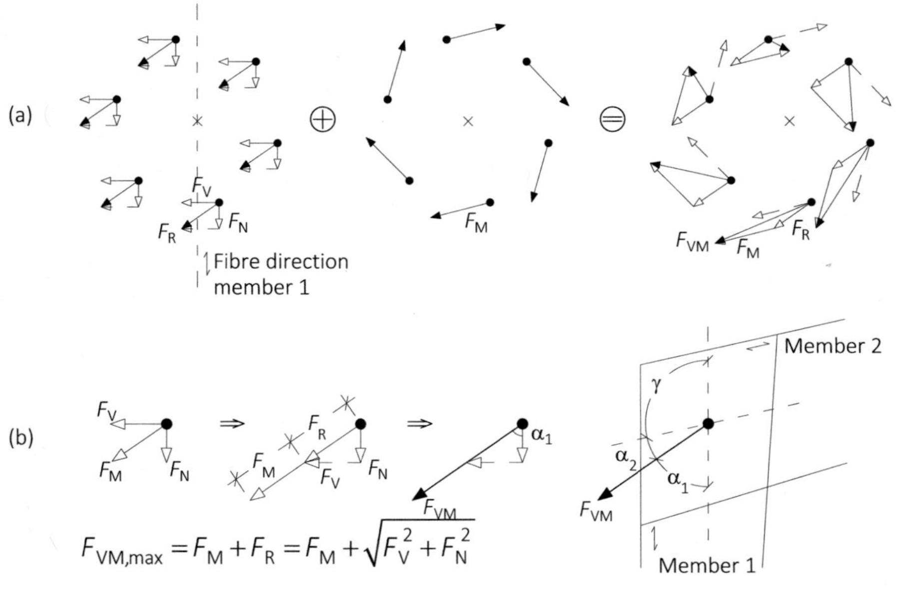

Figure E14-10 Loads on fasteners in a circular fastener arrangement with respect to member 1 = column. (a) vector addition of forces, (b) maximal force on fastener if F_R and F_M have the same direction.

Although the load-bearing capacity of dowel-type fasteners depends on the embedment strength of the members to be connected, the influence of the angle between the force and grain directions on embedment strength must be taken into consideration for fitted bolts, dowels and, where applicable, screws. This means that the load-bearing capacity of each fastener varies depending on its position or load direction and, in theory, each individual fastener of the joint would have to be verified. There is, however, scope to simplify this verification with the theoretical maximum fastener load. With this in mind, it is assumed that the resulting normal and shear forces and the force F_M generated from the moment act in the same direction (Figure E14-10 (*b*)). This maximum load is as follows for a circular fastener arrangement (index VM = fastener):

$$F_{VM,max} = F_M + \sqrt{F_N^2 + F_V^2} \tag{E14-13}$$

The angles between the force and grain directions are as follows in this case (Figure E14-10 (*b*)):

$$\alpha_1 = \arctan\left(\frac{F_V}{F_N}\right) \quad \text{and} \quad \alpha_2 = 180^\circ - \gamma - \alpha_1 \tag{E14-14}$$

where

α_1, α_2 Angle between the force direction of the theoretical maximum load and the grain direction of member 1 or 2

γ Angle between the axes of the beam and column

For a rectangular or trapezoidal fastener arrangement (Figure E14-9 (*b*)), the theoretical maximum value of F_{VM} should be verified for the outermost fastener:

$$F_{VM} = \sqrt{\left(F_N + \frac{x_{max}}{\sqrt{x_{max}^2 + y_{max}^2}} \cdot F_M\right)^2 + \left(F_V + \frac{y_{max}}{\sqrt{x_{max}^2 + y_{max}^2}} \cdot F_M\right)^2} \tag{E14-15}$$

The corresponding angle between the force and grain directions results as:

$$\alpha_1 = \arctan\left(\frac{y_{max} \cdot F_M + \sqrt{x_{max}^2 + y_{max}^2} \cdot F_V}{x_{max} \cdot F_M + \sqrt{x_{max}^2 + y_{max}^2} \cdot F_N}\right) \tag{E14-16}$$

E14.3 Timber stresses in the joint area

In addition to the loads on the fasteners, the stresses on the timber in the joint area must also be taken into consideration. Figure E14-11 shows two different fastener arrangements used to manufacture frame corners. The fastener arrangement marked with (*b*) reveals a far less favourable combination of shear stresses and stresses perpendicular to the grain compared to the circular arrangement (*a*), since higher tensile stresses perpendicular to the grain are generated over a larger member length and unfavourable interaction between tensile stresses perpendicular to the grain and shear stresses occurs throughout the member length in question. For fastener arrangements along the member edges, meanwhile, higher tensile stresses perpendicular to the grain are observed particularly in the vicinity of the end grain (Boult, 1988). The risk of splitting resulting from these observations can be mitigated by reinforcement measures (e.g. glued-on wood-based panels) or arranging fasteners with a high withdrawal capacity such as fully threaded screws to secure corners (see Section E14.4). For non-reinforced frame corners, the load-bearing capacity of the joint should be reduced by 15% if two circles of dowels are chosen as a fastener arrangement (Heimeshoff, 1977; Kolb, 1970). It is also advisable to select slender fasteners to further reduce the splitting risk.

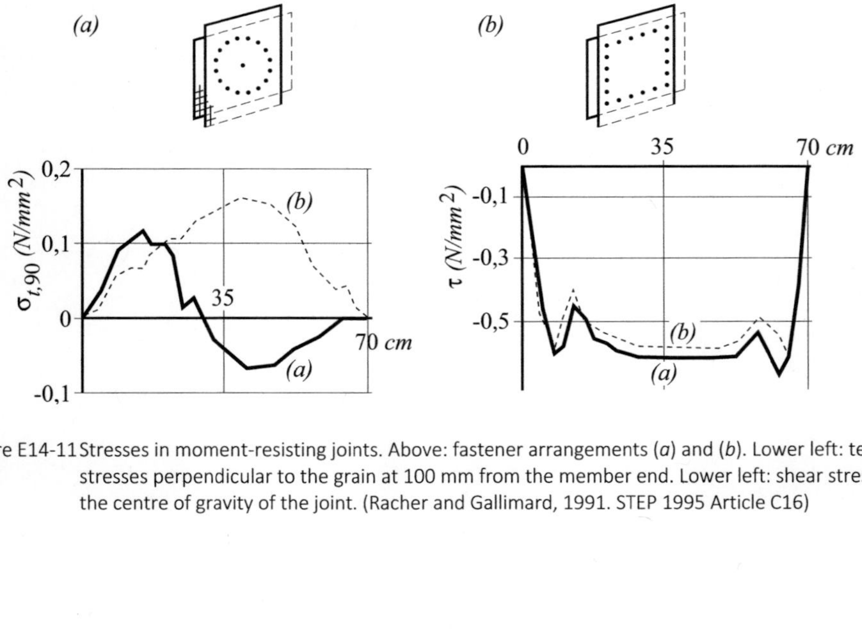

Figure E14-11 Stresses in moment-resisting joints. Above: fastener arrangements (*a*) and (*b*). Lower left: tensile stresses perpendicular to the grain at 100 mm from the member end. Lower left: shear stresses in the centre of gravity of the joint. (Racher and Gallimard, 1991. STEP 1995 Article C16)

To verify the **shear capacity of the timber members** in the joint area, shear stresses have to be derived, an example of which is shown using a joint with two circles of dowels. To do so, the components in *x*-direction of the forces transferred by the fasteners and above the centre of gravity are added, governing section see Figure E14-12 (*a*). For this purpose, the forces in the fasteners are transformed in a line load *f* and here, the following applies for the external circle of dowels (where $2 \cdot \pi \cdot r_1$ = circumference and $n_1 \cdot F_{M,1}$ = sum of the shear forces generated by the moment):

$$f_1 = \frac{n_1 \cdot F_{M,1}}{2 \cdot \pi \cdot r_1} \qquad \text{(E14-17)}$$

and for the internal circle of dowels:

$$f_2 = \frac{n_2 \cdot F_{M,2}}{2 \cdot \pi \cdot r_2} \qquad \text{(E14-18)}$$

As shown in Figure E14-12 (*b*), based on the line load at a circle segment of size $d\varphi$, the tangential force component $f \cdot r \cdot d\varphi$ and the force component perpendicular to the grain $f \cdot r \cdot d\varphi \cdot \sin\varphi$ (*x*-direction in Figure E14-12) can be determined. For both dowel circles, the sum of all force components perpendicular to the grain lying in the semicircle above the centre of gravity is calculated. This reveals the entire load component perpendicular to the grain generated by the corner moment:

$$V_M = \Sigma f_1 \cdot r_1 \cdot \sin\varphi \cdot d\varphi + \Sigma f_2 \cdot r_2 \cdot \sin\varphi \cdot d\varphi \qquad \text{(E14-19)}$$

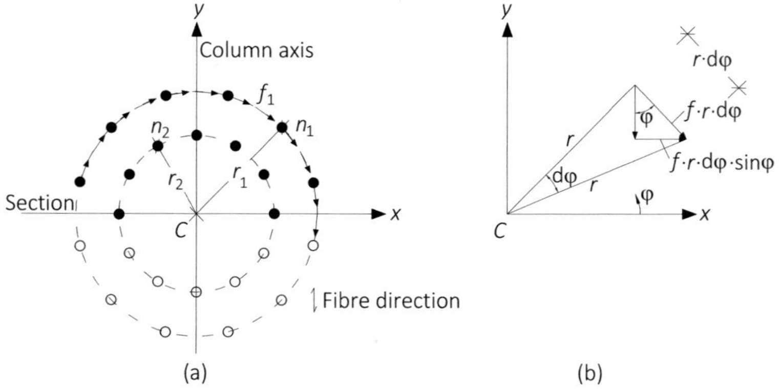

Figure E14-12 Derivation of stresses perpendicular to the grain in a circular arrangement of fasteners: (a) governing section, (b) circle segment with division of forces.

Equations (E14-17) and (E14-18) reveal that:

$$V_{\mathrm{M}}=\int_0^{\pi}\frac{n_1\cdot F_{\mathrm{M},1}}{2\cdot\pi}\cdot\sin\varphi\cdot\mathrm{d}\varphi+\int_0^{\pi}\frac{n_2\cdot F_{\mathrm{M},2}}{2\cdot\pi}\cdot\sin\varphi\cdot\mathrm{d}\varphi=\frac{1}{\pi}\cdot\left(n_1\cdot F_{\mathrm{M},1}+n_2\cdot F_{\mathrm{M},2}\right) \tag{E14-20}$$

In addition to the load component from the bending moment in accordance with equation (E14-20), the component from the shear force must also be taken into consideration to obtain the total shear force of the member:

$$V_{\mathrm{tot}}=V_{\mathrm{M}}-\frac{V}{2} \tag{E14-21}$$

For the circular fastener arrangement, the design shear load can be calculated from equation (E14-21) where the shear force component V_{M} generated by the moment, equation (E14-20), and the load F_{M} generated by the moment, equation (E14-10), are inserted:

$$V_{\mathrm{tot}}=\frac{M}{\pi}\cdot\frac{n_1\cdot r_1+n_2\cdot r_2}{n_1\cdot r_1^2+n_2\cdot r_2^2}-\frac{V}{2} \tag{E14-22}$$

For joint configurations with rectangular or trapezoidal fastener arrangements, the design shear load can be derived accordingly. Various fastener arrangements were systematically assessed by Kessel and Willemsen (1991).

For the rectangular arrangement shown in Figure E14-11 (*b*), the design shear load in the joint area can be calculated as follows:

$$V_{\mathrm{tot}}=V_{\mathrm{M}}-\frac{V}{2}=\frac{M}{2}\cdot\frac{1}{\sum x_1^2+\sum y_1^2}\cdot\sum_{\mathrm{i}=1}^{\mathrm{n}}\left|y_i\right|-\frac{V}{2} \tag{E14-23}$$

E14.4 Reinforcing measures for corners

The load direction of the fasteners varies according to their position, meaning the load-bearing capacity is not dependent on the number of fasteners arranged in a row. For circular fastener arrangements, minimum spacings and minimum end and edge distances of fasteners are not expressly regulated by design standards. Recommendations are given in Table E14-1 and the corresponding definitions are shown in Figure E14-13.

If no reinforcements are arranged in a corner joint with multiple dowel circles, the load-bearing capacity of the joint should be reduced by 15%:

$$n_{ef} = 0.85 \cdot n \tag{E14-24}$$

In corner joints featuring only a single circle of dowels or where reinforcement measures have been taken to prevent members splitting, $n_{ef} = n$ may be assumed.

Table E14-1 Recommendations for minimum spacings and edge and end distances for fasteners arranged in circles.

Minimum distances		Fitted bolts, dowels	Split ring and shear plate connectors	Toothed-plate connectors
To the loaded end	$a_{1,t}$	$7 \cdot d$	$2 \cdot d_c$	$1.5 \cdot d_c$
To the edge	$a_{2,t(c)}$	$4 \cdot d$	d_c	d_c
Spacing within a circle	a_1	$6 \cdot d$	$2 \cdot d_c$	$1.5 \cdot d_c$
Spacing between circles or rectangles	a_2	$5 \cdot d$	$1.5 \cdot d_c$	$1.5 \cdot d_c$

d, d_c = nominal diameter of the fasteners in mm

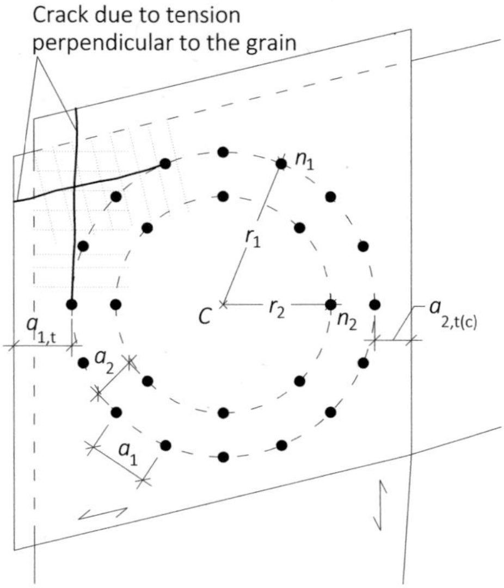

Figure E14-13 Definition of minimum distances and arrangement of the corner reinforcement. (Heimeshoff, 1977)

Reinforcing elements for beams and columns arranged to absorb tensile stresses perpendicular to the grain must be designed for a force $F_{ax,d}$, which corresponds to the lateral load $F_{M,d}$ of the fasteners of the external circle within a 30° sector (Heimeshoff, 1977):

$$F_{ax,d} = \frac{30}{360} \cdot n_1 \cdot F_{M,d} = \frac{1}{12} \cdot n_1 \cdot F_{M,d} \tag{E14-25}$$

The reinforcements against tension perpendicular to the grain are to be arranged in the external quarter of the corner in the column and beam as shown in Figure E14-13. The length of the reinforcing elements must be selected such as to prevent the risk of cracking tangential to the circle and parallel to the grain of the column or beam. Usual corner designs thus include reinforcing elements with lengths 10 to 12 times the diameter of the fasteners (dowels, fitted bolts) of the moment-resisting joint.

E14.5 Literature

P. Racher, original Article C16, STEP 1995.

Boult B.F. (1988). Multi-nailed moment resisting joints. International Timber Engineering Conference, Vol. 2:329-338, Seattle.

Heimeshoff B. (1977). Berechnung von Rahmenecken mit Dübelanschluss (Dübelkreis). Arbeitsgemeinschaft Holz, Holzbau-Statik-Aktuell, Folge 2, Düsseldorf.

Kessel H. and Willemsen T. (1991). Zur Berechnung biegesteifer Anschlüsse. Bauen mit Holz 93:342-352.

Kolb H. (1970). Festigkeitsverhalten von Rahmenecken. Bauen mit Holz 72:387.

Komatsu K. and Kawamoto N. (1992). Analysis of glulam semi-rigid portal frames under long-term load. Paper 25-8-1, CIB-W18 Meeting 25, Åhus.

Leijten A.J.M. (1988). Steel reinforced joints with dowels and bolts. International Timber Conference, Vol. 2:474-488, Seattle.

Ohashi Y. and Sakamoto I. (1989). Study on laminated timber moment resisting joint. 2nd Pacififc Timber Engineering Conference, Vol. 2:37-42, Auckland.

Racher P. and Gallimard P. (1992). Les assemblages de structures bois: a) comportement mécanique des principeaux types assemblages; b) analyse du fonctionnnement d'une couronne boulonnée. Annals ITBTP, No. 504, pp. 29-40.

E15 Joints with multiple shear planes

Original articles: H. Hartl, A. J. M. Leijten, B. O. Hilson

Trusses of various forms are primarily used in roof structures and bridges. Their advantages include a low dead weight and resulting material saving, while downsides include the significant construction height required in the centre of the truss compared to simple beams. Stand-out structural features include the in-plane shape of a truss and the joints between the chords, posts and diagonals, as well as any joints within the chords. Since high forces and sometimes moments occur in the joints, there is a need for particularly high-performance joints. In addition to joints with hinge bolts and steel nailing plates, those with slotted-in steel plates and connectors, double-shear joints with laterally loaded dowel-type fasteners and step or bonded joints, this is the main area of application for joints with multiple shear planes. Given the distribution of fastener loads over several shear planes, their compact joint areas make them the first choice to meet the construction requirements for joints in trusses. Multiple-shear steel-to-timber joints with covered dowels and individually adapted gusset plates even meet the dual requirements of visual appeal and effective fire resistance.

E15.1 Joints with multiple shear planes – construction types

In addition to the basic cases for fasteners in single- or double-shear, trusses also feature joints between members made of wood and wood-based materials which comprise more than three individual components (basic case of fastener in double-shear). Such joints arise, for instance, if the upper and lower chords comprise two or even three individual components, depending on the area involved. The posts and diagonals are laid between and/or outside the chord members. Furthermore, the limited length of sawn timber and the subsequent need for splice joints lead to tension joints within the two-component tension chords with internal and external wood members. This results in joints involving five or more different components arranged adjacent, while continuous fasteners have four or more shear planes accordingly. Such trusses would typically use what was previously a popular option but is now rarely seen, the nailed truss. The scope of trusses also includes steel-to-timber joints, within which two or more gusset plates are inserted into correspondingly tailored slots (or additionally externally attached), examples of which are included in Figure E15-1.

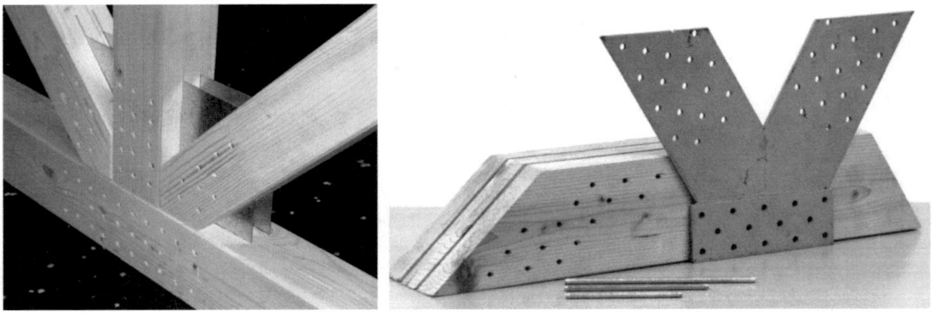

Figure E15-1 Left: GREIM system: truss joint between one post, two diagonals and lower chord, respectively made of glulam, nailed joint with two galvanised gusset plates. Right: connection BSB: slotted-in and predrilled glulam, galvanised predrilled 5 mm thick gusset plates and 6.3 mm thick dowels in lengths as required.

E15.2 Basis of design

To determine the total load-bearing capacity of joints with multiple shear planes, additional points must be considered when applying the basic Johansen cases discussed in Article E2. To calculate the total load-bearing capacity, the initial step involves considering plausible combination of possible individual failure modes (embedment failure in the wood, plastic hinges of the fastener) over the entire fastener length. Figure E15-2 provides an overview of the possible failure modes of a fastener with four shear planes for a timber-to-timber and a steel-to-timber joint. While the failure modes shown are practically feasible, they may not all be preferable as far as structural aspects are concerned. When using thin steel plates, an additional failure mode not shown in Figure E15-2 may also occur, since the thin steel plate cannot provide any clamping effect and the fastener may deform as shown in Figure E15-3 if the timber side members are thin.

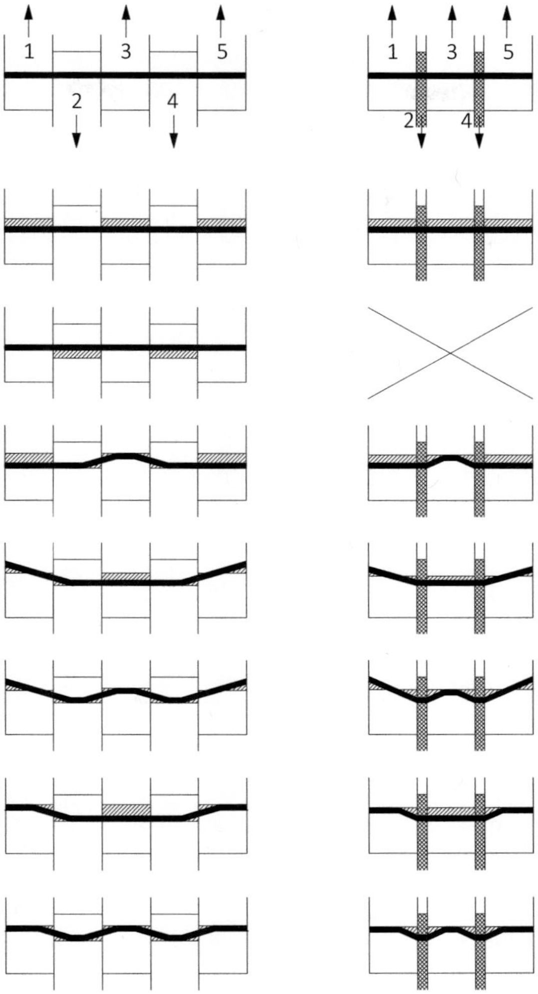

Figure E15-2 Failure modes of a fastener with four shear planes in a timber-to-timber (left) and a steel-to-timber connection (right).

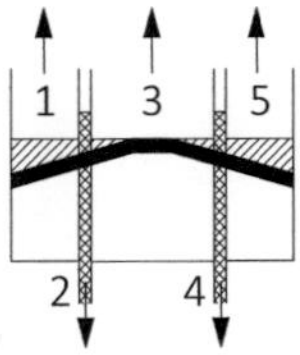

Figure E15-3 Possible deformation pattern of a fastener with four shear planes in a steel-to-timber connection with thin steel plates and thin timber side members (see case k in Figure E15-4).

The member thicknesses are constant in the qualitative representation in Figure E15-2. In reality, however, the mechanisms shown are strongly influenced by the individual member thickness and the corresponding embedment strength. The shaded areas indicate cavities which emerge after the respective displacement or rotation of a fastener relative to the member, i.e. where the wood matrix was "squeezed" by the fastener. It is assumed that the individual cross-sections 1, 3 and 5 are loaded in one direction and 2 and 4 in the other. The lower part of the illustration, meanwhile, shows the failure mechanisms with two plastic hinges per shear plane. This mechanism facilitates the exceptionally ductile nature of the joint, mitigates the overall splitting risk for solid and glulam timber and is ultimately a favoured option.

The **total load-bearing capacity of a joint with multiple shear planes is ultimately determined as the sum of the minimum load-bearing capacities of all shear planes**. As when designing the basic cases, likewise, ideal rigid-plastic behaviour can be assumed for fasteners under bending stress and the wood and wood-based materials under embedment stress when determining the load-bearing capacity per shear plane and fastener. The latter is calculated assuming that each shear plane is part of a double-shear joint. The relevant basic cases in accordance with Johansen, as already explained in Article E2, are shown in Figure E15-4. In shear planes which connect edge members (cross-sections 1 or 5), basically all failure modes are possible. The deformation pattern of a fastener, whose end is rotated (Figure E15-4: failure mode j for timber-to-timber joints and g and k for steel-to-timber joints, k only for thin steel plates, see also Figure E15-3), however, is only mechanically possible for an actual edge member. This deformation pattern, with rotated fastener ends, is not possible in centrally placed members (cross-sections 2, 3 or 4), which are "at the edges" only in the theoretical observation, since the deformed fastener must continually penetrate all individual members. When designing joints, another aspect to consider is the fact that total load-bearing capacity may be limited by shear failure along the external rows of fasteners or tensile failure of the wood (so-called block shear failure, Article E13). The step-by-step procedure used to determine the total load-bearing capacity is qualitatively shown in Figure E15-5.

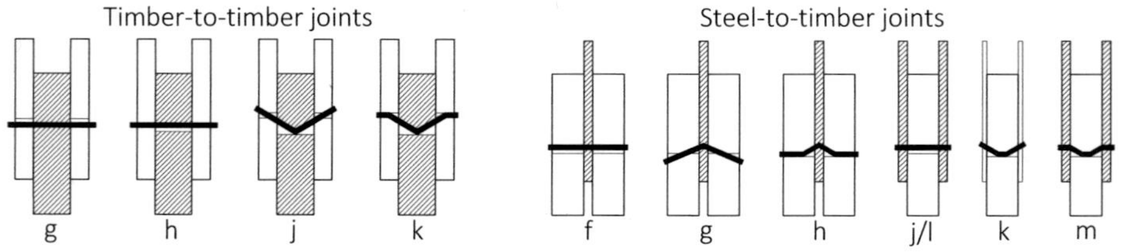

Figure E15-4 Basic cases in accordance with Johansen: failure modes of double-shear joints.

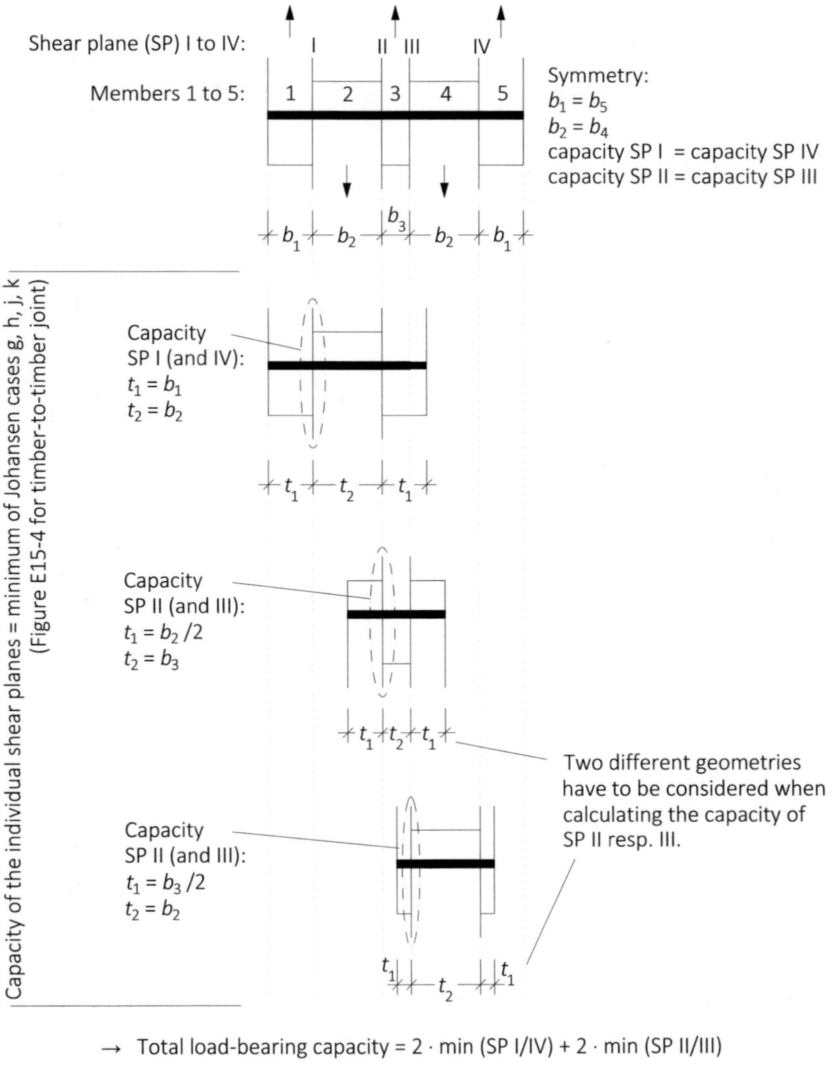

Figure E15-5 Step-by-step determination of the total load-bearing capacity of a symmetric timber-to-timber joint with four shear planes.

E15.3 Literature

H. Hartl, A.J.M. Leijten, B.O. Hilson, original Articles D1, 7 (volume 3), STEP 1995.

F

Serviceability

F1 Deformations

Original article: S. Thelandersson

The overall performance of structures should always meet two basic requirements. The first is structural safety, which is generally quantified by load-bearing capacity and the second is serviceability, which generally refers to the ability of the construction and its individual parts to perform satisfactorily under normal usage conditions. The serviceability limit state also encompasses the wellbeing of users, e.g. restricting unpleasant floor vibrations, and the visual appearance of the construction, e.g. impairments caused by excessive deformations. As a general rule, insufficient structural safety is a threat to human life and may also involve the risk of considerable damage costs. Conversely, exceeding the serviceability limit state rarely has consequences which impact on personal safety and any relevant economic risks are also likely to be low. Almost all cases of structural damage which occur in practice are attributable to a lack of serviceability, reflecting its importance when designing structures. For horizontal timber structures such as floors or roofs, serviceability requirements such as limiting deflections or vibrations are very often a decisive parameter when it comes to cross-sectional dimensions.

F1.1 Deformations over the lifetime of a structure

The portion of variable actions (e.g. imposed loads on floors or snow loads) in timber constructions tends to exceed the portion of permanent actions. This means that as the actions change considerably over time, the degree of deformations does the same and this fact must be taken into account appropriately when verifying serviceability. Figure F1-1 schematically shows the deflection history of a beam subject to permanent load and snow loads (Mårtensson, 1992; Mårtensson and Thelandersson, 1992). The total deflection is the sum of the deflection δ_1 from the permanent load directly after the load is applied and the deflection δ_2, which changes over time. δ_2 can be considered irreversible for the purpose of practical applications (Mårtensson, 1992). Short-term peak loads, such as those shown in Figure F1-1, emerge as either snow loads or imposed loads in the most common types of buildings.

With respect to the type of time-dependent behaviour shown in Figure F1-1, distinctions can be made between the following deflection portions (see Figure F1-2):

- δ_0 is the precamber in the unloaded state (state 0),
- δ_1 is the deflection due to permanent load directly after the load is applied (state 1),
- δ_2 is the deflection due to variable load $\delta_{2,inst}$ plus any time-dependent portions from permanent load δ_{creep} (state 2) and
- δ_{net} is the deflection (sag) of the beam relative to a straight line connecting the supports (= $u_{net,fin}$).

As a general rule, portions δ_0 and δ_1 are fixed as soon as the construction is complete. However, these portions may still be subject to change when the permanent load changes. δ_2 and δ_{net}, conversely, change over time.

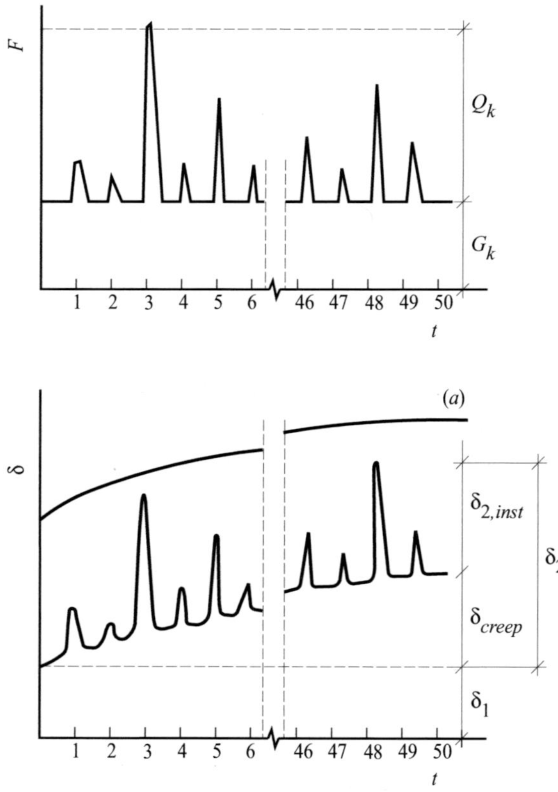

Figure F1-1 Illustration of the time-dependent deflection of a beam under permanent (*G*) and variable (*Q*) action. The curve (*a*) shows the resulting deflection δ when the beam is permanently loaded by the characteristic values $G_k + Q_k$. *F* is the load and *t* the time in years. (STEP 1995 Article A17)

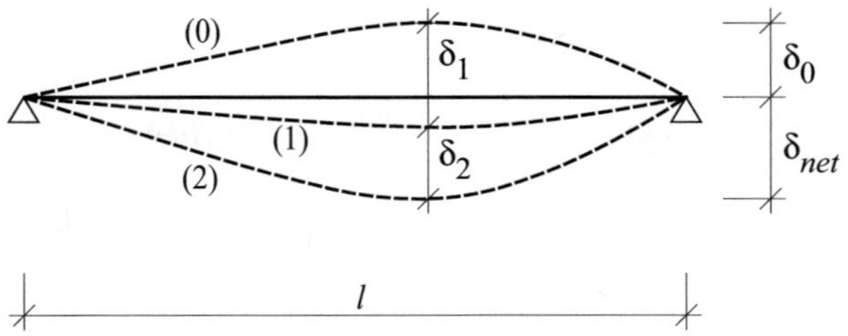

Figure F1-2 Deflection portions of a simply supported beam. (STEP 1995 Article A17)

F1.2 Load combinations for the serviceability limit state

The design principles valid for all Eurocodes define the various load combinations used to verify the serviceability limit state. For timber constructions, the characteristic and quasi-permanent combinations are particularly significant and the former is most relevant when exceeding the limit state triggers permanent damage or permanent and excessive deformations. This combination is used to determine the initial deformation and the symbolic notation of this combination is as follows (see also Equation (C2-5)):

$$\sum_{j\geq 1} G_{k,j} + Q_{k,1} + \sum_{i>1} \psi_{0,i} \cdot Q_{k,i} \qquad \text{(F1-1)}$$

The quasi-permanent combination applies for long-term effects and is used to determine the portion of creep deformation δ_{creep}. The symbolic notation of the quasi-permanent combination is as follows (see also Equation (C2-6):

$$\sum_{j\geq 1} G_{k,j} + \sum_{i\geq 1} \psi_{2,i} \cdot Q_{k,i} \qquad \text{(F1-2)}$$

In expressions (F1-1) and (F1-2), $G_{k,j}$ and $Q_{k,i}$ denote the characteristic values of the permanent or variable actions i, while values $\psi_{0,i} \cdot Q_{k,i}$ and $\psi_{2,i} \cdot Q_{k,i}$ represent the characteristic and quasi-permanent values of the variable action $Q_{k,i}$, respectively (see also Article C2).

F1.3 Limitation of deformations

The most common reasons for limitations applied to deformations in structures are:

- General usability and visual appearance (e.g. restricting disruptive visible deflections or avoiding uneven ground),
- Structural requirements (e.g. avoiding damage to non-load-bearing members such as partition walls or façade claddings and guaranteeing problem-free assembly, water- and airtightness or allowing water to discharge properly from roofs),
- Requirements concerning building equipment (e.g. guaranteeing the flawless functioning of machines, cables, exhaust air ducts and their supports).

Modern design standards such as EC 5 only cover general functional requirements, e.g. stipulating that structures must be designed and built to ensure the abovementioned serviceability perspectives can be taken into consideration. **Numeric values for deformation limits must be defined from case to case by the structural engineer in consultation with the building owner; EC 5 only provides recommendations.**

F1.4 Deflection limits to avoid damage

Deflections should be limited, when there is a risk of permanent damage to partition walls, installations, fittings and claddings if this is not done. In this case, it should be highly likely that the deflection limit will not be exceeded. For this purpose, the deflection is calculated with the characteristic load combination (F1-1) and any damage in this case is generally due to deformations which occur once the construction is complete. The formal condition applies to circumstances in which deflections can result in permanent damage:

$$u_{\text{inst}} \leq u_{\text{inst,crit}} \qquad \text{(F1-3)}$$

whereby u_{inst} represents the initial deformation for the characteristic combination, namely the portion δ_1 from Figure F1-2 and the portions of the variable loads on the initial deformation, the time-dependent nature of which is reflected in the combination coefficient $\psi_{0,i}$. $u_{\text{inst,crit}}$ is the limit value, which, if exceeded, may cause damage. The value of $u_{\text{inst,crit}}$ depends on the type and connection details of the members that can be damaged. In the absence of any more accurate details, $u_{\text{inst,crit}}$ is assumed to be either a fixed absolute value, e.g. 20 mm, or a percentage of the span ℓ, e.g. $\ell/300$ to $\ell/500$ for simply supported beams and $\ell/150$ to $\ell/250$ for cantilever beams (EC 5 Table 7.2). These values are often recommended for beams in floors and roofs, which are directly connected with partition walls or other non-load-bearing members.

F1.5 General usability and visual appearance

With regard to general usability and the visual appearance, limiting excessive deflections which may be permanent or occur over an extended period is often desirable. However, it may be permissible to exceed the limit values at times, if the deformations are reversible and/or limited to a short period. In this case, a higher probability of occurrence can be accepted for the limit values, whereupon additional deformation due to creep is calculated with the quasi-permanent load combination (F1-2) and added to the initial deformation u_{inst} (characteristic combination). The corresponding condition in this case is as follows:

$$u_{net,fin} \leq u_{fin,crit} \tag{F1-4}$$

Here, $u_{net,fin}$ denotes the sag of the beam relative to a straight line connecting the supports (see Figure F1-2, value δ_{net}), while $u_{fin,crit}$ is the deflection limit value with respect to the general visual appearance. The value of $u_{fin,crit}$ depends, among other things, on the type or style of construction and the attitude of building users as well as whether or not the beam is visible. The requirements for residential buildings tend to be far more stringent than those for industrial buildings. As a general recommendation for simply supported beams, the value $u_{fin,crit} = \ell/150$ to $\ell/300$ is specified; while for cantilever beams $u_{fin,crit} = \ell/75$ to $\ell/150$ applies (EC 5 Table 7.2).

F1.6 Calculation of deflection in accordance with EC 5

The elastic initial deformations caused by bending u_m can generally be calculated based on the simple beam theory and using formulas from corresponding textbooks. Given the low shear modulus of the wood compared to the modulus of elasticity, it is not always advisable to disregard shear deformations. Details used to calculate the shear deformation u_v can also be taken from textbooks and the overall initial deflection u_{ins} is ultimately revealed as the sum of u_m and u_v. To get an idea of the order of magnitude of the shear deformations, a simply supported beam with rectangular cross-section $b \cdot h$, span ℓ and subject to a uniformly distributed load q is observed. In this case, the ratio between deformation due to shear u_v and that due to bending u_m at mid-span of the beam can be approximated as:

$$\text{At mid-span:} \quad u_{\mathrm{m}} = \frac{5}{384} \cdot \frac{q \cdot \ell^4}{E \cdot I}$$

$$u'_{\mathrm{v}} = \frac{Q}{G \cdot A_s} = \frac{1}{G \cdot A_s} \cdot \left(\frac{q \cdot \ell}{2} - q \cdot x \right) \Rightarrow u_{\mathrm{v}}(x) = \frac{1}{G \cdot A_s} \cdot \left(\frac{q \cdot \ell}{2} \cdot x - \frac{q}{2} \cdot x^2 \right)$$

$$\text{At mid-span:} \quad u_{\mathrm{v}}\left(x = \frac{\ell}{2} \right) = \frac{q \cdot \ell^2}{8 \cdot G \cdot A_s} = \frac{q \cdot \ell^2}{8 \cdot G \cdot \frac{5}{6} \cdot A} \tag{F1-5}$$

$$\text{Ratio:} \quad \frac{u_{\mathrm{v}}}{u_{\mathrm{m}}} = 0.96 \cdot \frac{E}{G} \cdot \left(\frac{h}{\ell} \right)^2$$

The ratio E/G is around 16 for solid timber and glulam timber, which yields $u_v/u_m = 0.15$ for $\ell/h = 10$ and less than 0.04 for $\ell/h = 20$. For a concentrated load at mid-span, these ratios are 25% higher.

Long-term or creep deformations in wood under permanent load are significantly influenced by the surrounding climate. If, amid constant high moisture content, the level of creep deformations shows only a modest increase compared to that with low-level moisture content, this does not mean that moisture content has only a minor impact on creep deformations. Changes in moisture content are one of the primary causes of major creep deformations (mechano-sorptive creep, see Article B2). Accordingly, long-term deformations of members in outdoor climate, which include frequent and swift changes in relative humidity, exceed corresponding deformations of members located in air-conditioned buildings. For the same reason, members with large cross-sections have fewer creep deformations than those with small cross-sections, since the rate of change in moisture content tends to be far slower in the larger cross-sections. Surface treatments which prevent the exchange of moisture between wood and the surrounding air have the same effect (Mårtensson, 1992; Taylor et al., 1991).

As well as load-related deformations, the serviceability of timber structures is also impacted by the swelling and shrinking of the wood. The deformations in the wood triggered by changing moisture content may well be of the same order of magnitude as those caused by external loads. However, swelling and shrinking effects can be limited by employing a corresponding construction; particularly in terms of fine details and ensuring the use of dry wood. Deformations solely attributable to moisture changes will not be covered in any further detail in the current article.

Rules to calculate deflections are specified as non-binding in EC 5. In accordance with these rules, the elastic initial deformations u_{inst} are to be calculated based on the average values of the corresponding stiffness parameters. For structures made of members with the same creep properties, the final deformation u_{fin} can be simplified as follows, assuming a linear relation between load and deformation:

$$u_{fin} = u_{inst,G} \cdot \left(1 + k_{def}\right) + u_{inst,Q,1} \cdot \left(1 + \psi_{2,1} \cdot k_{def}\right) + \sum_{i>1}\left[u_{inst,Q,i} \cdot \left(\psi_{0,i} + \psi_{2,i} \cdot k_{def}\right)\right] \quad \text{(F1-6)}$$

where k_{def} is the creep coefficient, which takes into consideration the increase in deformations over time attributable to the surrounding climate and the load duration class. Values for k_{def} are specified in EC 5 Table 3.2 for various construction materials and differing service classes.

F1.7 Literature

S. Thelandersson, original Article A17, STEP 1995.

Mårtensson A. (1992). Mechanical behaviour of wood exposed to humidity variations. Report TVBK-1006, Department of Structural Engineering, Lund University.

Mårtensson A. and Thelandersson S. (1992). Control of deflections in timber structures with reference to EC 5. Paper 25-102-2, CIB-W18 Meeting 25, Åhus.

Taylor G.D., West D.J. and Hilson B.O. (1991). Creep of glued laminated timber under conditions of varying humidity. International Timber Engineering Conference, London.

F2 Vibrations

Original article: S. Ohlsson

The vibration verifications introduced in this article are those at the serviceability limit state, since most people perceive excessive vibrations as disruptive. The main focus in this case is on floor vibrations. Vibrations caused by machines and floor vibrations in rooms subject to rhythmical loading (e.g. dance or sports halls), must be examined in more detail and are not covered in this article. Vibrations are dynamic phenomena and the dynamic behaviour of floors is governed by the floors' mass, stiffness, damping and hence their natural frequency and eigenforms. To assess such complex dynamic behaviour cost-effectively, EC 5 replaces the actual dynamic load with equivalent vertical, static loads (unit impulse and equivalent static load). DIN 1052 stipulated that when verifying floor vibrations, the deflection due to quasi-permanent load should not exceed 6 mm, which limited the scope to a stiffness criterion and excluded any additional criteria, such as natural frequency limitations. EC 5, conversely, includes a more complex verification method, which means so-called unit impulse velocity responses are reviewed as well as stiffness and frequency criteria.

Although methods for a (simplified) vibration verification process already exist, in practice, numerous problems remain with floor vibrations. Users dislike the latter, even if the extent of such vibrations may be within the verification limit values specified in EC 5. Newer research projects (e.g. Hamm and Richter, 2009) have re-examined the vibration behaviour of timber and timber-composite floors and assessed the validity or relevance of the EC 5 verifications. The influence of the floor build-ups (e.g. with or without a screed) was also examined. New design and construction rules are discussed to round off this article, since they also pave the way to verify floor vibrations more reliably.

F2.1 Serviceability requirements

The fitness for purpose of a building depends on the extent to which it fulfils its prescribed functions in due form. All aspects of serviceability defined by the static system or the load-bearing members of a building can be summarised under the collective term "structural serviceability". Structural serviceability requirements are normally building-wide or on key building sections.

Most serviceability requirements are a factor of the following conditions:

- Acceptable human comfort (comfortable structural behaviour),
- Guaranteed functionality of the building, including installations and
- Acceptable visual building appearance.

These conditions are set out in detail in EC 5. Damage to claddings or partition walls may e.g. lead to a loss of functionality due to leaking floor decking in a bathroom or adversely affect the visual appearance due to visible cracks.

Serviceability requirements differ significantly from load-bearing capacity requirements at times. Unlike ultimate limit states, where there is no possibility of reversing the process when exceeded, when serviceability limit states are reached, such as the unpleasant feeling of excessive timber floor vibrations, reversal is possible. In both examples, the limit state is reached due to deviations from the expected straight and level flooring. For the serviceability limit condition, the key parameter is therefore the sum of all deviations, collectively comprising a portion caused by external loads and one caused by climatic influences. In this case, a precamber of the beams makes it possible to limit the total deflection to tolerable values. A consideration of precamber is included in EC 5. However, precamber has no influence on the beam vibration behaviour.

The limit values recommended in EC 5 for deflections and vibrations generally guarantee compliance with serviceability requirements. The following section covers these requirements with respect to vibrations in more detail. **For all verifications of the serviceability limit state, the rule is that the specified limit values are only recommended; the structural engineer must define limit values or required verifications on a case-by-case basis with the building owner.**

F2.2 Serviceability concerning vibrations

Vibrations which limit serviceability can be caused by wide-ranging loads and avoiding vibrations perceived as unpleasant by building users is particularly important. The sensitivity of the people concerned is thus the main factor dictating the limit of what are considered permissible vibrations. In timber frame constructions, human activities and installed machines both constitute key sources of vibrations. with the former usually comprising footfall from human traffic, running, or children playing around.

The key consequences of these causes are distinguished as follows:

- Discomfort due to vibrations caused by steps and
- Discomfort due to vibrations caused by machines.

The way different people react to vibrations is subjective and influenced by myriad factors. One comprehensive presentation of the relevant issues is included in Griffin (1990). The following relationships apply to most situations:

The human perception of vibration

- Depends on the vibration acceleration for frequencies of less than around 8 Hz,
- Depends on the vibration velocity for frequencies exceeding 8 Hz,
- Exhibits a logarithmic characteristic, resembling the subjective perception of volume,
- Increases with the vibration duration,
- Declines with proximity to and awareness of the vibration source, and
- Declines with increasing physical activity or familiarity.

Two design goals can be derived from the above relationships: first of all, the extent of vibrations in proximity to the vibration source should be limited, which can be ensured by designing the timber floor in accordance with the approach described in the following sections. In addition, transmission of vibrations to neighbouring construction sections should be prevented.

With this in mind, a suitable static system should be chosen, including consideration of the following points:

- Timber frame structures allow vibrations to be transferred to neighbouring storeys, which makes them less suitable for vibration-prone constructions. For the same reason, continuous floors between neighbouring residences are best avoided.
- Partition walls should be located above each other and constructed right down to the foundation. If a partition wall e.g. is only present in one storey and erected at the mid-span of a timber floor, this wall may structurally couple the two adjoining floors and generate reciprocal vibrations. While these vibrations may be acceptable within the storey in which the vibration source itself is located, they often cause discomfort in other storeys, where the vibration source is unknown.

F2.3 Human-induced vibrations

The initial section already explained that in DIN 1052, the scope for limiting vibrations only included deflection. This requirement included defining a stiffness criterion, since a low bending stiffness of a floor leads to the deflection limit value being exceeded. The greater the floor span, the higher the level of deflection and thus the lower the natural frequency. A floor with a smaller span will thus oscillate with a smaller period (= higher frequency) than one with a larger span. The natural frequency *f* can be represented depending on the stiffness and hence the static deflection u_{stat} of a floor, see equation (F2-1). Equation (F2-1) is based on a model of the floor as an undamped single degree of freedom system with stiffness *K*, mass *M* and load *F*:

$$f=\frac{\omega}{2\cdot\pi}=\frac{1}{2\cdot\pi}\cdot\sqrt{\frac{K}{M}}=\frac{1}{2\cdot\pi}\cdot\sqrt{\frac{F/u_{stat}}{F/g}}=\frac{1}{2\cdot\pi}\cdot\sqrt{\frac{g}{u_{stat}}}\approx\frac{5}{\sqrt{u_{stat}[cm]}} \qquad \text{(F2-1)}$$

By limiting deflection, the natural frequency of a timber floor can be minimised. The first natural frequency of a floor should be at least double the frequency of human-induced vibrations, since frequently repeated stimulation at a frequency of up to half that of the natural frequency can result in resonance (Hamm and Richter, 2008). One step per second corresponds to 1 Hz. Activities such as running have a frequency of up to 4 Hz. Accordingly, the desirable criterion is normally a natural frequency of at least 8 Hz.

The relationships observed shall now be explained in more detail. The vibration design of timber floors in accordance with EC 5 takes into consideration everyday human activities like the dynamic effects of footfall, with scope to determine the latter via tests (Ohlsson, 1982). Figure F2-1 shows the course of the load induced by a person treading in place over time. In residential buildings, the swiftly declining short-term reaction prevails and the load comprises two differing components:

- A low-frequency component (0 - 8 Hz), caused by the step frequency and its corresponding harmonic vibrations plus
- A high-frequency component (8 - 40 Hz), primarily due to the contact between the heel and floor, the so-called heel drop.

If the natural frequency of the floor exceeds 8 Hz, the low-frequency components generate quasi-static vibrations with amplitudes mainly determined by floor stiffness rather than mass.

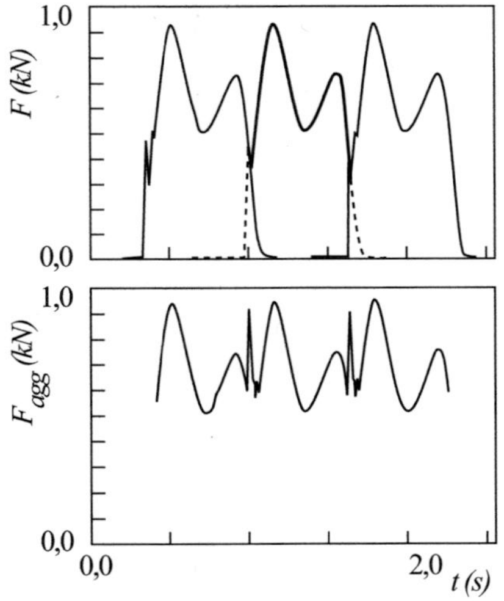

Figure F2-1 Course of the contact forces generated by the steps of a walking person (top) and corresponding load on the floor (below). (STEP 1995 Article A18)

The first natural frequency f_1 of a rectangular timber beam floor with pinned supports along all four edges can be determined with the approximate equation (F2-2). Equation (F2-2) shows how the "natural frequency" of a simply supported beam is determined in very simplified form; "natural frequency", since a beam, unlike a single degree of freedom system (simple harmonic oscillator), has not one but innumerable natural frequencies, the lowest thereof is designated as the first natural frequency:

$$f_1 = \frac{\pi}{2 \cdot \ell^2} \cdot \sqrt{\frac{(E \cdot I)_\ell}{M}} = f_{1,\text{beam}} \tag{F2-2}$$

where

M	Mass per unit of area in kg/m^2,
ℓ	Floor span in m
$(E \cdot I)_\ell$	Equivalent bending stiffness of the floor about an axis perpendicular to the beam axis in Nm^2/m

For biaxial floors (e.g. massive timber floors using cross-laminated timber) however, equation (F2-2) is overly simplified, since the effective bending stiffness in the perpendicular direction increases the first natural frequency.

The natural frequency of such floors can be determined using equation (F2-3):

$$f_{\text{plate}} = f_{\text{beam}} \cdot \sqrt{1 + \frac{1}{\alpha^4}} \quad \text{where} \quad \alpha = \frac{b}{\ell} \cdot \sqrt[4]{\frac{(E \cdot I)_\ell}{(E \cdot I)_\text{b}}} \tag{F2-3}$$

where

b	Floor width in m
$(E \cdot I)_\text{b}$	Equivalent bending stiffness of the floor about an axis parallel to the beam axis in Nm^2/m

From this, the initial two criteria of EC 5 can be derived:

- Frequency criterion:

  $$f_1 \geq 8\,\text{Hz} = f_{\text{lim}} \tag{F2-4}$$

- Stiffness criterion:

  $$\frac{u}{F} \leq a = a_{\text{lim}} \quad \text{in} \quad \frac{\text{mm}}{\text{kN}} \tag{F2-5}$$

To avoid resonance, a limit value is defined for the first natural frequency of the floor and for ratio $a = u/F$ from the vertical deflection u due to a vertical static load F. According to Kreuzinger and Mohr (1999), both criteria correspond to two of the three categories of vibration behaviour depending on the type of stimulation involved. The stiffness criterion limits the vibration due to a **one-time deflection caused by a footstep**, while the frequency criterion helps mitigate the risk of resonance caused by **repeated steps** and prevents any build-up of vibration amplitude. The final category, namely the impact of the high-frequency components caused by **heel drop**, is taken into consideration by a unit impulse of 1 Ns. The resulting vibration velocity v is a characteristic of the structure. If we take the floor as a freely movable and rigid body of mass M, it would be accelerated by the unit impulse to a velocity of 1 (m/s) /M.

According to Ohlsson (1988), the initial maximum value of velocity under a unit impulse, the so-called unit impulse velocity response v, can be determined in simplified form for the most common case in practice, a rectangular timber beam floor with pinned supports along all four edges. For a span ℓ in m, a floor width b in m and a uniformly distributed mass M in kg/m^2, the following approximation applies:

$$v = \frac{4 \cdot (0.4 + 0.6 \cdot n_{40})}{M \cdot b \cdot \ell + 200} \tag{F2-6}$$

In equation (F2-6), n_{40} denotes the number of eigenmodes with natural frequencies of less than 40 Hz and $M \cdot b \cdot \ell$ is the floor mass. An additional generalised mass of 50 kg is included in the quotient 4/200, which represents a vibrational portion of the body of the person supposedly disturbed by the vibration. n_{40} can be determined with the following approximate equation:

$$n_{40} = \left[\left(\left(\frac{40}{f_1} \right)^2 - 1 \right) \cdot \left(\frac{b}{\ell} \right)^4 \cdot \frac{(E \cdot I)_\ell}{(E \cdot I)_\mathrm{b}} \right]^{0.25} \tag{F2-7}$$

This simplifies the task of determining the maximum vibration velocity caused by a unit impulse, whereby the number of eigenmodes having to be taken into consideration is limited to natural frequencies below 40 Hz, since tests (Ohlsson, 1982) showed that natural frequencies over 40 Hz no longer contribute significantly to the initial maximum velocity v_{max}.

The favourable effect of a short vibration period is taken into consideration via a limit value depending on damping. The key damping parameter in this interaction is the damping coefficient σ_0, which indicates the decline in vibration amplitudes depending on time rather than the number of periods. The damping factor σ_0 is defined as:

$$\sigma_0 = f \cdot \zeta \qquad \text{(F2-8)}$$

In this case, for f, the natural frequency f_1 can be used, while the modal damping ratio ζ for common timber beam floors should be assumed at 0.01. The damping of timber beam floors shows considerable scatter with an average value of around 0.015 to 0.02 (Chui, 1988; Ohlsson, 1982). If a value exceeding 0.01 is used to verify vibrations, it is important to ensure this value remains valid throughout the service life of the floor.

The third criterion of EC 5 in accordance with equations (F2-4) and (F2-5) thus involves defining a limit value for the unit impulse velocity response:

$$v \le b^{(f_1 \cdot \zeta - 1)} = v_{lim} \quad \text{in} \quad \frac{\text{m}}{\text{Ns}^2} \qquad \text{(F2-9)}$$

The limit values of a from equation (F2-5) and b from equation (F2-9) are not independent. A low value for a requires a higher floor stiffness, although this results in an increase of the limit value of the unit impulse velocity response v_{lim} up to a theoretical maximum value of 1 m/(Ns^2) ($b^0 = 1$). If higher values are now assumed for the base b, the trend of v_{lim} to assume value 1 slows down and the vibration behaviour of the floor improves. Figure F2-2 symbolises this relationship: as b increases and a decreases simultaneously, the vibration behaviour of the floor improves whereas a decrease in b and a simultaneous increase in a leads to the vibration behaviour deteriorating. Common limit values are b = 120 and a = 1 mm/kN.

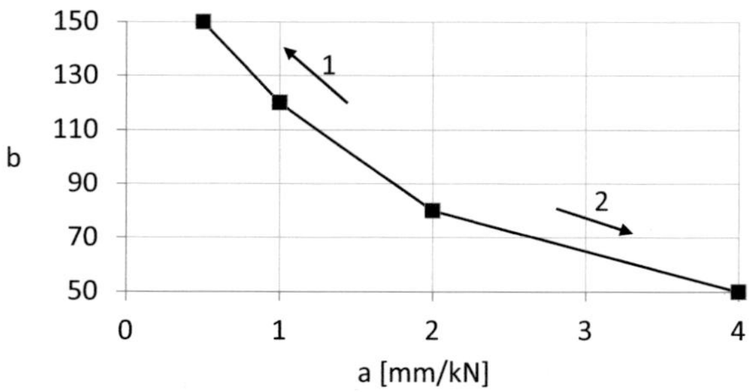

Figure F2-2 Relationship between a and b according to EC 5, 1 = better behaviour, 2 = worse behaviour.

F2.4 Recent research

All floors examined in Hamm and Richter (2009) met the criterion for unit impulse velocity response in accordance with equation (F2-9), although a portion of the floors in question were classified by users as unpleasant, hence the questionable significance of this criterion. Hamm and Richter were also able to show that floor build-ups such as screed use as well as the actual detailing, e.g. using a floating screed, have a key impact on the vibration behaviour. Moreover, to date, no distinction has been made between differing requirements for floors depending on their position, namely as a floor between or within units of use respectively. Hamm and Richter thus propose disregarding the verification of unit impulse velocity response and instead limiting the scope to the frequency and stiffness criteria, which, depending on usage, can be subdivided into two further categories. Moreover, requirements for structural detailing are also imposed and a value for the limit acceleration in the event of non-compliance with the frequency criterion. All details are summarised in Table F2-1. In Hamm (2012), a design example is included.

Table F2-1 Relationship between usage and requirements for design and structural detailing, from: Hamm and Richter, 2009, Table 3.

Installation position	Floors between different units of use	Floors within a unit of use	No requirements in terms of vibration behaviour
Assessment	Assessment: 1.0 to 1.5	Assessment: 1.5 to 2.5	Assessment: 2.5 to 4.0
Usage of rooms examined in this project	e.g. short-span corridors, apartment partitioning floors in multiple family units, office floors with PCs or meeting rooms	e.g. floors in normal single-family homes, existing floors with consent of the building owner	e.g. floors under unused rooms or undeveloped attic spaces
Description of the perceptions of the vibration behaviour	No vibrations at all or only very faintly perceptible, if focusing on it; not perceived as disruptive	Vibrations are perceptible, but not considered disruptive	Vibrations are perceived as perceptible to significantly perceptible, unpleasant and also disruptive at times
Frequency criterion $f_1 \geq f_{lim}$	f_{lim} = 8 Hz	f_{lim} = 6 Hz	-
Stiffness criterion u (2 kN) $\leq u_{lim}$	u_{glim} = 0,5 mm	u_{lim} = 1,0 mm	-
More accurate examination only if $f_1 < f_{lim}$	$f_{min} \leq f_1 < f_{lim}$ where f_{min} = 4.5 Hz and a_{lim} = 0.05 m/s^2	$f_{min} \leq f_1 < f_{lim}$ where f_{min} = 4.5 Hz and a_{lim} = 0.10 m/s^2	-
Structural requirements	Arrangement of floating heavy or light screed on grit fill	-	-

F2.5 Literature

S. Ohlsson, original Article A18, STEP 1995.

Chui Y.H. (1988). Evaluation of vibrational performance of light-weight wooden floors. International Conference on Timber Engineering 1:707-715, Forest Products Research Society, Madison.

Griffin M.J. (1990). Handbook for human vibration. Academic Press, London.

Kreuzinger H. and Mohr B. (1999). Gebrauchstauglichkeit von Wohnungsdecken aus Holz; Abschlussbericht. TU München, Fachgebiet Holzbau.

Hamm P. and Richter A. (2009). Bemessungs- und Konstruktionsregeln zum Schwingungsnachweis von Holzdecken. Fachtagungen Holzbau 2009. Leinfelden-Echterdingen, 15-29.

Hamm P. (2012). Schwingungen bei Holzdecken – Konstruktionsregeln für die Praxis. 2. Internationales Forum Holzbau, Beaune.

Ohlsson S. (1982). Floor vibration and human discomfort. Dissertation, Chalmers University of Technology, Göteborg.

Ohlsson S. (1988). Springiness and human-induced floor vibration. Document D12, Swedish Council for Building Research, Stockholm.

G

Accidental loads and additions

G1 Reaction to fire and fire design

Original article: H. Hartl

Describing the way materials behave in response to a fire simply is an almost impossible task. First of all, fire can be divided into two phases, the developing phase and the fully developed fire. Material behaviour, meanwhile, must be classified taking both these phases into consideration, where the reaction to fire involving a material describes its behaviour in the developing phase. The developing fire is influenced by a series of differing parameters, such as the combustibility of the material, the degree of inflammability, the propagation velocity of the fire over the surface and the rate at which heat is released. The fully developed fire represents the phase after the flash over, in which all combustible materials become involved in the fire. Desirable material properties during this phase include the ability to retain load-bearing capacity, restrict the spread of fire to the area in which it originated to avoid any propagation of flames, hot gases or excessive heat on the side of a wall or floor opposite a fire, which would indirectly help the fire spread to neighbouring areas. The ability to withstand the fully developed fire is generally known as fire resistance, but can only be attributed to a member and not a material. The behaviour of simple structural members such as columns or beams also depends on factors like the support conditions, magnitude of load and load distribution.

Observing how solid timber and wood-based materials behave in a developing fire, it becomes clear that wood-based materials burn and are thus classified as combustible. While this combustibility can be modified by applying a coating or impregnating with fire retardants, these measures cannot render wood-based materials non-combustible, despite the fact of more energy is required to make the material burn. Solid timber is difficult to ignite and there are few cases in which wood ignites earlier than other materials. Solid timber requires a surface temperature exceeding 400°C within a short to medium time frame for spontaneous combustion, namely without the presence of an ignition source. Even with an ignition source, the surface temperature must be maintained at more than 300°C for a specific period, whereupon the material will ignite. Wood is normally used as a benchmark material for comparison when classifying other materials, since wood is seen as a material, which, in most areas of use, presents an acceptable ignition risk. The actual values depend on the density, wood species, moisture content and the ratio of the circumference to the cross-section. The fire propagates over the surface of the combustible wood, and each ignition in the surrounding area triggers further ignitions. Since wood is difficult to ignite, the speed with which the flames spread is acceptably low for a combustible material. Virtually all countries allow the use of untreated wood in instances involving a low fire risk. The degree to which wood gives off heat depends on the type of temperature curve and the availability of oxygen as well as the density, circumference and size of the observed timber piece.

If wood-based materials are exposed to a fully developed fire, they respond with a series of favourable characteristics. The surfaces will ignite once the heat flow has reached a sufficient level and initially burn fairly vigorously; an insulating charcoal layer soon forms (see also Figure G1-9). However, since wood and charcoal in particular are poor thermal conductors, very little heat is transferred to the remaining unburnt material. For this reason, the construction is not damaged by any excess thermal expansions. If wood-based panels are used as load-bearing elements or as sheathing in partitioning constructions, meanwhile, the low thermal conductivity prevents the swift transfer of heat from the hot side of the construction to the colder side.

In tests, the fully developed fire is described in terms of the standard temperature-time curve in accordance with ISO 834 (see Figure G1-1) or in accordance with the corresponding national standard. The key (so-called *REI*) criteria include compliance with the

- Load-bearing capacity – criterion *R*,
- Integrity – criterion *E*,
- Insulation – criterion *I*.

These *REI*-criteria also govern design. Structures must be dimensioned such as to ensure they can retain their load-bearing capacity during fire of a certain duration (*R*) as well as, when required, their integrity (*E*) and insulation capacity (*I*).

The standard temperature-time curve in Figure G1-1 corresponds to a fictional fire, which differs significantly from a natural fire, since the latter also includes cooling phases. The favourable behaviour of wood in a cooling phase, however, is not taken into account when loading with the standard temperature-time curve. Moreover, a natural fire also tends to reach considerably higher temperatures much more rapidly and over a shorter time period, before quickly recooling. However, for most applications, a statement concerning reaction to fire under a standard temperature-time curve will suffice, which is far more feasible from an experimental perspective and facilitates comparisons between differing construction materials or components.

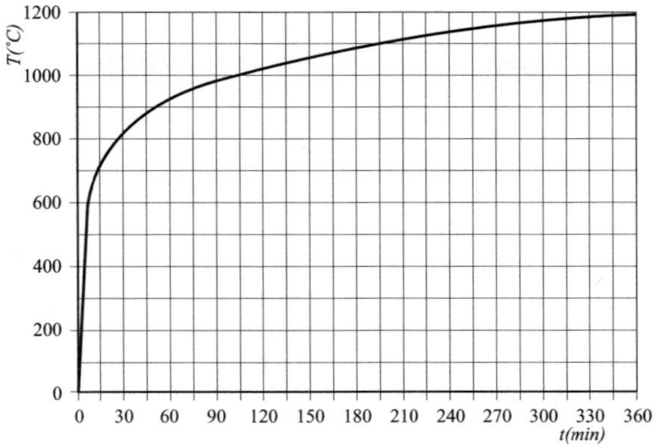

Figure G1-1 Standard temperature-time curve in accordance with ISO 834. (STEP 1995 Article A13)

Load-bearing capacity criteria apply in addition to the actual failure characteristics of critical deflection and deflection velocity. Integrity is generally determined using the formation of excessive cracks and splits (national definitions) or by the ignition of a ball of cotton wool placed on the side opposite the fire. The insulation capacity is deemed to be compromised when a mean temperature rise of 140°K is determined on the side opposite the fire or a maximum increase of 180°K is exceeded.

Wood only loses its load-bearing capacity when the remaining cross-section not yet destroyed by fire becomes so small that the stresses imposed by the loads attain the strength of the remaining cross-section. Wood-based materials will not rupture in the event of a fire or shrink to such an extent that cracks appear, but the surface will remain intact, until the wood is so thin that the fire ends up burning through. The increase in temperature will only exceed the limit value, if the thin and fire-stressed area reaches the opposite side, whereupon the burn-through soon follows. Under the conditions of a fully developed fire, the behaviour of wood can be predicted very reliably. In this article, the reaction to fire of wood and wood-based materials is explained, followed by the various design concepts for members and joints.

G1.1 Fire protection regulations

Before the actual "fire design" when exposed to fire (as opposed to "cold design" under normal temperature of the EC 5) is covered, the requirements and fire protection regulations will be summarised. On a European level, the impact in the event of fire is regulated in EC 1 Part 1-2, which provides the structural engineer with calculation methods to determine design fires and temperature distributions in structural members. However, it does not include requirements stating which building types must withstand a fire following the standard temperature-time curve and for how long. These requirements for buildings are regulated in Germany by the State Building Codes (LBO) (see also Article A1). The LBO classifies buildings into building classes (GK); of which there are five in Baden-Württemberg:

- Building class 1: Detached buildings up to 7 m high, comprising a maximum of two units of use covering a collective area not exceeding 400 m² as well as detached buildings used for agriculture and forestry purposes,
- Building class 2: Buildings up to 7 m high, comprising a maximum of two units of use covering a collective area not exceeding 400 m²,
- Building class 3: Other buildings up to 7 m high,
- Building class 4: Buildings up to 13 m high and comprising units of use each with an area not exceeding 400 m²,
- Building class 5: Other buildings, including underground buildings.

The LBO further stipulates the need to configure and erect structures, "which allow the formation of fires and the spreading of fire and smoke (fire propagation) to be prevented" (§ 15 LBO BW). Additional statements concern, among other things, escape routes, accessibility, spacing and lightning protection systems; more accurate provisions on building design (fire resistance, selection of construction material) are included in the "General design specification of the Ministry of Transport and Infrastructure for the State Building Code" (LBOAVO BW). The LBOAVO BW sets out the fire protection regulations for various building classes. For individual building components such as load-bearing walls and columns, floors, partition and fire walls, fire resistance classes are defined, see Table G1-1. Building components of fire resistance class F90 must be able to withstand a fire for at least 90 min without losing their load-bearing capacity and, if applicable, retain their integrity. This means, for example, according to LBOAVO BW, floors in buildings of building class (GK) 5 must be fire resistant, in GK 4 highly fire retardant and in GK 2 and 3 fire retardant. Additional definitions such as those of the fire resistance classes are included in the Construction Products List A (national document, published by DIBt Berlin). A selection of designations and classes on a European and federal level is included in Table G1-1.

Table G1-1 Fire resistance classes of building components, Construction Products List A Annex 0.1.

Fire resistance class in accordance with DIN 4102-2:1977	Fire resistance duration	Building authority terminology	Fire resistance class in accordance with EN 13501-2:2010		
			a	*b*	*c*
F30	≥ 30 min	Fire retardant	R30	REI30	EI30
F60	≥ 60 min	Highly fire retardant	R60	REI60	EI60
F90	≥ 90 min	Fire resistant	R90	REI90	EI90
F120	≥ 120 min	Highly fire resistant	R120	REI120	-
F180	≥ 180 min	Maximum fire resistant	-	-	-

a = load-bearing building components without compartmentation (no need to maintain integrity), *b* = load-bearing building components with compartmentation, *c* = non load-bearing internal walls

In addition to the required fire resistance of the building components, other requirements imposed on construction materials include those concerning reaction to fire as stipulated in the State Building Codes. The LBO BW (§ 26) divides construction materials into four classes, whereby class B3 is not generally used as a construction material (except when combined with other materials, whereupon it would have normal combustibility):

- Class A (A1 and A2): non-combustible,
- Class B1: difficult to ignite,
- Class B2: normal combustibility,
- Class B3: easily ignited.

The building components are classified in fire resistance classes in Table G1-1 and additionally linked with the reaction to fire requirements of their construction materials. Accordingly, LBO BW or the Construction Products List A distinguishes e.g. building components made of non-combustible or combustible materials (§ 26); concepts that are often linked: meaning building component F30-B1 is a fire retardant component made of construction materials that are difficult to ignite. A component of class F60-AB2, conversely, is a highly fire retardant component and mainly comprises non-combustible construction materials (meaning, consequently, that a portion of the construction materials may have normal combustibility). In addition, since not all combinations are permitted; a fire resistant component (F90) may not exclusively be made of combustible construction materials (B).

On a European level, EN 13501 stipulates the classification of construction products in classes A to E, which also include further sub-classifications specifying smoke development and burning droplets. The additional classification, discussed above, in line with *REI*-criteria closely resembles that of the State Building Codes (see Table G1-1). Accordingly, the European fire resistance class REI30 is comparable to the German class F30-B. For a non-load bearing but partitioning wooden wall, conversely, EI30 or F30-B may be required.

In general, the applicable regulations governing the requirements in the event of fire are very complex and relatively obscure, since European regulations are applied alongside those on a state or national level. The numerous relevant documents and frequent revisions made also hamper efforts to get a handle on which individual regulations apply. A good overview is provided by Scheer and Peter (2009) as well as the Construction Products List A. One stand-out feature of timber construction is the "Guideline on fire protection requirements for highly fire retardant components in timber construction", HFH-HolzR. This guideline paves the way for timber buildings of building class 4.

G1.2 Principles

Wood and wood-based materials mainly comprise cellulose and lignin, which, in turn, consist of carbon, hydrogen and oxygen (see Article B1). This means they are combustible and removing this property is practically impossible. However, absolute non-combustibility is only seldom required.

Influences on reaction to fire

The shape, surface, circumference and size of the building components made of wood and wood-based materials significantly influence their reaction to fire and the combustibility depends on the ratio of surface to volume. The greater this ratio, the more rapidly the flames will spread. Many sharp edges and rough surfaces increase this ratio and lead to unfavourable reaction to fire. Meanwhile, cracks and splits also increase the risk of fire breaking out, which explains why the charring rate of virtually crack-free glulam timber is lower than for solid timber.

The duration up to the point of the wood igniting and the fire spreading depends on the density (oven-dry density), meaning different wood species react differently to fire. The relation between density ρ and the rate of combustion *RC* is schematically shown in Figure G1-2. The relationship between density and ignition resembles the ρ-*RC* relationship: the higher the density, the later the point at which the wood will ignite.

The wood moisture content is another key influence parameter for the reaction to fire of wood. In timber structures, the moisture content generally ranges between 8 and 15%, which means 80 to 150 kg of water have to be extracted from each tonne of wood before it will burn. The influence of the moisture content on the charring rate can thus be disregarded within the low bandwidth for the equilibrium moisture content of between 8 and 15%.

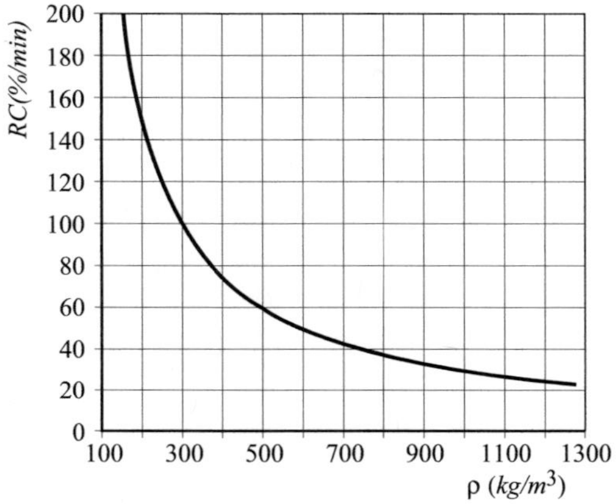

Figure G1-2 Relationship between density ρ and rate of combustion *RC*. (STEP 1995 Article A13)

Chemical and physical processes during the combustion of wood

When wood and wood-based materials burn, it sets off a chemical decomposition reaction, during which charcoal and combustible gases form. The spontaneous combustion of a thin strip of wood takes place at a temperature of between 340 and 430°C. The ignition temperature, however, may be far lower (e.g. 150°C), if the wooden piece has already been heated for an extended period. Temperatures under 100°C, but above room temperature, add heat to the wood and promote its drying out, whereupon the strength and modulus of elasticity decline.

At 100°C, the water begins to vaporise and the steam escapes via the route of least resistance, namely corners, edges, joints, open pores and cracks. These are places where the wood dries faster than normal. The temperature remains constant, until the water has completely vaporised. Figure G1-3 shows the temperature underneath the so-called pyrolysis layer depending on time, when the wood is exposed to fire corresponding to the ISO temperature-time curve. The curve clarifies the rise in temperature after the water vaporises at 100°C. The pyrolysis layer is the area between the charcoal and the unchanged wood. In this layer, although the wooden substance has already been chemically changed by the fire, it is not yet completely decomposed.

At between 150 and 200°C, gases are emitted, 70% of which comprise non-combustible carbon dioxide (CO_2) and 30% combustible carbon monoxide (CO). From 200°C meanwhile, more and more combustible gases form and the CO_2 portion declines. As soon as the gases ignite, the surface temperature rises significantly and the wood begins to decompose in the pyrolysis layer, which is around 5 mm thick. At temperatures exceeding 500°C, the gas formation declines very sharply and the charcoal formation rises, which explains the typical visual appearance of wood after a fire.

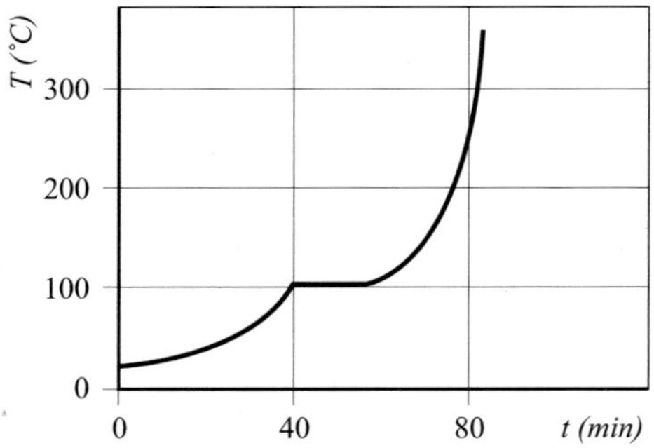

Figure G1-3 Temperature in the heated wood below the pyrolysis layer, wood is exposed to the ISO-temperature-time curve. (STEP 1995 Article A13)

The thermal conductivity of charcoal is only around a sixth of that of solid timber. The charcoal layer thus acts as a form of insulation and delays the decomposition of the deeper remaining cross-sectional areas. For this reason and due to the low thermal conductivity of wood, the temperature in the middle of the cross-section remains far lower than at the surface, making wood far more fire-resistant than is generally assumed. Figure G1-4 shows beams or columns, which have been exposed to fire on three or four sides.

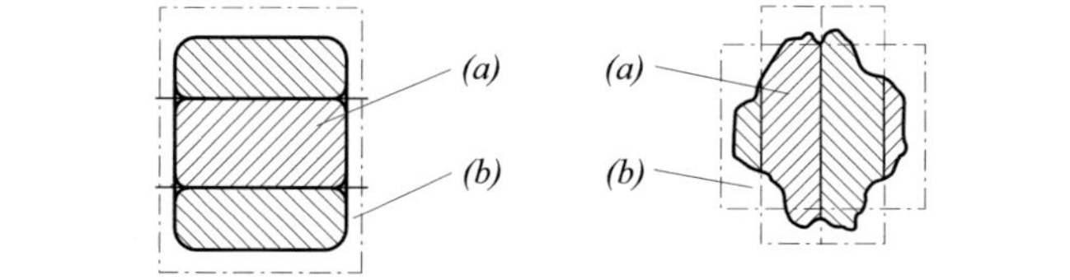

Figure G1-4 Beams and columns before and after exposure to fire. (*a*) remaining cross-section, (*b*) charcoal layer, (*c*) fire protection layer. (STEP 1995 Article A13)

Charring rates (rates of combustion)

Numerous tests with wood and wood-based materials reveal a linear dependency between the charring depth and fire duration. Accordingly, when calculating the fire resistance of a cross-section, a constant charring rate is assumed. Simple approaches to fire design without taking fillets into consideration include the use of one-dimensional charring rates β_0 in accordance with Table G1-2 (determined under standard exposure to fire in accordance with ISO 834). The remaining cross-section is assumed to be sharp-edged for the fire design. The effective charring rate β_n includes edge fillets and cracks and is therefore higher. For wood-based panels, the one-dimensional charring rates depend on density and panel thickness and must be calculated for characteristic densities ρ_k other than 450 kg/m³ and thicknesses h_p smaller than 20 mm:

$$\beta_{0,\rho,t} = \beta_0 \cdot k_\rho \cdot k_h \tag{G1-1}$$

where

$$k_\rho = \sqrt{\frac{450}{\rho_k}}$$

$$k_h = \sqrt{\frac{20}{h_p}}$$

and β_0 in accordance with Table G1-2.

Table G1-2 Charring rates β_0 and β_n. (β_n includes edge fillets and cracks), EN 1995-1-2, 2010.

Material	β_0 in mm/min	β_n in mm/min
Softwood and beech		
Glulam with $\rho_k \geq 290$ kg/m³	0.65	0.70
Solid timber with $\rho_k \geq 290$ kg/m³	0.65	0.80
Hardwood		
Solid timber and glulam with $\rho_k \geq 290$ kg/m³	0.65	0.70
Solid timber and glulam with $\rho_k \geq 450$ kg/m³	0.50	0.55
Laminated veneer lumber with $\rho_k \geq 480$ kg/m³	0.65	0.70
Panels		
Timber claddings	0.9[a]	-
Plywood	1.0[a]	-
Wood-based panels other than plywood	0.9[a]	-

[a] = values apply for ρ_k = 450 kg/m³ and a panel thickness of 20 mm

For multiple layers (cladding, wood-based panels) fitted seamlessly, the higher charring rate of subsequent layers must be taken into consideration following the failure of the initial layer, since these will already have heated up (see also Figure G1-6). For more precise design methods considering natural fires, the charring rate β_{par} in accordance with Annex A of the EC 5 Part 1-2 should be used.

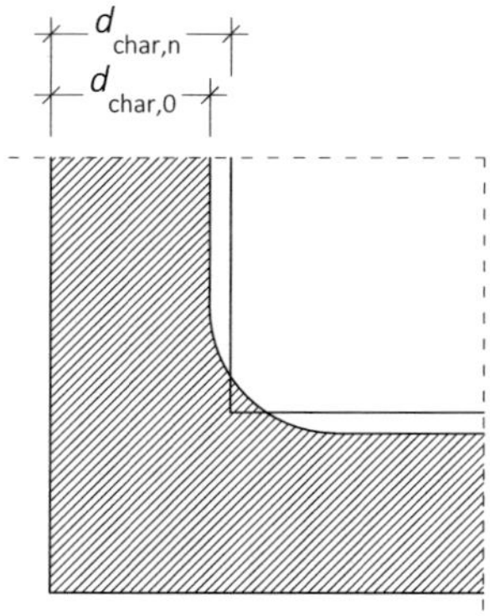

Figure G1-5 One-dimensional charring depth $d_{char,0}$ and effective charring depth $d_{char,n}$.

Using the charring rate β and the duration t of the fire (namely 30, 60 or 90 minutes), the charring depth d_{char} of an unprotected surface is determined, Figure G1-5.

One-dimensional charring depth:

$$d_{char,0} = \beta_0 \cdot t \qquad \text{(G1-2)}$$

Effective charring depth:

$$d_{char,n} = \beta_n \cdot t \qquad \text{(G1-3)}$$

If the one-dimensional charring depth is used, the corner fillets must be taken into consideration as shown in Figure G1-5, whereby $d_{char,0}$ is used as the fillet radius of the corners.

Normally, a simpler method using the effective charring depth is used (see Section G1.4) and the use of the one-dimensional or effective charring depth depends on the initial width of the building component. If the initial width b_{min} in equation (G1-4) applies, the one-dimensional charring depth may be used; but if the initial width is smaller than b_{min}, the effective charring depth must be used.

$$b_{min} = \begin{cases} 2 \cdot d_{char,0} + 80 & \text{for} \quad d_{char,0} \geq 13\text{mm} \\ 8.15 \cdot d_{char,0} & \text{for} \quad d_{char,0} < 13\text{mm} \end{cases} \qquad \text{(G1-4)}$$

For protected surfaces, the start of combustion of the wooden components is initially delayed. This only begins at time t_{ch}; for example at time t_f, after the protective layer has failed (t_{ch} need not be identical to t_f). The effect of the fire on the protective layer means the wood arranged behind is preheated, which, in turn, accelerates the rate at which it burns. Only after a charring depth of 25 mm or the charring depth of unprotected timber – whichever is lower – does the charring speed revert to the simple level, time t_a. Figure G1-6 shows this relationship and how components with initial protective cladding burn up faster once the protective layer fails, see curve 3. Accordingly, the protective layer is only of benefit when the failure time t_f of the protective layer exceeds the time required to reach a charring depth of 12.5 mm in an unprotected component.

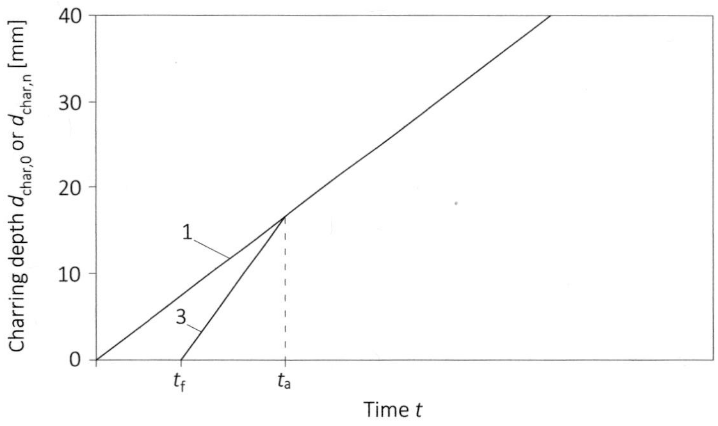

Figure G1-6 Illustration of charring depth depending on time for $t_{ch} = t_{f,1}$. 1: Course for unprotected surface; 3: course for initially protected surface; t_{ch} = delayed combustion of the wooden component, t_f = failure of protective layer, from time t_a = normal charring speed of wooden component.

The relationship from Figure G1-6 does not apply to all protected components; Figure G1-6 only applies to components with a protective layer that delays the combustion of the wooden component until the layer fails at t_f. For some types of protective layers, however, combustion of the wooden component may occur even before time t_f, if the layer e.g. cracks or lacks sufficient thermal insulation properties.

Temperature-dependent strength and stiffness values

The mechanical properties of wood are temperature-dependent, with stiffness and strength both declining with increasing temperature. The heated part of an unburnt remaining cross-section thus has lower strength or stiffness than at room temperature, while the magnitude of the decline varies for differing mechanical properties. The example shown in Figure G1-7 features tension strength parallel to the grain of solid timber depending on temperature, while Figure G1-8 shows compression strength parallel to the grain. The decrease in compressive strength under high temperatures is more prominent than the decrease in tensile strength. In addition, both figures clearly show the scattered nature of the test results. The dependency of mechanical properties on temperature declines at varying rates depending on the boundary conditions during testing, since among other things, ensuring constant temperature and humidity conditions in each case is far from easy, particularly with large test bodies.

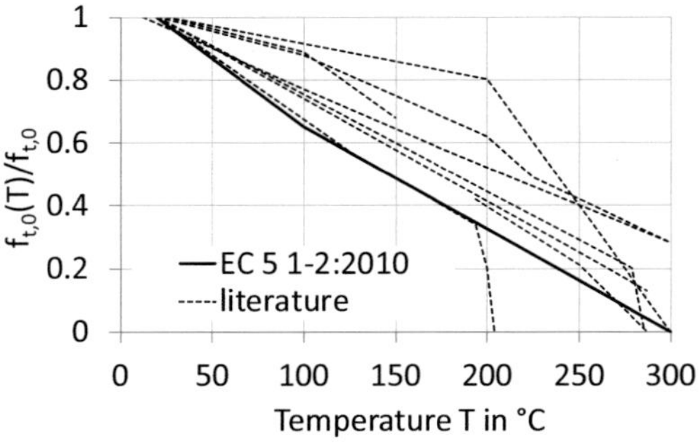

Figure G1-7 Relation between temperature and tensile strength $f_{t,0}$. (according to Scheer and Peter, 2009)

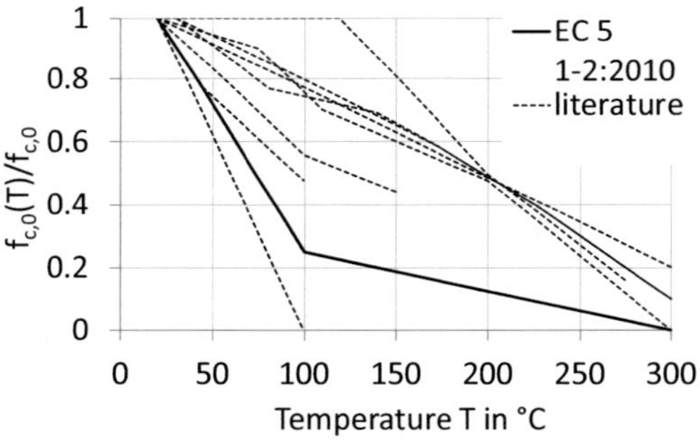

Figure G1-8 Relation between temperature and compressive strength $f_{c,0}$. (according to Scheer and Peter, 2009)

Temperature distribution in the cross-section

The temperature distribution in the cross-section depends on the charring speed, cross-sectional geometry, fire duration, fire exposure (on one or multiple sides) and the thermal conductivity of wood as well as on any existing metal parts (= high thermal conductivity). In the event of fire, surfaces ignite and initially burn quickly, but a heat-insulating charcoal layer soon forms, see Figure G1-9. Since wood, particularly in the form of charcoal, is a poor thermal conductor, very little heat is transmitted to the remaining unburnt material. In the solid timber, the temperature in the remaining cross-section in the proximity of the combustion area (base of the pyrolysis layer = area between charcoal and unchanged wood) is already unchanged. The temperature of 300°C at the combustion limit declines over a thickness a_0 (of around 25 to 40 mm) to the unchanged surrounding temperature level (e.g. 20°C), Figure G1-10. Even within the combustion area, the temperature declines, from that of the fire (800 to 1000°C) to around 300°C at the combustion limit. This effect of low thermal conductivity can be viewed positively, since among other things, the mechanical properties of the remaining cross-section, reduced by the heat-affected zone a_0, remain largely unchanged.

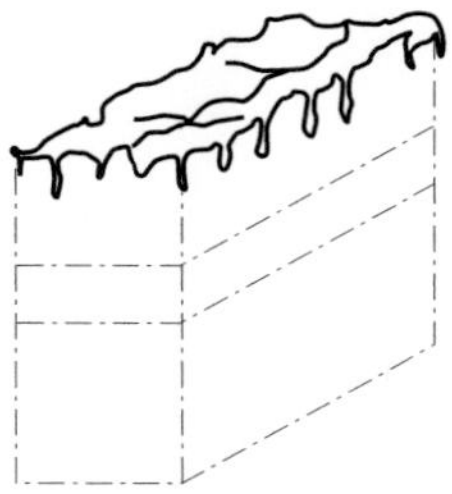

Figure G1-9 The changes in wood in the event of a fire: (*a*) charring layer, (*b*) pyrolysis layer, (*c*) unchanged wood. (STEP 1995 Article A13)

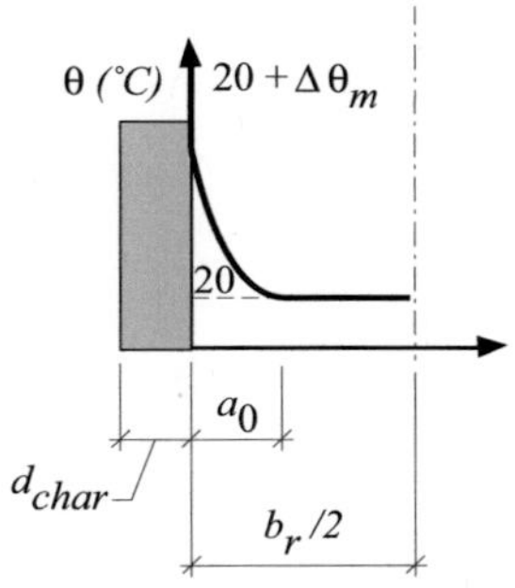

Figure G1-10 Temperature distribution for $b_r > 2 \cdot a_0$. (STEP 1995 Article B17)

Reaction to fire of joints with metal fasteners

The load-bearing capacity of metal fasteners, which are not protected against fire, is swiftly lost when heat is applied. In addition, the high thermal conductivity of the metal parts heats up the surrounding wood; causing both embedment strength and withdrawal capacity to decline. Covering the metal fasteners with wood or wood-based materials delays any heating of the metal parts, while the area of unprotected surfaces of metal parts is the decisive criterion governing how metal fasteners behave in the event of fire. Figure G1-11 shows the relevant yield point ψ of steel depending on temperature. Although steel is non-combustible, it features high thermal conductivity and a yield strength that declines swiftly with increasing temperature.

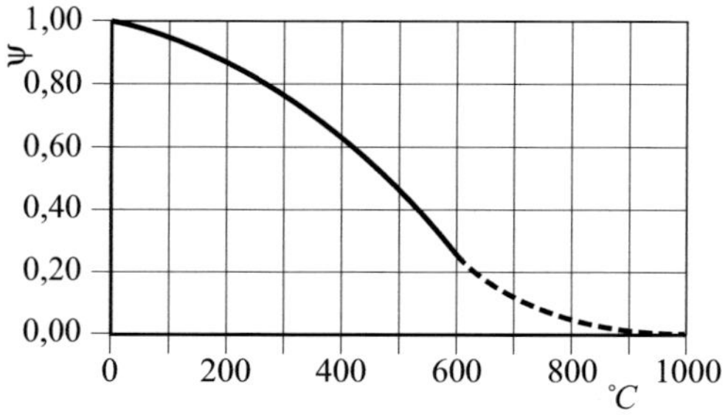

Figure G1-11 Relevant yield point ψ depending on temperature. (STEP 1995 Article C19)

G1.3 Safety concept for fire design

As a general rule, the applicable rules for fire design are the same principles as those applied for verifications in a "cold" state, whereby actions and material properties are determined using characteristic values. However, fire testing has usually been performed using deterministic methods, limiting the scope to average strength values. To reconcile test results with the calculation principles to be applied, EC 5 Part 1-2 contains specially tailored approximation methods for fire design. However, design for exposure to fire is not only possible via calculation but also via tests, while in Germany, classification via DIN 4102-4 is also possible. In Part 4 of DIN 4102, a set of classified construction materials and components is regulated, allowing a specific fire resistance class to be obtained without calculations or tests. For example, DIN 4102-4:1994 sets out that load-bearing, non-partitioning walls must be made of studs of a specific grading class and dimension and of sheathing with a minimum thickness and density to achieve the fire resistance duration F30-B.

If a fire design by calculation is performed, other partial safety factors and load combinations (actions and design material properties) apply compared to the case for normal design without exposure to fire. The effect of actions $E(t)$ and the resistance of timber members $R(t)$ during fire exposure are shown in Figure G1-12. The fire resistance is reached after time t_f, when $R(t)$ is smaller than $E(t)$.

Verification on the design level for the governing load duration t is thus as follows:

$$E_{d,fi} < R_{d,t,fi} \tag{G1-5}$$

where

$E_{d,fi}$ Design effect of actions in the event of fire

$R_{d,t,fi}$ Design resistance in the event of fire

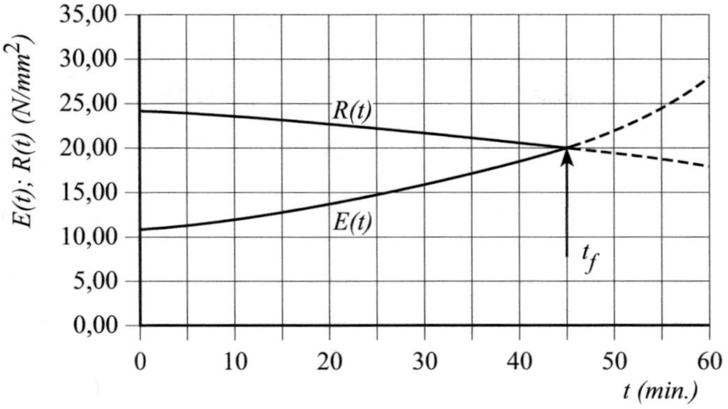

Figure G1-12 Effect of actions $E(t)$ and resistance of timber member $R(t)$ during exposure to fire. (STEP 1995 Article B17)

Design values for actions

According to EC 0 and EC 1 Part 1-2 or DIN 4102-22, the fire design should be implemented with a combination of actions for accidental design situations:

$$E_{dA} = \sum \gamma_{GA,j} \cdot G_{k,j} + \psi_{1,1} \cdot Q_{k,1} + \sum \psi_{2,i} \cdot Q_{k,i} \qquad \text{(G1-6)}$$

where

E_{dA} Actions in the event of fire

$\gamma_{GA,j}$ Partial safety factor for permanent actions in the event of fire, $\gamma_{GA,j} = 1.0$

Unlike combinations of actions for normal design, in this case, the partial safety factors for permanent and variable loads γ_{GA} and γ_Q are set at 1.0. However, for fire design purposes, the actions can be calculated from the normal design values by the following equation:

$$E_{dA} = \eta_{fi} \cdot E_d \qquad \text{(G1-7)}$$

η is the quotient of the actions from the accidental combination (fire) and the fundamental combination:

$$\eta_{fi} = \frac{\sum \gamma_{GA,j} \cdot G_{k,j} + \psi_{1,1} \cdot Q_{k,1} + \sum \psi_{2,i} \cdot Q_{k,i}}{\sum \gamma_{G,j} \cdot G_{k,j} + \gamma_{Q,1} \cdot Q_{k,1} + \sum \gamma_{Q,i} \cdot \psi_{0,i} \cdot Q_{k,i}} \qquad \text{(G1-8)}$$

There is scope to simplify in accordance with NA to EC 5 Part 1.2 to put $\eta = 0.60$ conservatively.

Design values of material properties

Values other than for normal design apply to the partial safety coefficients on the material side, since the material properties depend on temperature, resp. during an accidental design situation, such as in the event of fire, where the partial safety factor γ_M assumed is 1.0. To verify load-bearing capacity, the design strength $f_{d,fi}$ and stiffness $S_{d,fi}$ are as follows:

$$f_{d,fi} = k_{mod,fi} \cdot \frac{f_{20}}{\gamma_{M,fi}} \quad \text{or} \quad S_{d,fi} = k_{mod,fi} \cdot \frac{S_{20}}{\gamma_{M,fi}} \qquad \text{(G1-9)}$$

where

$k_{mod,fi}$ Modification factor in the event of fire, see below

$\gamma_{M,fi}$ Partial safety factor for material properties, $\gamma_{M,fi} = 1.0$

f_{20} or S_{20} 20%-quantiles of a strength or stiffness property at normal temperature

The 20%-quantiles are determined via the coefficient k_{fi} from the characteristic values:

$$f_{20} = k_{fi} \cdot f_k \quad \text{or} \quad S_{20} = k_{fi} \cdot S_{05} \qquad \text{(G1-10)}$$

k_{fi}-values are specified in Table G1-3 and vary for the different construction materials, since the scatter varies for the individual materials (namely the 20%-quantile may, depending on the scatter of a material property, differ to a greater or lesser extent from the 5%-quantile).

In the design case of fire, consequently, not only are the partial safety factors on the action and material side reduced, the modification factor k_{mod} is also replaced with $k_{mod,fi}$ and a 20%- rather than 5%-quantile value is used for the material properties, since it is assumed that the weakest cross-section is not simultaneously the one most stressed by fire.

Table G1-3 Values for k_{fi}, EC 5:2010 Part 1-2.

Material	k_{fi}
Solid timber	1.25
Glulam, wood-based materials	1.15
Laminated veneer lumber	1.1
Laterally loaded timber-to-timber joints	1.15
Laterally loaded joints with external steel plates	1.05
Joints with axially loaded fasteners	1.05

Coefficient $k_{mod,fi}$

As explained above, the mechanical properties depend on temperature to a greater or lesser extent, whereby different strength properties are influenced to varying extents by temperature. The most significant decline is in compressive strength, whereas tensile strength, for example, is less affected by temperature (see Figure G1-7 and Figure G1-8). This influence is considered with the coefficient $k_{mod,fi}$. The value of $k_{mod,fi}$ depends on the design method. With the reduced cross-section method, $k_{mod,fi}$ = 1.0 is assumed. With the reduced properties method, which can be applied for softwood cross-sections exposed to fire on three or four sides as well as for general fire exposure of softwood logs, other values for $k_{mod,fi} \leq 1.0$ must be determined, which take into consideration the reduction in mechanical properties of the construction material at elevated temperatures (for more details, see section "reduced properties method").

G1.4 Design methods for structural members

EC 5 Part 1-2 includes two methods for the fire design of members resp. to determine cross-sectional values in the event of a fire:

- Simplified method: Reduced cross-section method (RCSM),
- More accurate method: Reduced properties method (RPM).

In addition to both these methods, more complex methods can also be applied to calculate the charring depths (parametric fire exposure, EC 5 Part 1-2 Annex A) or determine all required design steps considering a parametric design fire (EC 5 Part 1-2 Annex B and EC 1 Part 1-2). However, in the vast majority of cases, these methods are infeasible for the structural engineer and simply represent alternatives for special constructions.

Reduced cross-section method

With this method, the fire resistance duration of a structural member depends on the load-bearing capacity of the remaining unburnt cross-section. Figure G1-13 shows this effective cross-section, where the following applies:

$$d_{ef} = d_{char,n} + k_0 \cdot d_0 \tag{G1-11}$$

where

$d_{char,n}$ Effective charring depth corresponding to equation (G1-3)

d_0 = 7 mm

k_0 See Table G1-4

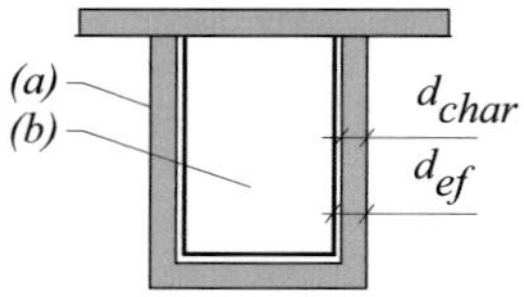

Figure G1-13 Effective cross-section. (STEP 1995 Article B17)

The additional layer thickness $k_0 \cdot d_0$ takes into consideration the fact that the wood near the combustion limit is very hot, and hence has only low strength and stiffness. The actual combustion layer is thus further increased by this "heat-affected zone" (a_0 in Figure G1-10).

d_0 was determined as follows:
The integration of the temperature distribution in accordance with Figure G1-10 reveals an average temperature of 80°C. If we assume the average strength at 80°C is around 70% of the initial strength and conclude that 70% of a_0 has wood strength unaffected by temperature and 30% can no longer bear any load, d_0 = 7 mm results in 30% of a_0 = 25 mm.

The coefficient k_0 corresponding to the required fire resistance duration t can be taken from Table G1-4. Using the remaining cross-sections, the cross-sectional values required for design can now be determined.

Table G1-4 Definition of coefficient k_0.

Unprotected surfaces	t < 20 min	$k_0 = t/20$
	t ≥ 20 min	$k_0 = 1.0$
Protected surfaces	$t_{ch} \leq 20$ min	$k_0 = t/20$
	$t_{ch} > 20$ min	k_0 rises in $t = 0$ to $t = t_{ch}$ linearly from 0 to 1
Interspaces in floor beams and wall studs etc:		
Cladding made of wood, wood-based materials or gypsum plasterboards type A	$t = t_f$	$k_0 = 0.3$
	$t_f < t < t_f + 15$ min	Linear interpolation of k_0 to $k_0 = 1.0$
Cladding made of gypsum plasterboards type F	$t \geq t_{ch}$	$k_0 = 1.0$
	$t < t_{ch}$	k_0 rises in $t = 0$ to $t = t_{ch}$ linearly from 0 to 1

t_{ch} = start of combustion delayed up until time t_{ch} due to protective cladding
t_f = time of failure of cladding

Reduced properties method

This method is also derived from the temperature distributions shown above. The load-bearing capacity is, in turn, calculated for a remaining cross-section, and here, the mechanical properties of the remaining cross-section are also reduced. This method is included in EC 5 Part 1-2 for softwood cross-sections exposed to fire on three or four sides as well as for general fire exposure of softwood logs. The reduction in mechanical properties is performed by the coefficient $k_{mod,fi}$, which is specified as depending on the load, on the circumference p and the area A_r of the remaining cross-section, see also Figure G1-14:

$$\text{Tension strength and modulus of elasticity: } k_{mod,fi} = 1.0 - \frac{1}{330} \cdot \frac{p}{A_r}$$

$$\text{Bending strength: } k_{mod,fi} = 1.0 - \frac{1}{200} \cdot \frac{p}{A_r} \qquad \text{(G1-12)}$$

$$\text{Compression strength: } k_{mod,fi} = 1.0 - \frac{1}{125} \cdot \frac{p}{A_r}$$

where

p Circumference of remaining cross-section subject to fire in m

A_r Area of remaining cross-section in m^2

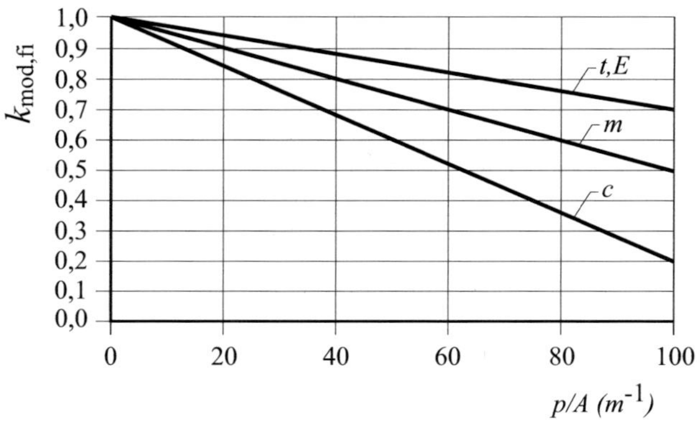

Figure G1-14 $k_{mod,fi}$ for tension (t), bending (m), compression (c) and modulus of elasticity (E). (STEP 1995 Article B17)

Figure G1-14 and equations (G1-12) show how the influence of temperature on the mechanical properties varies depending on the property. The compressive strength declines more than bending or tensile strength with increasing temperature.

With the above rules, cross-sectional values are determined to establish all relevant load-bearing capacity and stability verifications for the members exposed to fire. One key point which has to be noted is the applicable scope of the calculation methods introduced. Both simplified methods apply to members with constant cross-section and homogenous structure (namely no combined glulam). Combined glulam or beams with variable cross-section are not explicitly regulated in EC 5 Part 1-2, additional considerations are therefore required. Mechanically jointed beams, conversely, are regulated; the slip modulus K_{fi} can be determined from K_u using a conversion factor.

G1.5 Design methods for joints

Since metal has far higher thermal conductivity than wood, it is very effective at transmitting heat to the inside of members, which means yield strength declines very rapidly with increasing temperatures, Figure G1-11. Accordingly, metal fasteners should be protected. Such protection can be provided very easily for dowels, e.g. by applying wooden plugs.

Unprotected joints with side members made of wood

Presuming the proper design of joints and fasteners in accordance with EC 5 Part 1-1 (spacings and distances of the fasteners, etc.) and when complying with specific minimum thicknesses of joints and side members, unprotected joints with side members made of timber (also including steel-to-timber joints with slotted-in steel plates) have a fire resistance duration of $t_{d,fi}$ = 15 min, class R15 (20), without further verification. Higher fire resistance duration for the specified joints (not with bolts and fasteners in accordance with EN 912), but not exceeding R30, can be achieved by increasing the member thickness and the end and edge distances by the dimension a_{fi}:

$$a_{fi} = \beta_n \cdot k_{flux} \cdot (t_{req} - t_{d,fi}) \qquad \text{(G1-13)}$$

where

β_n Effective charring rate in accordance with Table G1-2

k_{flux} Coefficient for considering the elevated heat flow through the metal fasteners, generally $k_{flux} = 1.5$

t_{req} Required fire resistance duration

$t_{d,fi}$ Fire resistance duration in accordance with Table G1-5

Table G1-5 Fire resistance duration of unprotected joints with side members made of wood.

Fastener	Fire resistance duration $t_{d,fi}$ in min	Minimum dimension to be complied with
Nails	15	$d \geq 2.8$ mm
Screws	15	$d \geq 3.5$ mm
Bolts	15	$t_1 \geq 45$ mm
Dowels	20	$t_1 \geq 45$ mm
Fasteners corresponding to EN 912	15	$t_1 \geq 45$ mm

d = diameter fastener; t_1 = thickness of side member made of wood

Reduced stresses method

This method allows higher fire resistance durations to be achieved for unprotected joints, without having to increase cross-sections by the a_{fi}. A maximum fire resistance duration of 40 min can be achieved for timber-to-timber joints with dowels d ≥ 12 mm using the cross-sections for normal design. By combining the different methods (cross-sectional increase, protecting joints and reduced stresses method), a maximum fire resistance duration of 60 min can be achieved (Scheer and Peter, 2009).

Additional rules for joints with slotted-in steel plates

A fire resistance duration of 30 or 60 minutes (R30 or R60) for a joint with unprotected, non-protruding steel plates (plate thickness ≥ 2 mm) is achieved, when the widths b_{st} of the steel plates are complied with in accordance with Table G1-6.

Table G1-6 b_{st} of steel plates with unprotected edges.

	Unprotected edges	One edge or two opposite edges unprotected
R30	≥ 200 mm	≥ 120 mm
R60	≥ 280 mm	≥ 280 mm

Protected joints with side members made of wood

Cladding or wooden plugs can protect joints with side members made of wood. If wooden plugs are used, these should be of at least thickness a_{fi} in accordance with equation (G1-13). When using a cladding, it is important to ensure that the start of combustion of the protected member t_{ch} does not go below a certain time duration, which must be guaranteed by the cladding. Accordingly, the following applies for a cladding with wood-based panels or gypsum plasterboards of type A or H:

$$t_{ch} \geq t_{req} - 0.5 \cdot t_{d,fi} \qquad \text{(G1-14)}$$

and for cladding with gypsum plasterboards of type F:

$$t_{ch} \geq t_{req} - 1.2 \cdot t_{d,fi} \qquad \text{(G1-15)}$$

where

t_{ch} Time to the start of combustion, need not be identical to t_f, the time of failure of the protective cladding

t_{req} Required fire resistance duration

$t_{d,fi}$ Fire resistance duration of the unprotected joint in accordance with Table G1-5

Fire resistance classes of protected joints exceeding R30 can be reached in a simplified manner by increasing the end and edge distances by $2 \cdot a_{fi}$ in accordance with equation (G1-13).

Joints with external steel plates

Unprotected joints

Such joints require a fire design in accordance with EC 3 Part 1-2.

Protected joints

Joints are considered protected, if the external steel plates are completely equipped with cladding made of wood or wood-based materials of minimum thickness a_{fi} with $t_{d,fi}$ = 5 min in accordance with equation (G1-13).

Axially loaded screws

In the case of axially loaded screws exposed to fire, the design withdrawal resistance in accordance with EC 5 Part 1-1 is reduced by a factor η depending on the edge distances of the screw as well as the required fire resistance duration.

G1.6 Literature

H. Hartl, original Articles A13, B17, C19, STEP 1995.

Kollmann F.F.P and Coté W.A. (1968). Principles of wood science and technology. Volume I, Solid Wood. Springer Verlag, Berlin, 592 S.

Scheer C. and Peter M. (2009). Holz Brandschutz Handbuch. ed. Deutsche Gesellschaft für Holzforschung, Ernst und Sohn Verlag, Berlin.

G2 Joints subject to seismic loads

Original article: A. Ceccotti

Earthquake design of structures according to EC 8 takes two different earthquake loads into consideration. In the event of a light to moderate earthquake with a peak ground acceleration value corresponding to a return period of 95 years (corresponding to a probability of exceedance of 10% in 10 years), structures should not sustain any damage or excessive deformations. Structures should also be able to withstand a severe earthquake with a return period of 475 years (probability of exceedance of 10% in 50 years). In the event of a severe earthquake, although any structural collapse must be prevented, any larger scale of damage is acceptable. During a severe earthquake, a structure "softens", the natural frequency *i* declines and the kinetic energy conveyed by the earthquake is dissipated. Reducing the stiffness "slows down" the structural response to cyclic loading; the structure then has sufficient time to reverse its direction of motion prior to developing deformations triggering collapse.

The ability of a structure or parts thereof to withstand plastic deformations and hence allow energy to dissipate before reaching the load-bearing capacity is key to explaining why the impact of an earthquake does not lead to the structure collapsing (Ceccotti, 1989). Timber structures with plastic joints can withstand a stronger earthquake than corresponding constructions with rigid joints, which are prone to brittle failure. Wooden members generally behave in a linear-elastic manner when subject to cyclic load, whereby brittle failure ensues, due largely to growth irregularities like knots. Except for compressive stress perpendicular to the grain, very little energy is dissipated during this process. The same applies to glued joints, the behaviour of which can also be characterised as linear-elastic and brittle and which have similarly small impact on the energy dissipation within a timber construction. Timber constructions made of glued members or joints are therefore not considered dissipative, i.e. as linear-elastic up to failure and are assigned to a lower ductility class.

The ability to dissipate energy via plastic deformations, however, can be achieved by joints with mechanical fasteners, which behave "semi-rigidly", unlike rigid glued joints. Well-designed joints with mechanical fasteners generally show prominent plastic behaviour before the load-bearing capacity is reached. This plastic behaviour of joints is taken into consideration when verifying the structure, by classifying it in a ductility class which takes its energy dissipation ability into consideration. The more ductile the joints behave before the load-bearing capacity is reached, the greater the structure's resistance against earthquake effects.

If a structure exposed to an earthquake with a peak ground acceleration a_y reaches its elastic limit ("yield point"), it can also withstand a q-times stronger earthquake with a peak ground acceleration a_u without collapsing. The peak ground acceleration of the stronger earthquake is $a_u = q \cdot a_y$. The values for a_u correspond to the seismic action, which is defined in standards depending on the geographical location of the structure to be designed. In EC 8, the coefficient q is known as "a factor reducing seismic actions" or "behaviour factor". The above-specified ductility classes are identified by the value of the behaviour factor q.

For design purposes, therefore, it is sufficient to verify that a structure will remain in the elastic area during an earthquake with a peak ground acceleration a_y. The ability of the structure to withstand q-times stronger earthquakes, even when damaged, is then implicitly taken into consideration. For structures, which remain linear-elastic up to failure, $q = 1.5$. This value takes overstrength into consideration, meaning structures show higher strength under a dynamic load than under a static load. Larger values for q mean lower values for the calculated seismic actions, although this requires sufficient ductility and energy dissipation. If ductile behaviour cannot be guaranteed, the structure should be designed with $q = 1.5$.

Although $q = 1.5$ must only be assumed for structures such as three-hinged frames with glued frame corners, which lack any appreciable energy dissipation capability, there are times when designing even plastic structures with $q = 1.5$ makes sense. This may apply e.g. if a variable action other than the earthquake must be taken into consideration. If, for a comparatively lightweight structure, dead weight plus snow is the governing load combination, it is possible that all verifications can be complied with, even if the seismic action is considered with $q = 1.5$, since snow and earthquake actions need not be simultaneously considered with their characteristic values. Assuming $q = 1.5$ eliminates the need for the structural engineer to consider specific detailed rules or perform tests to determine ductility.

Apart from the described cases including specific load combinations, consideration of the behaviour factor q can also be a decisive factor dictating the economic efficiency of structures in seismic areas. When using a value $q > 1.5$, sufficient ductility of the joints with mechanical fasteners should be ascertained. This can be done either by corresponding tests or, when joint configurations with known plastic behaviour are used, by ensuring compliance with the detailing rules specified in the following section.

G2.1 Ductility

Provided sufficient distances are maintained between fasteners and the timber member edges, timber joints with mechanical fasteners, unlike glued joints, show ductile and plastic behaviour prior to failure. The plastic behaviour is caused both by plastic embedment deformations of the wood underneath the fasteners as well as the plastic behaviour of the metal fasteners subject to bending (see Article E2). The load-deformation curve of a joint under static load initially shows a steep rise (see Figure G2-1 (a), area I). As soon as either the steel or the wood under embedment stress can no longer show elastic behaviour, "yield point" F_y, the load-deformation curve flattens out, until a horizontal area finally emerges, which determines the load-bearing capacity F_{max} of the joint (see Figure G2-1 (a), area II). Subsequently, the load decreases with increasing deformations (see Figure G2-1 (a), area III). In this area, the joint has already failed due to splitting, withdrawal or failure of the fastener. This part of the load-deformation curve can only be recorded, if the test is performed in displacement-control.

The definition of ductility D_s is specified in Figure G2-1, where two cases are distinguished. In the initial case (a), the load-deformation curve can be approximated by two straight lines, while in the second case (b), the curve is completely non-linear. In case (b), the ratio of gradients of both areas was defined as 1/6. This definition allows the yield deformation v_y to be determined objectively and comparably and thus allows the so-called static ductility D_s to be specified at v_u/v_y. Even when the load-bearing capacity of the joint has been reached, it can still carry a portion of the maximum load. With decreasing load, the deformation v_u needed to define the static ductility may correspond to the deformation reached at a minimum of 80% of the maximum load.

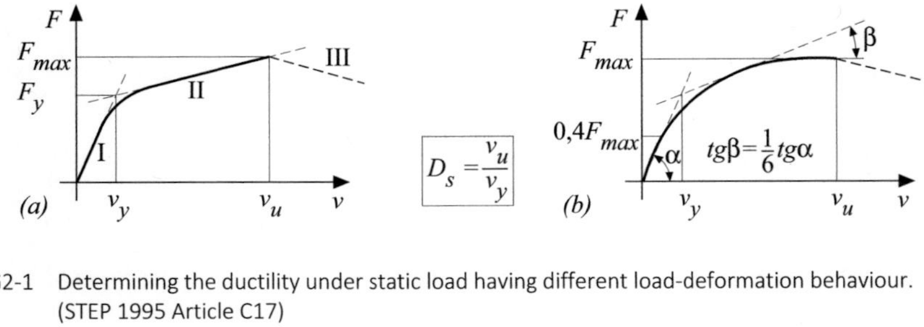

Figure G2-1 Determining the ductility under static load having different load-deformation behaviour. (STEP 1995 Article C17)

G2.2 Cyclic behaviour and energy dissipation

The cases shown in Figure G2-1 describe the behaviour under monotonic loading. In the following section, the behaviour under earthquake load is examined where the load direction changes within a short time. With this in mind, a nailed joint is observed, which is subject to a quasi-static load with changing direction (see Figure G2-2 (b)). When the elasticity limit is exceeded for the first time, the wood under the nail is compressed. This causes a cavity, within which the nail is unsupported during subsequent loading cycles within the same deformation range. The remaining load-bearing capacity is thus based on the nail acting as a cantilever over the length of the cavity or on a – mechanically identical – connection with a gap between the members. As soon as the deformation attained to date is exceeded, the nail, in turn, is supported by the surrounding wood and the load-deformation curve again follows the envelope curve, which would have formed had a monotonic load been applied. However, any withdrawal of the nail during alternate loading (see Figure G2-2 (b)) may reduce the load-bearing capacity.

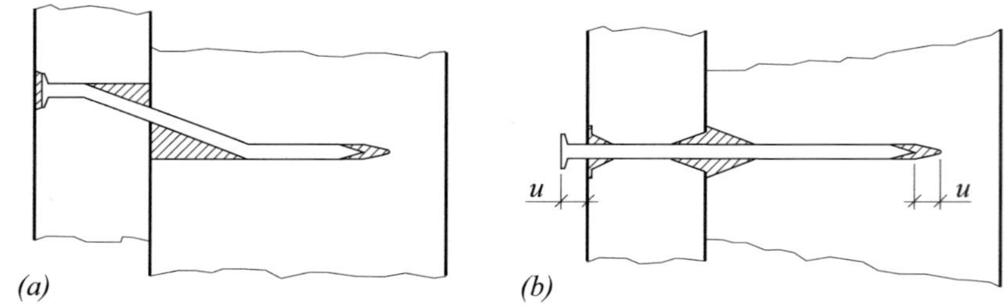

Figure G2-2 Cavities in the wood of a nailed joint, caused by cyclic load and plastic deformations. (STEP 1995 Article C17)

The typical hysteresis loops resulting from the behaviour described are shown for small, moderate and large deformations in Figure G2-3. The shape of the hysteresis loops varies significantly from those which emerge for an elastic-plastic material such as steel. For elastic-plastic material, the force required to restore plastic deformations is identical to the force originally required to generate the deformations in question (see Figure G2-4 (c)). In timber joints with mechanical fasteners, the cavities formed due to the wood being compressed mean hardly any hysteresis loops emerge in the second and fourth quadrant, leading to narrower and more "pinched" loops.

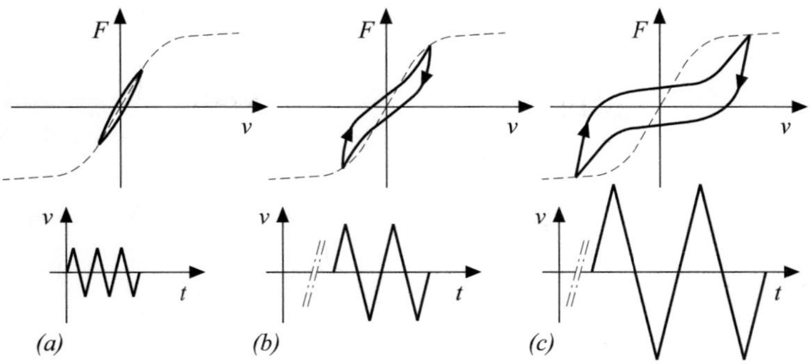

Figure G2-3 Load-deformation loops of dowelled joints at different load levels. (STEP 1995 Article C17)

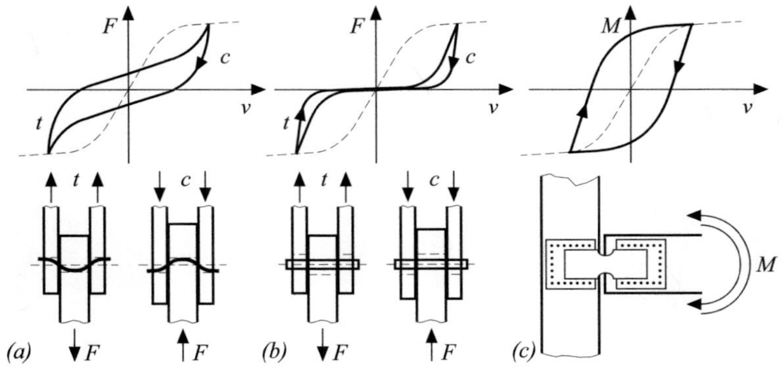

Figure G2-4 Different energy dissipation behaviour of various joints. *t* means tensile load, *c* compressive load. (STEP 1995 Article C17)

Figure G2-4 (a) shows load-deformation loops of a dowelled joint with slender dowels, in which both the embedment deformations of the wood and the plastic bending deformations of the dowels help dissipate energy. Sturdy dowels are not subject to plastic deformations and the energy dissipation is solely based on the plastic embedment deformation of the wood. Figure G2-4 (b) shows the corresponding load-deformation curves. As a general rule, the envelope curve for cyclic load matches the load-deformation curve for monotonic load, hence the load-deformation behaviour is assumed to be independent of the load history. Differences exceeding 10% can only be observed if the fastener arrangement in the joint changes, e.g. due to withdrawal or brittle fatigue failure of the fasteners. Unlike in masonry or concrete constructions however, timber joints are not usually prone to such low-cycle fatigue failures.

One exception are joints with punched metal plate fasteners, in which failure under cyclic load is often caused by the punched teeth being suddenly withdrawn from the timber or brittle failure of the metal plate itself. Other examples of the detrimental effect of cyclic load on the load-bearing capacity are joints with cold-formed steel connectors where nails can be withdrawn or joints in timber frame walls with brittle sheathing, where the cyclic embedment stresses can lead to sheathing material breaking out, which dramatically reduces the load-bearing capacity.

To obtain comparable information about how timber joints subject to cyclic loading behave, a CEN test standard was devised, which specifies a simple method to define the load-deformation loops. Tests should be implemented quasi-statically and under deformation control. The displacements are increased in steps with triple cycles of amplitudes, Figure G2-5. Any difference between the load achieved in the initial and third cycle subsequently represents the strength impairment under cyclic loading (see Figure G2-6).

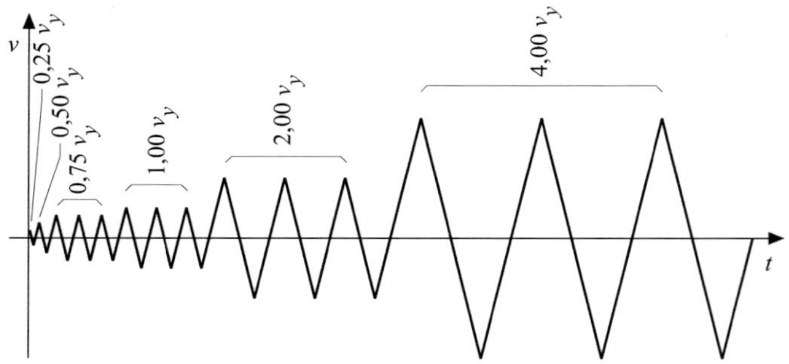

Figure G2-5 Test method for timber joint subject to cyclic loads, EN 12512. (STEP 1995 Article C17)

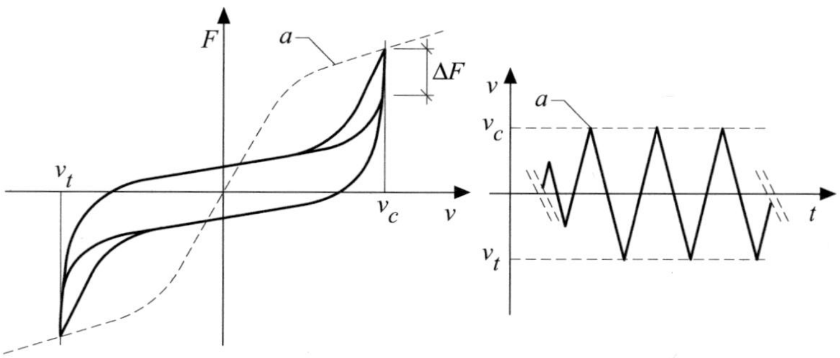

Figure G2-6 Strength impairment ΔF between the initial (*a*) and the third cycle at the same displacement level. (STEP 1995 Article C17)

The variable ΔF in Figure G2-6 clarifies the strength impairment between the envelope curve, which corresponds to the initial and third cycles at the same displacement level.

The energy dissipated by plastic deformations in a cycle is shown as a shaded area E_d in Figure G2-7, while the ratio between the dissipated energy E_d and the potential energy E_p is designated as the equivalent viscous damping ratio ν_{eq}. With increasing deformation, the energy dissipated in a hysteresis loop E_d increases, whereby ν_{eq} remains around the same level. Well-designed dowelled joints and nailed plywood-to-timber joints in timber frame constructions can achieve values of around 8 to 10% for ν_{eq}.

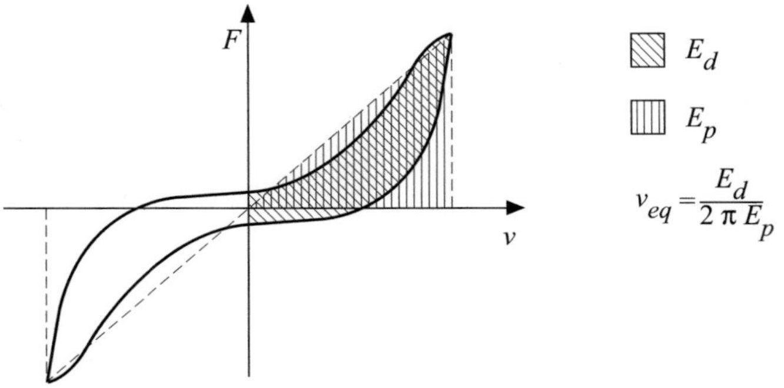

Figure G2-7 Energy dissipation through hysteresis. (STEP 1995 Article C17)

Provided the joints remain in the elastic area, the equivalent viscous damping ratio ν_{eq} is equal to zero (see Figure G2-3 (a)). However, still in the elastic area, when cyclic loading is applied, a certain amount of energy is dissipated. The "viscous" damping ratio of the construction material wood itself is in the order of magnitude of only 1%. Through friction between members or compression perpendicular to the grain, particularly in constructions with numerous non load-bearing members, e.g. timber frame constructions, the damping ratio in the elastic area may reach values of around 5%. Accordingly, for calculations in the elastic area, a damping ratio of 5% is often assumed.

G2.3 Behaviour of different joint typologies

The favourable behaviour of timber joints with mechanical fasteners under cyclic loads is characterised by a high level of ductility, the ability to withstand repeated loading and the ability to dissipate energy by plastic deformations. Premature failure due to timber splitting can often be avoided by ensuring sufficient spacings and end and edge distances. EC 5 specifies minimum values for fastener distances, which were selected to avoid the risk of the wood splitting under a static load. Despite the lack of indications that cyclic loads promote any splitting of the wood, greater fastener distances than the specified minimum distances in EC 5 may further reduce the tendency toward splitting and enhance the joint ductility as a result.

Splitting of the wood can also be prevented by gluing on wood-based panels or introducing fully threaded screws in the area of the joint. As well as preventing splitting, such measures may also enhance load-bearing capacity due to the generally higher embedment strength of the reinforcement materials (see Article E12).

The energy dissipation characteristics of a timber joint with mechanical fasteners can also be achieved using slender fasteners. Here, the slenderness ratio is defined as the ratio of member thickness to fastener diameter. Unlike sturdy fasteners which do not deform plastically, slender fasteners form plastic hinges, which dissipate energy during this process. In addition, the use of slender fasteners also helps guard against the splitting risk (see Figure G2-4).

The risk of losing load-bearing capacity due to cyclic loading can be largely prevented by using slender fasteners with a high withdrawal resistance, which are made of steel with a large deformation capacity. In addition, joints between wood and brittle materials should not be used to dissipate energy. As a general rule, the behaviour of normal fasteners in terms of the response under cyclic loading can be described as follows (see also Article G3 for indications of structural detailing).

Dowel-type fasteners

Nails, staples and screws generally show pronounced plastic behaviour in timber joints. The risk of withdrawal under repeated load means the length of the fasteners should be greater than for a joint under static load. For this reason, smooth shank nails should not be used for load-bearing joints in seismic areas. A slenderness ratio exceeding 8 generally already guarantees ductile behaviour under seismic loads (see Figure G2-8). As shown with tests on nailed wall panels (Yasumura, 1988), a high degree of ductility and favourable energy dissipation characteristics in nailed joints between plywood and wood in timber frame construction can already be observed with nails with a slenderness ratio of four in plywood (see Figure G2-9).

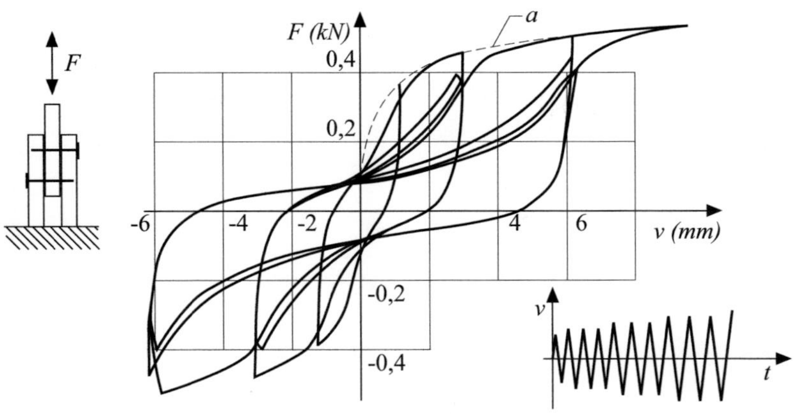

Figure G2-8 Typical behaviour of a nailed joint under repeated load (Slenderness ratio of the nails = 8.5). (STEP 1995 Article C17)

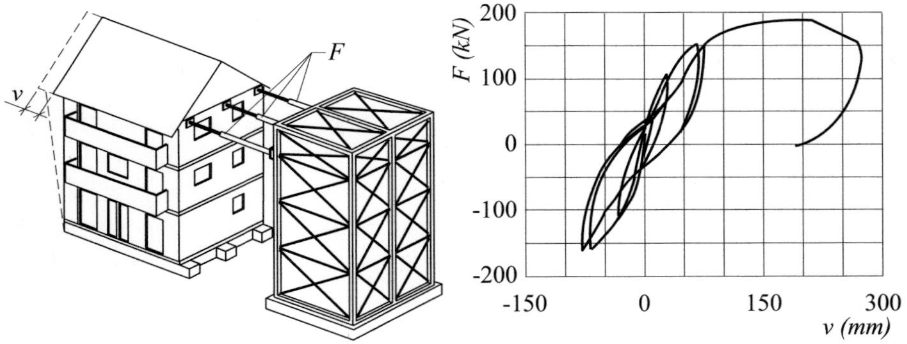

Figure G2-9 Ductile behaviour of a plywood sheathed timber frame house. (STEP 1995 Article C17)

Likewise, dowelled joints with minimum slenderness ratios of 8 show ductile behaviour under cyclic loading with plastic deformations of the dowels themselves as well as of the timber subject to embedment stress (provided, the steel grades of the delivered dowels correspond to the steel grades assumed in design). In this case, the failure mechanism 3 corresponding to Article E2 is the key mechanism, identified by the emergence of two plastic hinges per shear plane. While joints with sturdy dowels failing in mode 1 also show plastic behaviour due to the plastic embedment deformations of the wood, the scope for energy dissipation is far lower, given the lack of plastic deformations of the dowels themselves (Figure G2-4 (b)). Joints with sturdy dowels are also prone to splitting, making it particularly important to ensure sufficient edge distances and spacings in this case. If using dowelled joints with slenderness ratios of less than 8, tests to quantify their ductility will be required.

The load distribution in bolted joints is never uniform due to the oversized drilled holes. This may result in individual bolts in the joint being overloaded and potentially, splitting of the wood, preventing any redistribution which would allow more uniform loading of the individual bolts. Bolted joints in seismic areas are accordingly only recommended if slender bolts are used and the joints are carefully manufactured.

Connectors and punched metal plate fasteners

Split ring and shear plate connectors are not generally recommended for dissipating energy in structures in earthquake-prone areas due to their generally brittle failure (see Article E6). In contrast, toothed-plate connectors exhibit plastic behaviour. However, the prerequisite is ensuring sufficient end and edge distances and spacings, which guarantee embedment failure under the bolts and dowel teeth and prevent any timber splitting.

Although the load-deformation curves of joints with punched metal plate fasteners often show a certain degree of plastic deformations, given the risk of possible failure of the nail plate and withdrawal of punched nails, cyclic tests should be performed on the joints before considering any ductility.

G2.4 Seismic behaviour of joints with mechanical fasteners

Considerations to date relate to the behaviour of joints with mechanical fasteners under quasi-static and regularly changing loads. In the event of an actual earthquake, however, both the frequency and load cycles would vary from those of the above-described tests. Since both these variables differ for each earthquake, which makes them impossible to forecast, important statements on basic joint behaviour subject to seismic loads can be made although the tests cannot simulate the exact seismic behaviour. Since the stiffness and load-bearing capacity of timber and timber joints rises with decreasing load duration, we can assume that joints subject to seismic loads have a higher load-bearing capacity and stiffness than joints subject to quasi-static testing. Likewise for ductility, the same value can be assumed during seismic loads as was determined during quasi-static tests. Quasi-static tests thus suffice for the purpose of determining the key parameters outlining the response to seismic loads with sufficient accuracy. If the form of the hysteresis loops of the joint under cyclic load is known, the load-bearing capacity of the construction, taking into consideration the non-linear joint behaviour, can be calculated using a calculation program for any earthquake load (RILEM, 1994). Since load cycles from actual earthquakes occur very irregularly, unlike those of quasi-static tests, the number of cycles at which the deformation peaks is generally small, while numerous load changes occur at comparatively low deformation levels. Figure G2-10 shows the result of a non-linear calculation showing the moment-rotation curve of a frame corner joint using dowels under the impact of the El Centro earthquake, whose acceleration values were increased by 50% for this calculation.

If, when designing structures, the seismic action is reduced by dividing by the behaviour factor q and hence the ability to dissipate energy is taken into consideration, it is crucial to ensure that the joints in which energy dissipation should take place can actually undergo plastic deformation, before connected timber members or additional joints, such as anchors in concrete, fail. Timber members or additional joints must accordingly withstand greater loads before failing than those joints intended to function as dissipative areas. The higher load-bearing capacity of the members and remaining joints thus prevents brittle failure before plastic deformation in the dissipative joints.

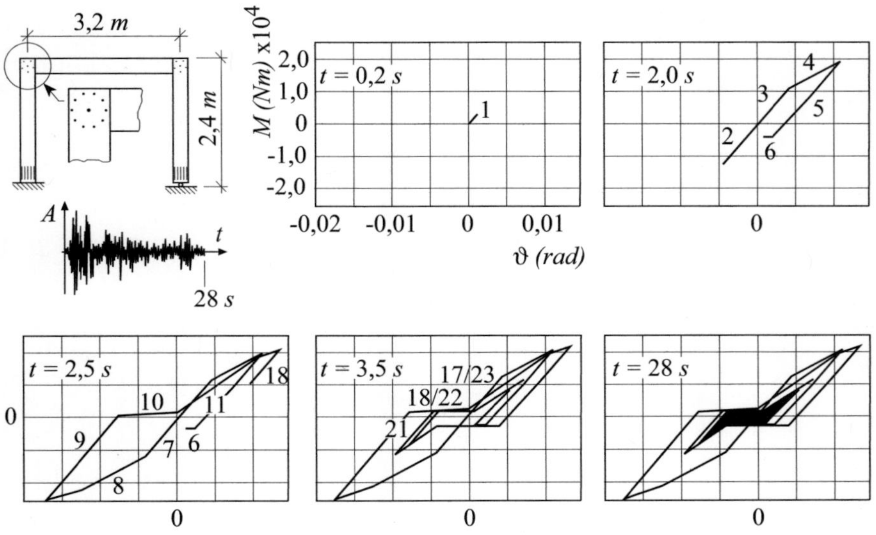

Figure G2-10 Moment-rotation curve of a frame corner joint using dowels under the impact of the El Centro earthquake. (STEP 1995 Article C17)

G2.5 Literature

A. Ceccotti, original Article C17, STEP 1995.

Ceccotti A. (Hrsg.) (1989). Structural behaviour of timber constructions in seismic zones. CEC DG III – University of Florence Workshop.

RILEM TC 109 TSA (1994). Timber structures in seismic regions: RILEM State-of-the-Art Report, Material and Structures 27:157-184.

Yasumura M. (1988). Experiment on a three-storied wooden frame building subjected to horizontal load. International Timber Engineering Conference, Seattle, pp. 262-275.

G3 Earthquake-compliant structural details

Original article: A. Ceccotti, P. Touliatos

When designing structures in areas at risk of earthquakes, one of the key tasks is to dimension load-bearing members: the corresponding rules set out in the provisions are particularly important. However, the structural engineer must also acknowledge that merely following the specified design rules to the letter alone will not ensure an earthquake-proof construction. The arrangement of members and structural details are also important to ensure a secure construction. As experience shows, for smaller structures with more regular plan and elevation, compliance with certain minimum member sizes and detailing rules of joints may suffice to achieve earthquake-proof construction without calculation. In addition, non-load-bearing components such as lightweight partition walls or floor build-ups impact on damping due to friction and hence affect the entire dynamic behaviour of the construction. This effect is only quantifiable in exceptional cases, but is normally favourable since it involves additional energy being dissipated. In many countries, code provisions for design in seismic areas hence include minimum dimensions and other requirements with corresponding examples, which eliminate the need to verify smaller structures.

In EC 8, this scope was not considered due to the diversity of construction techniques used among European countries. For the following reasons, in addition to calculation rules, EC 8 nevertheless includes a series of recommendations for structural details. There is a need to comply with specific details

- To guarantee the validity of the calculation rules, which are based on specific assumptions of structural details,
- To ensure that the assumed level of ductility is guaranteed, whereupon the correct behaviour factor q for calculating the inertial forces can be selected. One relevant example is detailing rules for joints with mechanical fasteners, which should show plastic behaviour before the maximum load is reached.

However, EC 8 does not impose any constraints on structural engineers in terms of possible timber construction designs. In principle, any type of timber joint can be used, provided it meets specific ductility requirements, which can be verified by tests (see Article G2). In most cases, few rules concerning structural detailing need to be considered which may eliminate the need to perform such tests. However, solely complying with a range of detailing rules in itself does not spawn a satisfactory design in most cases. A list including all possible details could never cover all conceivable cases. Accordingly, it is crucial to understand the significance of the recommendations on structural detailing and apply them accordingly. With this in mind, the current article addresses the key structural details of a timber construction subject to earthquakes in generic form. The structural engineer must transfer the solutions shown here to individual cases.

G3.1 Structural continuity

Earthquake actions can be considered horizontal, which, unlike vertical actions that only impact on a comparatively small part of the structure, generate loads throughout the entire structure (see Figure G3-1). Although the same also applies for wind loads, in Europe, earthquake loads are often decisive, particularly with heavier constructions (Ceccotti and Larsen, 1989).

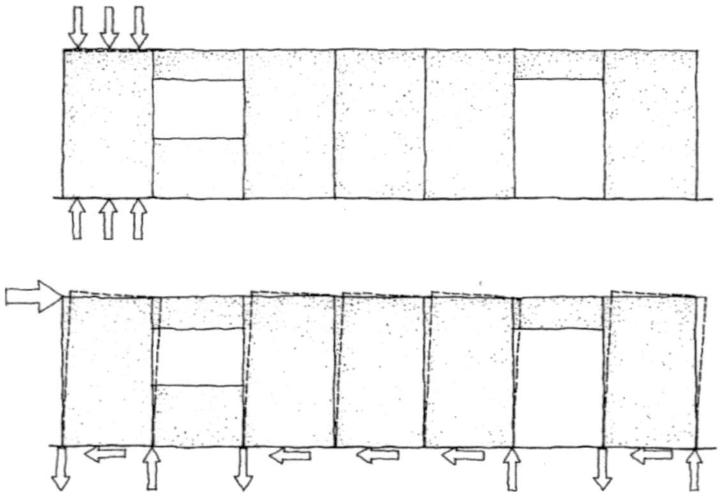

Figure G3-1 Different structural effects of vertical and horizontal loading. (STEP 1995 Article D10)

The continuous connection of the different members and their effectiveness in different load directions (e.g. tension and compression) are thus key prerequisites to ensure a favourable response in the event of an earthquake. For shear walls in timber frame constructions, the individual frames must be interconnected, to activate as many dissipative areas as possible and render the structure more effective overall. Figure G3-2 shows the key details and solution examples required to achieve interaction of the entire structure. Floors should always be designed as diaphragms. For this purpose a ring beam must be installed that absorbs the tension forces arising when the floor is loaded in plane (Figure G3-2b). The necessary continuity and force transfer at the floor corners can be guaranteed using a smaller spacing between the sheathing and ring beam (Figure G3-2a), where attention must be paid to avoid choosing overly small spacing as this may lead to premature failure. The vertically arranged members must also be interconnected through the floors, so tensile forces can be transferred, which may also be generated from the vertical accelerations in combination with the other internal forces (Figure G3-2c). The connection between upper and lower load-bearing timber frame walls by nailing the respective sheathing to the ring beam does not suffice when it comes to transmitting tensile forces, since the timber ring beam is subject to tensile stress perpendicular to the grain during an earthquake. The forces perpendicular to the grain can be transferred e.g. by using a nailed-on reinforcement of wood-based panels or crossbanded LVL for the ring beam.

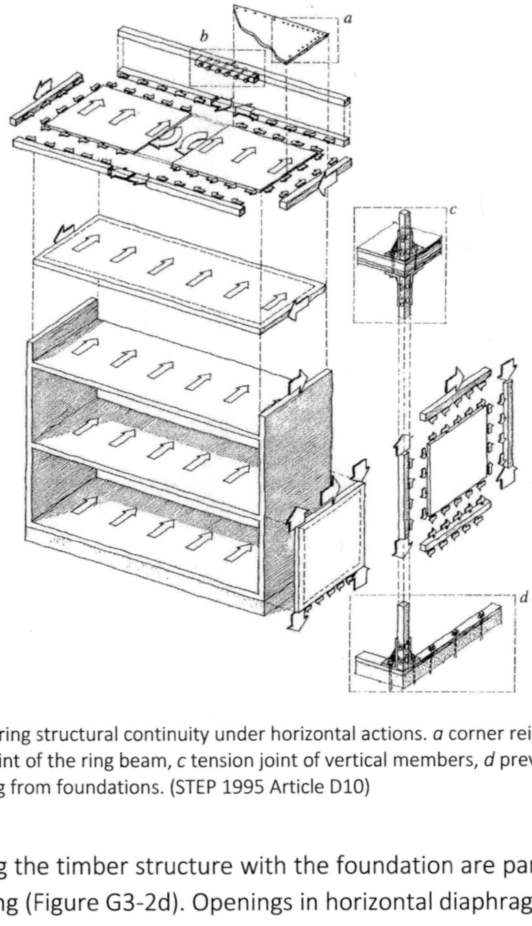

Figure G3-2 Details assuring structural continuity under horizontal actions. *a* corner reinforcement, *b* tension joint of the ring beam, *c* tension joint of vertical members, *d* prevention of sliding and uplifting from foundations. (STEP 1995 Article D10)

The joints connecting the timber structure with the foundation are particularly crucial, to avoid uplift and sliding (Figure G3-2d). Openings in horizontal diaphragms and shear walls are weak points and should be reinforced. Since diaphragms have to distribute loads, openings such as stairwells are particularly critical and their impact on the overall behaviour must also be taken into consideration. Figure G3-3 clarifies those diaphragm and shear wall areas which require specific focus. Tensile stresses perpendicular to the grain should be avoided at all costs. If tensile stresses perpendicular to the grain cannot be avoided, EC 8 specifies that "additional measures to avoid splitting must be incorporated" (e.g. punched metal plate fasteners or plywood panels, Figure G3-4, or a reinforcement with fully threaded screws inserted perpendicular to the grain).

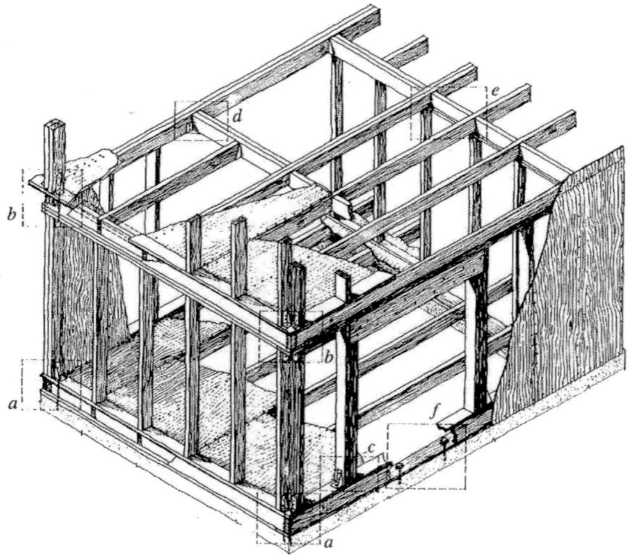

Figure G3-3 Detailing examples for a timber frame building. *a* prevention of uplift, *b* tensile joint in vertical members, *c* reinforcement of wall opening by adding studs and lintels, *d* reinforcement of diaphragm openings by adding girders and trimmer joists, *e* stiffening of floor beams by adding blocks, *f* prevention of sliding. (STEP 1995 Article D10)

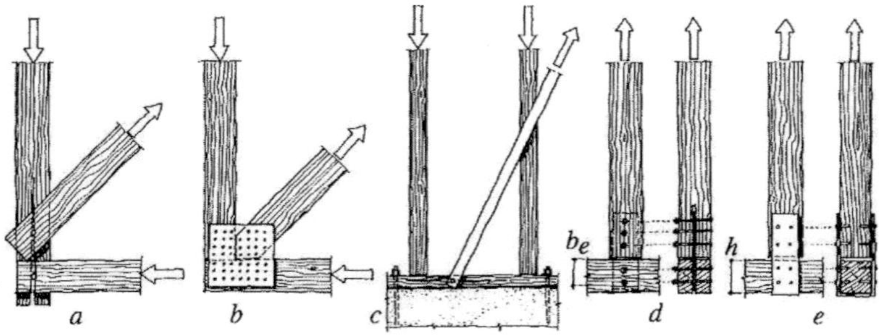

Figure G3-4 Detailing against tensile stresses perpendicular to the grain in joints. *a*, *c*: poor, *b*, *d*, *e*: good. (STEP 1995 Article D10)

In addition, joints must also be capable of withstanding alternating loads, since the load direction varies on multiple occasions during an earthquake. For these reasons, simple contact joints, which are only capable of transferring compressive forces, are unsuitable. In Figure G3-5, options for additional measures are shown to secure these joints. To prevent differing displacements of foundations, individual foundations should be avoided or interconnected. For homes constructed on softer ground in particular, the foundation must be configured such that it is sufficiently rigid and acts like a "raft" when exposed to earthquake loads.

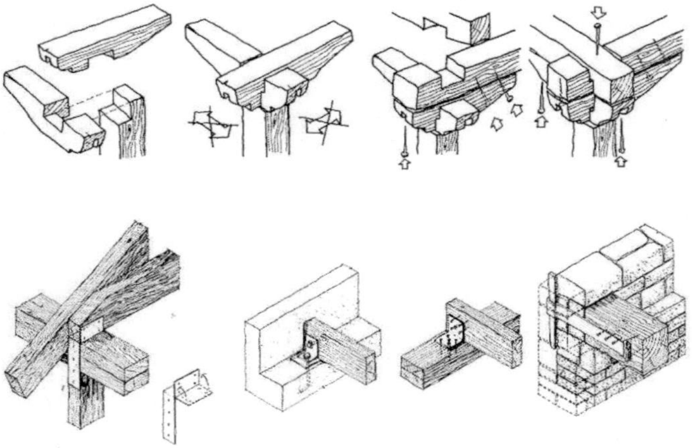

Figure G3-5 Possible provisions against the loosening of support in old and modern constructions. (STEP 1995 Article D10)

G3.2 Regularity of the building

Regularity in the plan and elevation of a building are crucial to ensure optimal behaviour in the event of an earthquake. Regular and symmetrical structures minimise torsional effects, which are difficult to estimate using normal calculation methods. Calculations of the effects of eccentrically exerted inertial loads on irregular structures are often unrealistic, due to the global action of earthquakes and impreciseness in stiffness distribution. Instead of such calculations, the building should be regularly constructed and the bracings designed to withstand earthquakes should be uniformly distributed (Figure G3-6).

This means torsional stresses can be virtually prevented and the design results obtained are also more reliable. If larger openings are unavoidable, the resulting eccentricity should be limited by additional bracing measures such as frames or truss-type components (see Figure G3-6f). Another option would be to reinforce existing shear walls choosing larger panel thicknesses, smaller fastener spacings or double- rather than single-sided sheathing. Non-load-bearing partition walls also respond positively in an earthquake, since they help dissipate energy via hysteresis and friction.

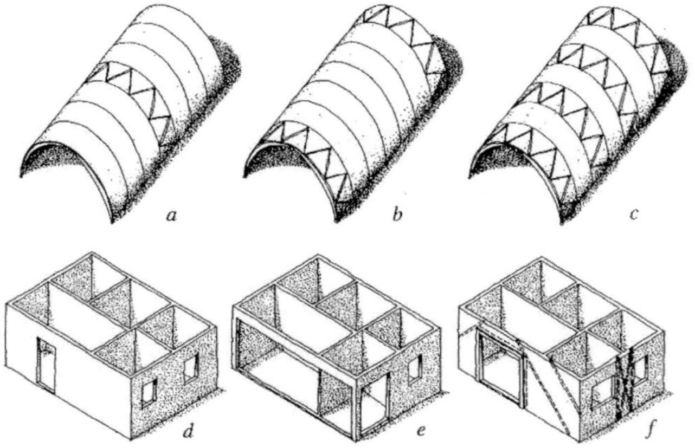

Figure G3-6 Schematic examples of distribution of bracings. *a*, *e*: poor, *b*, *f*: fair, *c*, *d*: good. (STEP 1995 Article D10)

G3.3 Structural compatibility

Problems may occur when members or structural parts of different stiffness are interconnected. Examples include joints between a timber construction and a chimney or a masonry wall. If larger differences in stiffness exist between connected members, both structures could either be constructed such that they could transfer their portion of seismic load independently of the other structure, or both structures could be connected to form a single structural unit. In the example shown in Figure G3-7, the glass façade is independent of the timber structure.

The joints between external masonry walls and an internal timber construction are often not designed to ensure the construction as a whole is capable of withstanding earthquake loads. Since less rigid timber constructions are generally subject to larger deformations than more rigid masonry constructions, the timber construction tends to rest against the masonry walls, which means they have to withstand the horizontal inertial forces of the entire construction and must be designed to take this into consideration (Figure G3-8). If the masonry is incapable of withstanding the entire load, the following approach is also possible: the masonry can be considered a mass without rigidity (damaged due to cracks) and the entire load is transferred by the timber construction. The construction in Figure G3-9 is designed such that in the event of an earthquake, the masonry would collapse without causing excessive damage to the timber construction. The failure of the masonry leads to a great deal of energy being dissipated, which means the timber construction and its far lower mass can withstand the earthquake without excessive damage.

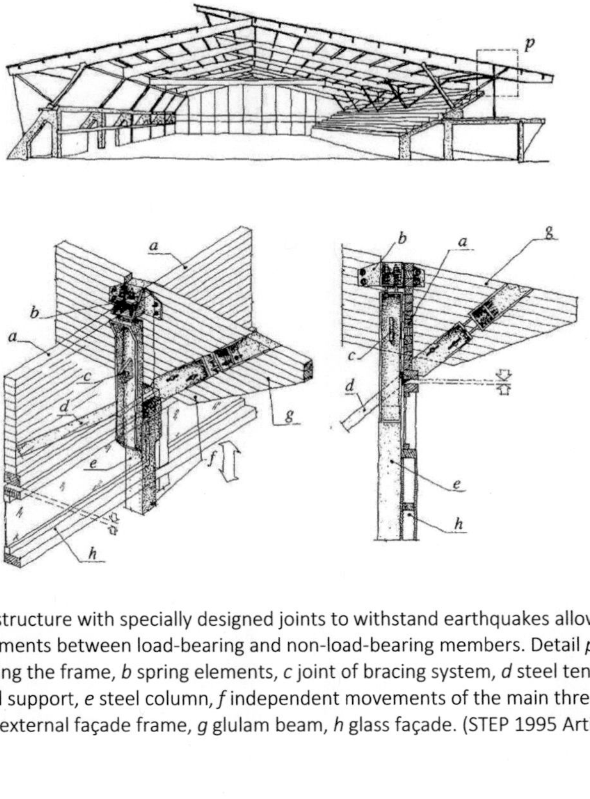

Figure G3-7 Timber structure with specially designed joints to withstand earthquakes allowing for relative displacements between load-bearing and non-load-bearing members. Detail *p*: *a* beam reinforcing the frame, *b* spring elements, *c* joint of bracing system, *d* steel tension rod with springed support, *e* steel column, *f* independent movements of the main three-hinged frame and the external façade frame, *g* glulam beam, *h* glass façade. (STEP 1995 Article D10)

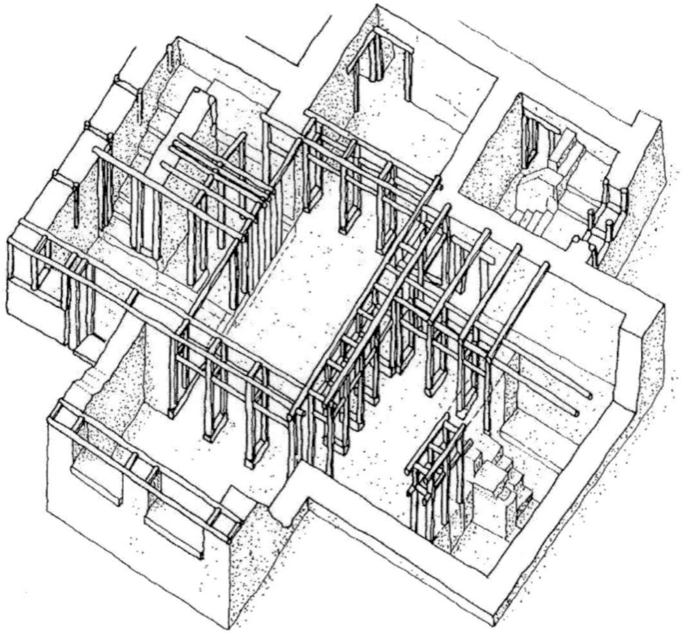

Figure G3-8 Mixed masonry-timber building (Greek islands, 1500 B. C.). The timber construction only carries vertical loads, while the masonry also accommodates horizontal loads. The floors cannot act as diaphragms, due to their workmanship. (according to Touliatos, 1993, STEP 1995 Article D10)

G3.4 Ductility and energy dissipation

In historic buildings, such that shown in Figure G3-9, the required energy dissipation was achieved by friction between wood and masonry as well as plastic compression deformations perpendicular to the grain in the timber. In modern, lighter structures, the energy dissipation is usually guaranteed by using plastic deformations in joints with mechanical fasteners (see Article G2).

When verifying structures with reduced inertia forces in accordance with EC 8 ($q > 1.5$; see Article G2) the required ductility and energy dissipation characteristics must be confirmed with tests. For specific configurations however, where experience suggests that ductile and plastic behaviour can be expected, no tests are required.

Timber frame panels with sheathing nailed to the studs show far more ductile behaviour than when using diagonals to ensure the lateral load-bearing capacity. According to EC 8, plastic behaviour of timber frame panels can be assumed, when, among other things, the sheathing comprises wood-based panels at least $4 \cdot d$ thick and the nail diameter d does not exceed 3.1 mm. This provision applies to particleboards with minimum density ρ_k of 650 kg/m^3, plywood at least 9 mm thick and particle- or fibreboards at least 13 mm thick.

Timber-to-timber and steel-to-timber joints with nails or dowels can be considered sufficiently ductile, if the timber thickness of the connected members is at least $8 \cdot d$ and the dowel diameter does not exceed 12 mm. This specification is imposed because joints respond favourably to cyclic loads when designed such that failure mechanism 3 in accordance with Johansen governs (two plastic hinges per shear plane).

Of course, the use of other mechanical fasteners and timber member sizes in constructions at risk of earthquakes is also possible. However, if a behaviour factor $q > 1.5$ is used in the design, the plastic behaviour of joints must be verified with tests (see Article G2). For joints with dowel-type fasteners, failure mechanism 3 should always be the goal, so that the plastic embedment deformations of the wood and the plastic bending deformations of the fasteners can boost the joint ductility.

Even when the structure is designed for $q = 1.5$, meaning the plastic behaviour is no longer required computationally, the use of slender fasteners with plastic deformation potential is recommended. This means a safety reserve, without additional costs for the structure. The recommendations for detailing timber constructions in accordance with EC 8 should thus be considered independently of the underlying behaviour factor q as far as possible.

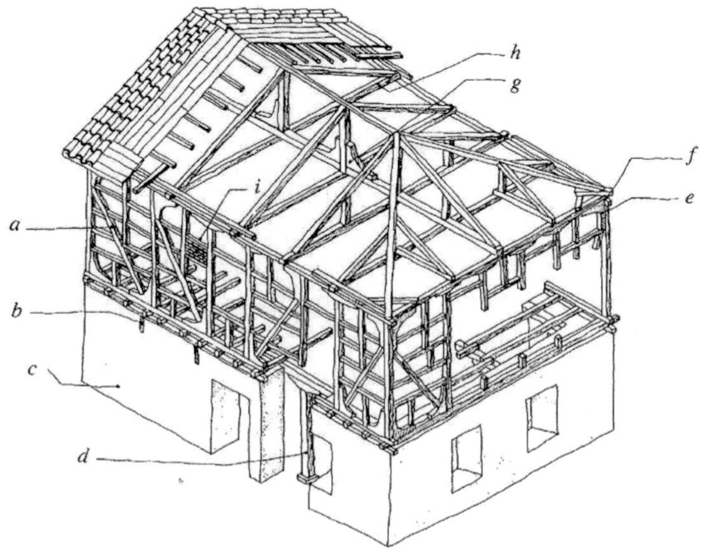

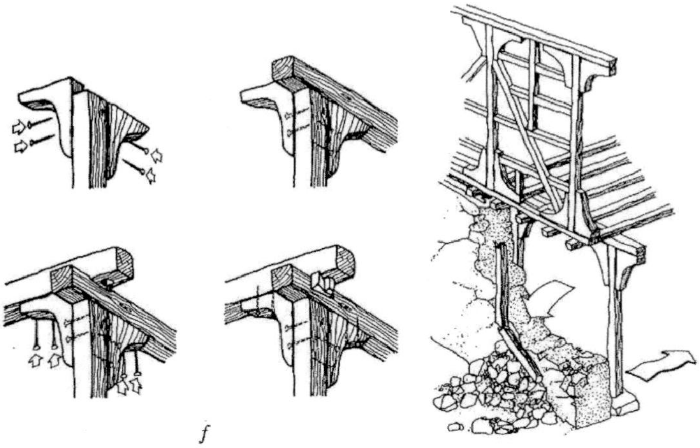

Figure G3-9 Example of a mixed masonry-timber building (Greek islands, 1800 A. D.). *a* timber diagonal, causing compression perpendicular to the grain in the lower corner, *b* anchoring of the timber construction in the masonry, *c* masonry as support for the upper timber construction, *d* timber columns to accommodate vertical loads after the masonry fails (see detail lower right), *e* bent timber corner piece to reinforce the timber frame wall, *f* joint that is simply to replace after damage (see detail lower left), *g* corner piece made of wood to reinforce the roof construction, *h* tie beam to prevent horizontal forces on walls, i brick infill. The floors cannot act as diaphragms, due to their workmanship. (according to Touliatos, 1993, STEP 1995 Article D10)

G3.5 Literature

A. Ceccotti, P. Touliatos, original Article D10, STEP 1995.

Ceccotti A. and Larsen H.J. (1988). Background Document for specific rules for timber structures in EC 8. Report EUR 12226 for the Commission of the European Communities, Brussels.

Touliatos P. (1993). Seismic disaster prevention in the history of structures in Greece. National Technical University of Athens.

G4 Damages in hall structures

Authors: Matthias Frese, Ann-Kathrin Grün and Hans Joachim Blass

In January 2006, Germany and neighbouring countries saw a series of collapses of what were, at times, decades-old timber roof structures. Although most failed due to the weight of snow, it was not possible to conclude that an exceptionally high snow load was the only possible cause of failure. Several recent damage cases have generally shown a combination of multiple causes to be considered or a need to differentiate between causes and catalysts, to explain structural failures; this is what applies to failures of technical systems in general, barring a few exceptions (Schmitt-Thomas, 2005).

At that time, there was no uniform statistical and systematic consideration of damage cases involving timber-hall structures, which could be used to draw conclusions for those planning and designing such structures. Damage cases, some of which are described in specialist literature, tend to be individual considerations, which makes it difficult to form concrete conclusions about their significance. This is largely what spurred on the research project in 2006 named "Damage analysis of timber-hall structures". It aimed to set out the basis for an overall consideration (constructing a database with damage cases, developing a system for classifying, describing and analysing damage), citing the causes of damage and formulating conclusions. The project results are shown in a research report (Blass and Frese, 2010). In the meantime, there was scope to expand the database and include certain damage cases and it seemed expedient to re-analyse data material compiling over 700 damage cases. The conclusions saw general requirements for reliable structures emerge, some of which had already resurfaced multiple times.

G4.1 Data collection

Origin of data

The data on the damage cases came from expert reports, an old database of the Studiengemeinschaft Holzleimbau e. V., the collections of the Technical Universities of Munich and Graz as well as the archive of renovation projects of a glulam company based in Northern Germany, Gebr. Schütt KG. Some cases were taken from literature or passed on to the authors directly. Accordingly, the data was inconsistent and exposed qualitative differences. The data compiled was the result of a non-controllable survey: the material used was consequently that which was available and accessible. This may lead to constraints concerning the representative nature of timber-hall structures in a general sense and the significance of the results. Most of the damage descriptions assessed concerned timber hall structures based in Germany.

Data classification

Structures generally comprise recurring individual members. These, in turn, are assembled based on recurring construction principles. Structures of a specific group, timber hall structures in this case, can thus be easily compared. This simplifies the data collection and common consideration of their characteristics, damages and their causes. A system was developed, allowing damage cases to be reliably recorded in a database. According to this, the features of a damage incident are assigned based into five thematic groups: construction, member, material and damage features as well as error sources. Figure G4-1 shows the systematic approach, with a selection of features and related or exemplary characteristics. Based on quantitative or metric parameters such as the height above sea level, qualitative features such as district or heating are collated, with a specific vocabulary consisting of keywords. The ongoing classification system, see Blass and Frese (2010), includes over 60 qualitative and quantitative features as well as multiple keywords, to describe a damage case consistently.

The building features convey an overview of the damaged hall structures. Member and material features are detail-oriented and directly linked to damage. They provide information on structural systems, damaged members and the materials used. Damage parameters describe the damage pattern, also in terms of consequences for the structural safety of members or the overall construction. Initial damage is first to occur. This is in contrast to secondary damage, when damage is e.g. inflicted through damage on another member. Not every incident of initial damage necessarily represents a type of failure; initial damage in the form of critical deformation, moisture penetration, rot, blue stain and mould fungi, corrosion or harmless cracks in the grain direction (cf. Figure G4-2) does not generally lead to a member or structure failing. Tensile failure includes bending failure, since this is normally triggered by a local tensile failure. Significant cracks in the grain direction (Figure G4-3 and Figure G4-8), shear (Figure G4-2) and tensile failures are particularly significant and allow the strength properties of solid timber and glulam to be demonstrated based on real member size. This “testing” independent of laboratory methods shows the strengths and weaknesses of the material against a background of changing climate, load history and member size.

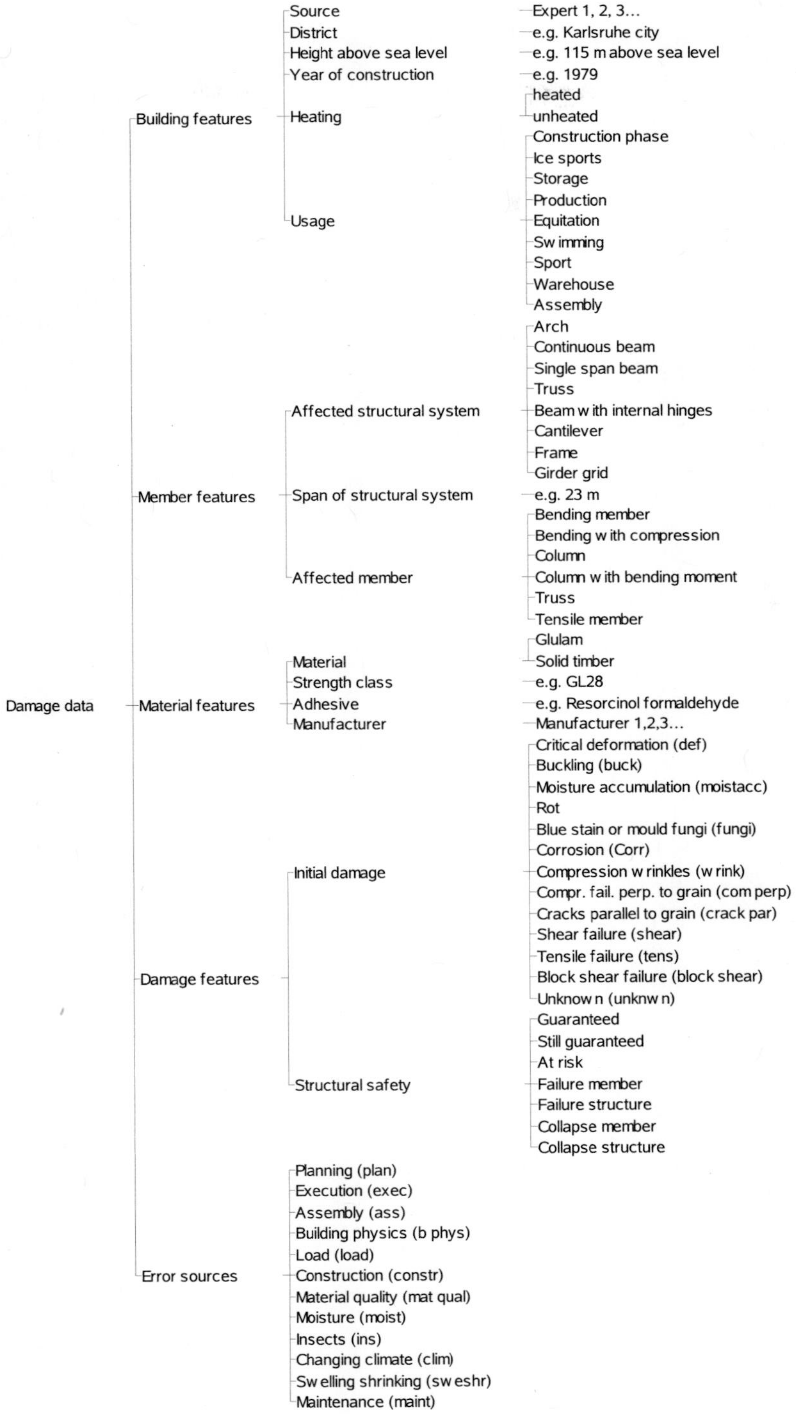

Figure G4-1 Selection of keywords and features. Abbreviations in brackets for Table G4-12.

Figure G4-2 Initial damages – rot (top left), blue stain or mould fungi (top right), cracks in grain direction (minor shrinkage cracks, lower left), shear failure (part of beam in compression zone protruding at end grain, lower right).

The characteristics of structural safety are assessments of experts in their reports. They relate to damaged members or entire structures. ***Guaranteed*** indicates sufficient structural safety; ***still guaranteed*** means that rehabilitation works are recommended soon; ***at risk*** are members or constructions including serious initial damage that leads to less structural safety than required. Significant repair works are required in this case. The keywords ***failure member*** or ***failure structure*** denote an obvious loss of structural safety, but not a collapse. The member or construction can no longer fulfil its function (Figure G4-3, left). To meet subjective safety requirements, e.g. propping or other safety measures must be implemented immediately. ***Collapse structure*** denotes the complete loss of a structure, ***collapse member*** a partial loss. A repair involves considerable expense, which is why the characteristics of structural safety are subject to an order of priority, identified by a spectrum between safe and collapse.

Figure G4-3 Cracks in pitched cambered beams, immediate propping/safeguarding required (left and centre), professionally retrofitted beams (right)

The characteristics of the error sources correspond to 12 categories, which are then allocated to the causes of initial damage. A selection of such causes is included in Table G4-1. The error source ***planning*** involves the technical planning of a construction or member. Typical for the error source ***building physics*** is generally an unfavourable thermal transport. ***Construction*** involves constructive characteristics, which lead to directly related circumstances in a damage case. Although originating in technical planning, ***building physics*** and ***construction*** are classed as sufficiently important to be individual error sources. ***Execution*** involves qualitative performance of works on the construction site and identifies unfavourable deviations from planning. ***Assembly*** involves defects when assembling members and constructions. ***Load*** refers to any type of overload. ***Moisture*** refers to the ingress of humidity, which is always due to external causes and thus excludes damaging humidity caused by e.g. thermal radiation (in indoor ice rinks) or a lack of detailing for durability. ***Changing climate*** is the cause of resulting residual stresses in glulam cross-sections (see Möhler and Steck, 1980; Häglund, 2010). ***Swelling*** or ***shrinking*** refers to the long-term directional volume changes affecting an entire member. ***Swelling*** or ***shrinking*** is only cited in combination with the error source ***construction***, when constructive characteristics prevent shrinking (a lack of shrinking does not lead to damage, even if shrinking is prevented). For certain error sources, responsible persons can be nominated: ***Planning***, ***building physics*** and ***construction*** concern the entire planning phase of a construction and hence the designer, ***execution*** and ***assembly*** the building contractor. ***Material quality*** is almost always associated with manufacturers of glulam and ***maintenance*** to owners or authorised representatives.

Table G4-1 Categories of error sources.

Keyword	Cause
Planning	Infringements of provisions/generally accepted technical rules; Failure to heed engineering expertise, e.g. in specialist literature; static calculation errors
Execution	Use of green wood; lack of drainage due to incorrectly installed outlets; incorrect number or diameters of fasteners; non-compliance with building execution provisions; unfavourable changes of the structural system, the cross-sectional dimensions or strength class
Assembly	Transport damage; insufficient member protection against weathering impacts; a lack of support measures during the construction phase
Building physics	Thermal radiation (indoor ice rinks); solar radiation; members exposed to indoor and outdoor climates (e.g. gable beams)
Load	Overload with regard to permanent loads, imposed loads, snow and wind loads as well as accumulation of water on flat roofs
Construction	Notched beams; dowel circles in frame corners; beams with openings; defective detailing for durability; tensile stresses perpendicular to the grain in joints; prevention of shrinking (e.g. slotted-in steel plates); curved or kinked members subject to bending; unwanted restraints; structures that are difficult to manage in terms of design and execution
Material quality	Defective quality of wood, finger joints or bond lines concerning required characteristics; preliminary damage
Moisture	High moisture content due to roof leaks or sprinkler systems in indoor riding arenas
Changing climate	Cyclic changes in moisture content in glulam due to changing climate; leading to changing humidity gradients in the cross-section
Swelling or shrinking	Long-term increase or decrease in moisture content leading to swelling or shrinking
Maintenance	Failures during inspection, maintenance and repair

G4.2 Presentation of the structures and their damage

General points

The database includes 709 datasets. Each concerns a single independent incidence of initial damage within a construction and concerning the construction, member and material features as well as the error sources to which the initial damage is attributed. Multiple attributions usually apply to error sources. Since many structures are subject to two or more incidents of damage, the 709 incidents of damage affect a total of 529 structures. In 44 cases, it was not possible to attribute a clear expression of characteristics to initial damage. In such cases, the initial damage was described as unknown. Due to the lack of details in the damage descriptions, not all the corresponding characteristics were defined for all features or error sources. The overall figures in the following tables and diagrams thus differ.

Building features

Most of the structures are located in the former West German states (Figure G4-4). Data on damage cases in the new German states was only available in individual cases. This is due to the divide up until 1989, which meant dialogue between experts was very limited up to the time of reunification and even for some considerable time thereafter. It is inaccurate, therefore, to conclude that timber-hall structures in the new German states did not suffer comparable damage. Most of the damage originated from the four most populous West German states: Baden-Württemberg, Bavaria, Lower Saxony and North Rhine-Westphalia as well as Schleswig-Holstein, the active sphere of the directly cooperating glulam construction company Gebr. Schütt KG (Table G4-2). More concerning is the fact that extending the database of 550 cases (values in brackets) in Blass and Frese (2010) with 159 cases has led to the damage tripling in Schleswig-Holstein as well as a clear increase in Lower Saxony. If a similarly rigorous survey examined damage incidents in the other German federal states, similar findings could be expected. Accordingly, the damage actually accessible or located in the course of the damage survey and recorded here is very likely to be less than a third of the actual extent of damage which could be described and analysed as part of such assessment. With the snow load in mind, the height above mean sea level of the structures was recorded (Figure G4-5). It ranges from -1 m in the North German lowlands up to 1145 m above sea level in the low mountain range, foothills of the Alps or the Alps themselves. The year in which the structures were constructed comprises the period from 1912 to 2006, the average year of which was 1980 (Figure G4-6). 81% of the timber-hall structures had a sealed building envelope. The distribution of the service classes shows that service class 1 prevails (Table G4-3). Most of the structures are heated. In terms of usage, the damaged structures frequently include sports halls, warehouses and production facilities and those used for conferences.

Table G4-2 Initial damage per federal state. In brackets: original analysis of 550 damage cases.

Federal state	Number/total	Portion [%]
Baden-Württemberg	91/709 (90/550)	12.8 (16.4)
Bavaria	113/709 (108/550)	15.9 (19.6)
Hesse	30/709 (30/550)	4.2 (5.5)
Lower Saxony	107/709 (70/550)	15.1 (12.7)
North Rhine-Westphalia	131/709 (120/550)	18.5 (21.8)
Schleswig-Holstein	101/709 (30/550)	14.3 (5.5)
Remainder (respectively < 20)	76/709 (42/550)	10.7 (7.6)
Not specified	60/709 (60/550)	8.5 (10.9)

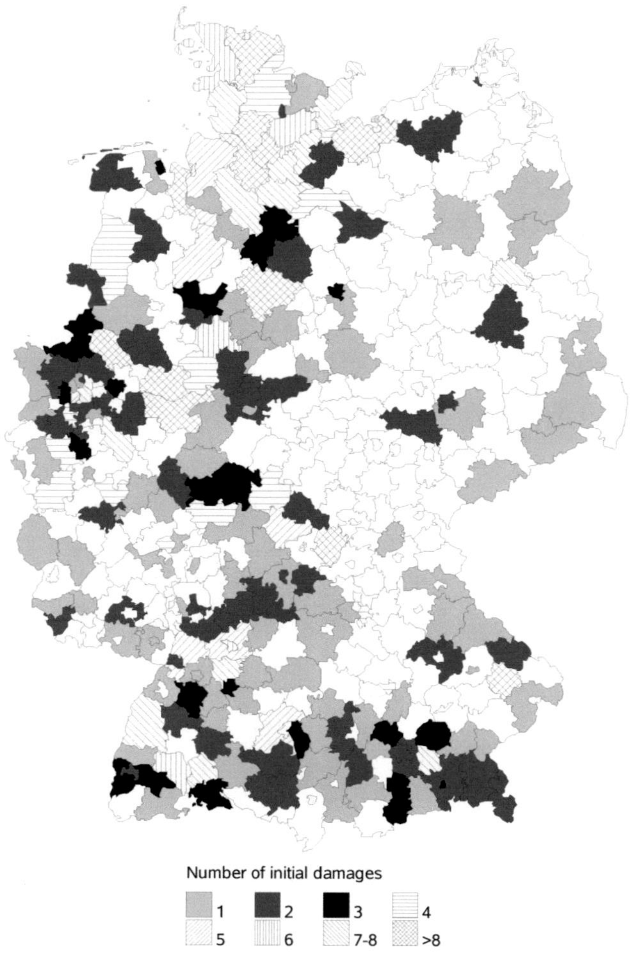

Figure G4-4 Distribution of initial damage over a German map with district borders.

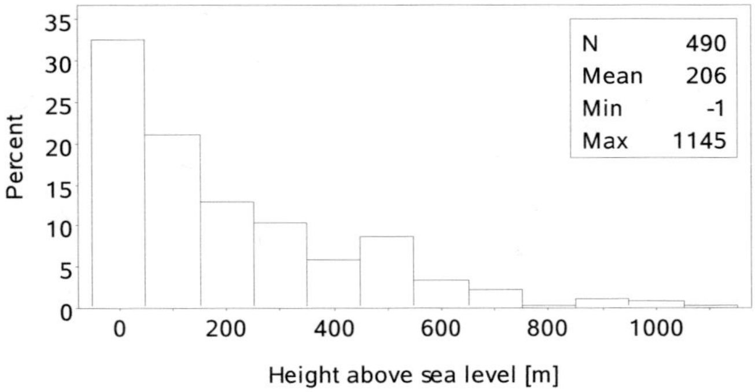

Figure G4-5 Frequency distribution of height above sea level.

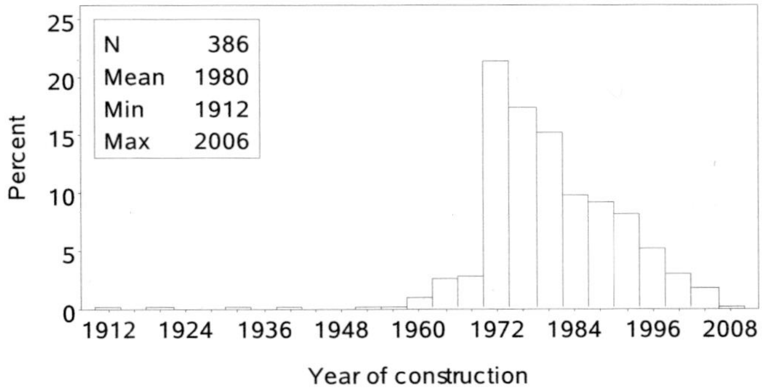

Figure G4-6 Frequency distribution of the year in which the structure was built.

Table G4-3 Service classes of the structures.

Service class	Number/total	Portion [%]
1	374/529	70.7
2	69/529	13.0
3	28/529	5.3
Not specified	58/529	11.0

Member and material features

Damage was mainly observed in single and multi-span beams (including those with hinges) as well as in frames (Table G4-4). The 19 listed trusses denote non-specific structural systems, in which only individual members like chords (beams), compression or tension members were affected. Most affected structural systems were statically determinate; there were rarely statically indeterminate systems (Table G4-5). The portions in both tables thus correspond to each other.

The distribution of the spans of the structures is shown in Figure G4-7. The modal value is 20 m and corresponds to the typical span for sports halls and indoor riding halls. Damage to very wide-spanned halls was relatively infrequently observed, but cf. Hansson and Larsen (2005).

Table G4-4 Affected structures.

Structural system	Number/total	Portion [%]
Single-span beam	370/601	61.6
Cantilever beam	16/601	2.7
Multi-span beam	36/601	6.0
Beam with internal hinges	17/601	2.8
Frame	56/601	9.3
Arch	13/601	2.2
Truss	19/601	3.2
Other	10/601	1.7
Not specified	64/601	10.7

Table G4-5 Statical determinacy of structures.

Statical determinacy	Number/total	Portion [%]
Determinate	477/601	79.4
Indeterminate (1 redundant)	19/601	3.2
Indeterminate (2 redundants)	2/601	0.3
Indeterminate (multiple redundants)	6/601	1.0
Not specified	97/601	16.1

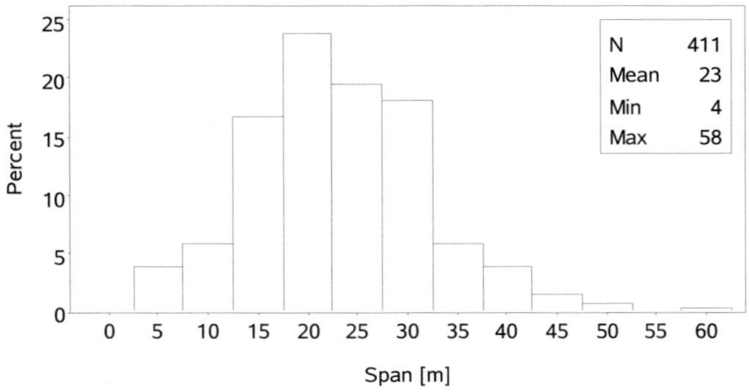

Figure G4-7 Frequency distribution of span.

Table G4-6 Affected members.

Affected member	Number/total	Portion [%]
Beam	425/601	70.7
Beam with compression	46/601	7.7
Column	9/601	1.5
Column with bending	47/601	7.8
Tension member	13/601	2.2
Truss	25/601	4.2
Not specified	36/601	6.0

The portions of the affected members are summarised in Table G4-6. Beams, which as expected often constitute the main structural member, are thus particularly damage-prone. By far the most common example is beams with compression (those of portal frames), columns with bending (columns of portal frames) and trusses, while columns and tension members were not significantly affected. The proportional composition of the member shapes in Table G4-7 shows that it is not only curved members and double-tapered beams, systematically subject to tensile stresses perpendicular to the grain, but also straight members that are affected. Virtually 90% of the damaged members are made of glulam (Table G4-8). In most cases, the glulam concerned is of quality class I. The clear leader is BS14 (nowadays GL28), followed by quality class II and finally, for portal frames included in more recent constructions, already a few members of European GL classes. The damaged glulam members are allocated among 40 different manufacturers. Here, it is assumed that the plant size or production output of the individual manufacturer and the frequency of damage to the members they manufacture are proportional. Damage frequencies in interaction with specific manufacturers must be securely eliminated. The number of manufacturers involved is a benchmark indicating that damage to glulam represents a product-specific problem.

Table G4-7 Member shapes.

Member shape	Number/total	Portion [%]
Straight	189/601	31.5
Kinked	10/601	1.7
Curved	59/601	9.8
Trapezoidal	31/601	5.2
Fish-bellied	13/601	2.2
Parallel*	3/601	0.5
Roof beam+	15/601	2.5
Pitched cambered	149/601	24.8
Double-tapered	80/601	13.3
Not specified	52/601	8.7

* Exclusively trusses. + Double-tapered or pitched cambered.

Table G4-8 Construction materials.

Material	Number/total	Portion [%]
Glulam	533/601	89.0
Glulam/solid timber*	2/601	0.3
Glulam/wood-based material*	7/601	1.2
Solid timber	47/601	7.8
Wood-based material	4/601	0.7
Not specified	6/601	1.0

* Combination of both.

Damage features and further considerations

Table G4-9 shows the allocation of initial damage to the defined damage patterns. Over 70% is due to cracks parallel to the grain, which were mainly discovered in glulam members (cf. Table G4-8). 5-6% is allotted to tensile or shear failure and rot. Overall, 5% affected the serviceability or appearance of the members. Buckling and corrosion were both in the region of 1%. Block shear and compression failure perpendicular to the grain are individual cases. Compression wrinkles are not included in the database. In 6% of cases, the damage descriptions lacked sufficient details to define the precise nature of the damage. The overwhelming frequency of cracks parallel to the grain in curved glulam beams with opening moments (Figure G4-3) triggered closer scrutiny into their origin (Blass and Frese, 2010, Frese, 2011). Considering the anisotropy of the degrees of swelling and shrinkage in the curved areas (longitudinal/tangential-radial ≈ 1:24), changes in moisture content lead to curvature changes and, under unfavourable support conditions, to stresses perpendicular to the grain. Different longitudinal shrinkage of the lamellae in the edge and core areas due to moisture content changes also generates stresses perpendicular to the grain in combined curved beams, which exacerbate the formation of cracks in curved glulam beams. To calculate the deformation of glulam members based on changes in moisture content, reference is made to Blass and Frese (2010), the American Institute of Timber Construction (1994) and Larsen and Riberholt (1983).

Expert reports in particular contain precise details of the relative crack depth (Figure G4-8) and moisture content of affected members. The diagram in Figure G4-9 shows the relations between both values. The illustration does not contradict the observation that cracks parallel to the grain occurred more frequently in members that were attributable to service class 1 or occurred at times when particularly low moisture content was recorded. Figure G4-10 shows the distribution of member depth (in the area of the supports) of glulam beams with shear failures. For 21 beams of over 29 registered shear failures, the member depths are known and ranged between 650 and 2400 mm. The distribution reflects that deep members with greater volume subject to shear have less effective shear strength than less deep members. In the damage descriptions shown, no shear failures of glulam members under 650 mm in depth were documented.

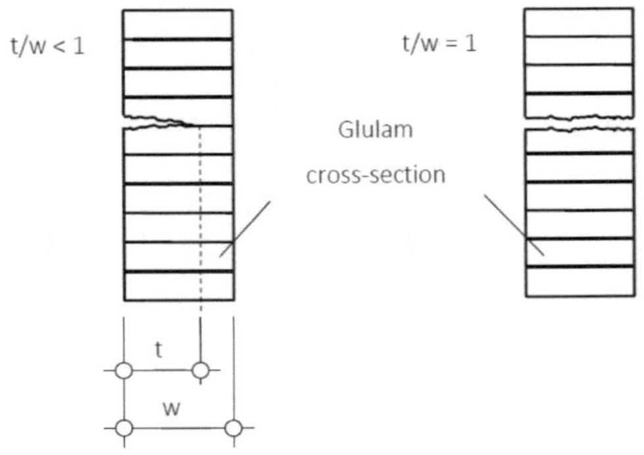

Figure G4-8 Relative crack depth in glulam: partial (left) and continuous crack (right).

Table G4-9 Distribution of initial damage.

Initial damage	Number/total	Portion [%]
Critical deformation	19/709	2.7
Buckling	6/709	0.9
Moisture accumulation	12/709	1.7
Rot	39/709	5.5
Blue stain or mould fungi	5/709	0.7
Corrosion	8/709	1.1
Compression wrinkles	0/709	0.0
Compression failure perpendicular to the grain	2/709	0.3
Cracks parallel to the grain	507/709	71.5
Shear failure	29/709	4.1
Tension failure	27/709	3.8
Tension or shear failure*	9/709	1.3
Block shear	2/709	0.3
Unknown	44/709	6.2

*No clear distinction is possible.

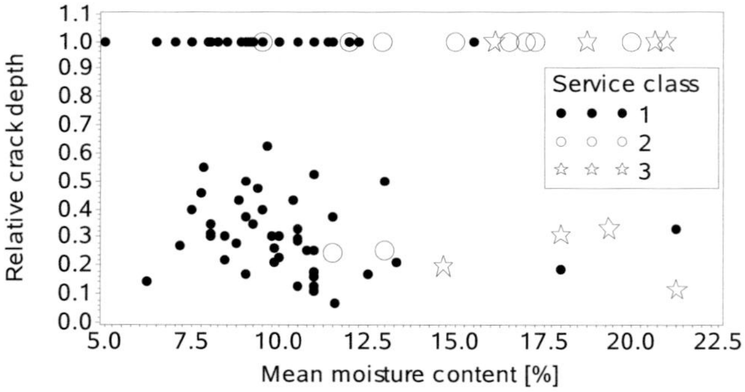

Figure G4-9 Relation between crack depth and moisture content.

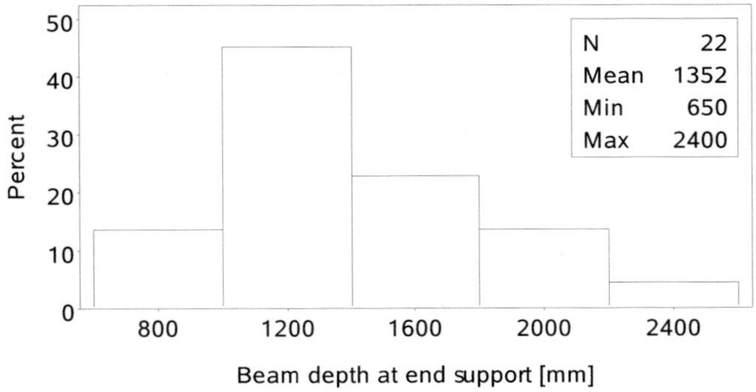

Figure G4-10 Frequency distribution of the depth of glulam members with shear failure at supports.

Table G4-10 shows the distribution of consequences inflicted as a result of the initial damage on the structural safety of the members or constructions. In just under a quarter, structural safety is ***guaranteed*** (including ***still guaranteed***), and ***at risk*** in another quarter. Another quarter succumbed to ***failure*** or ***collapse***. No assessments were available for the remainder. The characteristic values ***guaranteed*** and ***at risk***, the cumulative portions of which amounted to almost 50%, show that after professional inspections, apparently assessments can be made regarding the structural safety of members and constructions. This can be used as the basis for suitable measures, which ensure structural safety in future and extend the service life. The cumulative portion of the assessments ***still guaranteed***, ***threatened*** and ***failure member*** or ***structure*** (54%) shows that a large part of the timber-hall structures and their structural systems meet the basic requirements of structures in the case of failure (EC 0). A distribution without such assessments, which gradually subdivide the predominantly discontinuous transition between safe and collapsing structure, would indicate that structures would become unsafe and collapse without advance notice. As part of efforts to prevent a collapse, the above-specified assessments set out the scope for action within which prompt action can still make a difference (cf. Figure G4-3 on the left). Structures are subject to natural and man-made wear. The inspection, cf. Studiengemeinschaft Holzleimbau e. V. (2015), maintenance and repair of structures, including those made of timber, thus becomes highly significant. Accordingly, the planning must take into consideration the accessibility to load-bearing members for inspections as far as possible. However, the decisive factor dictating the success of a maintenance measure is not only the expertise, but also the independent nature of the individuals and companies commissioned to provide such service.

Table G4-10 Assessments of structural safety.

Structural safety	Number/total	Portion [%]
Guaranteed	58/709	8.2
Still guaranteed	109/709	15.4
At risk	178/709	25.1
Failure member	96/709	13.5
Failure structure	3/709	0.4
Collapse member	16/709	2.3
Collapse construction	41/709	5.8
Not specified	208/709	29.3

For 127 damage cases, the actual year – and for 100 the actual month – is known in which the initial damage relevant to the structural safety took place. These damage cases particularly include cracks parallel to the grain, tensile and shear failures. They occur in relation to internal forces and moments, the values of which generally peak under the impact of snow and accordingly show seasonal dependencies. While in Figure G4-11 above, the years in which damage occurred are uniformly distributed (except 1956), in Figure G4-11 below, there is a far higher likelihood of occurrence during the January to March period than for the remaining months. This therefore accurately reflects the assumption that a dry (internal) climate in the winter and early spring together with the impact of snow is at least part of the trigger for the initial damage affecting structural safety, particularly for cracks parallel to the grain. This fact notwithstanding, in the revised snow load standard of 2005 (DIN 1055-5) or 2010 (NA to EC 1 Part 1-3) – in contrast to the 1975 version (DIN 1055-5) – the characteristic value of the snow load to be applied for higher elevations had significantly increased case-by-case, cf. Schroeter (2007). If those timber-hall structures from the database are isolated for which an effective snow load according to present standards of more than 25% higher would apply than compared to the 1975 version, the portion of members and constructions suffering failure or collapse would be 2 to 3 times as many compared with the other hall structures. In this respect, excessive snow load from today's perspective was previously responsible for failure and collapses in certain cases. Increasing the characteristic values for the snow load at higher elevations is the right step to help eliminate damage concerning structural safety in specific geographic locations. In over 35% of the damage affecting structural safety, at least five years had elapsed between the year of construction and damage event, and even ten in most cases (Figure G4-12). Premature emergence of damage, dominated by cracks parallel to the grain, is what happens after timber is dried e.g. by swift heating, which results in unfavourable residual stresses or tensile stresses perpendicular to the grain.

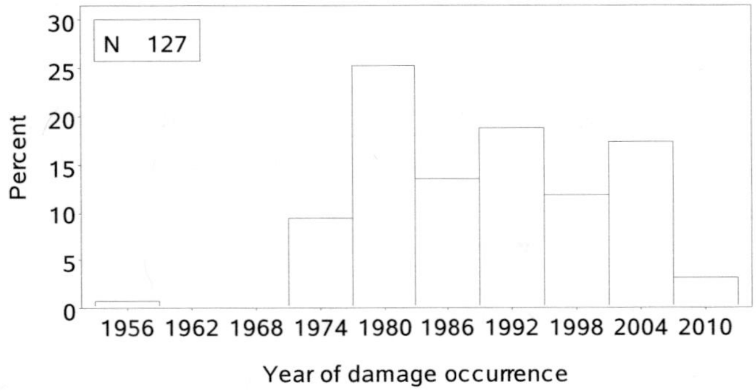

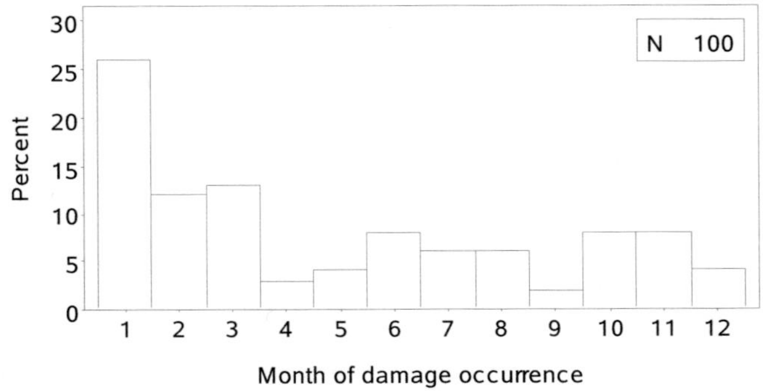

Figure G4-11 Frequency distribution of the year (above) and month (below), in which initial damage affecting structural safety took place.

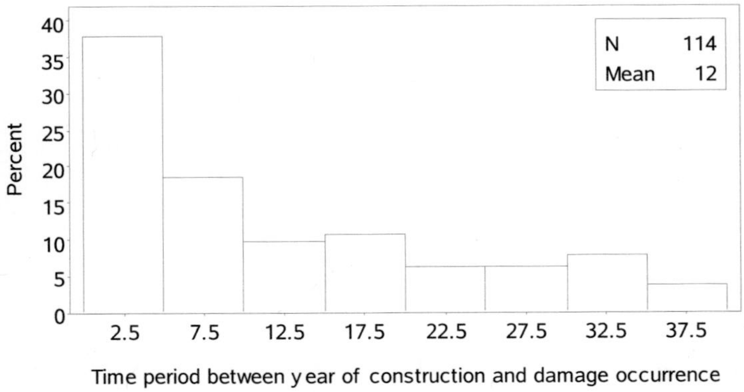

Figure G4-12 Frequency distribution of the time period between the year of construction and the incidence of damage affecting structural safety.

Error sources

A total of 1282 error sources could be allocated to 709 damage incidents. This means that each damage incident tends to be associated with two sources on average. If a causal link is established (cf. Figure G4-13), an error source was allocated without weighting of damage, namely it was either linked completely or not at all. Here too, there is no order of priority within the sources, which would reveal a specific quantitative significance or allow the individual sources to be compared. The individual portions of error sources are shown in Table G4-11. In this table, the key causes for the damage incidents analysed here are shown: The main damage occurs in relation with the ***construction*** (29%). These cases primarily involve incidents where stresses perpendicular to the grain are exerted or deficient detailing for durability becomes apparent. ***Changing climate*** (12.6%), ***swelling*** or ***shrinking*** (9.8%), only observed in isolation, ***load*** (9.5%), ***material quality*** (7.1%), ***planning*** (9.0%), ***building physics*** (6.3%) and ***execution*** (5.2%) are of moderate importance. Unfavourable influences of ***assembly***, ***moisture*** and ***maintenance*** play a minor role.

Figure G4-13 Circumstances, which may lead to damage. Error source planning/execution: torsional restraints at supports not built (top left), error source building physics: possible crack formation in glulam by local heating/drying behind planned glazing (top right), error source assembly: weathering of glulam (lower left) and construction of dowel circles, in which residual stresses are caused by shrinking (lower right).

Table G4-11 Distribution of the general error sources.

Error source	Number/total	Portion [%]
Planning	115/1282	9.0
Execution	67/1282	5.2
Assembly	27/1282	2.1
Building physics	81/1282	6.3
Load	122/1282	9.5
Construction	372/1282	29.0
Material quality	91/1282	7.1
Moisture	27/1282	2.1
Insects	0/1282	0.0
Changing climate	162/1282	12.6
Swelling or shrinking	125/1282	9.8
Maintenance	12/1282	0.9
Not specified	81/1282	6.3

Finally, all links between initial damage and error sources are broken down in Table G4-12. For damage which affects timber structures particularly seriously, such as rot, cracks parallel to the grain as well as shear and tension failure, Table G4-12 allows those error sources to be identified as the main causes, which include the most links:

- Construction (40%) + moisture (20%) → rot
- Construction (34%) + changing climate (16%) → cracks parallel to the grain
- Construction (19%) + load (17%) + material quality (14%) → shear failure
- Load (26%) + material quality (24%) → tension failure

The rounded percentage values in brackets correspond to those in Table G4-12. The criterion for the selection of the four error source combinations was that the sum of their respective portions should each comprise at least 50%, whereby the specified combinations will get the appropriate significance to be the main cause. For example, for rot, 40% of the links relate to the error source ***construction*** and 20% to ***moisture***. For cracks parallel to the grain and shear failures, ***construction*** was involved in 34% or 19% respectively of cases. The causes of damage were thus mainly associated with the selection or design of an unsuitable construction, as typified by deficient planning and detailing. Exceptionally high loads (as defined by the probabilistic safety concept) and poor material quality were the main causes of shear or tension failures recorded.

Consequences

Given the very limited scope for engineers to have a direct influence on exceptionally high loads, e.g. those exerted by snow loads far beyond characteristic values, and on material quality, the key to avoiding damage according to this analysis is linked to improving the construction process and design via technical calculations and considerations. The VDI guideline 2221 (1993) includes cross-industry indications for successful construction The key conclusions, which are individually drawn for those responsible for the construction process, are as follows:

- Designer: Avoiding tension stresses perpendicular to the grain or arranging reinforcements; consideration of detailing for durability and avoiding of exposed spruce glulam; awareness of the unfavourable influence of changing climates and high temperatures on glulam members, more in-depth details included in Möhler and Steck (1980), Häglund (2010) and Gamper et al. (2013); consideration of swelling and shrinking also concerning the anisotropy of the degree of swelling and shrinking artificially generated in glulam; design of structures with only limited scope for damage from impacts beyond standards, closer details, concerning general building construction, in Kersken-Bradley (1992), Pötzl (1996) and Harte et al. (2007); Ensuring accessibility to members of structural systems for inspections as well as consideration of the scope to replace members at risk and those subject to significant wear; planning of monitoring methods (Riedner, 2007; Fellmoser, 2011; Pawlowski et al., 2013); finally measures to offset time pressure during the planning and design phase.
- Glulam manufacturer: Timber drying up to the equilibrium moisture content to be expected in the building, see Gamper et al. (2013); where applicable, surface and especially end grain protection, to delay any exchange of humidity between wood and the surroundings, see Möhler and Steck (1980).
- Building contractor: Careful protection of the wood during the construction phase.

Table G4-12 Initial damage and attributed error sources. Abbreviations in Figure G4-1.

	plan	exec	ass	B phys	load	const	mat qual	moist	clim	sweshr	maint	not spe.+	Σ
def	1	9	2	2	9	2	1	0	0	1	0	1	28
	3.57*	32.1	7.14	7.14	32.1	7.14	3.57	0	0	3.57	0	3.57	
buck	2	2	0	0	3	0	0	0	0	0	0	1	8
	25	25	0	0	37.5	0	0	0	0	0	0	12.5	
moistacc	1	0	1	7	0	2	0	3	0	0	1	0	15
	6.67	0	6.67	46.7	0	13.3	0	20	0	0	6.67	0	
rot	2	1	0	6	1	22	1	11	0	0	7	4	55
	3.64	1.82	0	10.9	1.82	40	1.82	20	0	0	12.7	7.27	
fungi	0	1	0	5	0	0	0	2	0	0	0	0	8
	0	12.5	0	62.5	0	0	0	25	0	0	0	0	
corr	0	2	0	3	0	1	0	3	0	0	0	0	9
	0	22.2	0	33.3	0	11.1	0	33.3	0	0	0	0	
comp	4	0	0	0	1	0	0	0	0	0	0	0	5
perp	80	0	0	0	20	0	0	0	0	0	0	0	
crack	87	29	22	51	51	325	61	4	155	119	1	60	965
par	9.02	3.01	2.28	5.28	5.28	33.7	6.32	0.41	16.1	12.3	0.1	6.22	
shear	7	5	0	4	11	12	9	1	6	5	0	4	64
	10.9	7.81	0	6.25	17.2	18.8	14.1	1.6	9.38	7.81	0	6.25	
tens	8	10	0	3	14	2	13	2	0	0	1	2	55
	14.6	18.2	0	5.45	25.5	3.64	23.6	3.6	0	0	1.82	3.64	
shear or	1	0	0	0	8	3	2	0	1	0	0	0	15
tens	6.67	0	0	0	53.3	20	13.3	0	6.67	0	0	0	
block	0	0	0	0	0	0	0	0	0	0	0	2	2
shear	0	0	0	0	0	0	0	0	0	0	0	100	
unknwn	2	8	2	0	24	3	4	1	0	0	2	7	53
	3.77	15.1	3.77	0	45.3	5.66	7.55	1.89	0	0	3.77	13.2	
Σ#	115	67	27	81	122	372	91	27	162	125	12	81	1282
	8.97	5.23	2.11	6.32	9.52	29	7.1	2.11	12.6	9.75	0.94	6.32	100

Error sources (1st row) and initial damage (1st column); number (top) and percentages (bottom)

* Example: $1/28 \cdot 100 = 3.57\%$

+ Not specified

\# Totals and percentages in Table G4-11

G4.3 Summary

Around 70% of the independent damage incidents analysed here are cracks parallel to the grain. Rot, shear and tension failures respectively each comprise around 6%, while the remaining 12% concern serviceability. Damage is particularly observed in association with constructions subject to stresses perpendicular to the grain. Changing climates meanwhile, which trigger cyclical changes in moisture content, are significant for the formation of cracks parallel to the grain in glulam. The shrinking of entire cross-sections as well as error sources concerning load, material quality, planning, building physics and execution are of minor significance for damage. An even less important role is played by detrimental influences from assembly, maintenance and moisture. To avoid damage during construction, it is imperative to focus more closely on the design, calculation and detailing. Tension stresses perpendicular to the grain, as required for local or global equilibrium within members, should be avoided or accommodated by using reinforcements. The damage analysis supports the following findings in particular: Only glulam members with a large volume subject to shear were prone to shear failures; increasing the characteristic values for snow load in the National Annex of the EC 1 for higher elevations in comparison with the snow load values of 1975 is one correct step to ensure structural safety, independently of any geographical feature.

G4.4 Literature

This article has already been published (in German): Frese M, Grün A.-K. and Blass H.J. (2015). Schäden an Hallentragwerken aus Holz: Beschreibung – Ursachen – Vermeidung. KIT Scientific Working Paper No. 31, ISSN 2194-1629.

Schmitt-Thomas K.G. (2005). Integrierte Schadensanalyse – Technikgestaltung und das System des Versagens. 2. Aufl., Springer-Verlag, Berlin.

Blass H.J. and Frese M. (2010). Schadensanalyse von Hallentragwerken aus Holz. Karlsruher Berichte zum Ingenieurholzbau Band 16, KIT Scientific Publishing, Karlsruhe.

Möhler K. and Steck G. (1980). Untersuchungen über die Rißbildung in Brettschichtholz infolge Klimabeanspruchung. Bauen mit Holz 82:194-200.

Häglund M. (2010). Parameter influence on moisture induced eigenstresses in timber. European Journal of Wood and Wood Products 68:397-406.

Hansson M. and Larsen H.J. (2005). Recent failures in glulam structures and their causes. Engineering Failure Analysis 12:808-818.

Frese M. (2011). Wechselwirkung zwischen der Anisotropie der Schwind- und Quellmaße sowie Holzfeuchteänderungen in der Ebene von gekrümmtem Brettschichtholz. European Journal of Wood and Wood Products 69:359-367.

American Institute of Timber Construction (1994). Timber Construction Manual. 4th Ed., John Wiley & Sons, New York.

Larsen H.J. and Riberholt H. (1983). Trækonstruktioner, Beregning. SBI-Anvisning 135. Statens Byggeforskningsinstitut, Denmark.

Studiengemeinschaft Holzleimbau e.V. (2015). Leitfaden zu einer ersten Begutachtung von Hallentragwerken aus Holz. Online-Publikation, 9.4.2015, http://www.brettschichtholz.de/publish/binarydata/pdfs/aktuelles/stghb_leitfaden-hallentragwerke-2014_print_140218.pdf.

Schroeter H. (2007). Erläuterungen und Beispiele zur Lastnorm DIN 1055 neu. Bautechnik 84:559-571.

VDI-Richtlinie 2221 (1993). Methodik zum Entwickeln und Konstruieren technischer Systeme und Produkte.

Gamper A., Dietsch P., Merk M. and Winter S. (2013). Gebäudeklima – Langzeitmessung zur Bestimmung der Auswirkungen auf Feuchtegradienten in Holzbauteilen. Bautechnik 90:508-519.

Kersken-Bradley M. (1992). Unempfindliche Tragwerke – Entwurf und Konstruktion. Bauingenieur 67:1-5.

Pötzl M. (1996). Robuste Tragwerke - Vorschläge zu Entwurf und Konstruktion. Bauingenieur 71:481-488.

Harte R., Krätzig W.B. and Petryna Y.S. (2007). Robustheit von Tragwerken – ein vergessenes Entwurfsziel? Bautechnik 84:225-234.

Riedner W. (2007). Sicherheit und Überwachung von weitgespannten Hallensystemen. Bautechnik 84:78-80.

Fellmoser P. (2011). Monitoring von Holzkonstruktionen. Bauingenieur 86:541-543.

Pawlowski R., Henke K., Schregle P. and Winter S. (2013). Überwachung von Bauwerksverformungen mittels digitaler Figureverarbeitung. Bauingenieur 88:214-221.

H

Annexes

Annex 1: Dynamic modulus of elasticity

In the following section, exemplarily for the case of longitudinal vibrations, the equation is derived with which the dynamic modulus of elasticity can be calculated from the first Eigenfrequency measured. The equation (1) derived here corresponds to equation (B5-1).

Required fundamental equations, for definitions see Figure A1:

$$\sigma = \frac{F}{A} = E \cdot \varepsilon \;\Rightarrow\; F = A \cdot E \cdot \frac{\partial u}{\partial x}$$

$$F = F(x) - F(x+dx) = \frac{\partial F}{\partial x} \cdot dx = \frac{\partial}{\partial x} \cdot \left(A \cdot E \cdot \frac{\partial u}{\partial x} \right) \cdot dx \tag{a}$$

$$F = m \cdot a = m \cdot \frac{\partial^2 u}{\partial t^2} \tag{b}$$

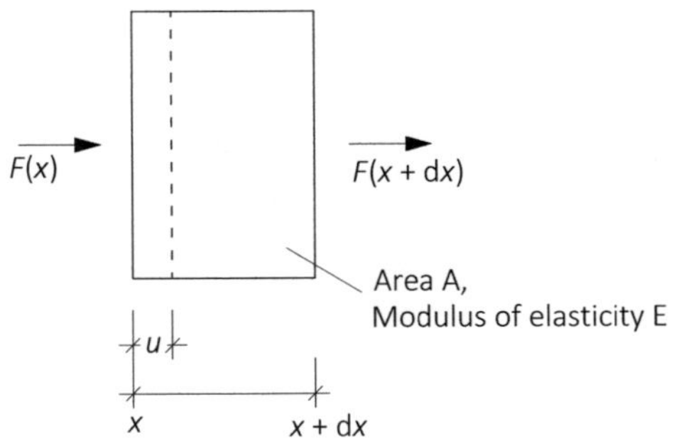

Figure A1 Definitions.

Assumption: Area A and Young's modulus E are constant → no transverse contraction.

Equating (a) and (b):

$$m \cdot \frac{\partial^2 u}{\partial t^2} = \frac{\partial}{\partial x} \cdot \left(A \cdot E \cdot \frac{\partial u}{\partial x} \right) \cdot dx$$

$$\frac{\partial^2 u}{\partial t^2} = \frac{A \cdot dx \cdot E}{m} \cdot \frac{\partial^2 u}{\partial x^2}$$

With $A \cdot dx$ = volume V, density ρ = m/V:

$$\frac{\partial^2 u}{\partial t^2} = \frac{E}{\rho} \cdot \frac{\partial^2 u}{\partial x^2} = c_{long}^2 \cdot \frac{\partial^2 u}{\partial x^2} \tag{c}$$

For long, thin bars (negligible transverse contraction), the following thus applies:

$$c_{long} = \sqrt{\frac{E}{\rho}} \tag{d}$$

where c_{long} = longitudinal sound velocity and the wavelength λ:

$$\lambda = c_{long} \cdot T \qquad \text{where} \quad T = \text{period} \tag{e}$$

Now, the case of resonance is considered:

$$\text{resonance} \Leftrightarrow \text{bar length } \ell = n \cdot \frac{\lambda}{2} \qquad n = 1,2,3,\ldots \tag{f}$$

From equations (e) and (f), the following first mode of vibration emerges:

$$T = \frac{\lambda}{c_{long}} = \frac{2 \cdot \ell}{n \cdot c_{long}} \quad \Leftrightarrow \quad c_{long} = \frac{2 \cdot \ell}{n \cdot T}$$

Inserting in equation (d) with $T = 1/f$:

$$\frac{2 \cdot \ell}{T} = \sqrt{\frac{E}{\rho}} \quad \Leftrightarrow \quad \frac{4 \cdot \ell^2}{T^2} = \frac{E}{\rho} \quad \Leftrightarrow \quad \boxed{E = 4 \cdot \ell^2 \cdot f^2 \cdot \rho} \tag{1}$$

Annex 2: Stress interactions

Multiaxial stresses

The anisotropy of the wood means stresses exerted at an angle to the grain lead to multiaxial stresses. This can be seen in Figure A2, where compression σ_{compr} exerted at an angle α to the grain generates normal stresses parallel and perpendicular to the grain $\sigma_{parallel}$ and $\sigma_{perpendicular}$ and shear loads τ. In EC 5, this fact is taken into consideration via interaction equations that depend on the angle α (see e.g. equations (D1-2) or (D1-6)). The following section shows the origin of these equations, to verify the equations given in EC 5 and improve understanding of their mechanical background. The basis is the assumption of a linear interaction criterion (see also Figure D1-7).

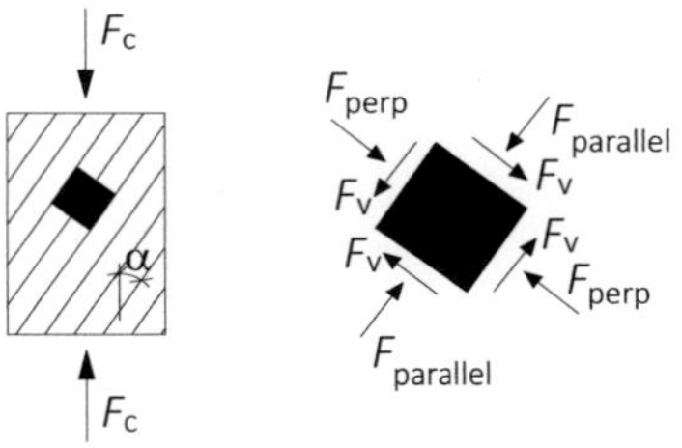

Figure A2 Multiaxial stresses in a specimen compressed at an angle to the grain.

Derivation Hankinson equation

In an initial step, only normal stresses are taken into consideration: $F_v = \tau = 0$.

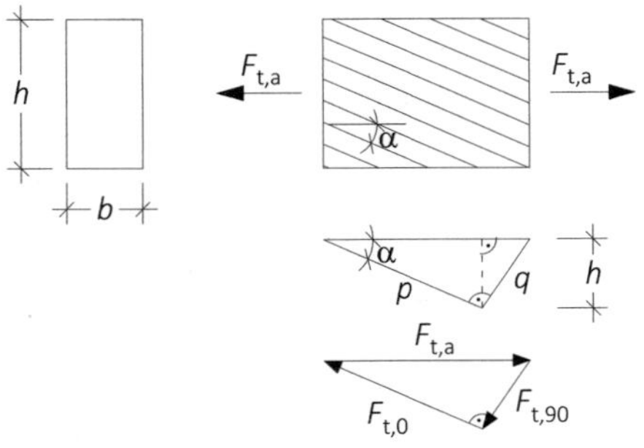

Figure A3 A piece of wood tensioned at an angle α to the grain.

Trigonometric relationships (Figure A3):

$$\sin\alpha = \frac{h}{p} \rightarrow p = \frac{h}{\sin\alpha} \tag{a}$$

$$\cos\alpha = \frac{h}{q} \rightarrow q = \frac{h}{\cos\alpha} \tag{b}$$

$$\sin\alpha = \frac{F_{\mathrm{t,90}}}{F_{\mathrm{t,\alpha}}} \rightarrow F_{\mathrm{t,\alpha}} = \frac{F_{\mathrm{t,90}}}{\sin\alpha} \tag{c}$$

$$\cos\alpha = \frac{F_{\mathrm{t,0}}}{F_{\mathrm{t,\alpha}}} \rightarrow F_{\mathrm{t,\alpha}} = \frac{F_{\mathrm{t,0}}}{\cos\alpha} \tag{d}$$

From forces to stresses:

$$\sigma_{\mathrm{t,0}} = \frac{F_{\mathrm{t,0}}}{b \cdot q} \overset{(b)}{=} \frac{F_{\mathrm{t,0}} \cdot \cos\alpha}{b \cdot h} \rightarrow F_{\mathrm{t,0}} = \frac{\sigma_{\mathrm{t,0}} \cdot b \cdot h}{\cos\alpha} \tag{i}$$

$$\sigma_{\mathrm{t,90}} = \frac{F_{\mathrm{t,90}}}{b \cdot p} \overset{(a)}{=} \frac{F_{\mathrm{t,90}} \cdot \sin\alpha}{b \cdot h} \rightarrow F_{\mathrm{t,90}} = \frac{\sigma_{\mathrm{t,90}} \cdot b \cdot h}{\sin\alpha} \tag{ii}$$

$$\sigma_{\mathrm{t,\alpha}} = \frac{F_{\mathrm{t,\alpha}}}{b \cdot h} \overset{(c)\,(d)}{=} \begin{cases} \dfrac{F_{\mathrm{t,0}}}{b \cdot h \cdot \cos\alpha} & \text{(iii)} \\ \dfrac{F_{\mathrm{t,90}}}{b \cdot h \cdot \sin\alpha} & \text{(iv)} \end{cases}$$

Failure criterion – assumption of a linear interaction without shear stresses:

$$\frac{\sigma_{\mathrm{t,0}}}{f_{\mathrm{t,0}}} + \frac{\sigma_{\mathrm{t,90}}}{f_{\mathrm{t,90}}} = 1 \tag{I}$$

Inserting:

(i) in (iii): $$\sigma_{\mathrm{t,\alpha}} = \frac{\sigma_{\mathrm{t,0}}}{\cos^2\alpha} \rightarrow \sigma_{\mathrm{t,0}} = \sigma_{\mathrm{t,\alpha}} \cdot \cos^2\alpha \tag{II}$$

(ii) in (iv): $$\sigma_{\mathrm{t,\alpha}} = \frac{\sigma_{\mathrm{t,90}}}{\sin^2\alpha} \rightarrow \sigma_{\mathrm{t,90}} = \sigma_{\mathrm{t,\alpha}} \cdot \sin^2\alpha \tag{III}$$

(II) + (III) in (I):

$$\frac{f_{\mathrm{t,\alpha}} \cdot \cos^2\alpha}{f_{\mathrm{t,0}}} + \frac{f_{\mathrm{t,\alpha}} \cdot \sin^2\alpha}{f_{\mathrm{t,90}}} = 1 \quad \text{where} \quad \sigma_{\mathrm{t,\alpha}} = f_{\mathrm{t,\alpha}} \tag{2}$$

The equation (2) thus obtained is now solved for $f_{t,\alpha}$:

$$f_{t,\alpha} = \frac{f_{t,0} \cdot f_{t,90}}{f_{t,0} \cdot \sin^2 \alpha + f_{t,90} \cdot \cos^2 \alpha} = \frac{f_{t,0}}{\frac{f_{t,0}}{f_{t,90}} \cdot \sin^2 \alpha + \cos^2 \alpha} \tag{3}$$

Equation (3) is the so-called Hankinson equation. The derivation applies to compressive or tensile stress; namely the index "t" for tension can be replaced with the index "c" for compression. Accordingly, Equation (D1-1) corresponds to the central expression in equation (3), while the last expression is in EC 5 in the interaction Equation (D1-6), where the coefficient $k_{c,90}$ is considered additionally.

Extension with shear

The following expression is added (for definitions see Figure A3):

$$\tau = \frac{F_{t,0}}{p \cdot b} = \frac{F_{t,\alpha} \cdot \cos\alpha}{b \cdot \frac{h}{\sin\alpha}} = \frac{F_{t,\alpha} \cdot \cos\alpha \cdot \sin\alpha}{b \cdot h} = \frac{(\sigma_{t,\alpha} \cdot b \cdot h) \cdot \cos\alpha \cdot \sin\alpha}{b \cdot h} = \sigma_{t,\alpha} \cdot \cos\alpha \cdot \sin\alpha \tag{IV}$$

Once again, assuming a linear interaction:

$$\frac{\sigma_{t,0}}{f_{t,0}} + \frac{\sigma_{t,90}}{f_{t,90}} + \frac{\tau}{f_v} = 1$$

Inserting (I), (II) and (IV)

$$\frac{f_{t,\alpha} \cdot \cos^2 \alpha}{f_{t,0}} + \frac{f_{t,\alpha} \cdot \sin^2 \alpha}{f_{t,90}} + \frac{f_{t,\alpha} \cdot \cos\alpha \cdot \sin\alpha}{f_v} = 1$$

and solving (where $\sigma_{t,\alpha} = f_{t,\alpha}$):

$$f_{t,\alpha} = \frac{f_{t,0}}{\cos^2 \alpha + \frac{f_{t,0}}{f_{t,90}} \cdot \sin^2 \alpha + \frac{f_{t,0}}{f_v} \cdot \cos\alpha \cdot \sin\alpha} \tag{4}$$

Equation (4) is used in Equation (D1-2) to verify tension at an angle α to the grain.

Quadratic interaction

If a quadratic rather than linear interaction is now assumed, equation (4) changes as follows:

$$\left(\frac{\sigma_{t,0}}{f_{t,0}}\right)^2+\left(\frac{\sigma_{t,90}}{f_{t,90}}\right)^2+\left(\frac{\tau}{f_v}\right)^2=1 \qquad \text{(quadratic interaction)}$$

$$\frac{f_{t,\alpha}^2\cdot\cos^4\alpha}{f_{t,0}^2}+\frac{f_{t,\alpha}^2\cdot\sin^4\alpha}{f_{t,90}^2}+\frac{f_{t,\alpha}^2\cdot\cos^2\alpha\cdot\sin^2\alpha}{f_v^2}=1 \qquad \text{(inserting (I), (II), (IV))}$$

Solving for $f_{t,\alpha}$:

$$f_{t,\alpha}=\frac{f_{t,0}}{\sqrt{\cos^4\alpha+\left(\frac{f_{t,0}}{f_{t,90}}\right)^2\cdot\sin^4\alpha+\left(\frac{f_{t,0}}{f_v}\right)^2\cdot\cos^2\alpha\cdot\sin^2\alpha}} \qquad (5)$$

Equation (5) is reflected in the verification of a step joint, equation (E9-3) (with compressive rather than tensile strengths). The difference between both equations exists in the form of factor 2, with which compression strength perpendicular to the grain and shear strength in equation (E9-3) is increased. This factor 2 is empirical in nature. Tests conducted with step joints show that the quadratic interaction criterion in accordance with equation (5) with doubled compression strength perpendicular to the grain and doubled shear strength correlates better with the test results.

Case of a tapered edge

In Figure A3, the stress used for derivations was parallel to the edge. In the case of tapered edges, however, the stress is parallel or perpendicular to the grain, as in Figure A4.

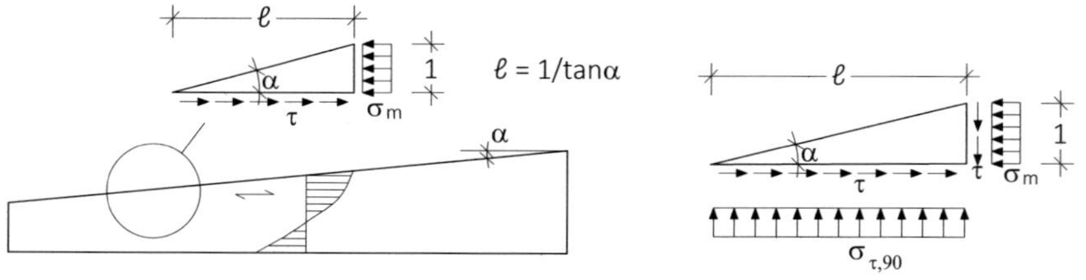

Figure A4 Tapered edge with stresses parallel and perpendicular to the grain. Left: Internal forces and moments at the considered infinitesimal element. Right: Shear stresses are at right angles to each other and of equivalent size, which means that due to equilibrium conditions, tension perpendicular to the grain $\sigma_{t,90}$ is developing.

The equilibrium of forces in both horizontal and vertical directions results in:

$$\Sigma H = 0 = \sigma_{\mathrm{m}} \cdot 1 - \tau \cdot \ell = \sigma_{\mathrm{m}} - \frac{\tau}{\tan\alpha} \quad \Rightarrow \quad \tau = \sigma_{\mathrm{m}} \cdot \tan\alpha \tag{x}$$

$$\Sigma V = 0 = \tau \cdot 1 - \sigma_{\mathrm{t,90}} \cdot \ell = \tau - \frac{\sigma_{\mathrm{t,90}}}{\tan\alpha} \quad \Rightarrow \quad \sigma_{\mathrm{t,90}} = \tau \cdot \tan\alpha \tag{y}$$

Inserting (x) in (y):

$$\sigma_{\mathrm{t,90}} = \sigma_{\mathrm{m}} \cdot \tan\alpha \cdot \tan\alpha = \sigma_{\mathrm{m}} \cdot \tan^2\alpha \tag{z}$$

Inserting in a quadratic interaction with shear:

$$\left(\frac{\sigma_{\mathrm{m}}}{f_{\mathrm{m}}}\right)^2 + \left(\frac{\sigma_{\mathrm{t,90}}}{f_{\mathrm{t,90}}}\right)^2 + \left(\frac{\tau}{f_{\mathrm{v}}}\right)^2 = \left(\frac{\sigma_{\mathrm{m}}}{f_{\mathrm{m}}}\right)^2 + \left(\frac{\sigma_{\mathrm{m}} \cdot \tan^2\alpha}{f_{\mathrm{t,90}}}\right)^2 + \left(\frac{\sigma_{\mathrm{m}} \cdot \tan\alpha}{f_{\mathrm{v}}}\right)^2 = 1$$

And solving for $\sigma_{\mathrm{t,0}}$:

$$\sigma_{\mathrm{m}}^2 \cdot \left(\frac{1}{f_{\mathrm{m}}^2} + \frac{\tan^4\alpha}{f_{\mathrm{t,90}}^2} + \frac{\tan^2\alpha}{f_{\mathrm{v}}^2}\right) = 1 \quad \Rightarrow$$

$$\sigma_{\mathrm{m}} = \frac{1}{\sqrt{\frac{1}{f_{\mathrm{m}}^2} + \frac{\tan^4\alpha}{f_{\mathrm{t,90}}^2} + \frac{\tan^2\alpha}{f_{\mathrm{v}}^2}}} = \frac{1}{\sqrt{\frac{1}{f_{\mathrm{m}}^2} \cdot \left(1 + \frac{f_{\mathrm{m}}^2}{f_{\mathrm{t,90}}^2} \cdot \tan^4\alpha + \frac{f_{\mathrm{m}}^2}{f_{\mathrm{v}}^2} \cdot \tan^2\alpha\right)}}$$

Additional transformations result in:

$$\sigma_{\mathrm{m}} = \frac{f_{\mathrm{m}}}{\sqrt{1 + \left(\frac{f_{\mathrm{m}}}{f_{\mathrm{t,90}}}\right)^2 \cdot \tan^4\alpha + \left(\frac{f_{\mathrm{m}}}{f_{\mathrm{v}}}\right)^2 \cdot \tan^2\alpha}} \tag{6}$$

Equation (6) is used to verify beams with a tapered edge, see equation (4-5) with equations (D4-6) and (D4-7). There too, additional factors are used to minimise or increase shear strength. For tensile stresses along the tapered edge, the factor is 0.75. Therefore, only 75% of the shear strength is considered, since the shear strength declines with simultaneously applied tension perpendicular to the grain (Equation (D4-6)). For simultaneously applied compression perpendicular to the grain, the opposite applies and shear strength increases. This is reflected in a factor of 1.5 for the shear strength, which is specified in the case of compressive stresses along the tapered edge (Equation (D4-7)).

Annex 3: Elastic buckling load and buckling lengths

Derivation of elastic buckling load for Euler case II

The elastic buckling loads for the four Euler cases shown in Figure D2-7 can be analytically derived. This is shown as an example for the Euler case II, which is shown in Figure A5; a slender column hinged at both ends.

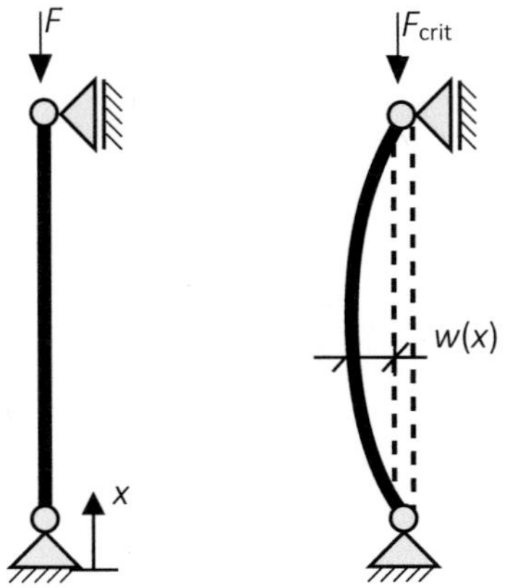

Figure A5 Column hinged at both ends and of length L, $E \cdot I$ = const. (STEP 1995 Article B6)

For the column shown in Figure A5 both the following equations apply:

$$M(x) = F \cdot w(x) \tag{a}$$

$$\frac{M}{E \cdot I} = -w''(x) \quad \text{(differential equation of deflection curve)} \tag{b}$$

In this case, it is assumed that the material concerned is homogeneous and elastic and the load is applied at the centre of gravity. Now, (a) is inserted in (b):

$$\frac{F \cdot w(x)}{E \cdot I} = -w''(x) \quad \Leftrightarrow \quad w''(x) + \frac{F}{E \cdot I} \cdot w(x) = 0$$

Substitution with $\alpha^2 = F/(E \cdot I)$ and the base function for the differential equation:

$$w(x) = A \cdot \cos(\alpha \cdot x) + B \cdot \sin(\alpha \cdot x)$$

With the boundary conditions w(x = 0) = w(x = L) = 0 it follows that:

$$w(x=0)=A\cdot 1+0=0 \Rightarrow A=0$$
$$w(x=L)=0+B\cdot\sin(\alpha\cdot L)=0 \Leftrightarrow \sin(\alpha\cdot L)=0 \Leftrightarrow \alpha\cdot L=n\cdot\pi \quad \text{mit} \quad n=1,2,3,\ldots$$

The smallest buckling load F_{crit} is thus obtained for $n = 1$, namely for $\alpha = \pi/L$. The reverse substitution yields the critical buckling load F_{crit} of the Euler case II:

$$\alpha^2=\frac{F}{E\cdot I}=\frac{\pi^2}{L^2} \Rightarrow F=F_{crit}=\frac{\pi^2\cdot E\cdot I}{L^2} \tag{7}$$

Equation (7) corresponds to equation (D2-19).

Buckling lengths of some structural systems

Connected columns

If columns hinged at both ends are braced by a column with a clamped support, the critical buckling load of the clamped column is reduced by the normal forces N_i in the hinged columns, since they generate horizontal forces in the deformed system. Considering the effect of rotation in the semi-rigid joint at the column base, the effective length factor β for buckling in the system plane (see Figure A6) is approximately:

$$\beta=\sqrt{\left(4+\frac{\pi^2\cdot E\cdot I}{\ell_r\cdot K_r}\right)\cdot(1+\alpha)}$$

In this case, α is specified in Figure A6. For the columns hinged at both ends, meanwhile, a buckling verification must be performed with a buckling length corresponding to their real length.

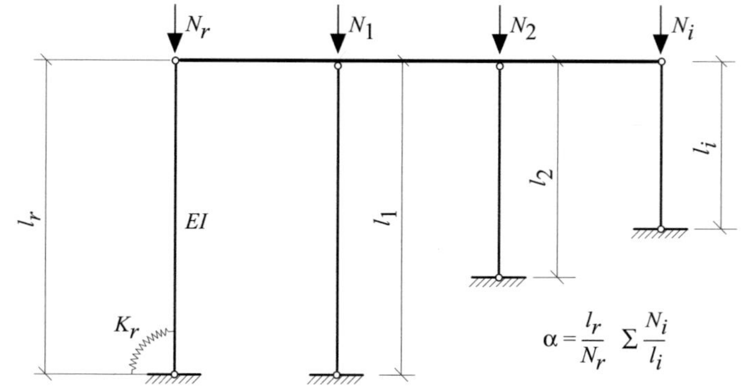

Figure A6 Connected columns. (STEP 1995 Article B7)

Arches

For three- and two-hinged arches (see Figure A7) with ratios h/ℓ between 0.15 and 0.5 and an essentially constant cross-section, the effective length for buckling in the arch plane may be assumed at:

$$\ell_{ef} = \beta \cdot s = 1.25 \cdot s$$

whereby s is equivalent to half the arch length. For the buckling verification, the normal force in the quarter point should be used.

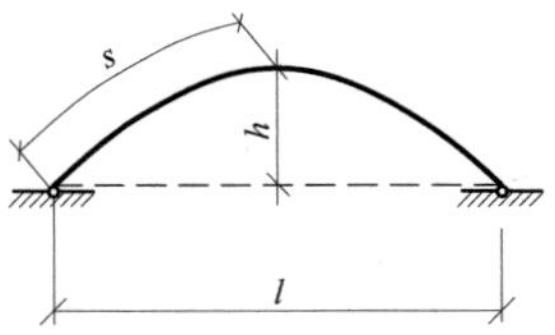

Figure A7 Two-hinged arch. (STEP 1995 Article B7)

Two- and three-hinged frames

For two- and three-hinged frames with normal forces N_S and N_R in columns (N_S) or rafters (N_R) and angles of inclination of $\alpha_S \leq 15°$ (see Figure A8), the following approximate equation can be applied for the buckling length of columns:

$$\ell_{ef} = \beta_S \cdot h \quad \text{where} \quad \beta_S = \sqrt{4 + \frac{\pi^2 \cdot E \cdot I_S}{h} \cdot \left(\frac{1}{K_\Phi} + \frac{s}{3 \cdot E \cdot I_R} \right) + \frac{E \cdot I_S \cdot N_R \cdot s^2}{E \cdot I_R \cdot N_S \cdot h^2}}$$

The corresponding buckling length of the rafter (if $\alpha_R \leq 20°$) is:

$$\ell_{ef} = \beta_R \cdot s \quad \text{where} \quad \beta_R = \beta_S \cdot \sqrt{\frac{E \cdot I_R \cdot N_S}{E \cdot I_S \cdot N_R}} \cdot \frac{h}{s}$$

If the second moments of area change (tapered rafters or columns), the cross-sections at $0.65 \cdot s$ or $0.65 \cdot h$ can be inserted in the above equations (for geometric definitions, see Figure A8).

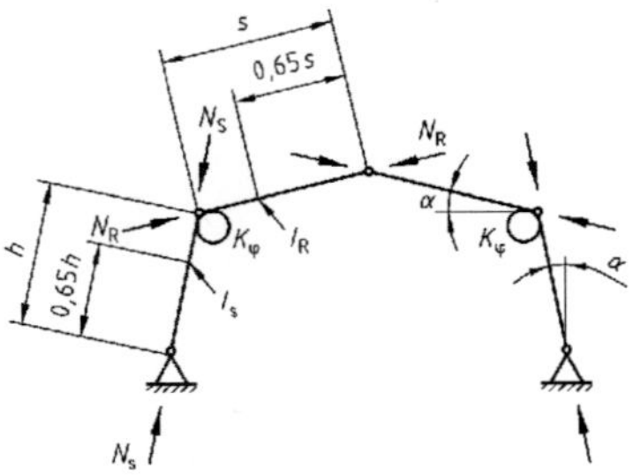

Figure A8 Three-hinged frame.

Columns or rafters with knee bracing

The effective lengths of the columns shown in Figure A9 on the left and of the rafters shown in Figure A9 on the right for buckling in the frame plane can be estimated as:

$$\ell_{ef} = 2 \cdot s_l + 0.7 \cdot s_o$$

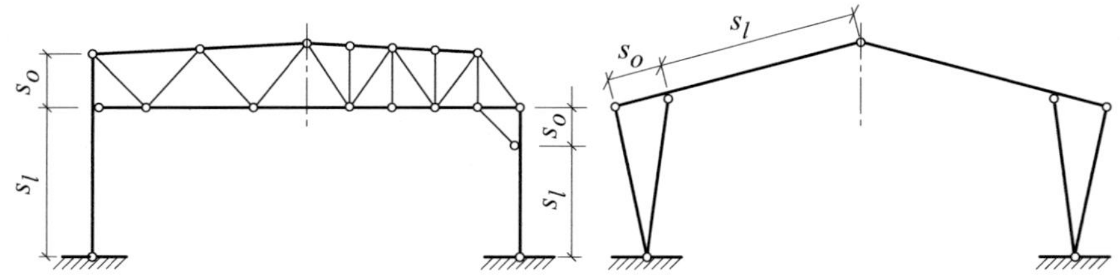

Figure A9 Frame with truss as rafter (left) and three-hinged frame with V-shaped columns (right). (STEP 1995 Article B7)

Torsional buckling of spatial frames

For rotationally symmetrical structures (see Figure A10), two types of buckling are to be examined in principle. As well as buckling within the plane of the half-frame, rotational buckling of the spatial structure represents one possible type of stability failure. The latter is characterised by a rotation of the compression ring about the vertical axis of symmetry. For $1 < \beta < 2$ and $a/s < 0.2$, the following approximate solution applies for the effective length factor β for the rafter:

$$\beta = \sqrt{1 + \frac{2 \cdot a}{s} + \frac{3 \cdot \pi^2 \cdot a \cdot E \cdot I}{4 \cdot s^2 \cdot \left(1 + a/s\right) \cdot K_r}}$$

Here, $E{\cdot}I$ is the bending stiffness of the rafter about the vertical axis and K_r the rotational stiffness of the connection between the compression ring and rafter, equally for bending about the vertical axis. Similar to the procedure for two- and three-hinged frames with variable second moments of area, the cross-sections at $0.65 \cdot s$ can be inserted if the rafters are tapered.

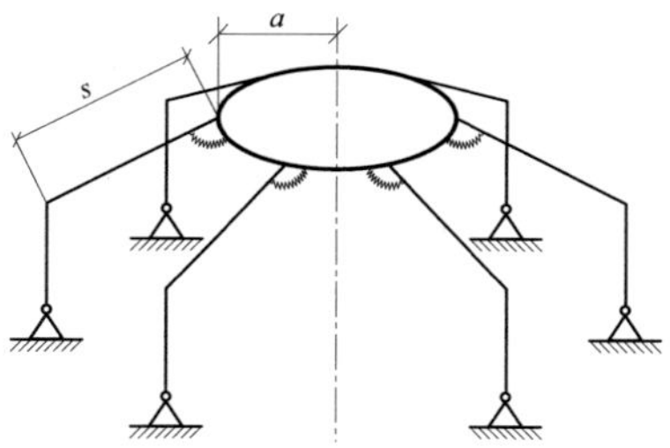

Figure A10 Rotationally symmetrical spatial frame. (STEP 1995 Article B7)

Annex 4: Derivations lateral torsional buckling

Critical moment M_{crit}

The equation used in EC 5 for the critical moment can be analytically derived. This is shown in the following section, whereby the following assumptions are important for understanding:

- The beam is subject to a constant moment and elastic (not plastic).
- The beam has a constant rectangular cross-section and constant stiffness.
- The beam cannot rotate at the supports (torsional restraints).
- The load application is in the centre of gravity of the beam cross-section.
- The beam cross-section does not warp and any twist ϕ is minimal.

Equations Euler-Bernoulli beam theory:

$$\frac{dQ}{dx}=-q,\quad \frac{dM}{dx}=Q,\quad \psi'=\frac{M}{E\cdot I},\quad -\psi=w',\quad -w''=\frac{M}{E\cdot I}$$

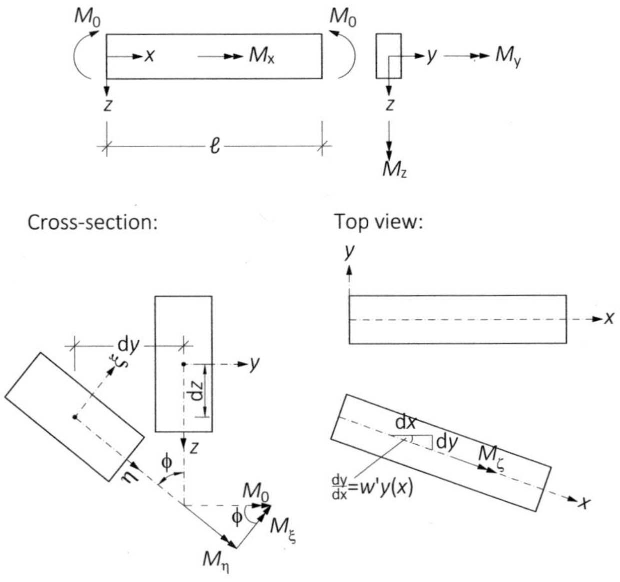

Figure A11 Definitions of laterally deviated beam with rectangular cross-section and constant moment M_0.

From Figure A11, the following relations can be derived, assuming a small angle of twist ϕ:

$$M_y : \ w''_z(x) = -\frac{M_\xi}{E \cdot I_y} \tag{a}$$

$$M_z : \ w''_y(x) = -\frac{M_\eta}{E \cdot I_z} \tag{b}$$

$$M_x(\text{no warping}) : \ \phi'(x) = \frac{M_\zeta}{G \cdot I_T} \tag{c}$$

$$\sin\phi = \frac{M_\eta}{M_0} \approx \phi \ \Rightarrow \ M_\eta = \phi \cdot M_0 \tag{d}$$

$$\cos\phi = \frac{M_\xi}{M_0} \approx 1 \ \Rightarrow \ M_\xi = M_0 \tag{e}$$

$$M_\zeta = w'_y(x) \cdot M_0 \tag{f}$$

insert (d), (e), (f) in (a), (b), (c):

$$w''_z(x) = -\frac{M_0}{E \cdot I_y} \tag{i}$$

$$w''_y(x) = -\frac{M_0}{E \cdot I_z} \cdot \phi \tag{ii}$$

$$\phi'(x) = \frac{M_0}{G \cdot I_T} \cdot w'_y(x) \tag{iii}$$

Differentiate equation (iii) with respect to *x*:

$$\phi''(x) = \frac{M_0}{G \cdot I_T} \cdot w''_y(x)$$

and insert in equation (ii):

$$\frac{G \cdot I_T}{M_0} \cdot \phi''(x) + \frac{M_0}{E \cdot I_z} \cdot \phi(x) = \phi''(x) + \frac{M_0^2}{E \cdot I_z \cdot G \cdot I_T} \cdot \phi(x) = 0$$

Now, solve the differential equations:

Initial value problem: $\phi(x = 0, x = \ell) = 0$ (→ torsional restraints and no twist)

Be: $\phi = A \cdot \sin(\alpha \cdot x) + B \cdot \cos(\alpha \cdot x)$ where $\alpha^2 = \dfrac{M_0^2}{E \cdot I_z \cdot G \cdot I_T}$ (iv)

$\phi(0) = A \cdot 1 + 0 = 0 \;\rightarrow\; A = 0$

$\phi(\ell) = B \cdot \cos(\alpha \cdot \ell) = 0 \;\leftrightarrow\; \cos(\alpha \cdot \ell) = 0 \;\leftrightarrow\; \alpha \cdot \ell = n \cdot \pi \;\leftrightarrow\; \alpha = (n \cdot \pi)/\ell$

Key criterion (smallest critical moment): $n = 1 \;\rightarrow\; \alpha = \pi/\ell$

Insert in (iv):

$$\alpha^2 = \frac{M_0^2}{E \cdot I_z \cdot G \cdot I_T} = \frac{\pi^2}{\ell^2}$$

Solve for $M_0 = M_{crit}$:

$$M_0^2 = \frac{\pi^2 \cdot E \cdot I_z \cdot G \cdot I_T}{\ell^2} \Rightarrow \boxed{M_{crit} = \frac{\pi}{\ell} \cdot \sqrt{E \cdot I_z \cdot G \cdot I_T}} \qquad (8)$$

Equation (8) corresponds precisely to the equation specified in EC 5 for M_{crit}, see also Equation (D2-21), which therefore only applies for systems under the requirements initially specified.

Textbooks such as Timoshenko (S.P. Timoshenko (1961). Theory of elastic stability. McGraw-Hill, New York.) provide examples of critical moments for other systems.

Derivation of the equation for k_{crit}

With lateral torsional buckling, other than bending about the major principal axis, bending about the minor principal axis and torsion also take place. The normal stresses can be taken into consideration here with a linear interaction:

$$\frac{M_y}{W_y \cdot f_{m,y}} + \frac{M_z}{W_z \cdot f_{m,z}} = \frac{\sigma_y}{f_{m,y}} + \frac{\sigma_z}{f_{m,z}} \leq 1$$

Transformation and definition k_{crit}:

$$\frac{\sigma_y}{f_{m,y}} \leq 1 - \frac{\sigma_z}{f_{m,z}} = 1 - \frac{M_z}{W_z \cdot f_{m,z}} = k_{crit} \qquad (9)$$

Design format:

$$\frac{\sigma_y}{k_{crit} \cdot f_{m,y}} \leq 1 \triangleq \frac{M_y}{k_{crit} \cdot W_y} \leq f_{m,y} \Rightarrow M_y = k_{crit} \cdot W_y \cdot f_{m,y} \qquad (10)$$

Using equations (9) and (10), coefficient k_{crit} can now be derived.

The base function in this case reads as follows:

$$M_z = \frac{\pi^2 \cdot E \cdot I_z}{\ell^2} \cdot \frac{\left(\frac{M_y}{M_{crit}}\right)^2}{1-\left(\frac{M_y}{M_{crit}}\right)^2} \cdot e \qquad (11)$$

The initial factor corresponds to the Euler buckling load about the z-axis (see equation (7)), the second factor is the so-called Dischinger factor. M_{crit} corresponds to the critical moment from equation (8) and e is the initial deformation about the minor axis of the system. The background to equation (11) lies in second order theory, its derivation is complex and will not be explained here in further detail[5].

Equations (11) and (10) are now inserted in equation (9):

$$k_{crit} = 1 - \frac{M_z}{W_z \cdot f_{m,z}} = 1 - \frac{\pi^2 \cdot E \cdot I_z}{\ell^2} \cdot \frac{\left(\frac{k_{crit} \cdot W_y \cdot f_{m,y}}{M_{crit}}\right)^2}{1-\left(\frac{k_{crit} \cdot W_y \cdot f_{m,y}}{M_{crit}}\right)^2} \cdot \frac{e}{W_z \cdot f_{m,z}}$$

Now, $\sigma_{crit} = M_{crit}/W_y$ is inserted:

$$k_{crit} = 1 - \frac{e \cdot \frac{\pi^2 \cdot E \cdot I_z}{\ell^2}}{W_z \cdot f_{m,z}} \cdot \frac{\frac{k_{crit}^2 \cdot f_{m,y}^2}{\sigma_{crit}^2}}{1-\frac{k_{crit}^2 \cdot f_{m,y}^2}{\sigma_{crit}^2}}$$

[5] Any readers who are interested can study the derivation in textbooks. A detailed derivation is also included in "O.P. Hörsting (2008). Zum Tragverhalten druck- und biegebeanspruchter Holzbauteile. Dissertation Universität Braunschweig (in German)"; equation (2.158) given there corresponds to equation (11) if initial rotation of the beam axis is disregarded.

The relative slenderness $\lambda_{rel}^2 = f_{m,y} / \sigma_{crit}$ is introduced:

$$k_{crit} = 1 - \frac{e \cdot \frac{\pi^2 \cdot E \cdot I_z}{\ell^2} \cdot k_{crit}^2 \cdot \lambda_{rel}^4}{W_z \cdot f_{m,z} \cdot \left(1 - k_{crit}^2 \cdot \lambda_{rel}^4\right)}$$

Now, multiply and substitute the expression in brackets for a better overview:

$$k_{crit}^3 \cdot \lambda_{rel}^4 - k_{crit}^2 \cdot \lambda_{rel}^4 \cdot \left(1 + \frac{e \cdot \pi^2 \cdot E \cdot b}{2 \cdot \ell^2 \cdot f_{m,z}}\right) - k_{crit} + 1 = k_{crit}^3 \cdot \lambda_{rel}^4 - a \cdot k_{crit}^2 \cdot \lambda_{rel}^4 - k_{crit} + 1 = 0$$

This equation is solved for λ_{rel} to be able to draw it. In EC 5, the resulting curve was simplified, see Figure A12 (for GL24h with $f_{m,z}$ = 24 MPa, $E_{0,g,05}$ = 9400 MPa, e/ℓ = 1/500 and $b = \ell /40$):

$$\lambda_{rel} = \sqrt[4]{\frac{k_{crit} - 1}{k_{crit}^3 - a \cdot k_{crit}^2}} \tag{12}$$

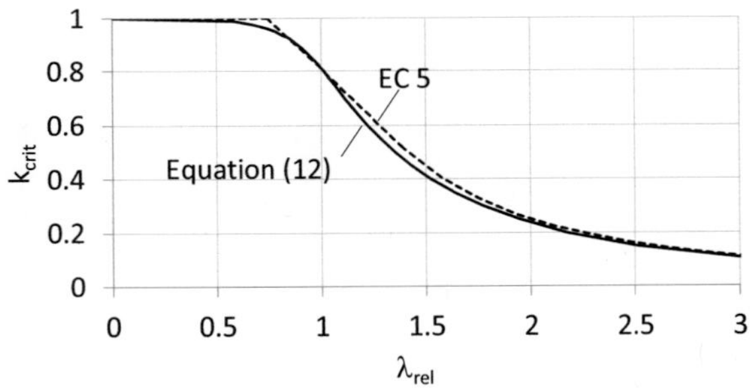

Figure A12 Coefficient k_{crit} depending on the relative slenderness λ_{rel}. See also Figure D2-10.

Annex 5: Mechanically jointed beams

Derivation γ-method

The derivation of the differential equations shown here is performed exemplarily for a T-cross-section comprising two components, see Figure A13. The following prerequisites apply: For each cross-section, Euler-Bernoulli beam theory applies and shear deformations are disregarded. The connection is assumed to be spread out along the axis of the beam, over which the cross-sections and joint stiffnesses are constant.

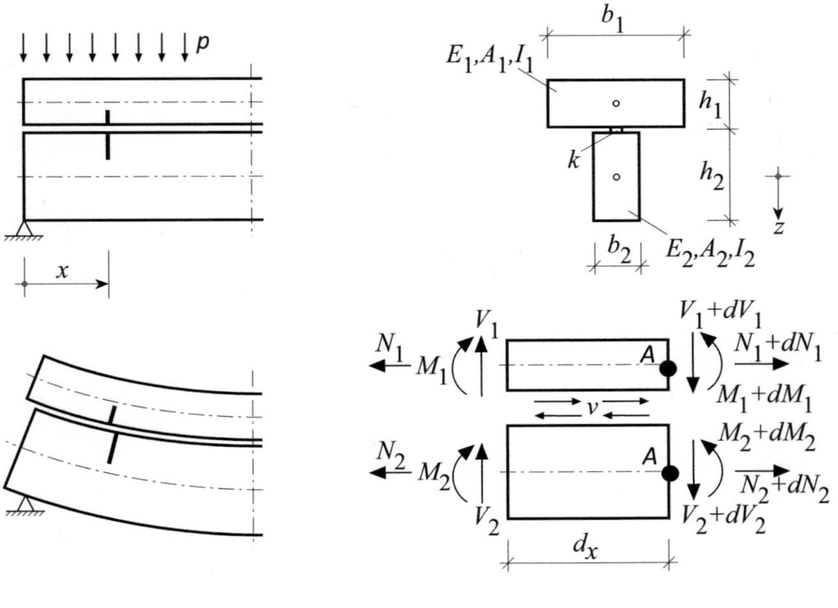

Figure A13 Mechanically jointed cross-section comprising two components: Dimensions, forces and moments. $A' = dA/dx$. (STEP 1995 Article B11)

Deformation relationships (see Figure A14):

$$u = u_2 - u_1 + w' \cdot \left(\frac{h_1}{2} + \frac{h_2}{2} \right) = u_2 - u_1 + w' \cdot a$$

u_1, u_2 Longitudinal displacements of cross-sections 1 and 2

w Common deflection

u Relative displacement of the individual cross-sections at the position of the fasteners

Here, we see that u is independent of the fastener position, while the distance of the axes a of the cross-sections is decisive instead. The derived equations thus apply not only to cross-sections arranged vertically, like the T-cross-section here, but also for those arranged adjacent to each other. However, this only applies if we disregard shear deformations of the individual cross-sections.

Elasticity relations according to beam theory ($'$ = differentiation with regard to x):

$$N_1 = E_1 \cdot A_1 \cdot u_1' \qquad N_2 = E_2 \cdot A_2 \cdot u_2'$$
$$M_1 = -E_1 \cdot I_1 \cdot w'' \qquad M_2 = -E_2 \cdot I_2 \cdot w''$$
$$V_1 = -E_1 \cdot I_1 \cdot w''' \qquad V_2 = -E_2 \cdot I_2 \cdot w'''$$

Shear flow in joint $v = k \cdot u = k \cdot \left(u_2 - u_1 + w' \cdot a\right)$

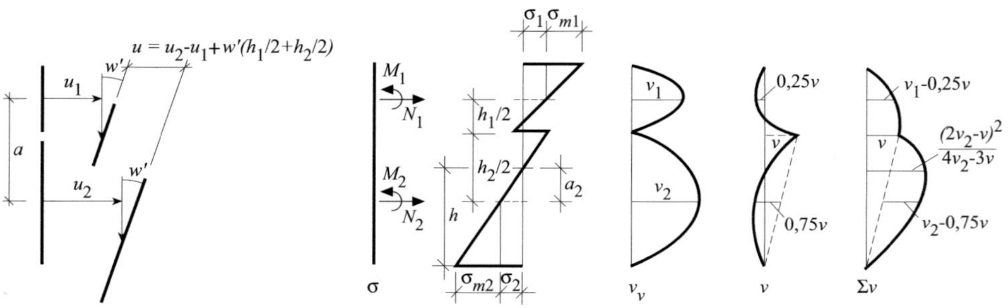

Figure A14 Cross-section from Figure A13: Deformations, longitudinal stresses, shear flow. (STEP 1995 Article B11)

Equilibrium of forces in x-direction (no external load p in x-direction), see Figure A13 ($N' = dN/dx$):

$$N_1 - \left(N_1 + \frac{dN_1}{dx}\right) - v \cdot dx = N_1' \cdot dx + v \cdot dx = N_1' + v = 0 \tag{a}$$

$$N_2 - \left(N_2 + \frac{dN_2}{dx}\right) + v \cdot dx = N_2' - v = 0 \tag{b}$$

in z-direction (p = external load in z-direction), Figure A13 ($V' = dV/dx$):

$$V_1 - \left(V_1 + \frac{dV_1}{dx}\right) + V_2 - \left(V_2 + \frac{dV_2}{dx}\right) - p \cdot dx = 0$$
$$-\frac{dV_1}{dx} - \frac{dV_2}{dx} - p \cdot dx = -V_1' - V_2' - p = 0 \tag{i}$$
$$V_1' + V_2' = V' = -p$$

Equilibrium of moments at point A, Figure A13 ($M' = dM/dx$):

$$\sum M^A = 0 = M_1 + V_1 \cdot dx - v \cdot dx \cdot \frac{h_1}{2} - M_1 - \frac{dM_1}{dx}$$

$$M_1' \cdot dx = V_1 \cdot dx - v \cdot dx \cdot \frac{h_1}{2} \tag{ii}$$

$$M_1' = V_1 - v \cdot \frac{h_1}{2}$$

$$M_2' = V_2 - v \cdot \frac{h_2}{2} \tag{iii}$$

Now, equations (ii) and (iii) are added, differentiated once with respect to x and $V'_1 + V'_2 = V'$ is expressed with $-p$ (equation (i)), where $h_1/2 + h_2/2 = a$ (Figure A14):

$$M_1'' + M_2'' + v' \cdot a + p = 0 \tag{c}$$

Accordingly, the three equilibrium conditions (a), (b) and (c) for the three deformations u_1, u_2 and w are formulated. If the internal forces and moments are replaced by the elasticity relations, we obtain a system of differential equations for the x-direction:

$$\begin{aligned} E_1 \cdot A_1 \cdot u_1'' + k \cdot (u_2 - u_1 + w' \cdot a) = 0 \\ E_2 \cdot A_2 \cdot u_2'' - k \cdot (u_2 - u_1 + w' \cdot a) = 0 \end{aligned} \tag{13}$$

and for the z-direction:

$$(E_1 \cdot I_1 + E_2 \cdot I_2) \cdot w'''' - k \cdot (u_2' - u_1' + w'' \cdot a) \cdot a = p \tag{14}$$

The variation in elastic energy

$$\Pi = \frac{1}{2} \cdot \int \left[E_1 \cdot A_1 \cdot u_1'^2 + E_2 \cdot A_2 \cdot u_2'^2 + (E_1 \cdot I_1 + E_2 \cdot I_2) \cdot w''^2 + k \cdot (u_2 - u_1 + w' \cdot a)^2 - 2 \cdot p \cdot w \right] \cdot dx$$

also elicits these equations.

In equation (14), the equilibrium in z-direction, elastic foundation k_w or the influence of second order theory can also be taken into consideration, whereby the expression $k_w \cdot w - N_0 \cdot w''$ is added to the left side.

The system of differential equations is particularly easy to solve for a single-span beam with sinusoidal load distribution (L = beam length):

$$p = p_0 \cdot \sin\left(\frac{\pi}{L} \cdot x\right)$$

Base function for the deformation:

$$u_1 = u_{10} \cdot \cos\left(\frac{\pi}{L} \cdot x\right)$$

$$u_2 = u_{20} \cdot \cos\left(\frac{\pi}{L} \cdot x\right)$$

$$w = w_0 \cdot \sin\left(\frac{\pi}{L} \cdot x\right)$$

Inserted in the system of differential equations (equations (13) and (14)) and cancelling out sine or cosine, we obtain:

$$\begin{array}{llll}
u_{10} & u_{20} & w_0 & = p_0 \\
-\frac{\pi^2}{L^2} \cdot E_1 \cdot A_1 - k & k & k \cdot \frac{\pi}{L} \cdot a & = 0 \\
k & -\frac{\pi^2}{L^2} \cdot E_2 \cdot A_2 - k & -k \cdot \frac{\pi}{L} \cdot a & = 0 \\
k \cdot \frac{\pi}{L} \cdot a & -k \cdot \frac{\pi}{L} \cdot a & -\frac{\pi^4}{L^4} \cdot (E_1 \cdot I_1 + E_2 \cdot I_2) - k \cdot \frac{\pi^2}{L^2} \cdot a^2 & = -1
\end{array}$$

The solving of the system of equations delivers:

$$w_0 = p_0 \cdot \frac{L^4}{\pi^4} \cdot \frac{1}{E_1 \cdot I_1 + E_2 \cdot I_2 + \dfrac{E_1 \cdot A_1 \cdot \gamma_1 \cdot a^2}{1 + \gamma_1 \cdot \dfrac{E_1 \cdot A_1}{E_2 \cdot A_2}}} = p_0 \cdot \frac{L^4}{\pi^4} \cdot \frac{1}{(E \cdot I)_{ef}}$$

$$u_{10} = w_0 \cdot \frac{\pi}{L} \cdot \frac{a \cdot \gamma_1 \cdot E_2 \cdot A_2}{\gamma_1 \cdot E_1 \cdot A_1 + E_2 \cdot A_2}$$

$$u_{20} = -w_0 \cdot \frac{\pi}{L} \cdot \frac{a \cdot \gamma_1 \cdot E_1 \cdot A_1}{\gamma_1 \cdot E_1 \cdot A_1 + E_2 \cdot A_2}$$

where:

$$\gamma_1 = \frac{1}{1 + \dfrac{\pi^2}{L^2} \cdot \dfrac{E_1 \cdot A_1}{k}} \qquad (15)$$

Equation (15) is coefficient γ.

Normal stresses

As was explained in Article D7, the effective bending stiffness $(E \cdot I)_{ef}$ and hence stress distribution is largely dependent on joint stiffness, which is recorded by the coefficient γ. For $(E \cdot I)_{ef}$ and where n = number of individual cross-sections, the following applies (equation (D7-1)):

$$\left(E \cdot I\right)_{ef} = \sum_{i=1}^{n}\left(E_i \cdot I_i + \gamma_i \cdot E_i \cdot A_i \cdot a_i^2\right) \tag{16}$$

In the following, the determination of stresses using the three cross-sections A, B and C shown in Figure A15 is explained.

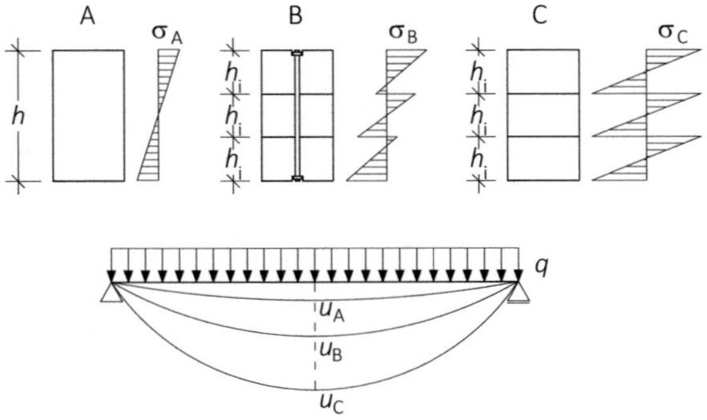

Figure A15 Deflection and bending stress distribution of a full cross-section (A), a cross-section comprising three mechanically jointed individual cross-sections (B) and a cross-section comprising three non-connected individual cross-sections (C). Corresponding to Figure D7-1.

Full cross-section A

Now we consider beam A as a rigidly connected beam with three components and with $E_1 = E_2 = E_3$ = const. and $h_i = h_1 = h_2 = h_3$ = heights of individual cross-sections. The entire external moment M is then distributed over the three individual cross-sections as shown in Figure A16.

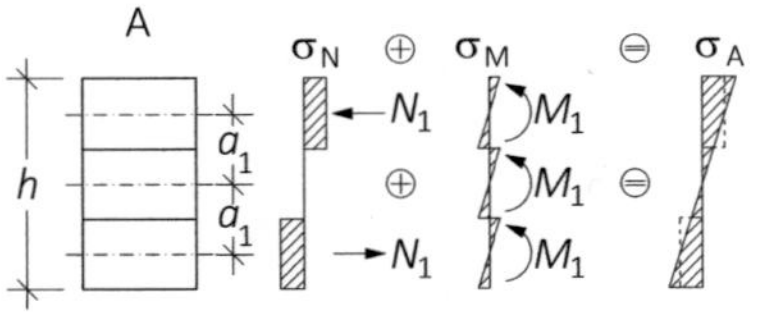

Figure A16 Beam A, comprising three rigidly connected individual cross-sections.

The external moment M must be in equilibrium with the internal moments and forces M_i and N_i of the individual cross-sections i:

$$M = \sum M_i + \sum N_i \cdot a_i$$

This means for the beam shown in Figure A16:

$$M = 3 \cdot M_1 + 2 \cdot N_1 \cdot a_1 \quad \text{or} \quad N_1 = \left(M - 3 \cdot M_1\right) \cdot \frac{1}{2 \cdot a_1} = \frac{M}{2 \cdot a_1} \cdot \left(1 - \frac{3 \cdot M_1}{M}\right) \tag{17}$$

The following applies to the stress σ_A at the edge of the beam:

$$\sigma_A = \frac{M_1}{I_1} \cdot \frac{h_1}{2} + \frac{N_1}{A_1} \tag{18}$$

For the considered „full cross-section“, the following also applies:

$$M_1 = M \cdot \frac{I_1}{I} \quad \text{and} \quad I = \sum_{i=1}^{3} \left(I_i + A_i \cdot a_i^2\right) = 3 \cdot I_1 + 2 \cdot A_1 \cdot a_1^2$$

Insert in equation (17) and considering the Steiner part (I - 3 · I_1):

$$N_1 = \frac{M}{2 \cdot a_1} \cdot \left(1 - \frac{3 \cdot I_1}{I}\right) = \frac{M}{2 \cdot a_1 \cdot I} \cdot \left(I - 3 \cdot I_1\right) = \frac{M}{2 \cdot a_1 \cdot I} \cdot 2 \cdot A_1 \cdot a_1^2 = \frac{M}{I} \cdot A_1 \cdot a_1$$

The stress component from the normal force N_1 is thus as follows:

$$\sigma_1 = \frac{N_1}{A_1} = \frac{M}{I} \cdot a_1$$

Accordingly, the overall stress σ_A at the beam edge from equation (18) evolves into the bending stress at the edge of a full cross-section of height h_{tot} = 3 · h_1:

$$\sigma_A = \frac{M_1}{I_1} \cdot \frac{h_1}{2} + \frac{M}{I} \cdot a_1 = \frac{M}{I_1} \cdot \frac{I_1}{I} \cdot \frac{h_1}{2} + \frac{M}{I} \cdot a_1 = \frac{M}{I} \cdot \left(\frac{h_1}{2} + a_1\right) = \frac{M}{I} \cdot \frac{h_{tot}}{2}$$

Cross-section C comprising three non-connected individual cross-sections

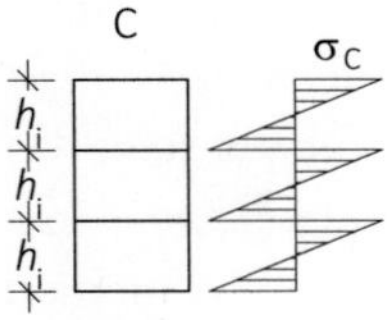

For beam C, the normal force is $N_i = 0$ and the bending stress σ_C at the edge of the beam is thus

$$\sigma_C = \frac{M_i}{I_i} \cdot \frac{h_i}{2} \qquad \text{where} \qquad M_i = M \cdot \frac{(E \cdot I)_i}{(E \cdot I)_{ges}} \quad \text{and} \quad I_{ges} = \sum I_i$$

Mechanically jointed cross-section B comprising three individual cross-sections

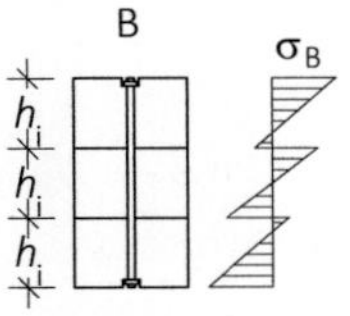

To simplify matters, we define $E_i = E_1 = E_2 = E_3$ = const. and $\gamma_1 = \gamma_3$. The derivation of the normal force is complex for the case in question, since the differential equations for deformations considering shear deformations in the joints need to be solved. This was done by Möhler (1956) and a reduction coefficient γ was defined, as explained in Article D7. In the case of semi-rigid joints, equation (17) can be transformed similarly to beam A, using $I_{ef} = 3 \cdot I_1 + 2 \cdot \gamma_1 \cdot A_1 \cdot a_1^2$ (E_i = const.) and $M_1 = M \cdot I_1 / I_{ef}$:

$$N_1 = \frac{M}{2 \cdot a_1} \cdot \left(1 - \frac{3 \cdot I_1}{I_{ef}}\right) = \frac{M}{2 \cdot a_1 \cdot I_{ef}} \cdot \left(I_{ef} - 3 \cdot I_1\right) = \frac{M}{2 \cdot a_1 \cdot I_{ef}} \cdot 2 \cdot \gamma_1 \cdot A_1 \cdot a_1^2 = \frac{M}{I_{ef}} \cdot \gamma_1 \cdot A_1 \cdot a_1$$

The stress component from the normal force N_1 is thus as follows:

$$\sigma_1 = \frac{N_1}{A_1} = \frac{M}{I_{ef}} \cdot \gamma_1 \cdot a_1$$

The overall stress σ_B at the beam edge can then be calculated as:

$$\sigma_B = \frac{M_1}{I_1} \cdot \frac{h_1}{2} + \frac{M}{I_{ef}} \cdot \gamma_1 \cdot a_1 = \frac{M}{I_1} \cdot \frac{I_1}{I_{ef}} \cdot \frac{h_1}{2} + \frac{M}{I_{ef}} \cdot \gamma_1 \cdot a_1 = \frac{M}{I_{ef}} \cdot \left(\frac{h_1}{2} + \gamma_1 \cdot a_1\right) \tag{19}$$

Equation (19) corresponds to equations (D7-6) and (D7-7) or the bending stress at the edge of the overall beam, whereby in the initial summand of equation (19), the bending stress at the edge of cross-section 1 is calculated and in the second summand, the normal stress in the centre of gravity of cross-section 1. While the effect of the semi-rigid joints (and the potential differing moduli of elasticity) on distribution of the internal forces and moments is considered via the effective stiffness when calculating the moment component $\sigma_{i,m} = M/(E \cdot I)_{ef} \cdot E_i \cdot h_i/2$, the normal force component depends directly on how the Steiner part contributes and hence the coefficient γ.

The normal forces N_i generated by an external moment in composed beams with rigid joints exceed those for beams with semi-rigid joints. For non-connected individual cross-sections, the external moment does not generate any normal force, $N_i = 0$. Consequently, the normal stress (tension t or compression c) is determined as in equations (19) and (D7-7) by considering γ_i:

$$\sigma_{i,t(c)} = \frac{M}{(E \cdot I)_{ef}} \cdot \gamma_i \cdot E_i \cdot a_i$$

One **reminder** here is that all internal forces and moments transferred via a semi-rigid joint must be reduced by the coefficient γ. This also applies to the shear stresses τ_{max}, since these are determined via the joint between cross-sectional parts 2 and 3, which is why γ_3 must be applied.

Shear stresses and loads on fasteners

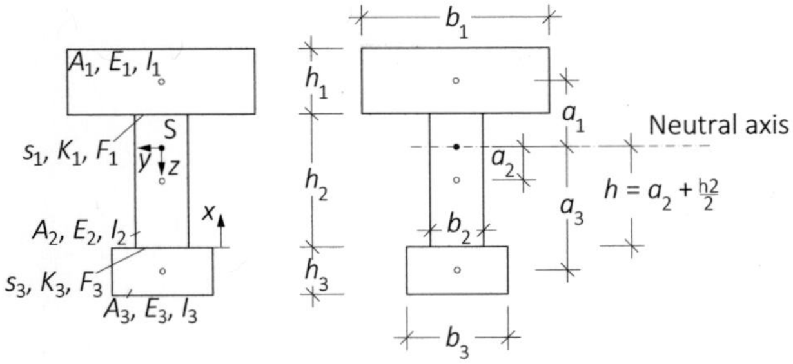

For the cross-section shown comprising three components, the maximum shear stress τ_{max} shall be determined. The shear stress peaks at the neutral axis at z = 0. To determine τ_{max}, the first moment of area S_{ef} of the lower beam part for z = [0, $h + h_3$] is needed, whereby the semi-rigidly connected cross-section 3 is to be reduced with γ_3 as explained above (z_{si} = distances from the centres of gravity of the individual cross-sections to the neutral axis):

$$S_{ef} = z_{s2} \cdot A_2 + z_{s3} \cdot A_3 = \frac{1}{2} \cdot h \cdot b_2 \cdot h + \gamma_3 \cdot a_3 \cdot b_3 \cdot h_3 = \frac{1}{2} \cdot b_2 \cdot h^2 + \gamma_3 \cdot a_3 \cdot A_3$$

S_{ef} can also be determined from the upper beam part for z = [0, $-(a_1 + h_1/2)$] and with γ_1. From S_{ef}, we obtain the following maximum shear stress, taking differing moduli of elasticity into consideration:

$$\tau_{max} = \frac{V_{max} \cdot (E \cdot S)_{ef}}{(E \cdot I)_{ef} \cdot b_2} = \frac{V_{max} \cdot \left(\gamma_3 \cdot E_3 \cdot A_3 \cdot a_3 + 0{,}5 \cdot E_2 \cdot b_2 \cdot h^2\right)}{(E \cdot I)_{ef} \cdot b_2}$$

This corresponds to equation (D7-8).

With τ_{max}, it is now possible to calculate the force exerted on the individual fasteners. The shear flow is:

$$t_{ef} = \tau_{max} \cdot b_2 = \frac{V_{max} \cdot E_i \cdot S_{joint}}{(E \cdot I)_{ef}} = \frac{V_{max} \cdot (\gamma_3 \cdot E_3 \cdot A_3 \cdot a_3)}{(E \cdot I)_{ef}}$$

whereby the first moment of area S_{joint} is now only determined for z = [h, $h + h_3$], namely for cross-section 3 (or 1 during verification of the fastener in the joint between cross-sections 1 and 2). Multiplying this by the fastener spacing s allows us to obtain the force per row of fastener and thus equation (D7-9):

$$F_3 = \frac{V_{max} \cdot \gamma_3 \cdot E_3 \cdot A_3 \cdot a_3 \cdot s_3}{(E \cdot I)_{ef}}$$

Annex 6: Derivation Johansen equations

The following section derives certain Johansen cases for single-shear timber-to-timber joints.

Failure mechanism b – embedment failure in member 2:

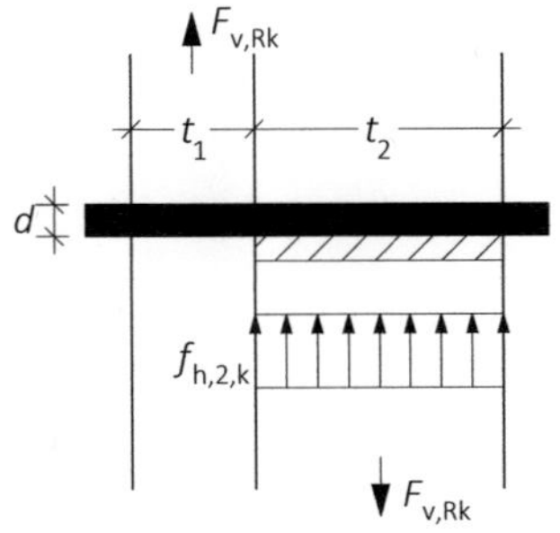

Figure A17 Failure mechanism b.

In the event of failure in accordance with mode b, as in Figure A17, the embedment strength in member 2 is decisive. The equilibrium of forces $\Sigma V = 0$ results in:

$$F_{v,Rk} = f_{h,2,k} \cdot d \cdot t_2 \tag{20}$$

Equation (20) precisely corresponds to the equation for failure mechanism b in EC 5.

Failure mechanism c – embedment failure in both members:

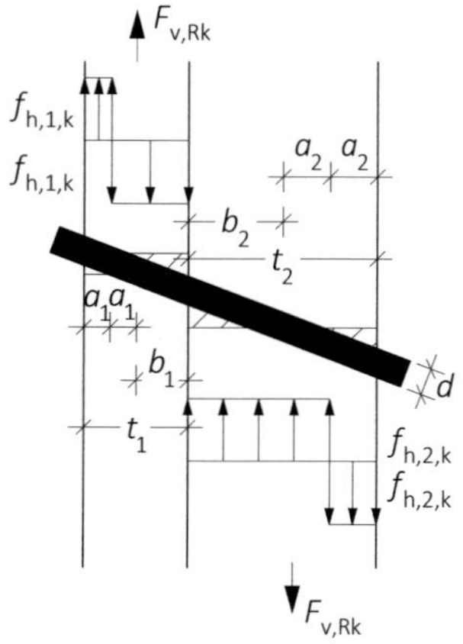

Figure A18   Failure mechanism c.

In the event of failure in accordance with mode c, as in Figure A18, the embedment strength has been reached in both members. The following applies:

$$\beta = \frac{f_{h,2,k}}{f_{h,1,k}}$$

The equilibrium of forces $\Sigma V = 0$ results in:

$$F_{v,Rk} = f_{h,1,k} \cdot d \cdot b_1 = f_{h,2,k} \cdot d \cdot b_2 = \beta \cdot f_{h,1,k} \cdot d \cdot b_2$$

Consequently:

$$\beta \cdot b_2 = b_1$$

The bending moment in the shear plane results in:

$$\text{Member 1:}\quad M_{SP} = f_{h,1,k} \cdot d \cdot \frac{(b_1 + a_1)^2}{2} - f_{h,1,k} \cdot d \cdot a_1 \cdot \left(b_1 + a_1 + \frac{a_1}{2} \right) = f_{h,1,k} \cdot d \cdot \left(\frac{b_1^2}{2} - a_1^2 \right)$$

$$\text{Member 2:}\quad M_{SP} = -f_{h,2,k} \cdot d \cdot \frac{(b_2 + a_2)^2}{2} + f_{h,2,k} \cdot d \cdot a_2 \cdot \left(b_2 + a_2 + \frac{a_2}{2} \right) = f_{h,2,k} \cdot d \cdot \left(a_2^2 - \frac{b_2^2}{2} \right)$$

Equating and replacing $f_{h,2,k}$ with $\beta \cdot f_{h,1,k}$:

$$f_{h,1,k} \cdot d \cdot \left(\frac{b_1^2}{2} - a_1^2 \right) = f_{h,2,k} \cdot d \cdot \left(a_2^2 - \frac{b_2^2}{2} \right) = \beta \cdot f_{h,1,k} \cdot d \cdot \left(a_2^2 - \frac{b_2^2}{2} \right)$$

Now replace b_2 with b_1/β:

$$f_{\mathrm{h,1,k}} \cdot d \cdot \left(\frac{b_1^2}{2} - a_1^2 \right) = \beta \cdot f_{\mathrm{h,1,k}} \cdot d \cdot \left(a_2^2 - \frac{b_2^2}{2} \right)$$

$$\frac{b_1^2}{2} - a_1^2 = \beta \cdot \left(a_2^2 - \frac{b_1^2}{2 \cdot \beta^2} \right)$$

$$\frac{b_1^2}{2} \cdot \frac{1+\beta}{\beta} = \beta \cdot a_2^2 + a_1^2$$

Replacing $a_1 = (t_1 - b_1)/2$ and $a_2 = (t_2 - b_2)/2 = (\beta \cdot t_2 - b_1)/(2 \cdot \beta)$:

$$\frac{b_1^2}{2} \cdot \frac{1+\beta}{\beta} = \frac{(\beta \cdot t_2 - b_1)^2}{4 \cdot \beta} + \frac{(t_1 - b_1)^2}{4}$$

$$(1+\beta) \cdot b_1^2 + 2 \cdot \beta \cdot (t_1 + t_2) \cdot b_1 - \beta \cdot \left(t_1^2 + \beta \cdot t_2^2 \right) = 0$$

The following is revealed when solving the quadratic equation:

$$b_1 = \frac{t_1}{1+\beta} \cdot \left[\sqrt{\beta + 2 \cdot \beta^2 \cdot \left[1 + \frac{t_2}{t_1} + \left(\frac{t_2}{t_1} \right)^2 \right] + \beta^3 \cdot \left(\frac{t_2}{t_1} \right)^2} - \beta \cdot \left(1 + \frac{t_2}{t_1} \right) \right] \qquad (21)$$

In addition, the following applies for the equilibrium of forces $\Sigma V = 0$ at member 1:

$$F_{\mathrm{v,Rk}} = f_{\mathrm{h,1,k}} \cdot d \cdot b_1$$

Inserting equation (21):

$$F_{\mathrm{v,Rk}} = \frac{f_{\mathrm{h,1,k}} \cdot d \cdot t_1}{1+\beta} \cdot \left[\sqrt{\beta + 2 \cdot \beta^2 \cdot \left[1 + \frac{t_2}{t_1} + \left(\frac{t_2}{t_1} \right)^2 \right] + \beta^3 \cdot \left(\frac{t_2}{t_1} \right)^2} - \beta \cdot \left(1 + \frac{t_2}{t_1} \right) \right] \qquad (22)$$

Equation (22) is included in the equation for failure mechanism c in EC 5, where equation (22) corresponds to the Johansen portion, while the rope effect is missing.

Failure mechanism e – embedment failure in both members and one plastic hinge in member 1:

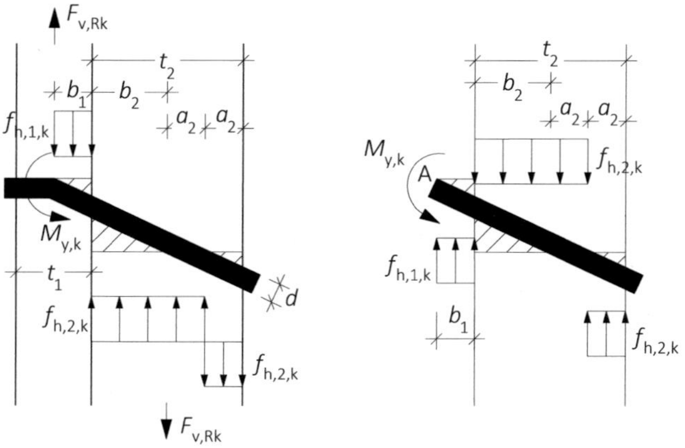

Figure A19 Failure mechanism e. Right: free-body diagram of fastener.

In the event of failure in accordance with mode e, as in Figure A19, both members reach the embedment strength and a plastic hinge forms. The following applies:

$$\beta = \frac{f_{h,2,k}}{f_{h,1,k}}$$

The shear force in the fastener is equal to zero at the point where the bending moment peaks:

$$f_{h,2,k} \cdot d \cdot b_2 - f_{h,1,k} \cdot d \cdot b_1 = \beta \cdot f_{h,1,k} \cdot d \cdot b_2 - f_{h,1,k} \cdot d \cdot b_1 = 0$$

Consequently:

$$\beta \cdot b_2 = b_1$$

The equilibrium of moments at the fastener $\Sigma M^A = 0$ is:

$$M_{y,k} = -f_{h,1,k} \cdot d \cdot b_1 \cdot \frac{b_1}{2} + f_{h,2,k} \cdot d \cdot (b_2 + a_2) \cdot \left(b_1 + \frac{b_2 + a_2}{2} \right) - f_{h,2,k} \cdot d \cdot a_2 \cdot \left(b_1 + b_2 + a_2 + \frac{a_2}{2} \right)$$

Now $f_{h,2,k}$ is replaced with $\beta \cdot f_{h,1,k}$, b_1 with $\beta \cdot b_2$ and $a_2 = (t_2 - b_2)/2$ is replaced and the equation is transformed:

$$b_2^2 + \frac{2 \cdot t_2}{2 \cdot \beta + 1} \cdot b_2 - \frac{t_2^2}{2 \cdot \beta + 1} - \frac{4}{2 \cdot \beta + 1} \cdot \frac{M_{y,k}}{f_{h,1,k} \cdot d \cdot \beta} = 0$$

Solving the quadratic equation reveals:

$$b_2 = -\frac{t_2}{2 \cdot \beta + 1} \cdot + \sqrt{\frac{t_2^2}{(2 \cdot \beta + 1)^2} + \frac{t_2^2}{2 \cdot \beta + 1} + \frac{4 \cdot M_{y,k}}{f_{h,1,k} \cdot d \cdot \beta \cdot (2 \cdot \beta + 1)}} \tag{23}$$

In addition, the following applies for the equilibrium of forces $\Sigma V = 0$ at member 2:

$$F_{v,Rk} = f_{h,2,k} \cdot d \cdot b_2 = \beta \cdot f_{h,1,k} \cdot d \cdot b_2$$

Inserting equation (23):

$$F_{v,Rk} = \frac{f_{h,1,k} \cdot d \cdot t_2}{2 \cdot \beta + 1} \cdot \left[\sqrt{2 \cdot \beta^2 \cdot (1 + \beta) + \frac{4 \cdot \beta \cdot (2 \cdot \beta + 1) \cdot M_{y,k}}{f_{h,1,k} \cdot d \cdot t_2^2}} - \beta \right] \tag{24}$$

Equation (24) is included in the equation for failure mechanism e in EC 5, where equation (24) corresponds to the Johansen portion, while the rope effect and the pre-factor 1.05 are missing (the pre-factor takes into consideration the differing partial safety factors and k_{mod} for steel and timber).

Failure mechanism f – embedment failure in both members and two plastic hinges:

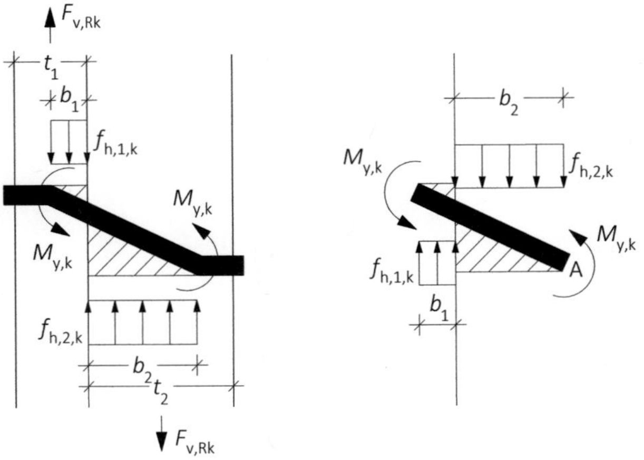

Figure A20 Failure mechanism f. Right: free-body diagram of fastener.

In the event of failure in accordance with mode e, as in Figure A20, both members reach the embedment strength and two plastic hinges form. The following applies:

$$\beta = \frac{f_{\text{h,2,k}}}{f_{\text{h,1,k}}}$$

and

$$\beta \cdot b_2 = b_1$$

The equilibrium of moments at the fastener $\Sigma M^{A} = 0$ is:

$$M_{\text{y,k}} + M_{\text{y,k}} = f_{\text{h,1,k}} \cdot d \cdot b_1 \cdot \left(b_2 + \frac{b_1}{2} \right) - f_{\text{h,2,k}} \cdot d \cdot b_2 \cdot \frac{b_2}{2}$$

Replacing b_2 with b_1/β and $f_{\text{h,2,k}} = \beta \cdot f_{\text{h,1,k}}$ and solving for b_1:

$$b_1 = \sqrt{\frac{2 \cdot M}{f_{h,1,k} \cdot d}} \cdot \sqrt{\frac{2 \cdot \beta}{1+\beta}}$$

The equilibrium of forces $\Sigma V = 0$ at member 1 results in:

$$F_{\text{v,Rk}} = f_{\text{h,1,k}} \cdot d \cdot b_1$$

Inserting b_1:

$$F_{\text{v,Rk}} = \sqrt{2 \cdot M \cdot f_{h,1,k} \cdot d} \cdot \sqrt{\frac{2 \cdot \beta}{1+\beta}} \tag{25}$$

Equation (25) is included in the equation for failure mechanism f in EC 5, again without rope effect and pre-factor 1.15.